DYNAMICS OF MATERIALS

DYNAMICS OF MATERIALS

Experiments, Models and Applications

LILI WANG
LIMING YANG
XINLONG DONG
XIQUAN JIANG

ELSEVIER

ACADEMIC PRESS
An imprint of Elsevier

Academic Press is an imprint of Elsevier
125 London Wall, London EC2Y 5AS, United Kingdom
525 B Street, Suite 1650, San Diego, CA 92101, United States
50 Hampshire Street, 5th Floor, Cambridge, MA 02139, United States
The Boulevard, Langford Lane, Kidlington, Oxford OX5 1GB, United Kingdom

Library of Congress Cataloging-in-Publication Data
A catalog record for this book is available from the Library of Congress

British Library Cataloguing-in-Publication Data
A catalogue record for this book is available from the British Library

ISBN: 978-0-12-817321-3

For information on all Academic Press publications visit our website at https://www.elsevier.com/books-and-journals

Publisher: Matthew Deans
Acquisition Editor: Brian Guerin
Editorial Project Manager: Peter Adamson
Production Project Manager: Mohana Natarajan
Cover Designer: Christian Bilbow

Typeset by TNQ Technologies

Contents

Preface for English Edition

Thanks to Elsevier and its editors, the English Edition of "Dynamics of Materials" is about to go to press.

Unlike the cases under quasistatic loading, the mechanical response of solids under dynamic (explosion/impact) loading is characterized mainly by two dynamic effects, namely the inertial effects of structures and the strain-rate effects of materials. The inertial effect of infinitesimal deformable element leads to the study of wave propagation in various (precise or simplified) forms, promoting the development of stress wave theory and structural impact dynamics, while the strain-rate effect leads to the study of all kinds of rate-dependent constitutive relations and failure criteria under high strain rates, promoting the development of material dynamics. They are different on the one hand and are interrelated on the other hand. It can be said that the "Foundation of Stress Waves" printed by Elsevier in 2007 and the "Dynamics of Materials" to be printed by Elsevier in 2019 are sister works to each other.

The study of the dynamic mechanical response of materials can essentially be attributed to how to quantitatively describe the whole process of dynamic flow/deformation until final failure. The flow/deformation of material is mathematically described by the constitutive relation, which can usually be decomposed into a spherical part and a deviator part. The former describes the law of volumetric change (volumetric law), while the latter describes the law of shape change (distortional law). Correspondingly, this book consists of the following three aspects. In the first part, the volumetric law is discussed from different views, including its macrothermodynamic basis, solid physics basis, and the related dynamic experimental study. When the loading stress σ_L is much higher than the material shear strength σ_S, the distortional part can be neglected, and the material can then be approximately dealt with as a fluid. In such a case, the volumetric law is usually referred to as the "Equation of State for Solids under High Pressure." In the second part, the distortional law is analyzed, wherein the rate-dependent macrodistortional law describing strain-rate effect, its micromechanism based on dislocation dynamics, and the dynamic experimental research based on the stress wave theory are discussed in turn, one after another. In the final part, the dynamic failure induced by dynamic damage evolution is examined, including the unloading failure of crack-free body taking "spalling" as a typical representative, the crack dynamics for cracked body describing the crack extension of a stationary or propagating crack under dynamic loading, the dynamic fragmentation due to dynamic growth of multicracks, and the dynamic evolution of meso-damage, first taking Adiabatic Shearing as the representative and then the general evolution law for different forms of meso-damage and macro-damage.

To help readers to understand the above contents more deeply and extensively, this book tries to follow a comprehensive approach, i.e., the approach characterized by the combination of macroscopic continuum mechanics and thermodynamics, the combination of macromechanics expression and microphysical mechanism, as well as the combination of theoretical analyses and experimental investigation.

Lectures on Dynamic Mechanical Properties of Materials at the University of Science and Technology of China (USTC) could be dated back to 1962. In that year, Academician Zheng Zhemin advised me to teach the course of Mechanical Properties of Metallic Materials Under Impact Load to college students, who were the first students majoring in the Explosion Mechanics Profession in the Department of Modern Mechanics. While lecturing, I wrote corresponding lecture notes, which were reviewed and approved by Academician Zheng and then printed for internal use (No. 07-58-b18). Different from traditional textbooks on the mechanical properties of metallic materials, this course is characterized by the combination of mechanical properties of materials and the knowledge of stress waves. In fact, the knowledge of stress waves has been taught since the second chapter. Two years later, it was changed to mainly focus on stress wave theory combined with some corresponding contents of dynamic constitutive relation of materials. A course on Plastic Dynamics was planned and corresponding lecture notes were printed (No. 07-60-E08) for internal use in USTC.

In 1978, I returned to the USTC from northwest China. It also coincides with the resumption of the national graduate enrollment system. The Head of the Explosion Mechanics Profession, Prof. Zhu Zhaoxiang, attached great importance to the curriculum and teaching material of postgraduates. He advised me to open two interrelated courses for postgraduates of the explosive mechanics profession, namely the Stress Wave Propagation and the Material Dynamics courses. In addition to the graduates, he took the lead and asked young teachers to listen to these two new courses. The heavy burden urges me to forge ahead. For me, lesson preparation is a process of deeply consulting new literature and restudying. The lecture is a process in which I give a presentation and then invite the audience to have a discussion. It's more like a seminar than a lecture. That kind of atmosphere with the tireless collective learning and lively discussion for seeking the truth is by now unforgettable!

When the course of Stress Wave Propagation was retaught again at the USTC, there were already old lecture notes, which, after being supplemented and edited, were published by National Defense Industry Press in 1985 under the title of "Foundations of Stress Waves." However, when I taught the new course of Material Dynamics, I only drew up a syllabus at the beginning, as I did not have enough time to write the lecture notes. Here I would like to thank my assistant, Lecturer Hu Shisheng. Using the detailed notes written by him when he was listening to my lecture and referring to the relevant materials, he sorted out the handout of "Material Dynamics," which was printed mainly for internal use. Until 2013, the Press of University of Science and Technology of China listed it as an Excellent Textbook, and then it was coauthored by Wang Lili, Hu Shisheng, Yang Liming, and Dong Xinlong and was officially published. During this time I was suddenly hospitalized for surgery, which delayed my due date. I would like to take this opportunity to express my gratitude to my wife, Lu Weixian, for her loving care, understanding, and support. I have not only regained my health but also finished the revision work together with my coauthors during my recovery. It lasted nearly 3 years.

When we revised this manuscript, although we still decided to emphasize the basic principles, basic concepts, and basic knowledge, we also hoped to reflect the new progress in this interdisciplinary field, as well as to reflect some research results from Chinese scholars in this field.

The publication of this book also entailed the support of many colleagues. Among them, my coauthors and I would like to express our particular appreciation to Associate Professor Miao Fuxing, Dr. Ding Yuanyuan, Dr. Le Yang, and Mr. Zhijin Jonathan Lin, who assisted us in the translations and the illustrations in this book.

Finally, I sincerely hope that this book, which incorporates more than 40 years of teaching and research experience and the inclusion of many people's painstaking efforts, can pave a way for the future young scholars and ensure them a better future in climbing the new scientific peak!

Wang Lili (Lili Wang, Li-Lih Wang)
Professor, Key Laboratory of Impact and Safety Engineering
(Ningbo University), Ministry of Education
March 2019

CHAPTER ONE

Introduction

Dynamics of Materials creates a new interdisciplinary branch, which studies the basic law of high-velocity flow/deformation and dynamic failure of materials under dynamic loading such as explosion and impact. It, together with the *Foundations of Stress Waves*, (Wang, 2007) constitutes the important foundation for studying Explosion Mechanics and Impact Dynamics. It is also one of the important development directions in Material Science research.

For people coming into contact with the *Dynamics of Materials* for the first time, the description above, a concise definition, is not clear enough. People will raise a series of specific questions: What is the relationship between material and mechanics? What is the relation between the *Dynamics of Materials* and other classical courses such as *Strength of Materials* and *Solid Mechanics*? What characteristics are expressed by the word "dynamics" in the *Dynamics of Materials*?

Human beings live in the **matter** world. The views of "matter is primary" and "matter is in motion" are the basic philosophical views for understanding and transforming the world. In order to know and transform the world, we must first know and study the matter that makes up the matter world. The matter to be studied and used is called **material**. Rock, for example, is a natural matter around us, when we use it in concretes as aggregates, or as a target for ground-penetrating projectiles, it becomes rock material when we study it.

Just as any science is closely related to matter, mechanics focuses on the mechanical movement of matter. Therefore, the mechanical properties of matter or materials play an irreplaceable important role in the study of mechanics. As we all know, the basic equations of continuum mechanics consist of the following aspects: first, a set of geometric equations (kinematic equations) that relates displacements, strains, and particle velocities, etc., reflects the **displacement continuity** or the **mass conservation** in a continuum. The second is the kinetic equation linking stress and particle acceleration, which represents the **momentum conservation**. Another set is the relation between various forms of energy, which embodies the **energy conservation**. The last set deals with the relationship between

Dynamics of Materials
ISBN: 978-0-12-817321-3
https://doi.org/10.1016/B978-0-12-817321-3.00001-2

stress, strain and their time-derivatives, reflecting the intrinsic mechanical properties of the material itself, called **material constitutive relations**, and the corresponding mathematical expressions are referred to as the **material constitutive equations**. Among those basic governing equations in continuum mechanics, three conservation equations reflect the common characteristics of each branch of mechanics, while the material constitutive relation reflects the special characteristics of each different branch. Reviewing the formation and classification of the basic branches of continuum mechanics, such as general mechanics (rigid body mechanics), fluid mechanics, and solid mechanics, and so on, are they mainly distinguished by the difference of constitutive relation of materials? It is no exaggeration to say that the importance of constitutive relations in the development of continuum mechanics cannot be overstressed.

In fact, the significance of the mechanical properties of materials is not only that all mechanical analysis aimed at determining the mechanical field (stress, strain, displacement, particle velocity distribution, etc.) of an object is based on the constitutive relation of materials, but is only half the task of solving practical problem. More importantly to further answer whether such stresses, strains, displacements, and particle velocities exceed the allowable limits so that will cause a failure of the object, it comes down to the establishment of generalized **strength criterion**.

Any generalized strength criterion for quantitative analysis can be reduced to the following simple

$$\Sigma \geq \Sigma_c \tag{1.1}$$

Once the formula is satisfied, it is judged either be of failure or invalid. The Σ on the left side of inequality is a mechanical characteristic quantity which can be computed from the basic governing equations mentioned previously under given conditions, and is solved by mechanics scientists. For solid structures, this is the task of "solid mechanics." The Σ_c on the right side is a critical parameter that can be experimentally measured to characterize the strength of materials, which is relied on material scientists to solve. Inequality (1.1) links the mechanical characteristic quantities with the critical parameters of material strength. It means that strength analysis is simultaneously based on mechanics and material science, and there is no lack of any one of them. If we do not have a comprehensive understanding of the mechanical properties of materials, we cannot establish the above strength criteria, and therefore cannot carry out any strength analyses.

Both the study of the constitutive relation of various materials required for the calculation of Σ and the study of the strength parameter Σ_c of various materials for the establishment of generalized strength criteria have formulated a new discipline branch, the "Mechanics of Materials" in true sense, which has nothing in common with the traditional university course "Strength of Materials." The latter, in fact, emphasizes the study not on the strength of the material itself, but on the analysis of the force and deformation of typical structural elements (such as rods, shafts, beams, etc.) under tension, compression, torsion, bending, and combined loads. The early traditional course "Strength of Materials" did not distinguish between two different concepts of **mechanical response of structures** and **mechanical response of materials**. Scholars at that time did not distinguish between structures and materials. Back to the inequality (1.1), it is not difficult to understand that the Σ on the left side of the inequality involves the structural mechanical response, while the Σ_c on the right side of the inequality, as well as the constitutive relation required for the calculation of the Σ, involves the material mechanical response.

Incidentally, in related literatures, besides the term **mechanical properties** of materials, people will encounter other terms, such as the term of **mechanical behavior** of materials and the term of **mechanical response** of materials. They are related to each other but have different expressions and meanings. Just as one's inner thought determines one's external behavior, it can be understood as that the inner mechanical properties of materials determine the external mechanical behavior of materials, while the mechanical behavior of materials describes the mechanical response of materials to various loads.

Studies on the mechanical properties of materials are becoming more and more important particularly because of all kinds of new engineering materials are constantly emerging, the application field of materials is increasingly extensive, and the service conditions of materials have become more extreme (high temperature, high pressure, high strain rate, strong magnetic field, strong corrosion, intense radiation, and so on). The theme of this book, namely the dynamic mechanical properties of materials under explosion/impact loading, is presently one of the most active directions in the material mechanics field.

The term "**dynamics**" in the "dynamics of materials" is mainly to emphasize the focus on the **dynamic mechanical behavior** of materials under dynamic loading such as explosion and impact. Such intensive dynamic loads are characterized by a short duration, high amplitude, and high variation rate over time. Therefore, unlike the problems under quasi-static loading, for the problems under dynamic loading, as will be discussed in this book as follows, the following characteristics should be taken into account.

1.1 Short duration

To characterize the short loading duration or process time of intensive dynamic loading, let us introduce a characteristic time T_L usually measured in the order of microsecond or even nanosecond. On the other hand, introduce a characteristic time $T_W = L_s/C_w$ to characterize the dynamic response of the structure, where L_s is the characteristic length of the structure and C_w is the characteristic wave velocity of the stress wave propagating in the structure. So the T_W indicates the time a stress wave needs to travel across a structural characteristic length. If the dimensionless time $\overline{T}$ is defined as the ratio between T_L and T_W and is less than 1 ($\overline{T} = T_L/T_W < 1$), namely the loading duration is shorter than the time it takes the stress wave to propagate across the structural characteristic length, then the problem cannot be treated as a quasi-static stress equilibrium problem, as the propagation of stress wave in the structure must be considered. For example, if L_s is in the order of a meter and C_w is in the order of 10^3 m per second, then T_W is in the order of milliseconds. Thus if the characteristic time of an explosion/impact load T_L is in the order of microseconds or even 10^{-1} milliseconds, then the propagation of the stress wave must be taken into account.

1.2 Intensive loading

For explosion/impact loading, the loading stress σ_L is characterized by high amplitude and a wide variation range. In order to express how high the relative amplitude of an intensive loading is, introduce a dimensionless stress $\overline{\sigma} = \sigma_L/\sigma_s$, where σ_s is the characteristic stress to describe the shear strength of materials. For example, in a nuclear explosion center, σ_L can rise to the order of 10^3—10^4 GPa within several microseconds, while the shear strength σ_s of metals is generally in the order of 10^{-1} GPa, then the dimensionless stress $\overline{\sigma}$ is about in the order of 10^4—10^5. Obviously, under such intensive loading, the shear strength of solid materials can be relatively ignored, like a fluid. Thus the solid material can be approximately treated using a hydrodynamic model. However, when $\overline{\sigma}$ was decreased to the order of 10^1—10^0, the shear strength effect of solid materials must be considered. Then it should be treated using the hydro-elastic-plastic model or the hydro-visco-elastic-plastic model.

Under the intensive load, the material will undergo a large strain (geometric nonlinearity), and the constitute relation is also obviously nonlinear (constitutive or physical nonlinearity); these nonlinearities greatly increase the complexity of problem.

1.3 High strain rate

The significant change of a high-amplitude load in a short duration must mean a high-loading rate or high strain rate. Generally, the strain rate under quasi-static loading is in the order of 10^{-5} to 10^{-1} s^{-1}, while the strain rate under explosion/impact loading is as high as 10^2 to 10^7 s^{-1} orders. So the latter is several orders higher than the former. A large number of experiments show that the mechanical behavior of materials is often different at different strain rates (Chou and Hopkins, 1972; Zukas et al., 1982; Meyers, 1994; Wang et al., 1992). From the viewpoint of a material deformation mechanism, only the ideal elastic deformation can be regarded as a transient response. All kinds of inelastic deformation and failure (such as that induced by the dislocation movement process, the stress-induced diffusion process, the damage evolution process, the crack extension and propagation process, etc.) are all nontransient responses taken place at a finite rate. In other words, the mechanical response of materials is essentially related to strain rates. Therefore, when we discuss the dynamic properties of materials, we must consider the effect of the strain rate.

Moreover, in the thermodynamic sense, the deformation or flow of materials at high strain rates is closer to an adiabatic process. Since the thermal softening caused by adiabatic temperature rises will affect the mechanical response of materials, this is a thermomechanical coupling problem. A typical example is the adiabatic shear instability of materials at high strain rates.

Furthermore, the material is often accompanied by phase transformation at intensive loads with high amplitude. Since a phase transformation can lead to a sudden change of the mechanical properties of a material, this is a deformation-phase transformation-coupling problem, which is another topic we are going to talk about.

In summary, compared to mechanical problems under quasi-static loads, mechanical problems under dynamic loads must take into consideration of two main dynamic effects, namely the **inertia effects** of structures and the **strain-rate effects** of materials. The inertia effect leads to the study of wave propagation in various (precise or simplified) forms, promoting the development of stress wave theory and structural impact dynamics; while the strain-rate effect leads to the study of all kinds of rate-dependent constitutive relations and failure criteria under high strain rates, promoting the development of material dynamics. There are differences but also close connections between these two aspects. In retrospect of the generalized strength criterion under quasi-static conditions (Formula 1.1), it should now be rewritten as (Wang, 2003)

$$\Sigma(t) \geq \Sigma_c(\dot{\varepsilon}) \tag{1.2}$$

The $\Sigma(t)$ on the left side of inequality is the characteristic quantity of the unsteady dynamic mechanical field taken into account of the stress wave propagation, and now is a function of time t, while the $\Sigma_c(\dot{\varepsilon})$ on the right side is a critical parameter characterizing the dynamic strength of materials taking into account the effect of the strain rate, and now is a function of the strain rate $\dot{\varepsilon}$.

The study on the stress wave effects of structures is the main objective of the *Foundations of Stress Waves* (Wang, 2007), while the study on the strain-rate effects of materials is the main objective of this book *Dynamics of Materials*. They can be regarded as sister-books to each other.

Dynamics of Materials formulates an interdisciplinary branch that concerns materials science, mechanics, and thermodynamics. In order to help readers to understand the contents more deeply and extensively, in the writing of this book, the authors try to follow such comprehensive approaches, that is, to combine macroscopic continuum mechanics with thermodynamics, to combine the macromechanics expression with the microphysical mechanism, and to combine the theoretical analyses with experimental investigation.

The main contents of this book can be briefly summarized as the study of dynamic constitutive relations and dynamic failure of materials. As we have learned in "elasticity" and "plasticity", the constitutive relation of materials can generally be decomposed into a spherical part and a deviator part. The former describes the law of volumetric change (**volumetric law**), while the latter describes the law of shape change (**distortional law**). When the dimensionless stress $\overline{\sigma}$ is much larger than one, $\overline{\sigma} = \frac{\sigma_L}{\sigma_S} >> 1$, then the shear strength of the material can be ignored, so the distortion law can be neglected accordingly, thus the constitutive relation is reduced to a single volumetric law. It is usually assumed that in the spherical stress (equal axial stress) state, no plastic volumetric deformation occurs and no volumetric strain-rate effect exists (no volumetric viscosity). In such cases the volumetric law is then reduced to the reversible nonlinear elastic law, equivalent to the state equation of inviscid compressible fluid. The volumetric law of solids in the case of $\overline{\sigma} >> 1$ and neglecting the volumetric plasticity and volumetric viscosity is often called the equation of the state of solids under high pressure, which is actually a particular form of the spherical component in a solid constitutive relation. Even when the shear strength of materials cannot be ignored, the state equation of solids under high pressure, as one component of the complete constitutive relation of solids, is indispensable, but it is usually assumed that the volumetric law and the distortional law are decoupled.

In this way, the book is divided into three parts. The first part deals with the volumetric law of materials; the second part deals with the distortional law, taken into account of the strain-rate effects; and the last part discusses the dynamic failure of materials.

PART ONE

Volumetric deformation law of dynamic constitutive relation of materials

The constitutive relation is a mathematical formulation of the mechanical behavior of materials. The strain rate—dependent dynamic constitutive relation of materials is usually expressed as a functional relation between stress tensor σ_{ij}, strain tensor $\boldsymbol{\varepsilon}_{ij}$, strain rate tensor $\dot{\boldsymbol{\varepsilon}}_{ij}$, and temperature T.

It's well known that the stress tensor σ_{ij} can be decomposed into the sum of the spherical stress tensor σ_m and the deviator stress tensor S_{ij}. Similarly the strain tensor $\boldsymbol{\varepsilon}_{ij}$ can be decomposed into the sum of the spherical strain tensor e_m and the deviator strain tensor e_{ij}, as well as the strain rate tensor $\dot{\boldsymbol{\varepsilon}}_{ij}$ can be decomposed into the sum of the spherical strain rate tensor $\dot{e}_m$ and the deviator strain rate tensor $\dot{e}_{ij}$.

$$\sigma_{ij} = \sigma_m \delta_{ij} + S_{ij}, \quad \boldsymbol{\varepsilon}_{ij} = e_m \delta_{ij} + e_{ij}, \quad \dot{\boldsymbol{\varepsilon}}_{ij} = \dot{e}_m \delta_{ij} + \dot{e}_{ij}, \tag{I.1}$$

Thus, the rate-dependent constitutive relation can be usually expressed as two parts: the spherical part, which describes volume changes (the volumetric deformation law in scalar form), and the deviator part, which describes shape changes (the distortional deformation law in functional form):

$$\text{Volumetric deformation law:} \quad \sigma_m = f_v(e_m, \dot{e}_m, T) \tag{I.2}$$

$$\text{Distortional deformation law:} \quad S_{ij} = \overline{\Phi}\left(e_{ij}, \dot{e}_{ij}, T\right), \tag{I.3}$$

Here, it has been supposed that the volumetric deformation law and the distortional deformation law are decoupled, namely they are not related to each other.

Under high pressures that allow the distortional deformation to be negligible, solids can be treated approximately as fluids, thus the constitutive relation degenerates into the volumetric deformation law. If it is supposed further that there is no plastic volumetric deformation under spherical stress (isoaxial stress) and there is no strain rate effect in volumetric deformation (no bulk viscosity), i.e., solids can be approximated to ideal compressible fluids, then the volumetric deformation law is reduced to a nonlinear elastic volumetric deformation law. From a thermodynamic point of view, the relation between mechanical quantities is independent of the "path" but only a function of the "state," which is often referred to as the "state equation for solids under high pressure."

The constitutive volumetric deformation law under the above assumptions will be discussed in this part, including three chapters: the macromechanics and thermodynamic foundation of the law will be discussed in Chapter 2, the microscopic physical foundation of the law in Chapter 3, and the dynamic experimental researches based on shock wave theory in Chapter 4.

CHAPTER TWO

State equation of solids under high pressures and its thermodynamic basis

2.1 Nonlinear elastic volumetric deformation law

From the view of constitutive theory of the continuum mechanics, the state equation of solids under high pressures is equivalent to a nonlinear volumetric deformation law if the plastic and viscous volumetric responses are negligible.

Let us start with the well-known linear generalized Hooke's law, and then generalize it to a nonlinear case.

According to the classical theory of elasticity, the linear constitutive relationship of homogeneous isotropic elastic materials can be expressed by the following generalized Hooke's law:

$$\sigma_{ij} = \lambda\delta_{ij}\varepsilon_{kk} + 2\mu\varepsilon_{ij}$$

where δ_{ij} is a unit tensor, λ and μ are the Lame's coefficients defined as below and both the stress and strain are assumed as positive in tension.

$$\lambda = \frac{E\nu}{(1+\nu)(1-2\nu)}, \tag{2.1}$$

$$\mu = G = \frac{E}{2(1+\nu)}, \tag{2.2}$$

where E, G, and ν are the Young's modulus, the shear modulus, and the Poisson ratio, respectively.

Dynamics of Materials
ISBN: 978-0-12-817321-3
https://doi.org/10.1016/B978-0-12-817321-3.00002-4

In the Lagrange system, introduce the volume strain Δ and the isoaxial pressure P (the negative of the average normal stress) as defined, respectively, by:

$$\Delta = \frac{V - V_0}{V_0} = (1 + \varepsilon_X)(1 + \varepsilon_Y)(1 + \varepsilon_Z) - 1$$

$$\approx \varepsilon_X + \varepsilon_Y + \varepsilon_Z = \varepsilon_{KK} \tag{2.3}$$

$$P = -\frac{1}{3}(\sigma_X + \sigma_Y + \sigma_Z) = -\frac{1}{3}\sigma_{KK}$$

where V and V_0 are the volume and its initial value, respectively.

Introduce further the deviator strain tensor e_{ij} characterizing the distortion (shape deformation), and the corresponding deviator stress tensor S_{ij}:

$$e_{ij} = \varepsilon_{ij} - \frac{1}{3}\Delta\delta_{ij}$$

$$S_{ij} = \sigma_{ij} + P\delta_{ij} \tag{2.4}$$

Then the generalized Hooke's law can be expressed by two parts: the volumetric deformation law (Eq. 2.5a) and the distortional deformation law (Eq. 2.5b).

$$-P = K\Delta, \tag{2.5a}$$

$$S_{ij} = 2Ge_{ij}, \tag{2.5b}$$

where K is the volumetric modulus (bulk modulus), which is related to other elastic coefficients by the following relation:

$$K = \lambda + \frac{2}{3}G = \frac{E}{3(1 - 2\nu)} \tag{2.6}$$

Eq. (2.5a) presents a linear elastic volumetric deformation law when the volumetric modulus K holds a constant, which is valid under the condition of small deformation. However, under the higher isoaxial pressure P much greater than the yield stress of materials Y, i.e., $P >> Y$, the deformation becomes larger, thus the volumetric modulus K is no longer a constant, but a function of the isoaxial pressure, $K(P)$ (or a function of the volume strain, $K(\Delta)$). In a differential form, the nonlinear elastic volumetric deformation law can be expressed as:

$$-dP = K(P)d\Delta \tag{2.7}$$

Note that Eq. (2.7) is a Lagrange expression of the nonlinear elastic volumetric deformation law, and $K(P)$ is called the Lagrange volumetric modulus (bulk modulus)

$$K(P) = -\frac{dP}{d\Delta} = -V_0\frac{dP}{dV} \tag{2.8}$$

If the law is described in terms of the Euler coordinates, the volumetric strain is defined in terms of logarithm strain (true strain) $\widetilde{\Delta}$ as:

$$\begin{aligned}\widetilde{\Delta} &= \int_{V_0}^{V}\frac{dV}{V} = \ln\frac{V}{V_0} = \ln(1+\varepsilon_X) + \ln(1+\varepsilon_Y) + \ln(1+\varepsilon_Z) \\ &= \widetilde{\varepsilon}_x + \widetilde{\varepsilon}_y + \widetilde{\varepsilon}_z\end{aligned} \tag{2.9}$$

then the corresponding volumetric modulus is

$$\widetilde{k}(p) = -\frac{dp}{d\widetilde{\Delta}} = -V\frac{dp}{dV}, \tag{2.10}$$

where p is the isoaxial stress defined in the Euler system. It is called the Euler volumetric modulus (bulk modulus).

The next key point is in what form the nonlinear compressive volumetric modulus depends quantitatively on the isoaxial stress (hydrostatic stress) P.

2.1.1 Bridgman equation

The studies on state equation of solids under high pressure can be firstly traced back to the experimental studies on the relationship between the volumetric modulus and the hydrostatic pressure under static high pressure. Bridgman (1949) studied such relationship for several tens of elements and compounds under the range of static high pressure P about 10^4 bars-10^5 bars $(1-10\text{ GPa})$. Based on his experimental results, he proposed an empirical equation:

$$\frac{V_0 - V}{V_0} = -\Delta = aP - bP^2 \tag{2.11}$$

where a and b are material constants. This equation is usually called the **Bridgman equation** (Bridgman, 1949) or the **isothermal state equation of solids** since the experiments were carried on in the isothermal and quasi-static conditions. For the most materials tested, when P is in the unit of bars, a is in the order of 10^{-6} bar^{-1} to 10^{-7} bar^{-1} and b is in the order of 10^{-12} bar^{-2}. For example, based on the data obtained from more than 400 times tests for iron at temperature 24°C, there is:

$$\frac{V_0 - V}{V_0} = 5.826 \times 10^{-7} P - 0.80 \times 10^{-12} P^2 \tag{2.12}$$

From Eq. (2.11), the Lagrange volumetric modulus K as a function of hydrostatic pressure P can be obtained:

$$K(P) = \frac{1}{a - 2bP} = \frac{1}{a\left(1 - \frac{2b}{a}P\right)}. \tag{2.13}$$

It is clear that K increases with the increase of P, namely the P—V curve is a concave curve as shown in Fig. 2.1. This indicates that the resistance of volumetric compression of solid materials increases with the increase of compressive deformation; in other words, it is more difficult to compress a solid as its volume is more compressed.

As b/a is about in the order of 10^{-5} bar^{-1} to 10^{-6} bar^{-1}, Eq. (2.13) shows that a pressure in the order of 10^3 bars—10^4 bars (0.1 - 1 GPa) is needed at least to obtain a change of 1% in K. This shows why the nonlinearity of elastic volumetric law is considered only in the problems under high pressures. When $2bP/a<1$, extending Eq. (2.13) into a power series and neglecting the small quantities of high orders, we obtain:

$$K(P) \approx \frac{1}{a}\left(1 + \frac{2b}{a}P\right) \tag{2.14}$$

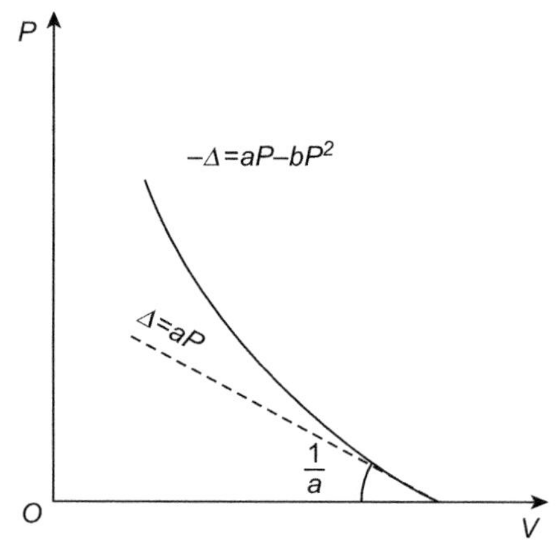

Figure 2.1 The *P*—*V* curve according to the Bridgman equation (Eq. 2.11).

Obviously, $1/a$ is equivalent to the volumetric modulus K_0 of the linear elastic volumetric law under low pressure; while $\frac{2b}{a^2}$ is the coefficient depicting the change of K with P when K is approximated by a linear function of P.

2.1.2 Murnaghan equation

Now let us start from the Euler volumetric modulus (bulk modulus) defined by Eq. (2.10) to discuss the relationship between $\widetilde{k}$ and p. Similar to Eq. (2.14), if assume that $\widetilde{k}$ is a linear function of p, namely

$$\widetilde{k} = k_0(1 + \eta p), \tag{2.15}$$

where k_0 and η are material constants, then the $p-V$ relationship can be written as an ordinary differential equation

$$-V\frac{dp}{dV} = k_0 + n_k p, \tag{2.16}$$

where $n_k = k_0\eta$, is the coefficient of the first-order derivatives of $\widetilde{k}$ with respect to pressure p. From the initial condition $V|_{P=0} = V_0$, we obtain

$$\left(p + \frac{1}{\eta}\right)\left(\frac{V}{V_0}\right)^{n_k} = \frac{1}{\eta} = const., \tag{2.17a}$$

or rewritten as

$$p = \frac{k_0}{n_k}\left\{\left(\frac{V_0}{V}\right)^{n_k} - 1\right\} \tag{2.17b}$$

In the one dimensional (1-D) strain condition ($V=(1+\varepsilon_X)V_0$), it is reduced to

$$p = \frac{k_0}{n_k}\left\{(1 + \varepsilon_X)^{-n_k} - 1\right\} = \frac{K_0}{n_k}\left\{\exp(-n_k\widetilde{\varepsilon}_x) - 1\right\}. \tag{2.18}$$

It should be noticed that Eq. (2.17a) is formally similar to the following isentropic state equation for ideal gases under the adiabatic condition.

$$p\left(\frac{V}{V_0}\right)^{\gamma} = \text{constant} \tag{2.19}$$

And the n_k in Eq. (2.17a) is similar to the adiabatic exponent γ in the above equation for ideal gases. For metals, n_k typically is taken as 4. Eq. (2.17) or (2.18) is usually called the **Murnaghan equation** or the

isentropic state equation of solids, where material constant K_0 and n_k can be obtained by experiments on isentropic wave propagation.

The Murnaghan equation was first derived from the research on the elastic theory of finite deformation by performing quite complicated mathematical calculations. As shown above, it is actually equivalent to the assumption that $\widetilde{k}$ is proportional to p (Eq. 2.15), namely the Euler volumetric modulus increases linearly with the increase of hydrostatic pressure. Obviously, the $p-V$ curve depicted by Eq. (2.16) is also a concave curve, formally similar to that in Fig. 2.1. This indicates also that the resistance of volumetric compression of solid materials increases with the increase of compressive deformation; in other words, it is more difficult to compress a solid as its volume is more compressed.

2.2 Thermodynamic equation of state

It was pointed out in the beginning of this part that solids can be treated approximately as fluids under high pressures that allow the distortional deformation of a material to be negligible. In this case, the nonlinear elastic volumetric deformation law is equivalent to the equation of state of solids under high pressures in the thermodynamics.

The equation of state (EOS) that the first time the reader learns may be the equation of state of ideal gas

$$P = nR\frac{T}{V} \tag{2.20}$$

where $R(= 8.314\,\mathrm{J/(mol\cdot K)})$ is called gas constant, n is the molar number of gas. The reason of calling the above formula as the equation of state is that the relation between P, V, and T is only dependent on their states but independent on the path of the state change. In a geometric sense, Eq. (2.20) can be regarded as a surface in a $P-V-T$ three-dimensional space, belonging to the more general type, the so-called temperature-type equation of state:

$$P = f_T(V, T) \quad \text{or} \quad f_T(P, V, T) = 0 \tag{2.21}$$

Please note that the Lagrange description is used here and below.

Reviewing the Bridgman equation and the Murnaghan equation, they are pure mechanical equation in form, where only mechanical parameters are involved. In fact, these two equations depict the relations between P and V in an isothermal process and an isentropic process, respectively.

Both are equations of state of solids in the specific thermodynamic conditions.

Similar to the temperature-type EOS described by Eq. (2.21), the entropy-type EOS described in the Lagrange system can be expressed by the following formula:

$$P = f_S(V, S) \text{ or } f_S(P, V, S) = 0 \tag{2.22}$$

Obviously, the Bridgman equation is nothing more than an intersection curve of the surface (Eq. 2.21) with an isothermal plane, and the intersection curve of the surface (Eq. 2.22) with an isentropic plane depicts an isentropic $P-V$ relation in the Lagrange description.

The slopes of these intersection curves determine, respectively, the Lagrange isothermal bulk modulus K_T

$$K_T = -V_0\left(\frac{\partial P}{\partial V}\right)_T \tag{2.23a}$$

$$k_T = -V\left(\frac{\partial P}{\partial V}\right)_T = (1+\Delta)K_T \tag{2.23b}$$

and the Lagrange isentropic bulk modulus K_S

$$K_S = -V_0\left(\frac{\partial P}{\partial V}\right)_S, \tag{2.23c}$$

$$k_S = -V\left(\frac{\partial P}{\partial V}\right)_S = (1+\Delta)K_S \tag{2.23d}$$

Note the two different expressions for the Lagrange bulk modulus K_T and k_T as well as K_S and k_S, as defined in Eq. (2.23), respectively. In addition, it should be emphasized to distinguish the Lagrange isentropic bulk modulus k_S with the Euler isentropic bulk modulus $\tilde{k}_s$ defined by Eq. (2.10).

In the classical elasticity, the distinction between the Lagrange description and the Euler description is usually negligible because of small deformation. And people also did not pay much emphasis on the distinction between the isothermal bulk modulus K_T and the isentropic bulk modulus K_S. But from thermodynamic viewpoint, particularly in the case of large deformation, people should pay attention to such distinction, which will be emphasized further in the following.

Eqs. (2.21) and (2.22) establish the relation between the mechanical quantities P, V and the thermal quantities T, S, reflecting the intrinsic intercoupling of mechanical properties and thermal properties of thermal elastic materials. Note that in the mechanical quantities, P is a quantity having intensive property and V is a quantity having extensive property, while in the thermal quantities, T is a quantity having intensive property and S is a quantity having extensive property.

However, starting from Eqs. (2.21) and (2.22), only a part of thermodynamic properties of materials, not all thermodynamic state parameters, can be determined in a thermodynamic equilibrium state (as will be discussed below). Thus, such equations can only be regarded as the **incomplete state equations**.

When P, V, T, and S are considered as the state variables in thermodynamics, there are four thermodynamic potential functions which constitute four complete equations of state. Matching one of the mechanical variables P, V and one of the thermal variables T, S as the independent state variables, respectively, four **complete equations of state** can be constructed based on the following four thermodynamic characteristic functions (or potential functions), see Fig. 2.2

$$
\begin{aligned}
E &= E(S, V) \\
A &= A(T, V) \\
H &= H(P, S) \\
G &= G(P, T)
\end{aligned}
\tag{2.24a}
$$

where E is the internal energy; A, the Helmholtz free energy; H, the enthalpy; and G, the Gibbs free energy.

The interrelations among them are described by the as-following relations, and graphically represented in Fig. 2.3.

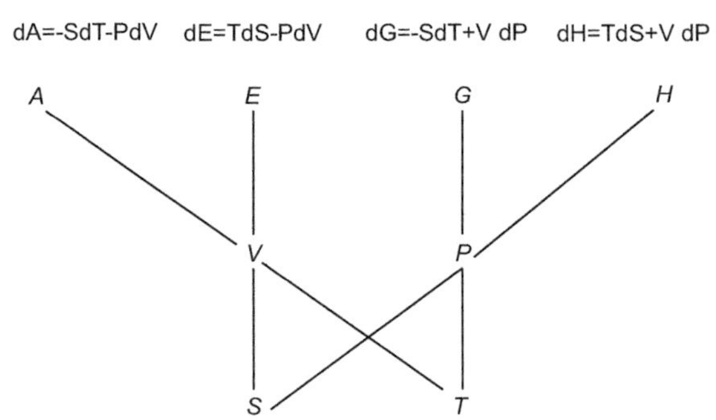

Figure 2.2 Four complete equations of state as the functions of P, V, T, and S, respectively.

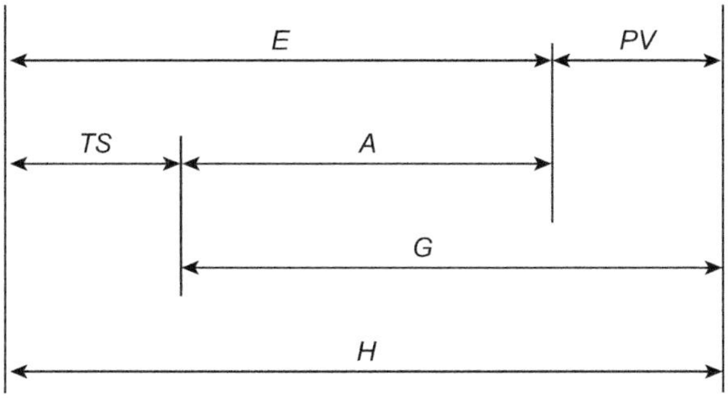

Figure 2.3 The interrelations among the *E*, *A*, *G*, and *H*.

$$\begin{aligned} A &= E - TS \\ E &= A + TS \\ G &= H - TS \\ H &= E + PV \end{aligned} \tag{2.24b}$$

As an example, the internal energy $E(S, V)$, one of the four thermodynamic characteristic functions, is firstly discussed below.

According to the first thermodynamics law (the energy conservation law), the increment in the internal energy of a system dE equals to the algebra sum of the heat increment gotten from outside δQ and the work increment δW that the system works toward outside.

$$dE = \delta Q - \delta W \tag{2.25}$$

Here the internal energy E is a state variable, but the heat Q and the work W are generally the process parameters which are dependent on the path.

For a reversible equilibrium process, on the one hand, the reversible work can be represented as $dW = PdV$; on the other hand, the state variable entropy S, which represents the confusion degree of the system, can be represented as $dS = dQ/T$ according to the second thermodynamics law, so we have:

$$dE = Tds - PdV. \tag{2.26}$$

If take a total differential of $E(S, V)$ directly (see Eq. 2.24a), we have

$$dE = \left(\frac{\partial E}{\partial S}\right)_V dS + \left(\frac{\partial E}{\partial V}\right)_S dV$$

Comparing Eq. (2.26) with above equation, we have:

$$T(S,V) = \left(\frac{\partial E(S,V)}{\partial S}\right)_V, \tag{2.27}$$

$$P(S,V) = -\left(\frac{\partial E(S,V)}{\partial V}\right)_S \tag{2.28}$$

Note that Eq. (2.28) is the right entropy-type EOS previously expressed by Eq. (2.22). It can be seen that the entropy-type EOS can be obtained by just taking the first partial derivative of internal energy characteristic function $E(S, V)$. More than that, once $E(S, V)$ is determined, other thermodynamic parameters can be determined too in turn. For example, by taking the second partial derivative, we have:

$$\text{isometric specific heat}: \ C_V \equiv T\left(\frac{\partial S}{\partial T}\right)_V = \frac{T}{\left(\frac{\partial T}{\partial S}\right)_V} = \frac{\left(\frac{\partial E}{\partial S}\right)_V}{\left(\frac{\partial^2 E}{\partial S^2}\right)_V}, \tag{2.29}$$

$$\text{isentropic bulk modulus}: \ k_S \equiv -V\left(\frac{\partial P}{\partial V}\right)_S = V\left(\frac{\partial^2 E}{\partial V^2}\right)_S, \tag{2.30}$$

$$\text{entropy stress function}: \ \phi_s \equiv \left(\frac{\partial P}{\partial S}\right)_V = -\frac{\partial^2 E}{\partial S \partial V} = -\left(\frac{\partial T}{\partial V}\right)_S, \tag{2.31}$$

and so on. Thus, the internal energy $E(S, V)$given in Eq. (2.24a) is a **complete EOS.**

Taking the Helmholtz free energy $A(T, V)$ as an example again, another type of complete EOS can be obtained. In fact, from Eq. (2.24), we have

$$dA = dE - SdT - TdS$$

Using Eq. (2.26), the above formula can be simplified to:

$$dA = -SdT - PdV \tag{2.32}$$

Comparing it with Eq. (2.26), it can be seen that in an isentropic process ($dS = 0$), the work done by the system is equal to the reduction of the internal energy, while in an isothermal process, the work done by the system is equal to the reduction of the Helmholtz free energy.

If taking a total differential of Eq. (2.24b) directly, we have:

$$dA = \frac{\partial A}{\partial T} dT + \frac{\partial A}{\partial V} dV$$

Comparing Eq. (2.32) with the above formula, it can be seen that the $P(T,V)$ and $S(T,V)$ can be obtained from the first partial derivative of A, namely

$$S(T, V) = -\left(\frac{\partial A}{\partial T}\right)_V, \tag{2.33}$$

$$P(T, V) = -\left(\frac{\partial A}{\partial V}\right)_T \tag{2.34}$$

Correspondingly, if transferring the functional form of the internal energy as $E\ (T,\ V)$, we get the following formula from the above formula:

$$E(T, V) = A + TS = A(T, V) - T\left(\frac{\partial A}{\partial T}\right)_V \tag{2.35}$$

Note that Eq. (2.34) is also a temperature-type EOS, as expressed by the previous Eq. (2.21). Thus the temperature-type EOS can be obtained by just taking the first partial derivative of the Helmholtz free energy characteristic function $A(T,\ V)$.

Similar to the previous example, other thermodynamic parameters can be also determined by taking the second partial derivate of A. For example:

$$\text{isometric specific heat}: \ C_V \equiv T\left(\frac{\partial S}{\partial T}\right)_V = -T\left(\frac{\partial^2 A}{\partial T^2}\right)_V, \tag{2.36}$$

$$\text{isothermal bulk modulus}: \ k_T \equiv -V\left(\frac{\partial P}{\partial V}\right)_T = V\left(\frac{\partial^2 A}{\partial V^2}\right)_T, \tag{2.37}$$

$$\text{temperature stress function}: \ \phi_T \equiv \left(\frac{\partial P}{\partial T}\right)_V = -\left(\frac{\partial^2 A}{\partial T \partial V}\right) = \left(\frac{\partial S}{\partial V}\right)_T, \tag{2.38}$$

Thus, the Helmholtz free energy $A(T,\ V)$ given in Eq. (2.23b) is a **complete EOS** too.

By similar approach, it is easy to prove that the enthalpy $H(S,P)$ and the Gibbs free energy $G(T,\ P)$ given in Eq. (2.24) are all also **complete EOS**s.

Correspondingly, by taking the first and the second partial derivates, we have the following relations:

2.2.1 Enthalpy: *H*(*S*,*P*)

$$dH = TdS + VdP \tag{2.39}$$

$$T = \left(\frac{\partial H}{\partial S}\right)_P, \tag{2.40}$$

$$V = \left(\frac{\partial H}{\partial P}\right)_S, \tag{2.41}$$

$$\text{constant pressure specific heat}: \ C_P \equiv T\left(\frac{\partial S}{\partial T}\right)_P = \frac{T}{\left(\frac{\partial^2 H}{\partial S^2}\right)_P} \tag{2.42}$$

$$\text{isentropic bulk modulus}: \ k_S \equiv -V\left(\frac{\partial P}{\partial V}\right)_S = -\frac{V}{\left(\frac{\partial^2 H}{\partial P^2}\right)_S} \tag{2.43}$$

$$\text{entropy expansion coefficient}: \ \alpha_S = \frac{1}{V}\left(\frac{\partial V}{\partial S}\right)_P = \frac{1}{V}\left(\frac{\partial^2 H}{\partial P \partial S}\right) \tag{2.44}$$

2.2.2 Gibbs free energy *G*(*T*,*P*)

$$dG = -SdT + VdP \tag{2.45}$$

$$S = -\left(\frac{\partial G}{\partial T}\right)_P, \tag{2.46}$$

$$V = \left(\frac{\partial G}{\partial P}\right)_T, \tag{2.47}$$

$$\text{constant pressure specific heat}: \ C_P \equiv T\left(\frac{\partial S}{\partial T}\right)_P = -T\left(\frac{\partial^2 G}{\partial T^2}\right)_P, \tag{2.48}$$

$$\text{isothermal bulk modulus}: \ k_T \equiv -V\left(\frac{\partial P}{\partial V}\right)_T = -\frac{V}{\left(\frac{\partial^2 G}{\partial P^2}\right)_T}, \tag{2.49}$$

$$\text{coefficient of thermal expansion}: \ \alpha_T = \frac{1}{V}\left(\frac{\partial V}{\partial T}\right)_P = \frac{1}{V}\left(\frac{\partial^2 G}{\partial P \partial T}\right), \tag{2.50}$$

It should be pointed out that the above thermodynamic parameters are not independent to each other. As we know in the classical elasticity, strain component $\boldsymbol{\varepsilon}_{ij}(i,j=1,2,3)$ is derived from the first partial derivates of the three displacement components u_k $(k=1,2,3)$. Thus, there should be some compatibility relationships between them in order to satisfy the continuity of displacements. Similarly, the above thermodynamic parameters are related with four thermodynamic characteristic functions represented in Eq. (2.24) and derived from the first, the second partial derivates of the functions. Thus, they are not independent to each other; some intrinsic compatibility relations must exist between them in order to satisfy the coordination requirements.

In the following, some intrinsic compatibility relations between the eight thermodynamic parameters which are derived from the second partial derivates of the above thermodynamic characteristic functions will be specifically discussed. On the one hand, these intrinsic compatibility relations are often used in the study on the EOS of materials, and on the other hand, these relations will help us to deepen our understanding of these thermodynamic parameters further.

It should be first pointed that the temperature stress function ϕ_T (Eq. 2.38), the isothermal bulk modulus k_T (Eq. 2.37), and the coefficient of thermal expansion α_T (Eq. 2.50) are related with each other. In fact, taking a total differential of temperature-type EOS, we have:

$$dP(T,V) = \left(\frac{\partial P}{\partial T}\right)_V dT + \left(\frac{\partial P}{\partial V}\right)_T dV \text{。} \tag{2.51}$$

For an isobaric process ($dP=0$), from the above equation we have:

$$0 = \left(\frac{\partial P}{\partial T}\right)_V + \left(\frac{\partial P}{\partial V}\right)_T \left(\frac{\partial V}{\partial T}\right)_P$$

According to Eqs. (2.37), (2.38), and (2.50), we have:

$$\phi_T \equiv \left(\frac{\partial P}{\partial T}\right)_V = -\left(\frac{\partial P}{\partial V}\right)_T \cdot \left(\frac{\partial V}{\partial T}\right)_P = -V\left(\frac{\partial P}{\partial V}\right)_T \cdot \frac{1}{V}\left(\frac{\partial V}{\partial T}\right)_P$$
$$= k_T \cdot \alpha_T, \tag{2.52}$$

It means that the temperature stress function ϕ_T equals to the product of the isothermal bulk modulus k_T and the coefficient of thermal expansion α_T.

Integrating above formula with respect to T, we have:

$$P(T, V) = P_K(V) + \int_0^T k_T \cdot \alpha_T dT = P_K(V) + P_T(T, V), \tag{2.53}$$

where $P_K(V)$ is only related to the specific volume V, and has nothing to do with the temperature T. So it is called the **cold pressure**. But $P_T(V,T)$ is determined by integrating ($k_T \cdot \alpha_P$) with respect to T and thus is called the **thermal pressure**. It can be seen that the pressure P in the temperature-type EOS is composed of two parts: the cold pressure $P_K(V)$ (had nothing to do with temperature) and the thermal pressure $P_T(V,T)$ (related with temperature), which in practice can be determined by $k_T, \alpha_P,$ and $P_k(V)$.

On the other hand, taking a partial derivative of $E(T, V)$ (expressed in Eq. 2.35) with respect to T under an isometric condition, and considering the definition of isometric specific heat (expressed in Eq. 2.36), we have:

$$\left(\frac{\partial E}{\partial T}\right)_V = \left(\frac{\partial A}{\partial T}\right)_V - \left(\frac{\partial A}{\partial T}\right)_V - T\left(\frac{\partial^2 A}{\partial T^2}\right)_V = -T\left(\frac{\partial^2 A}{\partial T^2}\right)_V = C_V \tag{2.54}$$

Integrating the above formula with respect to T, we have:

$$E(T, V) = E_K(V) + \int_0^T C_V dT = E_K(V) + E_T(T, V), \tag{2.55}$$

where $E_K(V)$ is only related to the specific volume V, and has nothing to do with T. So it is called the **cold energy**. But $E_T(V,T)$ is related to T and called the **thermal energy**. It can be seen that the internal energy $E(T, V)$ is also composed of two parts: the cold energy $E_K(V)$ (had nothing to do with temperature) and the thermal energy $E_T(V,T)$ (related with temperature), which in practice can be determined by C_V and $E_K(V)$.

Furthermore, under an isentropic condition ($dS = 0$), Eq. (2.51) can be rewritten as follow:

$$\left(\frac{\partial P}{\partial V}\right)_S = \left(\frac{\partial P}{\partial V}\right)_T + \left(\frac{\partial P}{\partial T}\right)_V \left(\frac{\partial T}{\partial V}\right)_S,$$

According to Eqs. (2.30), (2.31), (2.37), and (2.38), hence:

$$k_S = k_T + \phi_T \cdot \phi_S, \tag{2.56}$$

Due to $\phi_T \cdot \phi_S > 0$, the isentropic bulk modulus k_S must be greater than the isothermal bulk modulus k_T.

Similar to that by taking a total differential of a thermal EOS to get Eq. (2.51), now taking a total differential of an entropic EOS, *P(S, V)*, we have:

$$dP(S, V) = \left(\frac{\partial P}{\partial S}\right)_V dS + \left(\frac{\partial P}{\partial V}\right)_S dV \tag{2.57}$$

For an isobaric process ($dP = 0$), from the above equation, we have:

$$0 = \left(\frac{\partial P}{\partial S}\right)_V + \left(\frac{\partial P}{\partial V}\right)_S \left(\frac{\partial V}{\partial S}\right)_P$$

According to Eqs. (2.31), (2.30) and (2.44), hence:

$$\phi_S \equiv \left(\frac{\partial P}{\partial S}\right)_V = -\left(\frac{\partial P}{\partial V}\right)_S \cdot \left(\frac{\partial V}{\partial S}\right)_P = -V\left(\frac{\partial P}{\partial V}\right)_S \cdot \frac{1}{V}\left(\frac{\partial V}{\partial S}\right)_P = k_S \cdot \alpha_S \tag{2.58}$$

It means that the entropy stress function ϕ_S equals to the product of the isentropic bulk modulus k_S and the entropy expansion coefficient α_S.

If following the differential rule of composite functions, moreover noting the following relation

$$\left(\frac{\partial P}{\partial T}\right)_V = \left(\frac{\partial P}{\partial S}\right)_V \left(\frac{\partial S}{\partial T}\right)_V$$

and furthermore according to Eqs. (2.29), (2.31), and (2.38), then the ratio of the thermal stress function ϕ_T to the entropic stress function ϕ_S is obtained as:

$$\frac{\phi_T}{\phi_S} = \frac{C_V}{T} \tag{2.59}$$

Similarly, following the differential rule of composite functions, moreover noting the following relation

$$\left(\frac{\partial V}{\partial T}\right)_P = \left(\frac{\partial V}{\partial S}\right)_P \left(\frac{\partial S}{\partial T}\right)_P,$$

and furthermore according to Eqs. (2.44), (2.48), and (2.50), the ratio of the coefficient of thermal expansion to the entropic expansion coefficient is obtained as:

$$\frac{\alpha_T}{\alpha_S} = \frac{C_P}{T} \tag{2.60}$$

Divide Eq. (2.60) by Eq. (2.59), we have:

$$\frac{\phi_T}{\alpha_T} = \frac{\phi_S}{\alpha_S} \frac{C_V}{C_P} \tag{2.61}$$

Then by substituting Eqs. (2.52) and (2.58), we obtain finally:

$$\frac{k_T}{k_S} = \frac{C_V}{C_P} = \frac{\frac{\phi_T}{\alpha_T}}{\frac{\phi_S}{\alpha_S}} \tag{2.62a}$$

It is seen that the ratio of the thermal bulk modulus and the isentropic bulk modulus equals just to the ratio of the constant volume-specific heat and the constant pressure-specific heat, and equals also to the ratio of $\left(\frac{\phi_T}{\alpha_T}\right)$ and $\left(\frac{\phi_S}{\alpha_S}\right)$. On the other hand, it is also seen that from Eqs. (2.52), (2.56), and (2.59), there is the following relation between the isothermal volume modulus and the isentropic volume modulus:

$$\frac{k_S}{k_T} = 1 + \frac{k_T \alpha_T^2 T V}{C_V} \tag{2.62b}$$

From the above discussion, it can be seen that very close and mutual coupling relations exist between the mechanical quantities and the thermal quantities. These relations will be often used in further studies on the EOS of materials.

2.3 Grüneisen Equation of state

We have pointed out that the equations based on the four thermodynamics characteristic functions (Eq. 2.24a) are all complete equations of state. Once we can find any one of those equations, all the related thermodynamic quantities in the reversible equilibrium thermodynamics can be obtained.

But such complete state equations are difficult to be found. Their specific forms are also difficult to be determined by experiments. Thus, in practice, what can be used to describe the equations of state for solids under high pressures are still some incomplete equations of state. For example, as discussed previously, the so-called temperature-type EOS (e.g., Eq. 2.21) can be determined by a series of quasi-static and isothermal experiments or the so-called entropy-type EOS (e.g., Eq. 2.22) can be determined by a series of dynamic isentropic experiments (e.g., the stress wave propagation experiments).

However, if a shock adiabatic process should be dealt with, which generally occurs in the researches on "dynamics of materials", neither the temperature-type EOS (e.g., Eq. 2.21) nor the entropy-type EOS (e.g., Eq. 2.22) is suitable. Because if the distortion is negligible, the state parameters appeared in the conservation equations (the conservations of mass, momentum, and energy) across a shock wave front only contain the hydrostatic pressure P, the specific volume V, and the internal energy E, while the temperature T and the entropy S are not concerned directly. In such cases, it is convenient to use an internal energy-type state equation, which relates P, V, and E:

$$f_E(P, V, E) = 0, \tag{2.63a}$$

Of course, Eq. (2.63a) is still an incomplete equation of state, since it is impossible to determine all the thermodynamic state parameters by this type of EOS.

Such an internal energy-type state equation, in principle, can be derived from the internal energy-type complete state equation $E = E(S, V)$. In fact, the entropy S can be inversely solved from $E = E(S, V)$ as a function of E and V, i.e., $S = S(E, V)$, then substitute it into the entropy-type state equation $P = P(S, V)$, the following internal energy-type EOS is obtained:

$$P = P(S(E, V), V) = P(E, V) \tag{2.63b}$$

Now the question becomes what the specific form of this internal energy-type EOS is, and how to determine the material parameters involved in the equation.

The total differential of Eq. (2.63b) is:

$$dP(E, V) = \left(\frac{\partial P}{\partial E}\right)_V dE + \left(\frac{\partial P}{\partial V}\right)_E dV$$

The first partial derivative on the right side, $\left(\frac{\partial P}{\partial E}\right)_V$, representing the pressure increment per unit internal energy under the isometric condition, is of great significance in the study on internal energy-type EOS.

According to Eqs. (2.27), (2.29), and (2.38), it can be expressed as the quotient of the temperature stress function ϕ_T and the isometric specific heat C_V.

$$\left(\frac{\partial P}{\partial E}\right)_V = \frac{\left(\frac{\partial P}{\partial T}\right)_V}{\left(\frac{\partial E}{\partial T}\right)_V} = \frac{\left(\frac{\partial P}{\partial T}\right)_V}{\left(\frac{\partial E}{\partial S}\right)_V \left(\frac{\partial S}{\partial T}\right)_V} = \frac{\phi_T}{C_V} = \frac{k_T \alpha_T}{C_V} \tag{2.64}$$

For the last equal sign in the above equation, Eq. (2.52) has been taken into consideration.

From this, a new thermodynamic parameter is introduced:

$$\Gamma = V\left(\frac{\partial P}{\partial E}\right)_V = V \cdot \frac{\phi_T}{C_V} = \frac{k_T \alpha_T V}{C_V}, \tag{2.65}$$

which represents the change rate of the pressure with the internal energy change in unit volume under isometric condition, and is usually called the **Grüneisen coefficient**.

The last equal sign in Eq. (2.65) provides a very important relation, called the **Grüneisen's second law**, because the Grüneisen coefficient can be determined by several thermodynamic parameters k_T, α_T, V, and C_V; all are easy to be measured by experiments under normal conditions. Since the value of Γ obtained from such basic relation is determined indirectly by other thermodynamic experiments, is thus usually called it as the **thermodynamic Γ**, and is often represented by the symbol Γ_{th}.

Because the temperature stress function ϕ_T and the isometric specific heat C_V are all the functions of V and T, the Grüneisen coefficient Γ is in principle a function of V and T too, $\Gamma = \Gamma(V, T)$. But, a lot of experimental results show that the Γ is less sensitive to the temperature. Therefore in practice it can be assumed that the Γ is a function of only the specific volume, often represented by the symbol γ:

$$\gamma = \gamma(V) \tag{2.66}$$

This hypothesis is called the **Grüneisen's hypothesis** and it is a very important hypothesis in practice. Applying the hypothesis to Eq. (2.65) and taking integration, we have:

$$P(E, V) = \frac{\gamma(V)}{V} E + \Pi(V), \tag{2.67a}$$

Or taking into consideration of Eqs. (2.53) and (2.55), Eq. (2.67a) can be rewritten as:

$$P = P_K(V) + \frac{\gamma(V)}{V}\{E - E_K(V)\}, \tag{2.67b}$$

$$P_T(T, V) = \frac{\gamma(V)}{V} E_T(T, V) \tag{2.67c}$$

For convenience, the two terms in Eq. (2.67a) that contain only V are often expanded to be polynomials

$$P = A_0 + A_1\mu + A_2\mu^2 + A_3\mu^3 + \left(B_0 + B_1\mu + B_2\mu^2\right)E, \tag{2.67d}$$

where $\mu = \frac{\rho - \rho_0}{\rho_0}$, $\rho = \frac{1}{V}$, and the initial density $\rho_0 = \frac{1}{V_0}$. Note that the P_K (V) and P_T (T, V) in the equation are exactly the cold pressure and the thermal pressure given in Eq. (2.53), and the $E_K(V)$ and $E_T(T, V)$ are the cold energy and the thermal energy, respectively, given in Eq. (2.55). Eq. (2.67) is the so-called **Mie-Grüneisen EOS** (Mie, 1903; Grüneisen, 1912). Historically, Mie, G. in 1903 developed an intermolecular potential for deriving high-temperature EOS of solids (Mie, 1903). Grüneisen, E. then in 1912 extended Mie's model to temperatures below the Debye temperature at which quantum effects become important (Grüneisen, 1912).

Physically, the cold energy E_k which has nothing to do with temperature is the internal energy at temperature 0°K, and is also called lattice potential energy. In other words, it contains all the temperature-independent components of internal energy, such as the interaction energy (lattice binding energy) between molecules (ions, atoms, ...), the vibration energy at 0°K, and the compression energy of valence-electron gas, and so on. The temperature-dependent thermal energy E_T (T, V) is also called lattice kinetic energy, which contains all the temperature-dependent components of internal energy, such as the lattice thermal vibration energy and the thermal activation energy of electrons, and so on (see Chapter 3).

The cold pressure $P_K(V)$ and thermal energy $P_T(T, V)$ are corresponding to the cold energy $E_K(V)$ and the thermal energy $E_T(V, T)$, respectively. In fact, from Eq. (2.26)

$$dE = Tds - PdV$$

and by taking account of that the entropic at 0°K is 0 (Nernst theorem), we have:

$$P_K(V) = -\frac{dE_K(V)}{dV} \tag{2.68}$$

Furthermore, taking account of Eq. (2.64), the thermal pressure $P_T(T, V)$ in Eq. (2.53) becomes

$$P_T(V,T) = \int_0^T k_T \cdot \alpha_T dT = \int_0^T \frac{\Gamma C_V}{V} dT. \tag{2.69}$$

Using Gruneisen hypothesis and taking account of Eq. (2.55), the above equation becomes the Mie–Gruneisen in the form of Eq. (2.67c). This indicates that although the thermal pressure and thermal energy themselves are the functions of temperature T and specific volume V, but the ratio of them has nothing to do with temperature, and it is only the function of specific volume V. This is another manifestation of the physical meaning of Gruneisen hypothesis. And this is corresponding to the assumption that the lattice vibration frequency is only a function of atomic spacing (specific volume) in the crystal thermal vibration theory, called the simple harmonic vibration of lattice.

A further deeper understanding of the Gruneisen equation depends on the knowledge of microstructure of matter. Only in this way, the specific expressions of the cold energy $E_K(V)$, the cold pressure $P_K(V)$ and the Gruneisen coefficient $\gamma(V)$ can be determined further, and the physical meaning they represent can be understood well. These will be discussed in more detail in the next chapter.

The Grüneisen equation is the most commonly used EOS in the study on shock wave propagation in solids under high pressures. The first reason is that the Grüneisen equation is an internal energy-type equation. Thus it can be directly related to the P, V, and E in the three conservation equations of the governing equations for shock waves. The second reason is that all of the relevant parameters in the equation can be determined by experiments.

For any form of the Gruneisen equation, the most key issue is how to determine the Grüneisen coefficient Γ. Although it is usually assumed that the Grüneisen coefficient is a function of only the specific volume V, but what the most uncertain and the hardest to be determined is still the simplified Grüneisen coefficient $\gamma(V)$. Fortunately, it has been found that the error caused by the uncertainty of $\gamma(V)$ in the analyses of shock wave is not large, and the following Eq. (2.70) can be approximately used in practical engineering:

$$\frac{\gamma(V)}{V} = \frac{\Gamma_0}{V_0} = const. \tag{2.70}$$

Or in a more rough analysis, it is even approximately to take $\gamma = \Gamma_0 = const.$, and the value of Γ_0 can be determined by the known thermodynamic parameters as shown in Eq. (2.65). The calculated values of thermodynamic Grüneisen parameters for some typical materials are listed in Tables 2.1 and Table 2.2. (Tsien Hsue-shen, 1962)

The approximate hypothesis (Eq. (2.66) that the Grüneisen coefficient is only dependent on V is usually valid only in a certain pressure range. When the high pressure extends to a larger range, the Grüneisen hypothesis is no longer valid. In this case, the following form of internal energy-type EOS is commonly used in explosion/impact dynamics:

$$P = \left[a + \frac{b}{1 + \dfrac{E}{E_0\eta^2}}\right]\rho_0\eta E + A\mu + B\mu^2 \tag{2.71}$$

This equation is called the **Tillotson** equation (Tillotson, 1962), where $\eta = \mu+1 = \rho/\rho_0 = V_0/V$, ρ_0 and ρ are the initial density and current density, respectively; E is the specific internal energy (the internal energy in unit mass material); parameters a, b, A, and B are determined by fitting the experimental data of shock waves and the Thomas-Fermi-Dirac data under the maximum compression pressure (the monatomic gas state), see Table 2.3. As can be seen by comparing Eq. (2.67a) with Eq. (2.71) that if the Gruneisen coefficient is taken as a function of V and E, $\Gamma(V,E)$, and if the $\Gamma(V,E)$ and $\Pi(V)$ in Eq. (2.67a) are, respectively, taken as follows

$$\Gamma(V, E) = \left[a + \frac{b}{1 + \dfrac{EV^2}{E_0 V_0^2}}\right], \tag{2.72a}$$

Table 2.1 The calculated values of thermodynamic Gruneisen parameters for several elements (according to Eq. 2.65).

Element	Atomic weight	Density (g/cm^3)	Mole volume (cm^3/mol)	α ($\times 10^6\ ^\circ C^{-1}$)	k_T^{-1} ($\times 10^{12}\ cm^2/dyn$)	C_V ($\times 10^{-7} erg/(mol \cdot ^\circ C)$)	Γ_o
Li	6.94	0.546	12.7	180	8.9	22.0	1.17
Na	23.0	0.971	23.7	216	15.8	26.0	1.25
K	39.10	0.862	45.5	250	33	25.8	1.34
Rb	85.50	1.530	56.0	270	40	25.6	1.48
Cs	132.8	1.87	71.0	290	61	26.2	1.29
Cu	63.57	8.92	7.1	49.2	0.75	23.7	1.96
Ag	107.88	10.49	10.3	57	1.01	24.2	2.40
Au	197.2	19.2	10.3	43.2	0.59	24.9	3.03
Al	26.97	2.70	10.0	67.8	1.37	22.8	2.17
C (diamond)	12.00	3.51	3.42	2.91	0.16	5.66	1.10
Pb	207.2	11.35	18.2	86.4	2.30	25.0	2.73
P (white phosphorus)	31.04	1.83	17.0	370	20.5	24	1.28
Ta	181.5	16.7	10.9	19.2	0.49	24.4	1.75
Mo	96.0	10.2	9.5	15.0	0.36	25.2	1.57
W	184.0	19.2	9.6	13.0	0.30	25.8	1.62
Mn	54.93	7.37	7.7	63	0.84	23.8	2.42
Fe	55.84	7.85	7.1	33.6	0.60	24.8	1.60
Co	58.97	8.8	6.7	37.2	0.55	24.2	1.87
Ni	58.68	8.7	6.7	38.1	0.54	25.2	1.88
Pd	106.7	12.0	8.9	34.5	0.54	25.6	2.23
Pt	195.2	21.3	9.2	26.7	0.38	25.5	2.54

From Tsien Hsue-Shen. 1962. Notes on Physical Mechanics, Science Press, Beijing (in Chinese), p. 201.

Table 2.2 The calculated value of thermodynamic Gruneisen parameters for several ion crystal (according to Eq. (2.65)).

Crystal	Molecular weight	Density (g/cm^3)	Mole volume (cm^3/mol)	α ($\times 10^6\ ^\circ C^{-1}$)	k_T^{-1} ($\times 10^{12}\ cm^2/dyn$)	C_V ($\times 10^{-7} erg/(mol \cdot ^\circ C)$)	Γ_0
NaCl	58.46	2.16	27.1	121	4.2	47.6	1.63
NaBr	102.9	3.20	32.1	(120)	51	48.4	(1.56)
KCl	74.6	1.99	37.5	114	5.6	47.4	1.60
KBr	119.0	2.75	43.3	126	6.7	48.4	1.68
KI	166.0	3.12	53.2	128	8.6	48.7	1.63
RbBr	165.4	3.35	49.4	(107)	7.9	48.9	(1.37)
RbI	212.4	3.55	59.8	(102)	9.6	49.5	(1.41)
AgCl	143.3	5.55	25.8	99	2.4	50.2	2.12
AgBr	187.8	6.32	29.7	104	2.7	50.1	2.28
CaF_2	78.1	3.18	24.6	56.4	1.24	65.8	1.70
FeS_2	120.0	4.98	24.1	26.2	0.71	59.9	1.47
PbS	239.3	7.55	31.7	60	1.96	50	1.94

From Tsien Hsue-Shen. 1962. Notes on Physical Mechanics, Science Press, Beijing (in Chinese), p. 202–203.

Table 2.3 Parameters in the Tillotson equation for some materials.

	Steel	Aluminum	Thorium
ρ_0 (g/cm^3)	7.8	2.79	1.845
E_0 (erg/g)	9.5×10^{10}	5.0×10^{10}	17.5×10^{10}
a	0.5	0.5	0.55
b	1.5	1.63	0.62
A (dyn/cm^2)	1.28×10^{12}	0.75×10^{12}	1.17×10^{12}
B (dyn/cm)	1.05×10^{12}	0.65×10^{12}	0.55×10^{12}

1 erg $= 10^{-7}$ J, 1 dyn $= 10^{-5}$ N.
From China Encyclopedia, Mechanics, 1985

$$\Pi(V) = A\mu + B\mu^2, \tag{2.72b}$$

Eq. (2.67a) will become the Tillotson equation (Eq. (2.71)). This shows that the Tillotson equation is a specific form of the Gruneisen state equation in the case of $\Gamma = \Gamma(V,E)$.

CHAPTER THREE

Solid physics basis of the equation of state for solids under high pressures

3.1 Introduction

In the previous Chapter 2, the Volumetric Deformation Law of Dynamic Constitutive Relation of Materials is discussed from the view of macromechanics. Even though this kind of phenomenological study can be suitable for engineering application, its intrinsic mechanism has not been explored thoroughly. In terms of its physical essence, the equation of state (EOS) of solids is inseparable from the intrinsic microstructure characteristics of solids. With the development of modern science, it has become the consensus of the scientists in mechanics and materials science to study the mechanical properties of materials by combining macromechanics with microphysics. This is just the background of Mr. Qian Xuesen's advocacy of "Physical Mechanics" in the 1950s (Qian Xue-sen, 1962).

In general, matter has three states: gas, liquid, and solid. We are talking about solid matter, a state of matter with a certain macroscopic size and a relatively fixed shape. The solids are roughly divided into crystalline and amorphous. They are distinguished by the degree to which their microscopic particles are arranged remotely. Crystals are remotely ordered, i.e., the particles (atoms, ions, molecules) that make up crystals are periodically arranged in regular geometric order. But the particles in the noncrystal are disordered or only the neighboring particles are arranged in order (short-range ordering). Thus it is called amorphous, and also called glassy state or supercooled liquid. The "solid physics" what people usually say is in fact the "crystal physics". The solid physical foundation including the crystal structure, the binding between crystal particles (atoms, ions, molecules) and their thermal vibration, and so on will be briefly discussed in this chapter. Then the thermomechanical parameters related to the EOS of solids under high pressures and the Gruneisen EOS will be further derived from the viewpoint of solid physics.

Dynamics of Materials
ISBN: 978-0-12-817321-3
https://doi.org/10.1016/B978-0-12-817321-3.00003-6

3.2 Crystal structure

The remote order in the microstructure of crystal determines that the crystal has the following main characteristic distinguished from the amorphous: a regular geometrical shape on the macroscopic level, a definite melting temperature when changing from solid state to liquid state, anisotropy of physical properties, and the cleavage of the crystal surface along a certain orientation under the action of a force.

Crystal has regular geometry macroscopically that is just a reflection of a regular arrangement of crystalline particles (ions, atoms, or molecules) in the interior. One of the basic characteristics of the regularity of crystal microstructure arrangement is its **periodicity,** i.e., the arrangement of particles in a crystal repeats periodically in a certain way ceaselessly. The framework formed by the periodic arrangement of the mass center of particles in a crystal is called a **lattice**. The position of the particle's mass center is called the **lattice site** (Fig. 3.1).

The structure of the whole crystal can be seen as consisting of lattice sites along three different directions in space, which are periodically translated by a certain distance. The distance of each translation is called the period. There is a certain period in a certain direction, and different periods in different directions.

Due to the periodicity of the crystal, we can select a certain element with its edge length of a, b, c in three directions, respectively, and by translating repeatedly the element, the whole lattice can be formed. The vectors $\vec{a}$, $\vec{b}$, $\vec{c}$ which represent three element edge lengths and orientations are called the **translation vectors**. Such an element is called a **crystal cell**. There are several ways to select the crystal cell and the translation vector. They all have exactly the same lattice. The smallest cell is called a **unit cell**. The three translation vectors which form the edges of a unit

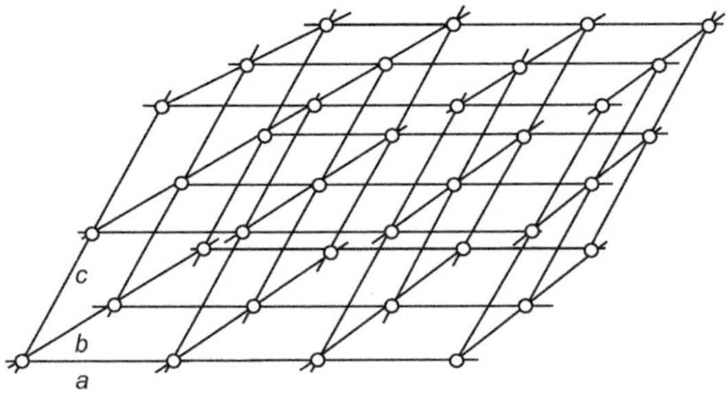

Figure 3.1 lattice site.

cell is called the **basic translation vector** (or called the basic vector). The size and shape of the crystal cell are often represented by the edge lengths *a*, *b*, and *c* on the axes $\vec{a}$, $\vec{b}$ and $\vec{c}$ and the angles α, β, and γ between the edges of the crystal cell. The edge lengths of the crystal cell are called lattice constants, and their measure unit is angstrom (Å).

If the crystal is composed entirely of exactly the same particles, the particles coincide with the lattice sites, and each lattice is surrounded by the same conditions, such a lattice is called a **single lattice**, or also known as the **Bravais lattice**. The Bravais lattice is a mathematical abstraction of a single lattice. If the crystal is composed of two or more particles, or if it is composed of the same particles, but the situation around each lattice is not the same, such lattice is called **compound lattice** or **non-Bravais lattice**, which can be considered to be constructed by two or more mutually nested Bravais lattices. According to their different edge lengths and angles, Bravais lattices have 14 types of seven crystal systems. Their shapes and names are shown in Fig. 3.2 (Gou, 1978). Of them, the face-centered cubic (in short, fcc), the body-centered cubic (in short, bcc), and the hexagonal close packing (in short, hcc) are three simplest and most typical crystal structures. The characteristics and symmetries of these seven crystal systems are listed in Table 3.1 (Gou, 1978).

Another basic characteristic of crystal microstructure is **symmetry**, i.e., the crystal can recover its original shape after some symmetrical operation, and the most basic operation of symmetry is rotation and reflection.

The lattice sites of a lattice can be separately arranged along a series of parallel straight lines. These sets of straight lines are called **crystal array** (Fig. 3.3). The sites of the same lattice can form different crystal arrays orientated in different directions. The orientation of a crystal array is called **crystal orientation**. The lattice sites of a lattice can also be placed on a series of planes parallel to each other, which are called **crystal planes**. In Fig. 3.4, several groups of different crystal planes are represented by bold lines. Once the crystal planes are determined, all the lattice sites should be contained in parallel and equidistant family of crystal planes.

To describe the direction of a crystal plane, **Miller indices** (*h*, *k*, *l*) are often used (Fig. 3.5). Select a set of translation vectors $\vec{a}$, $\vec{b}$, $\vec{c}$ as three coordinate axes; suppose that there is a crystal plane which intersects with these three axes at the M_1, M_2, and M_3, respectively; and the intercept is, respectively, equal to $h'a$, $k'b$, and $l'c$, accordingly (h', k', l') can be used to describe the direction of the crystal plane. However, if a crystal plane is parallel to one of these axes, e.g., the axis $\vec{a}$, then $h' \rightarrow \infty$. In order to avoid

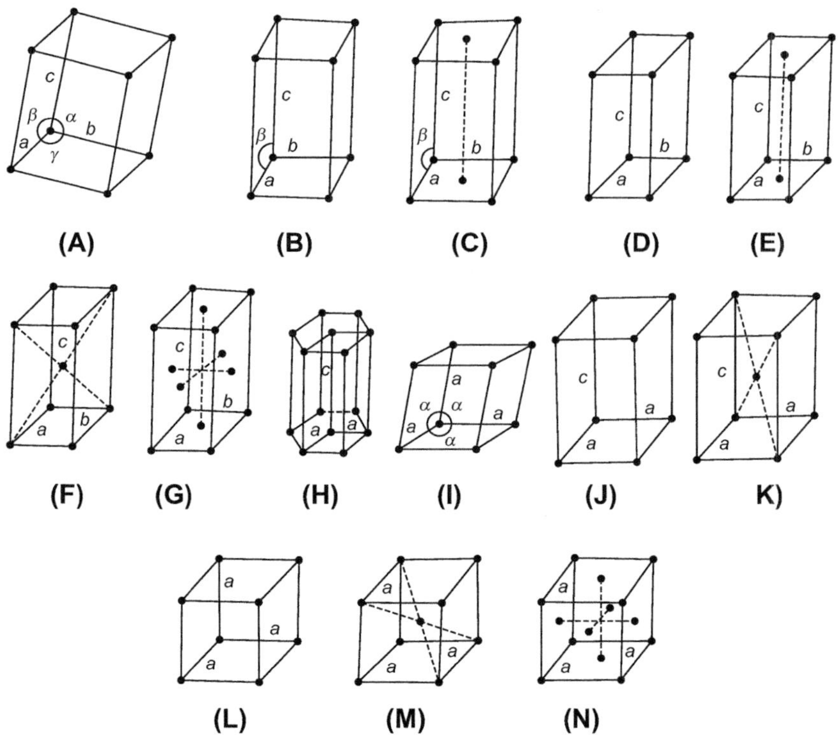

Figure 3.2 14 types of Bravais lattices (Gou. 1978), (A) simple triclinic; (B) simple monoclinic; (C) base-centered monoclinic; (D) simple orthorhombic; (E) base-centered orthorhombic; (F) body-centered orthorhombic; (G) face-centered orthorhombic; (H) hexagonal; (I) Trigonal; (J) simple tetragonal; (K) body-centered tetragonal; (L) simple cubic; (M) body-centered cubic; (N) face-centered cubic.

this case, Miller used the mutual prime numbers (h, k, l) of the reciprocals of (h', k', l') to indicate the crystal plane. The index (h, k, l) is called the **Miller index** of a crystal plane. For example, the Miller index of the crystal plane shown in Fig. 3.5 is (2, 3, 6), which is determined by the following formula:

$$\frac{1}{3}:\frac{1}{2}:1 = 2:3:6$$

In a crystal, all parallel crystal planes have the same Miller index. Because the number of atoms per unit volume in a crystal is constant, the greater the density of the atoms on the crystal planes in a particular direction is, the greater the spacing of the crystal planes is, and the smaller the binding force between the crystal planes is. Under the action of external forces, the property of a crystal that the brittle fracture occurs along the plane with the

Table 3.1 Characteristics of crystal systems (Gou, 1978).

Crystal family	Crystal system	Characteristics of unit cell	Basic symmetric elements
Advanced	Cubic	a = b = c, $\alpha = \beta = \gamma = 90°C$	4 threefold symmetry axes
Intermediate	Hexagonal	a = b ≠ c, $\alpha = \beta = 90°C$, $\gamma = 120°C$	1 sixfold symmetry axis
	Tetragonal	a = b ≠ c, $\alpha = \beta = \gamma = 90°C$	1 fourfold symmetry axis
	Trigonal	a = b = c, $\alpha = \beta = \gamma \neq 90°C$.	1 threefold symmetry axis
Low	Orthorhombic	a ≠ b ≠ c, $\alpha = \beta = \gamma = 90°C$	3 twofold symmetrical axes perpendicular to each other or 2 mirror planes perpendicular to each other
	Monoclinic	a ≠ b ≠ c, $\alpha = \gamma = 90°C$, $\beta > 90°C$	1 twofold symmetrical axis and 1 mirror plane
	Triclinic	a ≠ b ≠ c, $\alpha \neq \beta \neq \gamma$	Neither symmetrical axis nor mirror plane

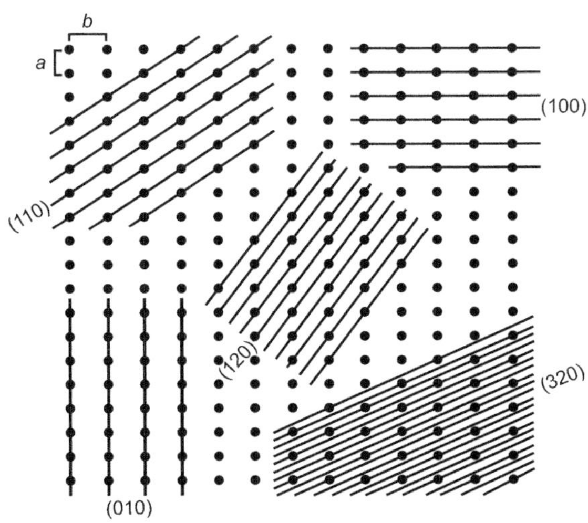

Figure 3.3 crystal arrays.

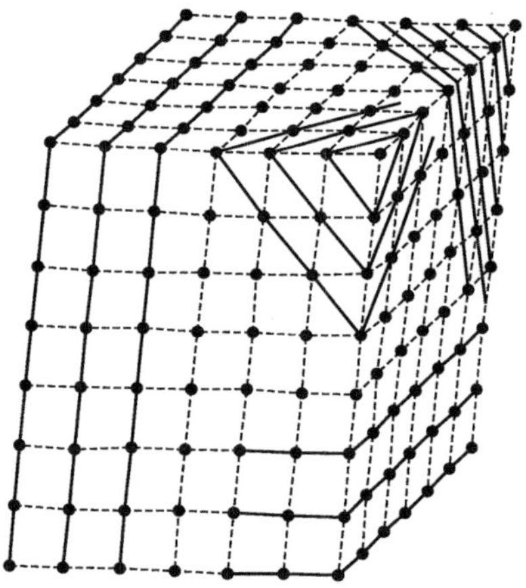

Figure 3.4 crystal plane.

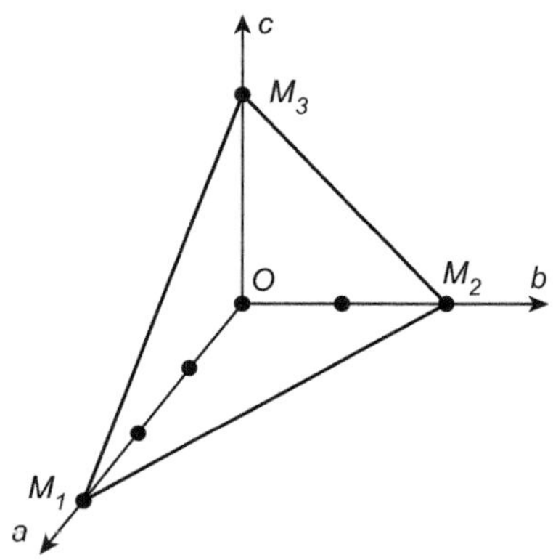

Figure 3.5 crystal plane index.

weakest binding force is called **cleavage**, and the corresponding smooth and flat plane formed in such process is called the **cleavage plane**; we therefore also call the crystal plane with the highest concentration of atoms as the cleavage plane.

The equilibrium position of the particles in the crystal corresponds to the minimum binding energy, so they are arranged as tightly as possible. In order to represent the compactness of the particles in a crystal, we can describe it by use of the number of particles closest to the particle, the **coordination number** (or **ligand number**). Sometimes the number of particles in the vicinity is also included, but at this time, it should be separately indicated. For example, the coordination number of metallic sodium is 8, the coordination number of metallic magnesium is 6 + 6, and so on.

3.3 The binding type of crystals

The reason why crystal structure presents regular and periodic arrangement and can be combined into a stable crystal with a certain geometry structure is due to the interaction force (binding force) between the crystal particles. And this binding force is related to the binding type of particles. The microgeometrical structures and physic-chemical properties of crystals are finally closely related to the binding types. Recall the cold pressure and the corresponding cold energy mentioned previously in the discussion of the EOS of solids under high pressures; they are mainly related to the crystal binding force and the corresponding binding energy, which will be discussed in the following.

The binding type of crystals depends mainly on the strength of atom's bound valence electron, and the result of reconfigurations of the interatomic electrons in the interaction. There are four basic forms of the reconfigurations of electrons. (1) Valence electrons transfer between interacting atoms to form ionic bonds. (2) Valence electrons are shared by a pair of interacting atoms (electron pairs) to form covalent bonds. (3) Valence electrons are shared by the all atoms (free-electron clouds), forming a metallic bond. (4) Electron configurations have no change, forming a molecular bond. Therefore, there are four basic binding types of crystals: ionic, atomic (covalent), metallic, and molecular.

The ability of a crystal atom to bind electrons depends on the electronegativity of the atoms that make up the crystal. The electronegativity of an atom consists of two parts: ionization energy and affinity energy. The so-called ionization energy is the energy necessary to make a neutral atom lose an electron to become a positive ion, i.e.,

$$\text{Neutral atom} + \text{ionization energy} \rightarrow \text{positive ion} + \text{electron}$$

It represents the ability of an atom to control its own valence electrons.

The so-called affinity energy is the energy released by making a neutral atom obtains an electron as a negative ion, i.e.,

$$\text{Neutral atom} + \text{electron} \rightarrow \text{negative ion} + \text{affinity energy}$$

It represents the ability of atoms to capture valence electrons.

Both the ability to control their own valence electrons (i.e., ionization energy) and the ability to capture other valence electrons (i.e., affinity energy) represent the binding capacity of atoms to valence electrons. This is the so-called atomic electronegativity.

The mathematical definition is:

$$\text{Electronegativity} = 0.18 \times (\text{ionization energy} + \text{affinity energy})$$

The selection of coefficient of 0.18 is only for the electronegativity of the metal Li to be 1, which has no practical significance.

Table 3.2 gives the ionization energy, affinity energy, and electronegativity of the three periodic elements in the periodic table of elements (Huang, 1966). Comparing all the data in the table, the following two trends can be seen:

(1) In the same period, elements from left to right show a significant trend of gradually increasing ionization energy, affinity energy, and electronegativity.

(2) In different period, elements from top to bottom show a significant tendency of gradually decreasing ionization energy, affinity energy, and electronegativity.

3.3.1 Ionic binding

When metal atoms that are less electronegativity (and thus less ionization energy) are close to nonmetallic atoms that are more electronegativity (hence more affinity), the former can easily release its valence electrons of the outermost layer to become positive ion, while the latter quickly absorbs the electrons released by the former to becomes negative ion. When positive and negative ions get close to each other to a certain extent under the effect of Coulomb force, the electron cloud of the shell of the positive and negative ions produces repulsive force due to overlapping according to the Pauli exclusion principle. When the absorbance and repulsion are equal, a stable

Table 3.2 Ionization energy and electronegativity of the three periodic elements in the periodic table of elements (Huang, 1966).

The second periodic	Li	Be	B	C	N	O	F	Ne
Ionization energy	5.392	9.322	8.298	11.260	14.534	13.618	17.422	21.564
Electronegativity	1.0	1.5	2.0	2.60	3.05	3.50	4.00	
The third periodic	Na	Mg	Al	Si	P	S	Cl	Ar
Ionization energy	5.139	7.646	5.986	8.151	10.486	10.360	12.167	15.759
Electronegativity	0.90	1.2	1.5	1.90	2.15	2.60	3.15	
The fourth periodic	K	Ca	Ga	Ge	As	Se	Br	Kr
Ionization energy	4.341	5.115	5.999	7.899	9.81	9.752	11.814	13.99
Electronegativity	0.8	1.0	1.6	1.90	2.00	2.45	2.35	

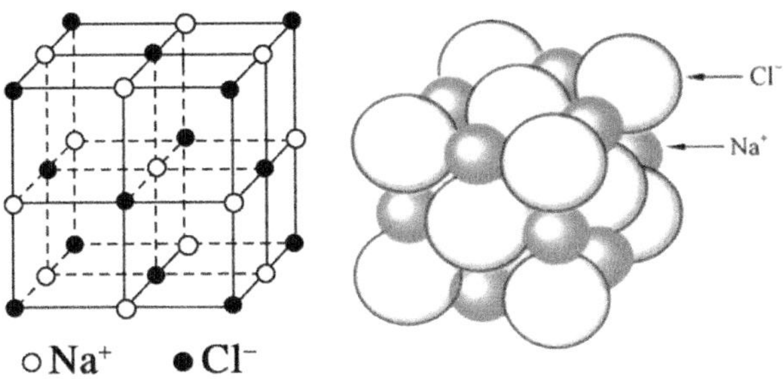

Figure 3.6 Ionic crystal structure of sodium chloride.

ionic bond can be formed, that is, the so-called ionic bond. Crystals bound together by these ionic bonds are called ionic crystals. All the two elements which differ greatly in electronegativity are mostly combined into ionic crystals, and sodium chloride is a typical ionic crystal. Its crystal structure, as shown in Fig. 3.6, is a complex lattice consisting of two mutually nested face-centered cubic crystals.

The formation of ionic crystals is mainly the role of Coulomb force, its binding force is strong, so its macroscopic properties express as more hard and high melting point, such as the melting point of sodium chloride is 801°C. On the other side, since there is no free electron in the ion crystal, its conductivity is poor.

3.3.2 Covalent binding

Covalent binding is also known as atomic binding of the two atoms that make up an atomic crystal that are highly electronegative, that is, they will not easily lose their own valence electrons and have a strong ability to capture the other's valence electron. When these two atoms are close to each other, they each contribute an electron and they share the two atoms, so it is to be the atomic binding, namely the covalent binding. Crystals bound by this covalent bond are called atomic (covalent) crystals. Diamonds made up of carbon atoms are typical atomic crystals whose crystal structure is shown in Fig. 3.7. Since the carbon atom in diamond has four valence electrons, it forms four covalent bonds with the adjacent four other carbon atoms, with an angle of 109°28′ between each covalent bond. Many important semiconductor materials, such as germanium and silicon, also have diamond-like structures. Atomic binding in atomic crystals is also mainly the role of Coulomb force, so its macroscopic properties also show high melting point, such as diamond melting point is above 3000°C. Because the atom (covalent) bond has a definite direction in the

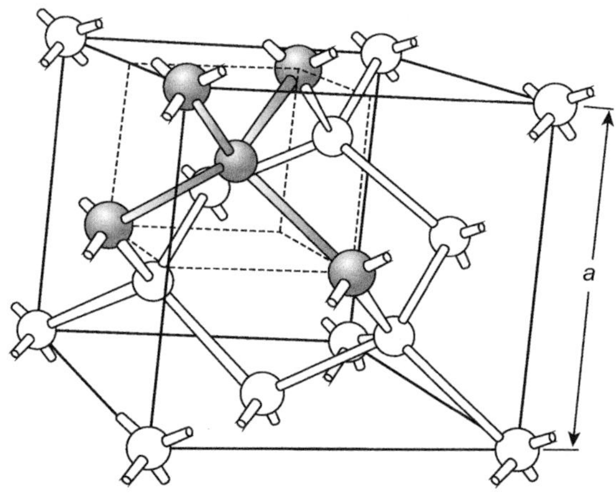

Figure 3.7 Atomic crystal structure of diamond.

crystal structure, it is not easy to change, so its macroscopic properties show both hard and brittle, and cannot be obviously bent. Moreover, the electrical conductivity of atomic crystals is very poor because there are no free electrons in their structure. However, when some impurities are added into the crystal, the electrical conductivity is improved greatly. The semiconductor materials such as germanium and silicon mentioned above are widely used by taking advantage of this characteristic.

3.3.3 Metallic binding

The basic characteristic of metallic binding is the communization of electrons. Metal atoms are less electronegative; they can easily lose their valence electrons and are difficult to capture valence electrons in other atoms. Thus, when a crystal is formed, the valence electrons originally belonging to each atom have been unbound and changed into free electrons moving in the whole crystal. This free electron is different from the valence electrons in an ionic crystal or atomic crystal, in which the electrons are tightly bound by their atoms. This basic difference makes the formation of metallic crystals different from that of ionic crystals and atomic crystals. In metal crystals, the Coulomb force acts to repel the electron-losing metal ions and is not conducive to the stability of the system. However, this Coulomb force is largely ineffective because the free electrons shield the metal positive ions so strongly that they become essentially neutralized particles without interaction, much like free atoms. The formation of metallic crystals must be explained by the theory of quantum mechanics. From quantum mechanics, we can see that the kinetic energy of particles such as free electrons decreases with the increase of the limited space volume. The energy is proportional to $V^{-2/3}$, where V is the space volume. When metal atoms are in a free state (gaseous state), the movement of valence electrons is confined to a very small atomic volume, so their kinetic energy is high. But in the crystal state, the valence electron can move freely throughout the crystal volume, this makes its kinetic energy reduces greatly. Thus, the total energy of a metal in a crystal state is much smaller than that in a gaseous state, which is the reason why metallic binding forms. The free electrons with negative electricity are like glue, holding the metal ions together.

Metallic binding is weaker than ion binding or atom (covalent) binding, and their strength is usually inversely related to the metal ion radius and positively related to the free electron density inside the metal. The melting point of metals generally increases with the strength of the metallic binding, so some simple metals formed by pure metallic binding have a lower melting point, for example, sodium has a melting point of only 97.8°C. Also, because the free electrons in metal crystals can easily move under the influence of electric field, the metal is highly electroconductive. The reason of that metals have good heat conductivity may be similarly explained too. In addition, the principle of minimum total energy also causes metal positive ions to pile up as tightly as possible, so metals have a high density. Metal bonds have no directional properties, so the metal has good ductility.

3.3.4 Molecular binding

For atoms or molecules with a stable electronic structure, the atomic electronegativity is very strong, that is, their ability to bind electrons is very strong, and they do not lose electrons to become ions. Then crystals bind by a weak attraction between molecules. It generally consists of three parts: (1) intermolecular forces between polar molecules by inherent dipole action, known as dipole—dipole forces or Keesom force; (2) nonpolar molecules, acting under the inherent dipole of polar molecules, polarize and generate induced dipoles, and the resulting intermolecular forces are called induced force or the Debye force; and (3) intermolecular forces resulting from the interactions of transient dipole between no polar molecules is called dispersion forces (named for their expression of action energy similar to the dispersion formula of light), or the London forces. These three are collectively known as Van der Waals force. In turn, this binding is called a molecular binding or a Van der Waals binding, and the resulting crystal is called a molecular crystal. Because the binding effect of molecular binding is very weak, its macroscopic properties show very low melting point. For example, neon crystals have a melting point of −248.7°C, very low hardness, and of course their electrical conductivity is very poor. All solids formed by inert gases at low temperatures are molecular crystals.

The above discussion is about the four most basic types of crystal binding. The actual crystal can be a combination of two or three types. For example, in a graphite crystal, a carbon atom has four valence electrons, three of which are covalently bonded to the surrounding three carbon atoms, almost on the same plane. Another valence electron appears as a free electron in a free state, forming a metallic binding, while the molecular binding is formed between the layers. The above comprehensive types determine that the graphite crystal is flaky, easy to peel from layer to layer, and good lubrication performance. In addition, due to free electrons in the crystal, its electrical conductivity is good. Some relatively complex metals (for example, transition elements iron, nickel, and so on) can form covalent binding in addition to metallic binding, so the melting point and strength of these metals are much higher than that of simple metals like sodium. The melting point of iron is 1535 °C and that of nickel is 1455°C.

3.4 The binding force and binding energy of crystals

Although crystals have several basic types of binding, the binding energy and binding force of different types of crystals have a common characteristic, that is, the interaction energy $u(r)$ and the interaction force $f(r)$ between two microparticles vary with the spacing r of the particles, and $f(r)$ can be derived from the derivative of the binding energy $u(r)$ with respect to r as its potential function, as shown in Fig. 3.8 and (3.1)

Fig. 3.8A shows a schematic diagram of the change rule of the interaction force $f(r)$ with distance r between two particles. When two particles are very far apart, the interaction force is zero. As the two particles get closer, the attraction between them increases with the shortening of the distance. The binding force reached its maximum at $r = r_m$. When the distance is further shortened, the binding force between the particles decreases gradually. At $r = r_0$, the interaction force is zero and the particle is in equilibrium. When the distance is less than the equilibrium distance r_0, the repulsion force is generated between the particles and it increases rapidly with the shortening of the distance. The extreme value of the binding force at $r = r_m$ means that the interaction forces between the particles include both the force of attraction and the force of repulsion. When $r < r_0$, the repulsion

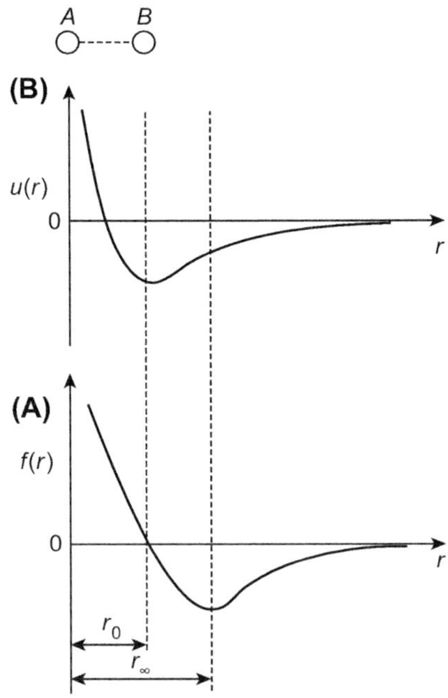

Figure 3.8 Interaction between two crystal particles. (A) relationship of interaction force with particle spacing; (B) relationship of interaction energy with particle spacing.

force is greater than the attraction force, while when $r > r_0$ the attraction force is greater than the repulsion force; When $r = r_0$, the attraction force and repulsion force cancel out, and the particle is in equilibrium.

Fig. 3.8B shows the change rule of the interaction energy $u(r)$ with distance r, which is the potential function of the interaction force $f(r)$ between the two particles. The relationship between the $f(r)$ and $u(r)$ is:

$$f(r) = -\frac{du(r)}{dr}。 \tag{3.1}$$

When $r = r_0, f(r_0) = 0$,

$$f(r_0) = -\frac{du(r)}{dr}\bigg|_{r=r_0} = 0 \tag{3.2}$$

which denotes the minimum value of the interaction energy $u(r)$, corresponding to the stable state of the crystal.

When $r = r_m$, the effective attraction force between particles is the greatest, that is,

$$\left.\frac{df(r)}{dr}\right|_{r=r_m} = \left.\frac{d^2u(r)}{dr^2}\right|_{r=r_m} = 0 \tag{3.3}$$

It corresponds to the turning point of the energy curve.

The following is a case study of ionic crystals. Ionic crystals consist of positively charged positive ions and negatively charged negative ions. Assuming that the charge distribution of each ion is spherically symmetrical, the mutual attraction between the ions is mainly the Coulomb electrostatic attraction of the opposite charge, while the repulsion between the ions is mainly the Pauli incompatibility repulsion force due to the overlap of the closed-shell electron cloud and the Coulomb electrostatic repulsion of the same charge.

At first, consider the mutual attraction between a pair of positive and negative ions, according to Coulomb's law:

$$f_1(r) = \frac{z_1 z_2 e^2}{r^2} \tag{3.4}$$

The corresponding mutual attraction energy is

$$u_1(r) = -\frac{z_1 z_2 e^2}{r} \tag{3.5}$$

where e is the charge of an electron (1.602×10^{-19} coulombs); z_1 and z_2 are the valence numbers of positive and negative ions, respectively; and r is the distance between two ions.

When two ions are very close, the effect of repulsion increases significantly, so Born assumed that the mutually exclusive energy can be

$$u_2(r) = \frac{b}{r^n} \tag{3.6}$$

where b and n in the equation are constants greater than zero, which can be determined by experiments. The corresponding repulsion force is

$$f_2(r) = -\frac{nb}{r^{n+1}} \tag{3.7}$$

When the above two interactions are combined, the interaction force and energy between the two ions can be obtained. That is,

$$f(r)=\frac{z_1 z_2 e^2}{r^2}-\frac{nb}{r^{n+1}}, \tag{3.8}$$

$$u(r)=-\frac{z_1 z_2 e^2}{r}+\frac{b}{r^n} \tag{3.9}$$

It should be pointed out that the influence of ion thermal vibration on the action force and action energy is not considered in the above derivation.

The binding energy U of an ionic crystal shall include the interaction energy between all the ions in the crystal, which could be written as:

$$U=\frac{1}{2}\sum_{i=1}^{N}\sum_{j=1}^{N}u(r_{ij}) \quad i\neq j \tag{3.10}$$

where N is the number of ions in the ionic crystal. The (1/2) term appears because the interaction occurs twice between a pair of ions. Because the number of ions contained in the crystal is so large that the surface effect can be ignored; the interaction energy of each ion is the same as that of all other ions in the crystal. So Eq. (3.10) can be written as following:

$$\begin{aligned} U &= \frac{1}{2}N\sum_{j\neq 1}^{N}u(r_{1\mathrm{j}}) \\ &= -\frac{1}{2}N\sum_{j\neq 1}^{N}\left\{\pm\frac{z_1 z_\mathrm{j} e^2}{r_{1\mathrm{j}}}-\frac{b}{r_{1\mathrm{j}}^{\mathrm{n}}}\right\} \end{aligned} \tag{3.11}$$

The positive and negative sign of the first item in brackets is determined by the charge sign of the ion. If the signs of charge symbol are the same, take the minus sign; if the two ion's charge signs are different, take the plus sign. The crystalline binding energy per unit mass (1 mol) is called **lattice energy**, which is equal to the energy released when a mole of positive and negative ions combine from the separated gaseous state into ionic crystals under 0K and without external pressure.

If r is the shortest distance between ions, the r_{1j} in the above equation can be represented as:

$$r_{1\mathrm{j}} = a_j r \tag{3.12}$$

where a_j is determined by the geometric structure of the crystal, which indicates how many multiples of r are the distance from the first ion to the jth ion. In this way, Eq. (3.11) can be rewritten as:

$$U = -\frac{1}{2}N\left\{\frac{z_1 e^2}{r}\sum_{j\neq 1}^{N}\left(\frac{\pm z_j}{a_j}\right) - \frac{1}{r^n}\sum_{j\neq 1}^{N}\left(\frac{b}{a_j^n}\right)\right\} \tag{3.13}$$

Let

$$\sum_{j\neq 1}^{N}\frac{b}{a_j^n} = B \tag{3.14}$$

In the case of only two ions, we further let

$$\sum_{j\neq 1}^{N}\left(\frac{\pm z_j}{a_j}\right) = \alpha z_2 \tag{3.15}$$

By substituting it into Eq. (3.13), the binding energy of two ion crystals can be obtained:

$$U = -\frac{1}{2}N\left\{\frac{\alpha z_1 z_2 e^2}{r} - \frac{B}{r^n}\right\} \tag{3.16}$$

This is consistent with the interaction energy of a pair of ions (Eq. 3.9) in form, where α is the Madelung constant and can be determined by the crystal structure. For example, for NaCl ion crystals, both z_1 and z_2 are equal to one; hence from Eq. (3.15), it is known that:

$$\alpha = \sum_{j\neq 1}^{N}\left(\pm\frac{1}{a_j}\right) \tag{3.17}$$

After some specific calculations (Xie, 1961; Gou, 1978), it can be gotten that $\alpha = 1.747{,}565$.

Eq. (3.16) indicates that if the parameters B and n are known, the binding energy U can be calculated. In fact, B and n are not independent of each other. The relationship between them can be determined by the equilibrium condition, that is, when $r = r_0$

$$\left.\frac{dU}{dr}\right|_{r=r_0} = -\frac{N}{2}\left\{-\frac{\alpha z_1 z_2 e^2}{r^2} + \frac{nB}{r^{n+1}}\right\}_{r=r_0} = 0 \tag{3.18}$$

From which we can get

$$B = \frac{\alpha z_1 z_2 e^2}{n} r_0^{n-1} \tag{3.19}$$

By substituting it into Eq. (3.16), the binding energy at equilibrium can be obtained:

$$U(r_0) = U_0 = -\frac{1}{2} N \frac{\alpha z_1 z_2 e^2}{r_0} \left\{ 1 - \frac{1}{n} \right\} \tag{3.20}$$

where n is related to the repulsion force between ions; this repulsion is shown in macroscopic terms as the compression resistance of solid materials, thus the relation between n and the elastic bulk modulus can be established. According to Eq. (2.8) in Chapter 2, the Lagrangian bulk modulus K is defined as:

$$K = -V_0 \left. \frac{\partial P}{\partial V} \right|_{V=V_0} = V_0 \left. \frac{\partial^2 U}{\partial V^2} \right|_{V=V_0} \tag{3.21}$$

where V_0 is the crystal volume at equilibrium and it can be expressed as:

$$V_0 = \beta N r_0^3 \tag{3.22}$$

where β is a constant related to the crystal structure. By substituting Eq. (3.16) into Eq. (3.21) and by using Eq. (3.19) and Eq. (3.22), we can obtain:

$$K = \frac{\alpha z_1 z_2 e^2}{18\beta\ r_0^4} (n - 1) \tag{3.23a}$$

or

$$n = 1 + \frac{18\beta\ r_0^4}{\alpha z_1 z_2 e^2} K \tag{3.23b}$$

where α and β in the equation can be calculated by the crystal geometric structure, r_0 can be determined by X-ray experiments, and K can be determined by the elastic wave propagation experiments under the isentropic condition. Thus n can be obtained directly by using Eq. (3.23b). For most ionic crystals, n is between 5 and 9. Table 3.3 lists the values of K and n for several ionic crystals.

The exponential term n is much greater than 1, which reflects the dramatic increase in the repulsion force after the ionic spacing is less than the equilibrium position r_0. Eq. (3.23) shows that the value of the bulk modulus depends mainly on the repulsion force, the steeper the repulsion change (the

Table 3.3 The values of K and n for several ionic crystals (Xie, 1961).

Crystals	NaCl	NaBr	NaI	KCl	ZnS
K(GPa)	24.0	19.9	15.1	17.4	77.6
n	7.90	8.41	8.33	9.62	5.4

Table 3.4 The binding energy U_0 of several ionic crystals (kcal/mol) (Huang, 1966).

Crystal	NaCl	NaBr	KCl	KBr
Calculated value	182	172	165	159
Experimental value	185	176	168	161

larger the n) is, the more difficult to compress the crystal (the larger the K) is. Eq. (3.20) shows that the binding energy of crystal is mainly supplied by the Coulomb force, and the energy supplied by the repulsion force is only $1/n$ of the Coulomb force. By substituting the calculated n into Eq. (3.20), the crystal binding energy U_0 at equilibrium can be accurately calculated which is very close to the measured results. Table 3.4 shows the calculated and experimental values of the binding energy U_0 of several ionic crystals at equilibrium.

It is not difficult to generalize the above discussion of ionic crystals to other types of crystals. Similar to Eq. (3.9), the binding energy of two particles in other types of crystals can be regarded as the sum of attraction terms and exclusion terms:

$$u(r) = -\frac{a}{r^m} + \frac{b}{r^n}, (\mathrm{m} < \mathrm{n}) \tag{3.24}$$

which is called the Mie potential expression, suitable for low temperature and pressure is not high ($\rho \leq 2\rho_0$, where ρ is the density after compressed and ρ_0 is the initial density). In the formula, r is the distance between particles, which is related to the lattice constant; the first term represents the attraction energy, the Coulomb attraction between the opposite charges, showing the long-term effect; the second term represents the repulsion energy, caused by the Coulomb repulsion force between the same charge and by the Pauli principle, showing the short-term effect; and the constants a, b, m, n are all positive constants depending on the crystal type, and $n > m$ means that exclusion varies faster with distance than attraction does. Accordingly, the interaction force between the two particles is:

$$f(r) = \frac{-du(r)}{dr} = \frac{ma}{r^{m+1}} - \frac{nb}{r^{n+1}} \circ \tag{3.25}$$

Similar to the binding energy of ionic crystals (Eq. 3.11), the binding energy of other types of crystals can be written:

$$\begin{aligned} U &= -\frac{N}{2}\left\{\sum_{j\neq 1}^{N}\left(\frac{a}{r_{1j}^{m}}-\frac{b}{r_{1j}^{n}}\right)()\right\} \\ &= -\frac{N}{2}\left\{\frac{a}{r^{m}}\sum_{j\neq 1}^{N}\left(\frac{r}{r_{1j}}\right)^{m}-\frac{b}{r^{n}}\sum_{j\neq 1}^{N}\left(\frac{r}{r_{1j}}\right)^{n}\right\} \\ &= -\frac{Na}{r^{m}}\sigma_m+\frac{\mathrm{Nb}}{r^{n}}\sigma_n \end{aligned} \tag{3.26}$$

where σ_m and σ_n are, respectively:

$$\begin{aligned} \sigma_m &= \frac{1}{2}\sum_{j\neq 1}^{N}\left(\frac{r}{r_{1\mathrm{j}}}\right)^{m} \\ \sigma_n &= \frac{1}{2}\sum_{j\neq 1}^{N}\left(\frac{r}{r_{1\mathrm{j}}}\right)^{n} \end{aligned} \tag{3.27}$$

It should be emphasized that σ_m and σ_n are only the functions of crystal structures, that is, the functions of crystal binding types and the functions of exponential terms of m and n, and have nothing to do with the volume of crystals. The role of surface particles has also been neglected in this analysis, which is entirely permissible when the crystal volume is not too small.

By using Eq. (3.22), the volume of a single particle is:

$$v=\frac{V}{N}=\beta r^{3}$$

Assume further:

$$A=a\beta^{m/3}\sigma_m,\ B=b\beta^{n/3}\sigma_n$$

By substituting it into Eq. (3.26), we have:

$$U=-\frac{NA}{v^{m/3}}+\frac{NB}{v^{n/3}}。 \tag{3.28}$$

In this way, the problem of how to determine the binding energy of crystals comes down to how to determine A, B, m, and n.

Set v_0 to be the volume of each particle in equilibrium position at the temperature $T = 0°K$ and the external pressure $P = 0$, then the following condition shall be satisfied for U:

$$\left.\frac{dU}{dv}\right|_{v=v_0} = -\frac{N}{3v_0}\left\{-\frac{Am}{v_0^{m/3}} + \frac{Bn}{v_0^{n/3}}\right\} = 0 \tag{3.29}$$

So there is

$$\frac{Am}{v_0^{m/3}} = \frac{Bn}{v_0^{n/3}} \tag{3.30}$$

Since v_0 can be inferred experimentally, the above formula provides a relation between A, B, m, and n.

The binding energy U_0 in the equilibrium position can be obtained by substituting Eq. (3.30) into Eq. (3.28):

$$U_0 = -\frac{NA}{v_0^{m/3}}\left(1 - \frac{m}{n}\right)。 \tag{3.31}$$

Since the U_0 is the same thing as the sublimation heat Q_0 at $T = 0°K$ and $P = 0$, and it can be inferred experimentally, the above formula provides another relation between A, m, and n.

Similar to Eq. (3.21) of ionic crystals, the bulk modulus K of other binding types of crystals at equilibrium is

$$K = -V_0\left.\frac{\partial P}{\partial V}\right|_{V=V_0} = V_0\left.\left(\frac{\partial^2 U}{\partial V^2}\right)\right|_{V=V_0} = \frac{v_0}{N}\left.\left(\frac{\partial^2 U}{\partial v^2}\right)\right|_{v=v_0} \tag{3.32a}$$

where $V_0 = Nv_0$, Substitute Eq. (3.28) into the above equation,

$$K = v_0\left\{\left(\frac{m}{3}\right)\left(\frac{m}{3}+1\right)\frac{A}{v_0^{\frac{m}{3}+2}} - \left(\frac{n}{3}\right)\left(\frac{n}{3}+1\right)\frac{B}{v_0^{\frac{n}{3}+2}}\right\} \tag{3.32b}$$

Taking account of Eq. (3.30), after finishing, we have

$$K = \frac{A}{9v_0}\cdot\frac{m(n-m)}{v_0^{m/3}} \tag{3.32c}$$

Substituting it into Eq. (3.31), we have

$$U_0 = -K\frac{9Nv_0}{mn} = -K\frac{9V_0}{mn} \tag{3.33}$$

It is called the **first rule of Grüneisen**.

Since K can also be inferred from experiments, Eqs. (3.30–3.32) provide three relations relating A, B, m, and n, as long as there is another relation, all four can be determined. This fourth relation can be provided by establishing the relation between the Grüneisen coefficient γ and n after the thermal vibration of lattices is discussed below. It can also be provided by other relationships that can be experimentally determined. For example, let us examine the rate of change of volume modulus K with regard to pressure P; from Eq. (3.32b), we have

$$\left(\frac{\partial K}{\partial V}\right)_T=\frac{v_0}{N}\left\{\left(\frac{m}{3}\right)\left(\frac{m}{3}+1\right)\left(\frac{m}{3}+2\right)\frac{A}{v_0^{\frac{m}{3}+3}}-\left(\frac{n}{3}\right)\left(\frac{n}{3}+1\right)\left(\frac{n}{3}+2\right)\frac{B}{v_0^{\frac{n}{3}+3}}\right\} \tag{3.34a}$$

Due to

$$\begin{aligned}\left(\frac{\partial K}{\partial P}\right)_T&=\left(\frac{\partial K}{\partial V}\right)_T\left(\frac{\partial V}{\partial P}\right)_T\\&=\frac{v_0}{N}\left\{\left(\frac{m}{3}\right)\left(\frac{m}{3}+1\right)\left(\frac{m}{3}+2\right)\frac{A}{v_0^{\frac{m}{3}+3}}-\left(\frac{n}{3}\right)\left(\frac{n}{3}+1\right)\left(\frac{n}{3}+2\right)\frac{B}{v_0^{\frac{n}{3}+3}}\right\}\\&\cdot\left(-\frac{V_0}{K}\right)\end{aligned} \tag{3.34b}$$

Substituting Eq. (3.32b) into the K in the above formula, and considering Eq. (3.30), after finishing, we have

$$\left(\frac{\partial K}{\partial P}\right)_T=\frac{\left(\frac{m}{3}+1\right)\left(\frac{m}{3}+2\right)-\left(\frac{n}{3}+1\right)\left(\frac{n}{3}+2\right)}{\left(\frac{m}{3}+1\right)-\left(\frac{n}{3}+1\right)}=\frac{9+m+n}{3} \tag{3.34c}$$

Since $\left(\frac{\partial K}{\partial P}\right)$ can also be inferred experimentally, Eq. (3.34) together with Eqs. (3.30–3.32) provide four relations relating A, B, m, and n. In other words, once the v_0, U_0, K, and $\left(\frac{\partial K}{\partial P}\right)$ are known experimentally, we can determine A, B, m, and n, and thus the binding energy of crystals expressed by Eq. (3.28) can be determined.

3.5 Lattice thermal vibration

In the previous section, when we discussed the binding force and the binding energy of crystals, we assumed that the particles were in a state of static equilibrium. In fact, even at zero absolute temperature, the particles in the crystal will vibrate slightly near the equilibrium position, known as the **zero-point vibration**. At room temperature, these particles make thermally vibrate at different frequencies near their respective equilibrium positions.

In general, the energy of a crystal is the sum of potential energy and kinetic energy. The potential energy is the cold energy corresponding to the cold pressure discussed in chapter 2, including the crystal binding energy, the vibration energy at zero ^{0}K, and the compression energy of valence electron gas in metal. Among those, the crystalline binding energy discussed in the preceding section is predominant. The kinetic energy is the lattice vibration energy taking off the vibration energy at zero ^{0}K, as well as the electron thermal activation energy at sufficiently high temperature. That is, the thermal energy corresponding to the thermal pressure discussed in chapter 2.

The thermal vibration of the crystal particles more and more increases with the increase of temperature. Once the melting-point temperature is reached, the particles of the whole crystal move irregularly, and the lattice structure disintegrates, corresponding to the melting of the solid–liquid phase transition. However, at not too high temperature, constrained by the strong binding force (Eq. 3.25) between the lattice particles, the particles can only make thermal vibration at the equilibrium position, known as **lattice thermal vibrations**. Many of the physical properties of crystals, especially the mechanical and thermal properties associated with temperature, are actually macroscopic representation of this microscopic thermal vibration of the lattice. This is exactly what we are going to discuss next.

The thermal vibration characteristics of the crystal particles depend on the binding force or energy between the particles. As can be seen in Fig. 3.8, the binding energy curve is asymmetric relative to the equilibrium point r_0, so the vibration of particles near the equilibrium point is actually nonlinear, enharmonic vibration. It is quite difficult to discuss such a problem, so we usually assume: The frequency of vibration is a function of the particle spacing (i.e., the density), but the amplitude is small enough (weak vibration) to be approximated by the harmonic vibration at each density, known as the **quasi-harmonic approximation**.

According to Eq. (3.24), the binding energy $u(r)$ between two particles is an asymmetric function of the particle spacing r. Expand the $u(r)$ to a power series near the equilibrium point r_0, we have:

$$u(r) = u(r_0) + \left(\frac{du}{dr}\right)_{r_0} X + \frac{1}{2!}\left(\frac{d^2u}{dr^2}\right)_{r_0} X^2 + \frac{1}{3!}\left(\frac{d^3u}{dr^3}\right)_{r_0} X^3 + \cdots$$

The X in the formula is the displacement of the particle from its equilibrium position, that is, $X = r\text{-}r_0$. The first term in the formula $u(r_0) = u_0$ is constant; the second term is zero according to Eq. (3.2). Thus the formula can be rewritten as:

$$u(r) = u(r_0) + \frac{1}{2}\alpha X^2 - \frac{1}{3}\beta X^3 + \cdots \tag{3.35a}$$

where

$$\alpha = \left(\frac{d^2u}{dr^2}\right)_{r_0}, \quad \beta = -\frac{1}{2}\left(\frac{d^3u}{dr^3}\right)_{r_0}$$

If the X^3 term and higher terms in the above formula are omitted, the interaction potential energy becomes:

$$u(r) = u(r_0) + \frac{1}{2}\alpha X^2。 \tag{3.35b}$$

It corresponds to a parabola symmetric to the equilibrium r_0 in geometric description. The corresponding interaction force is reduced to linear Hooke elastic force:

$$f(r) = -\frac{du(r)}{dr} = -\alpha X。 \tag{3.35c}$$

where α is the Hooke constant. Thus, the **quasi-harmonic approximation** is equivalent to that the asymmetric energy curve is approximated by a parabola symmetric to the equilibrium position r_0, as shown in Fig. 3.9.

Of course, some of the properties of crystals (such as thermal expansion, heat conduction, etc.) are closely related to enharmonic vibration, and then the higher order terms of X^3 in Eq. (3.35a) should be taken into account. However, the quasi-harmonic vibration analysis of crystals is sufficient to explain many basic properties of crystals, so we will mainly discuss the quasi-harmonic vibration of crystals in the following sections. The effect of enharmonic vibration is roughly discussed only if necessary.

1. Quasi-harmonic vibration of a one-dimensional simple lattice

First, we discuss the one-dimensional longitudinal lattice thermal vibration of a simple cubic crystal consisting of N identical atoms. For the one-dimensional longitudinal motion, the lateral binding force has no effect, and the problem is simplified as the one-dimensional vibration of an infinite chain of atoms ($N \rightarrow \infty$), as shown in Fig. 3.10. Let the mass of the atom is m, the spacing of the atoms in equilibrium is a (for a simple cubic crystal, i.e.,

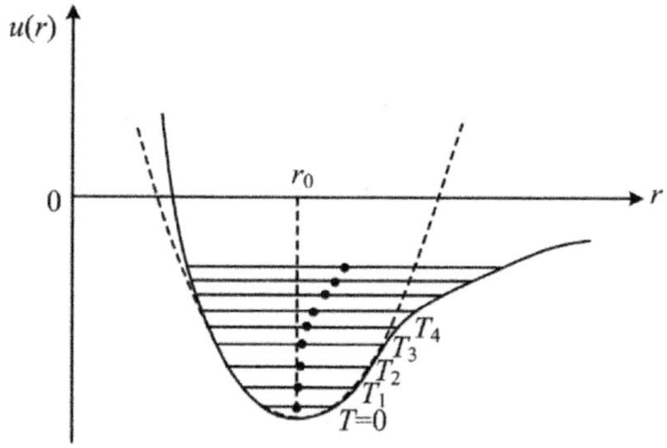

Figure 3.9 Energy curve near equilibrium r_0.

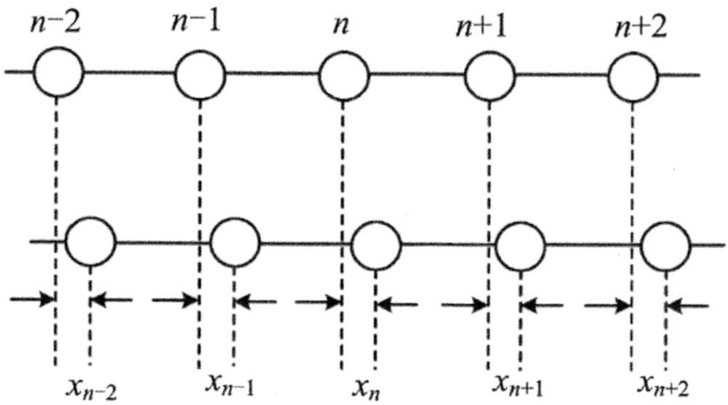

Figure 3.10 Vibration of one-dimensional infinite atomic chain.

the lattice constant), and X_n expresses the displacement of the nth atom from its equilibrium position due to thermal vibration.

Thus the relative displacement between the nth and (n+1)th atoms is $X_n - X_{n+1}$. In the quasi-harmonic approximation, the force between atoms is reduced to linear Hooke elastic force according to formula (3.35c), and the action force of the nth atom by the (n+1)th atom is:

$$F_{n,n+1} = -\alpha(X_n - X_{n+1})$$

while the action force of the nth atom by the (n-1)th atom is

$$F_{n,n-1} = -\alpha(X_n - X_{n-1})$$

As what discussed in this section is the weak vibration when the temperature is not too high, the thermal vibration of the particle is mainly affected by adjacent particles and the affections from other particles can be ignored.

So the total force of the nth atom is approximately equal to the sum of the $F_{n,\ n+1}$ and the $F_{n,\ n-1}$, then the equation of motion of nth atom is:

$$m\ddot{X}_n = \alpha(X_{n+1} + X_{n-1} - 2X_n), \mathrm{n} = 1, 2, 3, \ldots \mathrm{N} \tag{3.36}$$

This formula is true for all atoms, and the solution has a simple harmonic form:

$$X_n = Ae^{i(\omega t - 2\pi nak)} \tag{3.37}$$

where A is the amplitude, ω is the angular frequency, na is the equilibrium position of the nth atom relative to the origin, k $(=1/\lambda)$ is the wave number, and λ is the wavelength. Eq. (3.37) shows that the thermal vibration propagates in the lattice in the form of simple harmonic wave when each atom vibrates in its equilibrium position, and is called the **lattice wave**. Noting that the relationships exist between the crystal density ρ, Young's modulus E, and the one-dimensional stress elastic wave velocity C_0 with the m, a, α, respectively, as shown below:

$$\rho = m/a^3, E = \alpha/a, C_0 = \sqrt{\frac{E}{\rho}} = a\sqrt{\frac{\alpha}{m}} \tag{3.38}$$

Eq. (3.36) is consistent with the difference scheme after discretization of the one-dimensional elastic wave motion equation in continuum mechanics.

Substituting the harmonic solution (3.37) into Eq. (3.36), we have

$$\omega^2 = \frac{4\alpha}{m}\sin^2(\pi ak) = \omega_{\max}^2 \sin^2(\pi ak) \tag{3.39a}$$

Eq. (3.39a) gives the correspondence between the angular frequency ω and the wave number k (or wavelength λ). As shown in Fig. 3.11, it describes the dispersion phenomenon of lattice wave propagation dependent on frequency. Similar to the dispersion phenomenon in optics, it is called **dispersion or diffusion relation**. When k changes from 0 to $\pm 1/(2a)$, ω changes from 0 to its maximum $\omega_{\max}$. When $|k| \geq 1/(2a)$, there is periodic repetition. Therefore, in order for the ω to be a single-value function of the k, limit k to the following range:

$$-\frac{1}{2a} < k \leq \frac{1}{2a} \tag{3.39b}$$

The positive k corresponds to the positive propagation of lattice waves, while the negative k corresponds to the negative propagation of lattice waves.

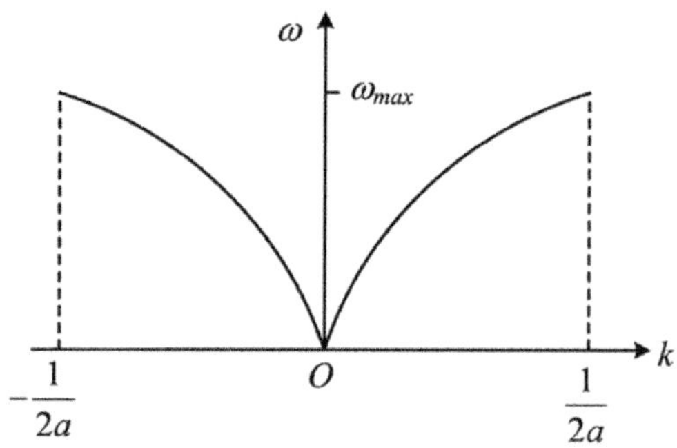

Figure 3.11 Correspondence relationship between the angular frequency ω and the wave number k.

Eq. (3.39) does not contain n, indicating that all particles have the same ω-k relationship; each k has a definite frequency ω, called a **vibration mode**. Different vibration modes correspond to lattice waves with different wavelengths. Eq. (3.39a) also shows that each mode of the lattice wave is independent of each other, known as the normal mode. It is worth emphasizing here that Eq. (3.36) originally represents the motion equation of N-coupled harmonic oscillator, while its solution (Eq. 3.39a) is the decoupling vibration in the normal mode. This shows that the discussion for the vibration problem under the interaction of n atoms can be reduced to a discussion for the vibration problem of n independent normal mode by means of simple harmonic approximation. This is of great significance for the following quantization analysis of lattice vibration energy.

The formula also shows that there is a maximum $\omega_{\max}$ in the angular frequency of lattice waves, that is,

$$\omega_{\max} = \sqrt{\frac{4\alpha}{m}} \tag{3.40}$$

Corresponding to this is that the lattice wave has a minimum wavelength $\lambda_{\min}$.

$$\lambda_{\min} = 2a \tag{3.41}$$

This conclusion is reasonable when the microstructure of the medium is considered, while in the continuum mechanics, the angular frequency of the plane wave is not limited, and the wavelength can be arbitrarily small. This is because in continuum mechanics the microstructure scale (such as lattice constant a) has been continuous at the macroscale.

In addition, the wave velocity of lattice wave can be obtained by using Eqs. (3.38)–(3.40).

$$C = \frac{1}{k}\frac{\omega}{2\pi} = a\sqrt{\frac{\alpha}{m}}\cdot\frac{\sin(\pi ak)}{\pi ak} = C_0\frac{\sin(\pi ak)}{\pi ak} \tag{3.42}$$

This shows that the lattice wave is a dispersive wave, its wave velocity C depends on ak $(=a/\lambda)$, that is, it depends on the ratio of the lattice characteristic size (lattice constant a) to the wavelength λ. This is similar to the elastic stress wave in bars in continuum mechanics, its wave velocity depends on the ratio of the characteristic size of the bar (bar diameter d) to the wavelength λ (Wang, 2007).

When $ka<<1$ or $\lambda>>a$, i.e., in the case of long waves

$$\sin(\pi ak) \approx \pi ak,$$

so we have

$$\omega \approx \omega_{\max}\pi ak \tag{3.43}$$

The corresponding wave velocity is reduced to

$$C = \frac{1}{k}\frac{\omega}{2\pi} \approx a\sqrt{\frac{\alpha}{m}} = \sqrt{\frac{E}{\rho}} = C_0 \tag{3.44}$$

At this time, the wave velocity of the lattice wave is a constant that is independent of the wave number but only related to the properties of the material. This conclusion is consistent with the results of elastic long wave in one-dimensional slender bar of continuum mechanics. When the wavelength is greater than the lattice constant, the crystal can be regarded as a continuous medium.

The above discussion is applicable to the vibration of a one-dimensional infinite chain ($N \rightarrow \infty$) because for a finite chain (N is finite), the particles at both ends are different from the internal particles, so the assumption that only affected by neighboring particles (Eq. 3.36) is no longer holds. It can be imagined, however, that if a finite chain consisting of N enough particles is connected head to tail to form a ring chain, then all particles are still under exactly the same condition, avoiding the consideration of the end effect This model was proposed by Born and Karman (1912). At this point, Eq. (3.36) is still valid, but the **periodic condition** (or called **Born-Karman condition**) of the end connection for a limited-length loop chain should be added.

$$X_i = X_{i+N}$$

After making the above assumption, we can treat the finite-length lattice approximation as the infinite lattice with periodic repetition. By substituting the harmonic solution (Eq. 3.37) into the additional periodic condition (Eq. 3.44), we have:

$$e^{-i2\pi Nak} = 1$$

Thus we obtain

$$Nak = S$$

where S is an integer. Taking into account Eq. (3.39b), the integer S must be limited to the following range:

$$-\frac{N}{2} < S \leq \frac{N}{2}$$

Thus, the possible values of S are

$$S = -\left(\frac{N}{2} - 1\right), \ \cdots, \ -2, \ -1, \ 0, \ 1, \ 2, \ \cdots, \ \frac{N}{2}$$

S has N different integer values, so $k = \frac{S}{Na}$ can only have N different values. Each value of k corresponds to an independent mode, so a finite-length lattice of N atoms can have N independent modes. In a one-dimensional lattice, the vibration degree of freedom of each atom is 1, so we can also say that the number of independent vibration modes of the lattice is equal to the number of degrees of freedom of the crystal. This means that the lattice vibrations are quantized. This conclusion also applies to three-dimensional lattices.

2. Quasi-harmonic vibration of one-dimensional compound lattice

The above discussion deals with the thermal vibration of a one-dimensional simple lattice consisting of N identical atoms. The thermal vibration of a one-dimensional compound lattice consisting of two different atoms is discussed further below.

Suppose that on an infinite chain, two different atoms, P_1 and P_2, are arranged periodically, and the distance between the neighboring atoms is 2a, as shown in Fig. 3.12. P_1 atoms with mass m_1 are located at the positions of 2n-1, 2n+1, 2n+3 ... and P_2 atoms with mass m_2 are located at the positions of 2n-2, 2n, 2n+2, ... In this way, P_2 atom is next to P_1 atom and P_1 atom is next to P_2 atom, and the two are coupled to each other.

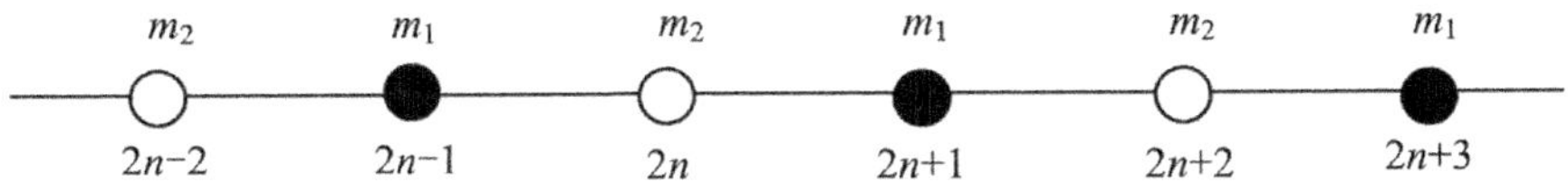

Figure 3.12 Vibration of a one-dimensional infinite chain consisting of two different atoms.

Similar to Eq. (3.36), the corresponding equations of motion are

$$\begin{aligned} m_1\ddot{X}_{2n+1} &= \alpha(X_{2n+2} + X_{2n} - 2X_{2n+1}) \\ m_2\ddot{X}_{2n} &= \alpha(X_{2n+1} + X_{2n-1} - 2X_{2n}) \end{aligned} \tag{3.45}$$

For the convenience of discussion, suppose $m_2 > m_1$.

Suppose the tentative solutions of the set of Eq. (3.45) are the following harmonic solutions.

$$\begin{aligned} X_{2n+1} &= Ae^{i[\omega t - 2\pi k(2n+1)a]} \\ X_{2n} &= Be^{i[\omega t - 2\pi k(2n)a]} \end{aligned} \tag{3.46}$$

where A and B are the vibration amplitudes of the atom with mass m_1 and m_2, respectively, ω is the angular frequency, k $(=1/\lambda)$ is the wave number, and λ is the wavelength. To substitute the tentative solutions into Eq. (3.45), we have

$$\begin{aligned} -m_1\omega^2 A &= \alpha\left[\left(e^{-i2\pi ka} + e^{i2\pi ka}\right)B - 2A\right] \\ -m_2\omega^2 B &= \alpha\left[\left(e^{-i2\pi ka} + e^{i2\pi ka}\right)A - 2B\right] \end{aligned} \tag{3.47}$$

or can be arranged as

$$\begin{aligned} \left(m_1\omega^2 - 2\alpha\right)A + (2\alpha\cos 2\pi ka)B &= 0 \\ (2\alpha\cos 2\pi ka)A + \left(m_2\omega^2 - 2\alpha\right)B &= 0 \end{aligned} \tag{3.48}$$

To have a nonzero solution to A and B, the coefficient determinant must be equal to zero, that is,

$$\begin{vmatrix} m_1\omega^2 - 2\alpha & 2\alpha\cos 2\pi ka \\ 2\alpha\cos 2\pi ka & m_2\omega^2 - 2\alpha \end{vmatrix} = 0 \tag{3.49}$$

So we can work out the following two solutions.

$$\begin{aligned} \omega^2 &= \frac{\alpha}{m_1 m_2}\left[(m_1 + m_2) \pm \left(m_1^2 + m_2^2 + 2m_1 m_2\cos 4\pi ka\right)^{\frac{1}{2}}\right] \\ &= \frac{\alpha(m_1 + m_2)}{m_1 m_2}\left\{1 \pm \left[1 - \frac{4m_1 m_2}{(m_1 + m_2)^2}\sin^2(2\pi ka)\right]^{\frac{1}{2}}\right\} \end{aligned} \tag{3.50a}$$

Take ω_+ to express the solution with the plus sign in Eq. (3.50a), and ω_- to express the solution with the minus sign in Eq. (3.50a), then the corresponding amplitude ratio $\left(\frac{A}{B}\right)$ of the two particles can be obtained after substituting Eq. (3.50a) into Eq. (3.48).

$$\left(\frac{A}{B}\right)_+ = \frac{2\alpha\cos(2\pi ka)}{2\alpha - m_2\omega_+^2}, \tag{3.50b}$$

$$\left(\frac{A}{B}\right)_- = \frac{2\alpha\cos 2\,\pi ka}{2\alpha - m_1\omega_-^2} \tag{3.50c}$$

Thus, for each value of k, there are two kinds of independent vibration with frequency ω_+ and ω_-, respectively. The relationship curve between ω and k is divided into two branches, as shown in Fig. 3.13. The one corresponding to ω_+ is called the **optical branch**, and the one corresponding to ω_- is called the **acoustic branch**. The name comes from the fact that their long-wave limits are related to light waves and sound waves, respectively (as described below). From Eq. (3.50a) and Fig. 3.13, we can see that ω is a periodic function of k, with a period of $\frac{1}{2a}$, and when k increases

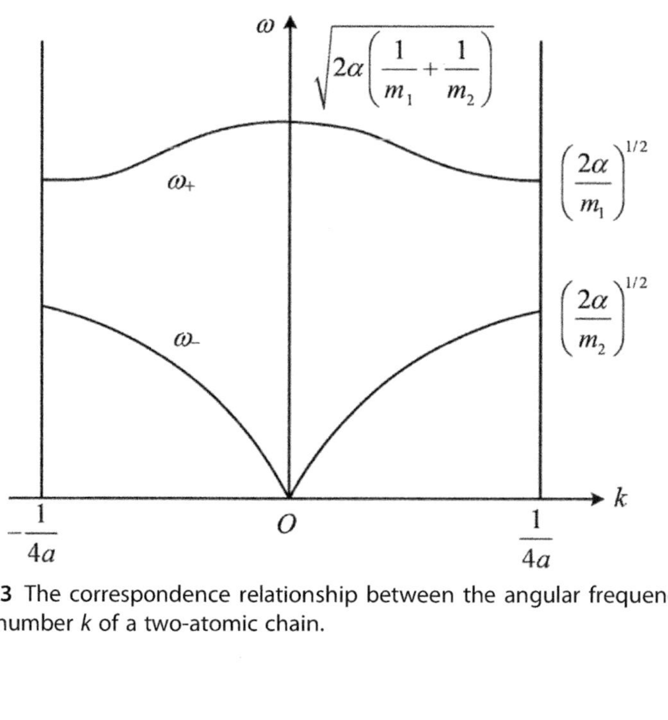

Figure 3.13 The correspondence relationship between the angular frequency ω and the wave number k of a two-atomic chain.

or decreases by $\frac{1}{2a}$, the value of ω remains unchanged. Therefore we can choose k to change in the range of

$$-\frac{1}{4a} < k \leq \frac{1}{4a}.$$

The optical and acoustic branches are discussed further below, respectively.

3.5.1 Optical branch

For optical branches, when $k = 0$ (long-wave limit), from Eq. (3.50), we have

$$\omega_+ = \left[2\alpha\left(\frac{1}{m_1} + \frac{1}{m_2}\right)\right]^{\frac{1}{2}}, \left(\frac{A}{B}\right)_+ = -\frac{m_2}{m_1} \text{。} \tag{3.51a}$$

At this point, ω_+ has a maximum value (Fig. 3.13), and $\left(\frac{A}{B}\right)$ is negative which means that the vibration of the two particles has a completely opposite phase (Fig. 3.14a). This corresponds to that the Bragg lattice formed by particles having mass m_1 and the Bragg lattice formed by particles having mass m_2 make a relative vibration as a whole, but the center of mass remains stationary ($m_1A + m_2B = 0$). For ionic crystals, the vibration direction of positive ions and negative ions is opposite, which will polarize the positive and negative ions and generate a certain electric dipole moment, so the long optical wave is also called the **polarized wave**. The angular frequency of the polarized wave in actual crystals is in the range of 10^{13}–10^{14} s^{-1} (far-infrared light wave). Polarized waves can interact with light waves, and when the wave number and the frequency are the same, resonance occurs, and the light can be used to excite this vibration. The resonance excited by the optical wave in ionic crystals can cause the intense absorption of far-infrared light, which is an important effect in infrared spectroscopy. ω_+ lattice wave is hence named as optical branch.

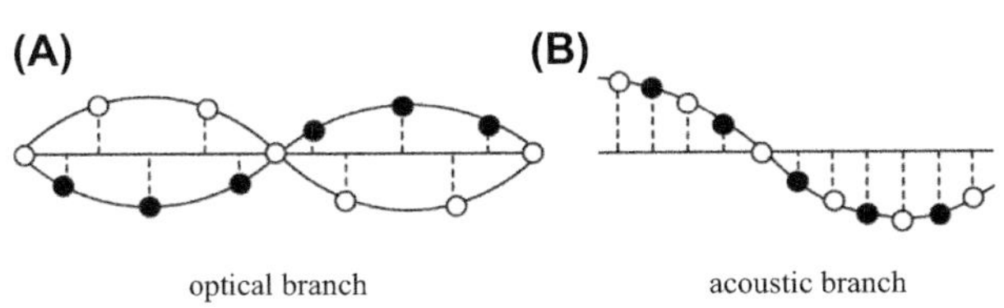

Figure 3.14 Vibration characteristics of optical and acoustic branches (the vertical axis represents the vibrating phase) (A) optical branch and (B) acoustic branch.

For an optical branch when $k = \pm\frac{1}{4a}$, from Eq. (3.50), we have

$$\omega_+ = \left(\frac{2\alpha}{m_1}\right)^{\frac{1}{2}}, \left(\frac{B}{A}\right)_+ = 0 \tag{3.51b}$$

In such case, the angular frequency ω_+ of the optical branch reaches the minimum, and $\left(\frac{B}{A}\right)$ equals to zero which indicates the particles having mass m_2 ($> m_1$) do not move, only the particles having mass m_1 vibrate.

3.5.2 Acoustic branch

For the acoustic branch, when $k = 0$ (long-wave limit), from Eq. (3.50), we have

$$\omega_- = 0, \left(\frac{A}{B}\right)_- = 1。 \tag{3.52a}$$

In such case, ω_- has a minimum value (Fig. 3.13), while $\left(\frac{A}{B}\right) = +1$ which means that the vibration of the two different particles has exactly the same amplitude and phase, representing the vibration of the mass center of a unit cell (Fig. 3.14B).

For the acoustic branch, when$k = \pm\frac{1}{4a}$, it can be calculated from Eq. (3.50).

$$\omega_- = \left(\frac{2\alpha}{m_2}\right)^{\frac{1}{2}}, \left(\frac{A}{B}\right)_- = 0 \tag{3.52b}$$

In such case, the angular frequency ω_- of the acoustic branch reaches its maximum, and $\left(\frac{A}{B}\right)$ equal to zero which indicates that the particles having mass m_1 ($< m_2$) do not move, only the particles having mass m_2 vibrate.

When $ka<<1$ or $\lambda>>a$, namely in the case of long waves

$$\sin(2\pi ak) \approx 2\pi ak$$

Thus from Eq. (3.50a), we have

$$\omega_- \approx \left(\frac{2\alpha}{m_1 + m_2}\right)^{\frac{1}{2}} 2\pi ak \tag{3.52c}$$

and the corresponding wave velocity is

$$C = \frac{1}{k}\frac{\omega}{2\pi} \approx a\sqrt{\frac{2\alpha}{m_1 + m_2}} = C_0$$

which is consistent with Eq. (3.44), and the last equal sign of the above formula is obtained in accordance with Eq. (3.38) and replace m with $\frac{m_1+m_2}{2}$. It could be seen that in the case of $ka<<1$ (long waves), the ω_- long wave is an elastic sound wave, and the wave velocity is a constant which is independent of the wave number and only related to the properties of materials. ω_- lattice wave is hence named as **acoustic branch**.

The other vibration properties of the acoustic branch are similar to the quasi-harmonic thermal vibrations of the one-dimensional simple lattice discussed previously, and are not repeated here.

As a generalization of the above discussion in the three-dimensional lattice, if the crystal consists of N unit cells and each unit cell contains S particles, there are SN particles in total. Now let's discuss the vibration in three dimensions. Each particle can vibrate in three directions, with a total of 3 SN degrees of freedom. Similar to the previous conclusions, it can be proved that for the quasi-harmonic small vibration, the three-dimensional lattice vibrations of the SN particles can be replaced by the superposition of the 3 SN independent simple harmonic vibrations (equivalent to 3 SN independent harmonic oscillators). It can also be proved that 3 SN independent vibrations can be divided into 3S different nodes of vibration, and each mode has N different frequencies of resonance. Only three of all these vibration modes are acoustic branches. For these three acoustic branches, the vibration directions of the S particles in the unit cell are consistent, which are a kind of the vibration between the mass centers of the unit cell. In the long-wave limit, the three acoustic waves coincide with the elastic waves in continuum mechanics. The remaining 3(S-1) branches are optical branches, and the vibration directions of the S particles in the unit cell of the optic branch are inconsistent, which are a kind of vibration between the particles in the unit cell.

The lattice vibration discussed above is in the category of classical mechanics, and the results show that as long as it is a small-amplitude linear vibration (harmonic vibration), the arbitrary complex small-amplitude vibration of a crystal composed of N particles can be considered as the superposition of 3N normal vibrations. In other words, a system of N interacting particles is equivalent to 3N independent harmonic oscillators. In this way, the above discussion can be easily transited to the quantum mechanics of lattice vibrations. Then it will be not difficult to know the vibration energy of the system by just studying the simplified system composed of 3N independent harmonic oscillators.

According to the theory of quantum mechanics, at temperature T, the vibration energy level ε_{ni} of a harmonic oscillator with basic frequency ν_i is

$$\varepsilon_{ni} = \left(\frac{1}{2} + n\right) h\nu_i \quad (n = 0, 1, 2, \cdots) \tag{3.53}$$

where h is the Planck constant (6.62×10^{-27} erg s); n_i is the ordinal number of quantum energy levels or the master quantum number, indicating that the energy of the harmonic oscillator has discrete values with uniform spacing (quantization); $\frac{1}{2} h\nu_i$ is the so-called zero-point energy which means that the motion does not stop even in the lowest energy state. The internal energy is the sum of the vibration energy ε_{ni} and the potential energy u_i (the interaction energy between the first oscillator and other oscillators in the system), that is, the internal energy level E_{ni} of the ith oscillators is

$$E_{ni} = u_i + \varepsilon_{ni} = u_i + \left(\frac{1}{2} + n\right) h\nu_i \quad (n = 0, 1, 2, \cdots) \tag{3.54}$$

For the macroscopic system composed of arbitrary particles in a thermal equilibrium state, as long as the particles of the system conform to the laws of classical mechanics, and the particles are completely statistically independent, then the statistical distribution of particles between energy levels satisfies the Boltzmann distribution law. Thus, the average internal energy $\overline{E}_{ni}$ of the ith oscillator, namely the thermodynamic internal energy E_i of the ith oscillator, is

$$\overline{E}_{ni} = E_i = \sum_{n=0}^{\infty} P_n E_{ni} \tag{3.55a}$$

where P_n is the most probable distribution rate of the oscillator at the E_{ni} level:

$$P_n = \frac{\exp\left(-\dfrac{E_{ni}}{k_B T}\right)}{\sum\limits_{n=0}^{\infty} \exp\left(-\dfrac{E_{ni}}{k_B T}\right)} \tag{3.55b}$$

where k_B is the Boltzmann constant (1.38×10^{-23} J/K^0), the numerator term on the right side of the equal sign represents the number of the effective state of the particle at the E_{ni} level, and the denominator term, which represents the sum of the numbers of the effective state of particles at each level of energy, is called **partition function**, and for convenience it is denoted as:

$$Z_i = \sum_{n=0}^{\infty} \exp\left(-\frac{E_{ni}}{k_B T}\right)。 \tag{3.55c}$$

So the internal energy of the ith oscillator can be written as

$$E_i = \sum_{n=0}^{\infty} P_n E_{ni} = \frac{\sum\limits_{n=0}^{\infty} E_{ni} \exp\left(-\frac{E_{ni}}{k_B T}\right)}{\sum\limits_{n=0}^{\infty} \exp\left(-\frac{E_{ni}}{k_B T}\right)} = \frac{\sum\limits_{n=0}^{\infty} E_{ni} \exp\left(-\frac{E_{ni}}{k_B T}\right)}{Z_i}。 \tag{3.55d}$$

If the particles in the same energy level ν_i may have several different microscopic states, called the degeneracy of this energy level, expressed in terms of g_i, then the exponential function term of the numerator and the denominator (partition function) on the right side of the equation should be multiplied by g_i. For example, there are two different states of spin quantum numbers in the same energy level state, then the degeneracy $g_i = 2$, and so on. However, for simplicity, $g_i = 1$ is set here and below, which does not affect the basic substance of what is discussed.

Let $x = \frac{1}{k_B T}$, $dx = -\frac{dT}{k_B T^2}$, then the above formula can be expressed in the following form:

$$E_i = \frac{\sum\limits_{n=0}^{\infty} E_{ni} \exp(-E_{ni} x)}{\sum\limits_{n=0}^{\infty} \exp(-E_{ni} x)} = -\frac{\frac{\partial Z_i}{\partial x}}{Z_i} = -\frac{\partial \ln Z_i}{Z_i} = k_B T^2 \frac{\partial \ln Z_i}{\partial T} = k_B T \frac{\partial \ln Z_i}{\partial \ln T} \tag{3.55e}$$

After the determination of internal energy E_i, it is not difficult to determine other relevant thermodynamic characteristic parameters by the thermodynamic relations provided in chapter 2. For example, the constant volume-specific heat C_{Vi}, which is closely related to thermal energy, can be obtained from Eq. (2.54):

$$C_{vi} = \left(\frac{\partial E_i}{\partial T}\right)_V = k_B T \left[\frac{\partial}{\partial T}\left(\ln Z_i + \frac{\partial \ln Z_i}{\partial \ln T}\right)\right] \tag{3.56}$$

Then by using the relation $C_V \equiv T\left(\frac{\partial S}{\partial T}\right)$ (Eq. 2.36), $\left(\frac{\partial S_i}{\partial T}\right)$ can be first obtained from Eq. (3.56) and then the entropy S_i can be further obtained:

$$S_i = k_B\left(\ln Z_i + \frac{\partial \ln Z_i}{\partial \ln T}\right) \tag{3.57}$$

Similarly, the Helmholtz free energy A_i can be written out from Eq. (2.24) and, accordingly, pressure P_i can be written out from Eq. (2.34):

$$A_i = E_i - TS_i = -k_B T \ln Z_i \circ \tag{3.58}$$

$$P_i = -\left(\frac{\partial A_i}{\partial V}\right)_T = k_B T\left(\frac{\partial \ln Z_i}{\partial \ln V}\right)_T \tag{3.59}$$

Furthermore, the enthalpy H_i can also be written out from Eq. (2.24):

$$H_i = E_i + P_i V = k_B T\left(\frac{\partial \ln Z_i}{\partial \ln T} + \frac{\partial \ln Z_i}{\partial \ln V}\right) \tag{3.60}$$

Thus, the key problem is how to determine the distribution function Z_i of the oscillators.

For this reason, we substitute the internal energy-level expression (Eq. 3.54) of the ith oscillator into the partition function expression (Eq. 3.55c), and use the following relation of the geometric series summation.

$$\sum_{n=1}^{\infty} a_1 q^{n-1} = \frac{a_1}{1-q}$$

then we have

$$\begin{aligned} Z_i &= \sum_{n=0}^{\infty} \exp\left[-\frac{u_i + \left(\frac{1}{2} + n\right) h\nu_i}{k_B T}\right] \\ &= \exp\left[-\frac{u_i + \frac{1}{2} h\nu_i}{k_B T}\right] \sum_{n=0}^{\infty} \exp\left[-\frac{n h\nu_i}{k_B T}\right] \\ &= \exp\left[-\frac{u_i + \frac{1}{2} h\nu_i}{k_B T}\right] \cdot \frac{1}{1 - \exp\left[-\frac{h\nu_i}{k_B T}\right]} \end{aligned} \tag{3.61}$$

or

$$\ln Z_i = -\frac{\left(u_i + \frac{1}{2} h\nu_i\right)}{k_B T} - \ln\left[1 - \exp\left(-\frac{h\nu_i}{k_B T}\right)\right] \tag{3.62}$$

Substituting the above results into the expressions of thermodynamic state quantities and after finishing, we can obtain:

$$E_i = \left(u_i + \frac{1}{2}h\nu_i\right) + \frac{h\nu_i}{\exp\left(\frac{h\nu_i}{k_B T}\right) - 1} \tag{3.63}$$

$$= E_{Ki}(V) + E_{Ti}(V, T)$$

$$A_i = \left(u_i + \frac{1}{2}h\nu_i\right) + k_B T \ln\left[1 - \exp\left(-\frac{h\nu_i}{k_B T}\right)\right] \tag{3.64}$$

and so on. Note that the first and second terms on the right side of Eq. (3.63) equal to cold energy and heat energy, respectively. It is not difficult to further find out the isometric specific heat C_{Vi}, entropy S_i, and pressure p_i by the partial differential of E_i and A_i. (The reader is advised to deduce it as an exercise.)

For a crystal composed of N particles, each particle has three degrees of freedom, which is equivalent to 3N oscillators under quasi-harmonic approximation, whose internal energy can be summed up by Eq. (3.54):

$$E_n = \sum_{i=1}^{3N} E_{ni} = \sum_{i=1}^{3N}\left(u_i + \left(\frac{1}{2} + n\right)h\nu_i\right)$$

Thus, similar to Eq. (3.55c), the partition function Z of the entire crystal system can be written as

$$Z = \sum_{n=0}^{\infty} \exp\left(-\frac{E_n}{k_B T}\right)$$

$$= \sum_{n=0}^{\infty} \exp\left[-\frac{\sum_{i=1}^{3N}\left[u_i + \left(\frac{1}{2} + n\right)h\nu_i\right]}{k_B T}\right] = \sum_{n=0}^{\infty}\prod_{i=1}^{3N} \exp\left[-\frac{u_i + \left(\frac{1}{2} + n\right)h\nu_i}{k_B T}\right]$$

$$= \prod_{i=1}^{3N}\sum_{n=0}^{\infty} \exp\left[-\frac{u_i + \left(\frac{1}{2} + n\right)h\nu_i}{k_B T}\right] = \prod_{i=1}^{3N} Z_i \tag{3.65a}$$

Or rewrite as

$$\ln Z = \sum_{i=1}^{3N} \ln Z_i \tag{3.65b}$$

With the expression of the partition function Z of the system, compared with Eqn. (3.55 e) to Eq. (3.59), the related specific forms of the thermodynamic state quantities of the system can be written, respectively, as

$$\begin{aligned} E &= k_B T \frac{\partial \ln Z}{\partial \ln T} = k_B T \cdot \sum_{i=1}^{3N} \frac{\partial \ln Z_i}{\partial \ln T} \\ &= U + \frac{1}{2} \sum_{i=1}^{3N} h\nu_i + k_B T \sum_{i=1}^{3N} \frac{\frac{h\nu_i}{k_B T}}{\exp\left(\frac{h\nu_i}{k_B T}\right) - 1} \\ &= E_K(V) + E_T(V, T) \end{aligned} \tag{3.66}$$

where $E_K(V)$ and $E_T(V, T)$ are cold energy and heat energy, respectively.

$$\begin{aligned} A &= -k_B T \ln Z = -k_B T \sum_{i=1}^{3N} \ln Z_i \\ &= U + \frac{1}{2} \sum_{i=1}^{3N} h\nu_i + k_B T \sum_{i=1}^{3N} \ln\left[1 - \exp\left(-\frac{h\nu_i}{k_B T}\right)\right] \end{aligned} \tag{3.67}$$

$$\begin{aligned} C_V &= k_B T \left[\frac{\partial}{\partial T}\left(\ln Z + \frac{\partial \ln Z}{\partial \ln T}\right)\right] \\ &= k_B \sum_{i=1}^{3N} \left(\frac{\frac{h\nu_i}{k_B T}}{\exp\left(\frac{h\nu_i}{k_B T}\right) - 1}\right)^2 \cdot \exp\left(\frac{h\nu_i}{k_B T}\right) \end{aligned} \tag{3.68}$$

$$\begin{aligned} S &= k_B \left(\ln Z + \frac{\partial \ln Z}{\partial \ln T}\right) \\ &= -k_B \sum_{i=1}^{3N} \ln\left[1 - \exp\left(-\frac{h\nu_i}{k_B T}\right)\right] + k_B \sum_{i=1}^{3N} \left(\frac{\frac{h\nu_i}{k_B T}}{\exp\left(\frac{h\nu_i}{k_B T}\right) - 1}\right) \end{aligned} \tag{3.69}$$

About the isometric specific heat C_V, it is worth comparing Eq. (3.68) of the quantum mechanics theory with that of the classical theory, the Dulong–Petit law.

According to the classical theory, the energy is divided equally by degrees of freedom, and the average energy of each degree of freedom is $k_B T$ (half of kinetic energy and half of potential energy). For a crystal with N particles and with 3N degrees of freedom, the total average vibration energy is $3Nk_B T$. Thus we have:

$$C_V = \left(\frac{\partial E}{\partial T}\right)_V = 3Nk_B \tag{3.70}$$

which is called the Dulong–Petit law.

On the other hand, when the temperature is very high, namely $(h\nu_i/k_B T) << 1$, expand Eq. (3.68) into a series with regard to $h\nu_i/k_B T$, we have

$$C_V = 3Nk_B\left\{1 - \frac{1}{36N}\sum_{i=1}^{3N}\left(\frac{h\nu_i}{k_B T}\right)^2 + \cdots\right\} \doteq 3Nk_B \tag{3.71a}$$

In such case, Eq. (3.68) is reduced to the Dulong–Petit law, and is also consistent with the experimental results.

However, when the temperature is very low, namely $h\nu_i/k_B T >> 1$, Eq. (3.68) can be simplified as

$$C_V = k_B\sum_{i=1}^{3N}\left(\frac{h\nu_i}{k_B T}\right)^2 \exp\left(-\frac{h\nu_i}{k_B T}\right) \tag{3.71b}$$

The results show that C_V decreases rapidly with the temperature T, and when $T \to 0$, $C_V \to 0$, which is consistent with the experimental results, while the Dulong–Petit law cannot explain this phenomenon.

We have applied the quantum theory to derive the specific forms of thermodynamic state quantities by using the partition function, and the key of the next step comes down to determine the energy-level distribution; in other words, the key is how to determine the 3N normal frequencies ν_i. This is a difficult and complex problem for real crystals, so it is usually approximated by simplified models, such as the commonly used Einstein model and Debye model These models are discussed below.

3.5.3 The Einstein model

Einstein (1906) assumed that all particles in a crystal vibrate at the same frequencyν_0, i.e.,

$$\nu_1 = \nu_2 = \cdots \nu_i = \cdots \nu_0$$

This is equivalent to ignoring the interaction between particles at thermal equilibrium. Thus, the internal energy E (Eq. 3.66) and the isometric specific heat C_V (Eq. 3.68) are reduced to, respectively:

$$E = U + 3Nk_BT\left(\frac{1}{2}\frac{h\nu_0}{k_BT} + \frac{\frac{h\nu_0}{k_BT}}{\exp\left(\frac{h\nu_0}{k_BT}\right) - 1}\right)$$

$$C_V = 3Nk_BT\left(\frac{h\nu_0}{k_BT}\right)^2 \frac{\exp\left(\frac{h\nu_0}{k_BT}\right)}{\left(\exp\left(\frac{h\nu_0}{k_BT}\right) - 1\right)^2}$$

Let $\theta_E = h\nu_0/k_B$, called the **Einstein characteristic temperature**, thus

$$E = U + 3Nk_BT\left(\frac{1}{2}\frac{\theta_E}{T} + \frac{\frac{\theta_E}{T}}{\exp\left(\frac{\theta_E}{T}\right) - 1}\right) \tag{3.72}$$

$$C_V = 3Nk_B\left(\frac{\theta_E}{T}\right)^2 \frac{\exp\left(\frac{\theta_E}{T}\right)}{\left[\exp\left(\frac{\theta_E}{T}\right) - 1\right]^2} = 3Nk_Bf_E\left(\frac{\theta_E}{T}\right) \tag{3.73a}$$

where

$$f_E\left(\frac{\theta_E}{T}\right) = \left(\frac{\theta_E}{T}\right)^2 \frac{\exp\left(\frac{\theta_E}{T}\right)}{\left[\exp\left(\frac{\theta_E}{T}\right) - 1\right]^2} \tag{3.73b}$$

which is called the Einstein specific heat function.

The formula of isometric specific heat will be further discussed particularly below, because it is closely related to the thermal energy (see Eq. 2.57). What is more, it already has corresponding classical theoretical model, and it is also easier to be measure experimentally. Besides, there are a large number of experimental results, which is convenient to compare and analyze each other. It will also be used in determining the Grüneisen parameter γ in the future.

According to the formula of isometric specific heat (3.73a) based on the Einstein's model, when the temperature is very high, that is, when $\frac{\theta_E}{T} << 1$, we have

$$C_V = 3Nk_B, \quad \frac{\theta_E}{T} << 1 \tag{3.73c}$$

This is consistent with the Dulong–Petit law (Eq. 3.70) and consistent with the experimental results.

When the temperature is very low, that is $\frac{\theta_E}{T} >> 1$, from Eq. (3.73a), we have

$$C_V = 3Nk_B\left(\frac{\theta_E}{T}\right)^2 \exp\left(-\frac{\theta_E}{T}\right), \quad \frac{\theta_E}{T} >> 1 \tag{3.73d}$$

As can be seen from the above equation, when $T \rightarrow 0$, $C_v \rightarrow 0$, which is consistent with the experiment. At low temperatures, however, the experiment indicated that C_v is proportional to T^3, while the C_v calculated by the above Eq. (3.73d) (including exponential decay term) goes to zero at a faster rate, that is, when the temperature goes to zero, the calculated C_v decreases faster than the experimental value. The reason for this deviation is that the Einstein model assumes that all particles in a crystal vibrate at the same frequency while ignores the effect of other frequencies in a real lattice.

When the Einstein model is adopted, the frequency ν_0 or the Einstein characteristic temperature θ_E is generally determined by experiments, making the theoretical curve given by Eq. (3.73a) as consistent as possible with the experimental results in a larger temperature range. According to this principle, the Einstein characteristic temperature θ_E of most solids is in the range of 100–300°K, and the corresponding frequency ν_0 is about 10^{13} Hz, belonging to the high-frequency range of infrared light. And this is equivalent to replacing the optical branch mode with a constant frequency ν_0 (a horizontal dotted line) as shown in Fig. 3.15. Since optical branch is a high-frequency branch, its frequency changes relatively smoothly, so the Einstein model basically reflects the characteristics of the optical branch. Its error at lower temperatures is mainly due to the omission of the role of the low-frequency part (acoustic branch). At higher temperatures, that is, when $\frac{\theta_E}{T} << 1$, the low-frequency part has little effect (see Eq. 3.73c), so it can be treated approximately by the Einstein model.

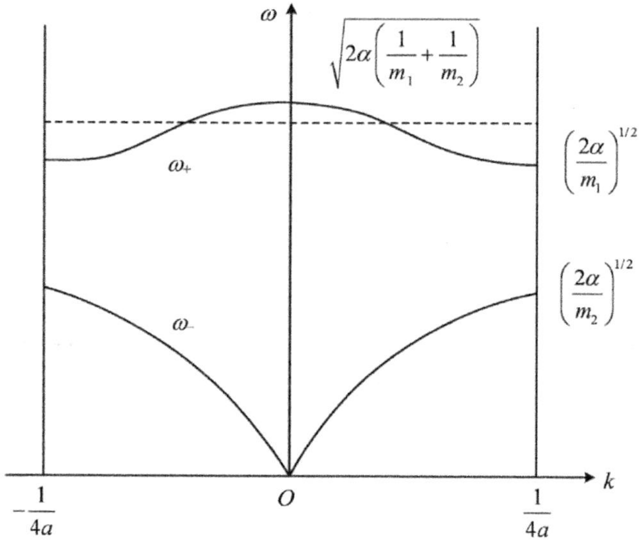

Figure 3.15 The relation between the ω and k of the Einstein model (a horizontal dotted line), comparing with the curves given in Fig. 3.13.

3.5.4 The Debye model

Debye (1912) believes that at low temperatures the contribution of low-frequency vibration to the specific heat and thermal energy is too great to ignore. As described in "§3.5 the thermal vibration of a lattice", the low-frequency lattice wave is a long acoustic wave with elastic wave properties. Therefore, he regarded crystals as isotropic continuous media, and long lattice waves produced by the crystal vibration as elastic waves propagating in continuous media.

The frequency ν_i of the lattice wave is originally discrete; if the frequency value is very dense, ν can be viewed as continuous, the modulus of vibration between ν and $\nu + d\nu$ is $g(\nu)d\nu$, and $g(\nu)$ is called the frequency distribution function. Because the total number of independent oscillators is 3N, there is

$$\int_0^{\nu_m} g(\nu)d\nu = 3N \tag{3.74}$$

ν_m is the maximum frequency in the formula. Accordingly, the original formula expressed in the form of discrete terms sum can be replaced by integral form. For examples, Eq. (3.66) and Eq. (3.68) can be, respectively, rewritten into

$$E = U + k_B T \int_0^{\nu_m} \left(\frac{1}{2} + \frac{1}{\exp\left(\frac{h\nu}{k_B T}\right) - 1} \right) \left(\frac{h\nu}{k_B T} \right) g(\nu)\ d\nu \tag{3.75}$$

$$C_V = k_B \int_0^{\nu_m} \left(\frac{h\nu_i}{k_B T} \right)^2 \frac{\exp\left(\frac{h\nu_i}{k_B T}\right)}{\left(\exp\left(\frac{h\nu_i}{k_B T}\right) - 1\right)^2} \cdot g(\nu)\ d\nu \tag{3.76}$$

So the key to the Debye model is how to calculate the frequency distribution function $g(\nu)$.

We study a cubic crystal with length of $\boldsymbol{L}$ on each side. According to the Debye model, it can be regarded as isotropic elastic continuous medium. By the classical elastic wave theory (Wang, 2007), the elastic waves propagating in isotropic elastic continuous medium consist of three components: a longitudinal wave component propagating with the longitudinal wave velocity C_l and two transverse wave components propagating with the transverse wave velocity C_t. The relationship between the wave velocity C, the frequency ν, and the wave number k or the wavelength λ is:

$$\nu = \frac{C}{\lambda} = Ck \tag{3.77}$$

where C may be C_l or C_t. Constrained by the free boundary condition at $x = 0$, $x = L$, $y = 0$, $y = L$ $z = 0$, and $z = L$, the half-wavelength $\frac{\lambda_x}{2}$, $\frac{\lambda_y}{2}$, $\frac{\lambda_z}{2}$ of the wave propagating along the three axes must be the integer divisor of side length L.

$$\frac{\lambda_x}{2} = \frac{L}{n_x} \quad \frac{\lambda_y}{2} = \frac{L}{n_y}, \quad \frac{\lambda_z}{2} = \frac{L}{n_z} \tag{3.78}$$

where n_x, n_y, and n_z are positive integers, that is, $n_x, n_y, n_z = 0, 1, 2, \cdots$. For waves propagating in any direction $\vec{n}$, in order to meet the free boundary conditions, the wavelength must satisfy:

$$\lambda = \lambda_x \cos\theta_x, \quad \lambda = \lambda_y \cos\theta_y, \quad \lambda = \lambda_z \cos\theta_z \tag{3.79}$$

where $\cos\theta_x$, $\cos\theta_y$, $\cos\theta_z$ are the direction cosines of $\vec{n}$. Because

$$\cos^2\theta_x + \cos^2\theta_y + \cos^2\theta_z = 1,$$

from Eq. (3.78) and Eq. (3.79), we can get the relations between the n_x, n_y, and n_z:

$$n_x^2 + n_y^2 + n_z^2 = \frac{4L^2 v_l^2}{C_l^2} \tag{3.80a}$$

$$n_x^2 + n_y^2 + n_z^2 = \frac{4L^2 v_t^2}{C_t^2} \tag{3.80b}$$

For given v_l and v_t, the value on the right side of Eq. (3.80) has been determined, so Eq. (3.80) provides two equations for determining the possible sets of n_x, n_y, n_z on the left side of Eq. (3.80). The set number of the positive integer n_x, n_y, n_z is the same thing as the number of independent oscillators corresponding to the given frequency, while the set number of n_x, n_y, n_z can be obtained by using the following geometric relations.

Take (n_x, n_y, n_z) as coordinate system, each equation in Eq. (3.80) represents a spherical surface with a half diameter of $R = 2Lv/C$, and any set of positive integers (n_x, n_y, n_z) that satisfy the equation is equivalent to a point on an eighth spherical surface. The number of these points is the number of independent oscillators whose frequency is v. We can thus assume that the number of oscillator with frequencies between v and $v + dv$ should be equal to one-eighth of the number of points in a shell with small radius $R = 2Lv/C$ and large radius $R + dR = 2L(v + dv)/C$. Or it is equal to one-eighth of the volume of this shell because in the coordinates we choose, the average unit volume contains one point. Therefore, the number of independent oscillators with frequencies between v and $v + dv$ is

$$g(v)dv = \frac{1}{8} \cdot 4\pi R^2 dR = 4\pi \frac{V}{C^3} v^2 dv$$

where $V = L^3$ is the volume of the crystal. Considering that for each vibration frequency v, there are one longitudinal wave and two transverse waves, the above formula can be used to the longitudinal wave and the transverse waves, respectively, and then add the results from the longitudinal wave and the transverse waves to get the number of independent oscillators with frequencies between v and $v + dv$:

$$g(v)dv = 4\pi V \left(\frac{1}{c_l^3} + \frac{2}{c_t^3} \right) v^2 dv = \frac{12\pi V}{c^{*3}} v^2 dv = Bv^2 dv \tag{3.81}$$

where

$$\frac{3}{c^{*3}} = \frac{1}{c_l^3} + \frac{2}{c_t^3}, \quad B = 4\pi V\left(\frac{1}{c_l^3} + \frac{2}{c_t^3}\right) = \frac{12\pi V}{c^{*3}} \tag{3.82}$$

Both c^* and B are constants determined by the numerical values of the longitudinal wave velocity C_l and the transverse wave velocity C_t of elastic waves.

In order to equalize the number of oscillators with the degrees of freedom of the crystal, Debye introduced an upper limit of frequency ν_D, making

$$\int_0^{\nu_D} g(\nu)d\nu = 3N$$

Substituting Eq. (3.81) into the above formula, we have

$$\int_0^{\nu_D} B\nu^2 d\nu = 3N$$

Thus we obtain

$$B = \frac{9N}{\nu_D^3} \tag{3.83a}$$

or

$$\nu_D = \left(\frac{9N}{B}\right)^{\frac{1}{3}} \tag{3.83b}$$

Note that ν_D is only a function of V, independent of temperature. The introduction of the upper limit of frequency ν_D means that in the Debye model, the short wave with the frequency larger than ν_D does not exist, while the vibration with the frequency smaller than ν_D can be approximated by elastic waves.

With the frequency distribution function $g(\nu)$ and the integral upper limit ν_D, the corresponding thermodynamic state function can be determined. For example, substituting Eq. (3.81) into Eqs. (3.75) and (3.76), the internal energy E and the isometric specific heat C_V can be determined, respectively:

$$E = U + 9Nk_BT\left(\frac{T}{\theta_D}\right)^3 \int_0^{\theta_D/T} \left(\frac{1}{2} + \frac{1}{e^x - 1}\right)x^3 dx \tag{3.84a}$$

or

$$E = U + \frac{9}{8} N k_B \theta_D + 3 N k_B T D\left(\frac{\theta_D}{T}\right) \tag{3.84b}$$

and

$$C_V = 9 N k_B \left(\frac{T}{\theta_D}\right)^3 \int_0^{\theta_D/T} \frac{e^x}{e^x - 1^2} x^4 dx \tag{3.85a}$$

or

$$C_V = 3 N k_B f_D\left(\frac{\theta_D}{T}\right) \tag{3.85b}$$

where

$$\begin{aligned}
\theta_D &= \frac{h\nu_D}{k_B} = \left(\frac{h}{k_B}\right)\left(\frac{3N}{4\pi V}\right)^{1/3} c^* = \left(\frac{h}{k_B}\right)\left(\frac{3N}{4\pi V}\right)^{1/3}\left(\frac{3}{1/c_l^3 + 1/c_t^3}\right)^{1/3}, \\
f_D\left(\frac{\theta_D}{T}\right) &= 3\left(\frac{\theta_D}{T}\right)^3 \int_0^{\theta_D/T} \frac{e^x}{e^x - 1} x^4 dx, \\
D(x_D) &= \frac{3}{x_D^3} \int_0^{x_D} \frac{x^3}{e^x - 1} dx, \\
x &= \frac{h\nu}{k_B T}, \quad x_D = \frac{h\nu_D}{k_B T} = \frac{\theta_D}{T}
\end{aligned} \tag{3.86}$$

where θ_D is called the **Debye characteristic temperature**, $D(x_D)$ is called the **Debye function**, and $f_D\left(\frac{\theta_D}{T}\right)$ is called the **Debye specific heat function**.

Similarly, expressions of such as the Helmholtz free energy A, entropy S, and pressure P can be written (it is suggested to derive them as exercises). Thus, in the Debye model, the thermodynamic quantities associated with thermal vibration are determinately determined by the parameter ν_D or Debye temperature θ_D; in other words, the characteristics of the thermal energy, thermal pressure, entropy and specific heat, and so on of the crystal are completely determined by ν_D and θ_D.

With regard to isometric specific heat (Eq. 3.85), when the temperature is very high, that is, when $x<<1$, e^xcan be expanded to $e^x \approx 1 + x$, and substitute them into Eq. (3.86), we can get $f_D\left(\frac{\theta_D}{T}\right) = 1$, so we obtain

$$C_V = 3Nk_B$$

This is consistent with the Dulong–Petit law.

On the contrary, when the temperature is very low, that is, when $x>>1$, the upper limit of the integral x_D in Eq. (3.85a) can be regarded as ∞, and Eq. (3.85a) can be rewritten as

$$C_V = 9Nk_B\left(\frac{T}{\theta_D}\right)^3 \int_0^{\infty} \frac{e^x}{(e^x - 1)^2} x^4 dx \tag{3.87a}$$

The integrand in the formula can be expanded into

$$\begin{aligned} \frac{e^x}{(e^x - 1)^2} x^4 &= x^4 e^{-x}\left(1 - e^{-x}\right)^{-2} \\ &= x^4 e^{-x}\left(1 + 2e^{-x} + 3e^{-2x} + \cdots\right) \\ &= x^4 \sum_{n=1}^{\infty} ne^{-nx} \end{aligned}$$

Substituting it into Eq. (3.87a), taking into.account the operation of the Riemann zeta function when integral, we have

$$C_V = \frac{12}{5}\pi^4 Nk_B\left(\frac{T}{\theta_D}\right)^3 \tag{3.87b}$$

That is, at low temperatures, the isometric specific heat is proportional to the cubic of temperature, called the **Debye law**, which is consistent with the experimental results.

For the atomic crystal and some simple ion crystals, the estimated value of C_V by the Debye theory is in good agreement with the experimental results in a wide temperature range. So the Debye theory goes a step further than the classical and Einstein theory. The limitations of Debye theory are mainly: (1) not suitable for compounds. Because the compound vibrates not only at a lower frequency but also at a higher frequency, while the Debye model just ignores the contribution of the optical branch higher than ν_D. (2) By the definition shown in Eq. (3.86), the θ_D is a temperature-independent constant, but the experimental results show that it is temperature dependent. (3) According to this model, the wave velocity is constant, that is,

$$C = \frac{v}{k} = const.$$

This is equivalent to that the acoustic branch mode is replayed by a dotted line, in which frequency ν (or angular frequency $\omega = 2\pi\nu$) is proportional to the wave number k, as shown in Fig. 3.16. This approximation is possible if ν is lower and the wavelength is much larger than the lattice constant. For those parts of the frequency distribution close to the maximum frequency ν_D, the corresponding wavelength is only one order of magnitude larger than the lattice constant, and the approximation of the continuous medium will lead to error.

There are two ways to determine the Debye characteristic temperature θ_D by experiments. The first method is the **specific heat method**, which is similar to the determination of Einstein characteristic temperature θ_E, that is, the selected θ_D can make the theoretical curve of isometric specific heat C_V as consistent as possible with the experimental results in a larger temperature range. The second method is the **elastic wave method**, which is to measure the longitudinal wave velocity c_l and the transverse wave velocity c_t

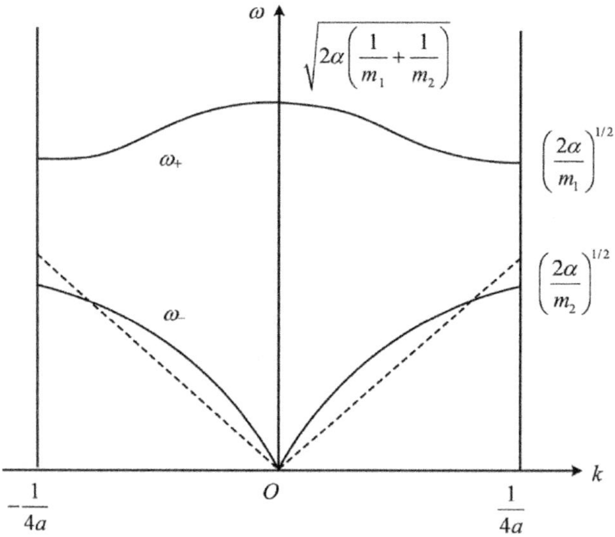

Figure 3.16 The relation diagram of ω and k of the Debye model.

Table 3.5 The θ_D determined by the specific heat method and the elastic wave method (Qian, 1962; Xie, 1961).

Material		Pb[a]	Ag[a]	Cu[a]	Al[a]	KCl[b]	NaCl[b]
θ_D	Specific heat method	88	215	315	398	230	308
	Elastic wave method	73	214	332	402	246	320

[a]From Qian (1962), p. 158.
[b]From Xie (1961), p. 110.

of the crystal, and then calculate the θ_D by Eqs. (3.82), (3.83) and (3.86). Table 3.5 shows the θ_D determined by these two methods and the results are satisfactory.

In conclusion, under the quasi-harmonic approximation, any complex small vibration of a crystal composed of N particles can be regarded as the superposition of 3N simple normal vibrations. At this time, the problem of determining the thermodynamic state quantities related to the thermal vibration of the lattice comes down to how to determine the partition function Z of the oscillator. While the determination of the partition function Z further comes down to how to determine the 3N normal frequencies ν_i. In order to solve such a difficult and complicated problem, the simplified Einstein model or Debye model is usually used for approximate treatment. The former is suitable for higher temperature $\left(\frac{\theta_E}{T} << 1\right)$, reflecting the optical branch of lattice vibration. At this point, the thermodynamic state quantities associated with thermal vibration depend on the Einstein characteristic temperature θ_E. The latter is suitable for lower temperature $\left(\frac{\theta_D}{T} >> 1\right)$, reflecting the acoustic branch of lattice vibration. At this point, the thermodynamic state quantities associated with thermal vibration depend on the Debye characteristic temperature θ_D or the corresponding upper frequency ν_D.

In fact, as far as the core of this chapter is concerned, what we are most concerned about is the EOS of solid under high pressure. The previous analyses provide a basic condition for us to further understand and study the EOS of solid under high pressure, especially the Grüneisen state equation, from the viewpoint of solid physics. This will be the focus of our discussion in the next section.

3.6 The solid physical foundation of the Grüneisen equation of state

In this section, we will discuss the contribution of thermal vibration to pressure in the same way as in the previous section, and derive the Grüneisen EOS of crystal; the equation and the related Grüneisen coefficient will be discussed in depth in the following.

From Eqs. (3.55e) and (3.59) in the previous section, it can be seen that the contribution of the thermal vibration of the ith oscillator to the internal energy E_i and the pressure P_i are respectively:

$$E_i = k_B T \frac{\partial \ln Z_i}{\partial \ln T},$$

$$P_i = k_B T \left(\frac{\partial \ln Z_i}{\partial \ln V} \right)_T 。$$

where the partition function Z_i of the ith oscillator, according to Eqs. (3.55c) and (3.61), is expressed as:

$$\begin{aligned}
Z_i &= \sum_{n=0}^{\infty} \exp\left(-\frac{E_{ni}}{k_B T} \right) \\
&= \sum_{n=0}^{\infty} \exp\left[-\frac{u_i + \left(\frac{1}{2} + n \right) h\nu_i}{k_B T} \right] \\
&= \exp\left[-\frac{u_i + \frac{1}{2} h\nu_i}{k_B T} \right] \sum_{n=0}^{\infty} \exp\left[-\frac{n h\nu_i}{k_B T} \right] \\
&= \exp\left[-\frac{u_i + \frac{1}{2} h\nu_i}{k_B T} \right] \cdot \frac{1}{1 - \exp\left[-\frac{h\nu_i}{k_B T} \right]}
\end{aligned}$$

Or in accordance with Eq. (3.62), it can be expressed in the following logarithmic form:

$$\ln Z_i = -\frac{\left(u_i + \frac{1}{2}h\nu_i\right)}{k_B T} - \ln\left[1 - \exp\left(-\frac{h\nu_i}{k_B T}\right)\right]$$

Substituting the above equation into Eq. (3.55) and after rearranging, the following equation given in the previous section as Eq. (3.63) can be obtained again.

$$\begin{aligned} E_i &= \left(u_i + \frac{1}{2}h\nu_i\right) + \frac{h\nu_i}{\exp\left(\frac{h\nu_i}{k_B T}\right) - 1} \\ &= E_{Ki}(V) + E_{Ti}(V, T) \end{aligned}$$

Substituting the above equation into Eq. (3.59) for P_i and after rearranging, we have

$$\begin{aligned} P_i &= k_B T\left(\frac{\partial \ln Z_i}{\partial V}\right)_T \\ &= -\frac{d\left(u_i + \frac{1}{2}h\nu_i\right)}{dV} - \frac{\frac{\partial \ln \nu_i}{\partial \ln V}}{V} \cdot \frac{h\nu_i}{\exp\left(\frac{h\nu_i}{k_B T}\right) - 1} \\ &= -\frac{dE_{Ki}(V)}{dV} + \frac{\gamma_i}{V} \cdot E_{Ti} \\ &= P_{Ki}(V) + \frac{\gamma_i}{V}[E_i - E_{Ki}(V)] \end{aligned} \tag{3.88}$$

This is the Grüneisen equation of the *i*th oscillator; in the equation

$$\gamma_i = -\left(\frac{\partial \ln \nu_i}{\partial \ln V}\right)_T \tag{3.89}$$

is called the Grüneisen coefficient of the *i*th oscillator.

The pressure P of a three-dimensional crystal shall be the superposition of the contribution of 3N oscillators, that is,

$$P = \sum_{i=1}^{3N} P_i$$

$$= -\sum_{=1}^{3N} \frac{dE_{Ki}(V)}{dV} - \sum_{i=1}^{3N} \frac{\frac{\partial \ln \nu_i}{\partial \ln V}}{V} \cdot \frac{h\nu_i}{\exp\left(\frac{h\nu_i}{k_B T}\right) - 1} \tag{3.90}$$

$$= -\frac{dE_K}{dV} + \sum_{i=1}^{3N} \frac{\gamma_i(V)}{V} E_{Ti}$$

$$= P_K(V) + \sum_{i=1}^{3N} \frac{\gamma_i(V)}{V} E_{Ti}$$

Under the assumption of quasi-harmonic vibration, $\gamma_i = \gamma(V)$, above form can be further reduced to:

$$P = P_K(V) + \sum_{i=1}^{3N} \frac{\gamma_i(V)}{V} E_{Ti}$$

$$= P_K(V) \tag{3.91}$$

$$= P_K(V) + \frac{\gamma(V)}{V}[E - E_K(V)]$$

This equation can also be derived by another way, namely directly using the following partition function of the system (Eq. 3.65)

$$Z = \prod_{i=1}^{3N} Z_i, \quad \ln Z = \sum_{i=1}^{3N} \ln Z_i$$

substituting it into the relation of $P = k_B T \left(\frac{\partial \ln Z}{\partial V}\right)_T$, and assuming $\gamma_i = \gamma(V)$, then after finishing, Eq. (3.91) can also be obtained. This is the **Grüneisen equation** of crystals. In chapter 2, we have derived this equation from the view of thermodynamics, and now we derived the same equation on the basis of the microstructure of crystals.

It should be emphasized that Eq. (3.91) is derived from the quasi-harmonic vibration approximation assumption, that is, it is assumed that the normal mode vibration frequency ν_i is a function of the particle spacing (i.e., the density ρ or the volume V). For the strict simple harmonic vibration, the normal frequency ν_i does not change with the volume V, and if so, according to Eq. (3.89), there will be $\gamma_i = 0$, which means there is no Grüneisen EOS. So the Grüneisen assumption of $\gamma_i(T, V) = \gamma(V)$ is just an approximation of the nonlinear vibration effect of lattices. If start directly from Eq. (3.90) to formulate an EOS with a form similar to the Grüneisen EOS but without the limitation of Grüneisen assumption, then it needs to introduce a coefficient $\Gamma(V, T)$ defined by Eq. (3.92a). Different from the Grüneisen coefficient $\gamma(V)$, $\Gamma(V, T)$ is generally a function of V and T.

$$\Gamma = \Gamma(V, T) = \frac{\sum_{i=1}^{3N} \gamma_i E_{Ti}(V, T)}{\sum_{i=1}^{3N} E_{Ti}(V, T)} = \frac{\sum_{i=1}^{3N} \gamma_i E_{Ti}(V, T)}{E_T(V, T)} \tag{3.92a}$$

Thus Eq. (3.90) can be rewritten as:

$$P = P_K + \frac{\Gamma(V, T)}{V} E_T \tag{3.92b}$$

It should emphasize again that the Γ (V, T) and the Grüneisen coefficients γ (V) are different, it is usually a function of V and T. However, in exceptional circumstances such as high and low temperatures, it is not necessary to rely on the Grüneisen assumption and Γ (V, T) would be reduced to a function only of V. In fact, at high temperatures, that is, $\frac{h\nu_i}{k_B T} << 1$, the thermal energy of the ith oscillator is reduced to.

$$E_{Ti} = \frac{h\nu_i}{\exp\left(\frac{h\nu_i}{k_B T}\right) - 1} \approx \frac{h\nu_i}{\left[1 + \frac{h\nu_i}{k_B T} + \frac{1}{2}\left(\frac{h\nu_i}{k_B T}\right)^2 + \cdots\right] - 1} \approx k_B T$$

This means that the thermal energy of all oscillator in the crystal is $k_B T$, which is equivalent to that the energy is equally distributed among all the vibration modes, independent of ν_i. Substituting it into Eq. (3.92a), we have

$$\Gamma(V, T) = \frac{\sum_{1}^{3N} \gamma_i}{3N} = \gamma(V) \tag{3.93}$$

This shows that at high temperatures, we do not need to rely on the Grüneisen assumption; we can also get the Grüneisen equation.

At low temperatures, that is $\frac{h\nu_i}{k_B T} >> 1$, the thermal energy of the ith oscillator is reduced to

$$E_{Ti} = \frac{h\nu_i}{\exp\left(\frac{h\nu_i}{k_B T}\right) - 1} \approx 0(\nu_i)$$

where $0(\nu_i)$ is a small quantity of ν_i, substituting it into Eq. (3.92a), we have

$$\Gamma(V,T) = \frac{\sum_{i=1}^{3N} \gamma_i \left(\frac{h\nu_i}{\exp\left(\frac{h\nu_i}{k_B T}\right) - 1}\right)}{\sum_{i=1}^{3N} \frac{h\nu_i}{\exp\left(\frac{h\nu_i}{k_B T}\right) - 1}} \approx \frac{\sum_{i=1}^{3N} \gamma_i \cdot 0(\nu_i)}{\sum_{i=1}^{3N} 0(\nu_i)} = \gamma(V) \tag{3.94}$$

Thus, at low temperatures, the Grüneisen equation can be derived directly without the Grüneisen assumption.

The Grüneisen EOS establishes the relationship between the cold pressure P_K, the cold energy E_K, and the Grüneisen coefficient γ (V). The relationships between the cold energy E_K and the thermal energy E_T as well as the cold pressure P_K and the cold energy E_K etc. have been discussed in the section of "Section 3.4, the bonding force and binding energy of crystals" and the section of "Section 3.5, lattice thermal vibration" of this chapter, respectively. For example, the cold energy E_K has been given in Eq. (3.66) as

$$E_K = U + \sum_{i=1}^{3N} \frac{1}{2} h\nu_i \cong U$$

When the zero-point energy $\sum_{i=1}^{3N} \frac{1}{2} h\nu_i$ is negligible relative to the crystal binding energy U, the cold energy is equal to the crystal binding energy; according to Eq. (3.28), we have

$$U = -\frac{NA}{v^{m/3}} + \frac{NB}{v^{n/3}}\text{。}$$

Further, the differential relationship between the cold pressure and the cold energy, i.e., $P_K = -\frac{dE_K}{dV}$, can determine the cold pressure:

$$P_K = -\frac{dE_K}{dV} \approx -\frac{dU}{dV} = -\frac{dU}{Ndv} = -\frac{m}{3} \cdot \frac{A}{v^{\frac{m+3}{3}}} + \frac{n}{3} \cdot \frac{B}{v^{\frac{n+3}{3}}}$$

So, the remaining key lies in how to determine the Grüneisen coefficient $\gamma\ (V)$ as a function of V.

We will discuss the Grüneisen coefficients approximated by the Einstein model and the Debye model, respectively, in the following.

According to the Einstein model, the vibration frequency ν_i of all oscillators in the crystal is the same, that is, $\nu_i = \nu_0$, and thus the Grüneisen coefficient γ_E of the Einstein approximation is obtained.

$$\gamma_E = -\frac{d \ln \nu_0}{d \ln V} = -\frac{d \ln \theta_E}{d \ln V} \tag{3.95}$$

Obviously, if the ν_0 is a constant, or $\theta_E (= h\nu_0/k_B)$ is a constant, it will result in a zero Grüneisen coefficient. Thus, the corresponding Grüneisen equation can be derived only by assuming $\nu_0 = \nu_0(V)$ or $\theta_E = \theta_E(V)$. This is equivalent to the Grüneisen assumption.

On the other hand, according to the Debye model, the vibration frequency ν_i of all oscillators in the crystal is distributed between $0 \sim \nu_D$ (the upper frequency limit), and its frequency distribution function $g(\nu)$, according to Eqs. (3.81) and (3.83), is:

$$g(\nu) = 4\pi V\left(\frac{1}{c_l^3} + \frac{2}{c_t^3}\right)\nu^2 = \frac{12\pi V}{c^{*3}}\nu^2 = B\nu^2 = \frac{9N}{\nu_D^3}\nu^2 \tag{3.96}$$

From this we can determine the expression of internal energy, then by referring to Eq. (3.84), we obtain

$$\begin{aligned} E &= k_B T\left(\frac{\partial \ln Z}{\partial \ln T}\right)_V \\ &= E_K(V) + 3Nk_B T D\left(\frac{\theta_D}{T}\right) \end{aligned} \tag{3.97}$$

where $\theta_D\left(=\frac{h\nu_D}{k_B}\right)$ is the Debye characteristic temperature, $D\left(\frac{\theta_D}{T}\right)$ is the Debye function, which according to Eq. (3.86) is expressed as

$$D\left(\frac{\theta_D}{T}\right) = 3\left(\frac{\theta_D}{T}\right)^{-3}\int_0^{\theta_D/T}\frac{x^3}{e^x - 1}dx$$

Similarly, the expression of the pressure P can be determined, by referring to Eq. (3.59), we have

$$\begin{aligned} P &= P_K - \frac{1}{V}\frac{d\ln\theta_D}{d\ln V}\cdot 3Nk_B TD\left(\frac{\theta_D}{T}\right) \\ &= P_K + \frac{\gamma_D(V)}{V}[E - E_K(V)] \end{aligned} \tag{3.98}$$

where γ_D is the Grüneisen coefficient of the Debye theory. Taking into account the different expressions of the Debye characteristic temperature θ_D (see Eq. 3.86)

$$\theta_D = \frac{h\nu_D}{k_B} = \left(\frac{h}{k_B}\right)\left(\frac{3N}{4\pi V}\right)^{1/3} c* = \left(\frac{h}{k_B}\right)\left(\frac{3N}{4\pi V}\right)^{1/3}\left(\frac{3}{\frac{1}{c_l^3}+\frac{1}{c_t^3}}\right)^{1/3}$$

The Grüneisen coefficient under the Debye approximation γ_D can also be expressed in different forms:

$$\gamma_D(V) = -\frac{d\ln\theta_D}{d\ln V} = -\frac{d\ln\nu_D}{d\ln V} = \frac{1}{3} - \frac{d\ln c^*}{d\ln V} \tag{3.99}$$

Therefore, the Grüneisen equation can also be derived by the Debye model. There is already an implicit assumption, $\gamma_D = \gamma_D(V)$. Notice the second equal sign of Eq. (3.96), $g(\nu) = \frac{12\pi V}{c^{*3}}\nu^2$, which is equivalent to regarding lattice waves as elastic waves with average wave velocity c^*; therefore, according to Eq. (3.99), the Grüneisen coefficient γ_D under the Debye approximation is determined by the change of Debye temperature θ_D with V, that is, the change of average wave velocity c^* with V. The wave velocity c^* is the average wave velocity obtained from the elastic longitudinal wave velocity c_l and the transverse wave velocity c_t according to the following definition given by Eq. (3.82):

$$\frac{3}{c^{*3}} = \frac{1}{c_l^3} + \frac{2}{c_t^3}$$

Pastine (1965) pointed out that a more logically solid expression of Debye theory is to consider the longitudinal mode and the transverse mode separately, so as to obtain two different Debye temperatures and two corresponding Grüneisen coefficients γ_l and γ_t. Royce (1971) further proposed the third Grüneisen coefficient γ_v corresponding to the volume expansion vibration (bulk wave). The wave velocities corresponding to these three values of γ are c_l, c_t, and c_v, expressed, respectively, as:

$$c_l = \sqrt{\frac{k_l}{\rho}} = \sqrt{\frac{k + \frac{4}{3}G}{\rho}} = \sqrt{\frac{(1-\mu)}{(1-2\mu)(1+\mu)} \cdot \frac{E}{\rho}}$$

$$c_t = \sqrt{\frac{k_t}{\rho}} = \sqrt{\frac{G}{\rho}} = \sqrt{\frac{1}{2(1+\mu)} \cdot \frac{E}{\rho}} \tag{3.100}$$

$$c_v = \sqrt{\frac{k_v}{\rho}} = \sqrt{\frac{k}{\rho}} = \sqrt{\frac{1}{3(1-2\mu)} \cdot \frac{E}{\rho}}$$

Substituting the above results into Eq. (3.99), we have

$$\gamma_v = -\frac{1}{6} - \frac{1}{2}\frac{d \ln k_v}{d \ln v} = -\frac{1}{6} - \frac{V}{2}\frac{k'_v}{k_v} \tag{3.101a}$$

$$\gamma_l = \gamma_v + \frac{V\mu'}{1-\mu^2} \tag{3.101b}$$

$$\gamma_t = \gamma_v + \frac{3V\mu'}{2(1-2\mu)(1+\mu)} \tag{3.101c}$$

where k is the bulk modulus, G is the shear modulus, E is the Young's modulus, μ is the Poisson's ratio, $\mu' = \frac{d\mu}{dV}$, and $k'_v = \frac{dk_v}{dV}$.

When $T \geq \theta_D$, by using the energy equivalence theorem, the energy of the longitudinal mode and the transverse mode are equal, then the Grüneisen coefficient γ_T, which considers both the longitudinal and transverse mode, is:

$$\gamma_T = \frac{\gamma_l + 2\gamma_t}{3} = \gamma_v + \frac{V(4-5\mu)\mu'}{3(1-\mu^2)(1-2\mu)} \tag{3.101d}$$

The difficulty in applying this formula is that μ' is unknown.

If the variation of Poisson's ratio with the specific volume can be ignored, $\mu' = 0$ thus $\gamma_l = \gamma_t = \gamma_T = \gamma_v$. This is equivalent to assuming that all modes have the same value of γ, all are equal to γ_v; that is, the average wave velocity is $c_v = (Vk_v)^{1/2}$. This is correspondent to a fluid model with negligible shear strength of solid under high pressures and only bulk waves with volume compression/expansion.

Since k_v is the bulk modulus of compression defined on the cold pressure curve $P_K(V)$, i.e., $k_v = -V\frac{dP_K}{dV} = -VP'_k$, substituting it into Eq. (3.101a), we have

$$\gamma_s = -\frac{2}{3} - \frac{V}{2}\cdot\frac{P_K''}{P_K'} \tag{3.102}$$

This result was first derived by Slater (1939), known as the **Slater relation**, written as γ_s. Obviously, the formula is based on the Debye model and assumes that the Poisson's ratio of the medium is a constant. The Slater formula establishes the relation between $\gamma(V)$ and the curvature of cold pressure curve $P_k(V)$ for the first time, which is of great significance. Once the cold energy $E_K(V)$ is known, the cold pressure $P_K(V) = -\frac{dE_K(V)}{dV}$ can be determined, and then γ_s can be determined. But the Slater formula also has disadvantages, that is, γ_s does not approach zero at simple harmonic vibrations.

Dugdale and MacDonald (1953) pointed out that, according to the lattice vibration theory, γ should be equal to zero at the pure harmonic vibration, while in this case $\gamma_s \neq 0$, there appears an abnormal thermal expansion. Therefore, they proposed a new formula:

$$\gamma_{DM} = -\frac{1}{3} - \frac{V}{2}\cdot\frac{\left(P_k(V)V^{\frac{2}{3}}\right)''}{\left(P_k(V)V^{\frac{2}{3}}\right)'} \tag{3.103}$$

called the **Dugdale–MacDonald** relationship. The formula can ensure that γ_{DM} approaches zero at harmonic vibrations.

Ващенко and Зубарев (1963) based on the free volume theory, assuming that $\frac{1-\mu}{1+\mu}$ is proportional to $\frac{P_K}{k_v}$, proposed the following formula:

$$\gamma_f = -\frac{V}{2}\cdot\frac{\left(P_k(V)V^{4/3}\right)''}{\left(P_k(V)V^{4/3}\right)'} \tag{3.104}$$

called the **free volume relations**, written as γ_f.

The formulas of the above three γ may be unified and written as

$$\gamma = -\frac{(2-n)}{3} - \frac{V}{2}\frac{\left(P_k(V)V^{2n/3}\right)''}{\left(P_k(V)V^{2n/3}\right)'} \tag{3.105}$$

where n is 0, 1, 2, respectively, for the three different models, that is

$$\gamma = \begin{cases} \gamma_s & n = 0 \\ \gamma_{DM} & n = 1 \\ \gamma_f & n = 2 \end{cases}$$

This is a unified form for the three theoretical models: the Slater model, the D–M model, and the free volume model, respectively.

However, in nature, the actual solids are varied, and the three simple models given above do not give a good description of any of them. It is known from the study of the state equations of a large number of solid materials that γ_s,γ_{DM} and γ_f are models only applicable to different solids. For example, Royci (1974) pointed out that under normal conditions, the Dugdale–MacDonald gamma γ_{DM} fits best with the thermodynamic gamma Γ_{th} for ordinary metals, while the free volume gamma γ_f is more suitable for alkali metals and their halides. However, for many solids, no matter which one is chosen, it is impossible to get a good agreement with the experiment.

In view of the above situation, Migault (1971, 1972), in the case that the unproven Eq. (3.105) is universally valid, proposed that the formula can be generalized as follows. That is, assuming that n is a characteristic parameter of the material, which can be changed continuously without being restricted by the above three kinds of γ. For specific materials, n value can be determined by making the experimental Hugoniot curve consistent with the theoretical calculation of the Hugoniot curve. For example, in the literature of Urtiew and Grover (1977), $n = 0.55$ is derived from Eq. (3.105) in the study of magnesium alloys. The theoretical calculation obtained is quite consistent with the experimental data, and the normal γ given by the formula is very close to the thermodynamic γ_{th}.

CHAPTER FOUR

Dynamic experimental study of equation of state of solids under high pressures

Since at the pressures as high as the shear strength of solid can be negligible, solids can be treated as inviscid compressible fluids, as mentioned earlier. At this time, the results of many studies on isentropic waves and shock waves in gas dynamics can be directly applied to the study on the stress wave propagation in solids under high pressures and on the equation of state of solids under high pressures. It is required only to pay attention to the difference between the equation of state of solids under high pressures and the equation of state of gas. This approximate method is often referred to as the hydrodynamic approximation.

Such high pressures which are sufficient to ignore the shear strength of solids are usually generated under intensive dynamic loading by explosions/impacts and are transmitted by intensive shock waves. Therefore, the study of the equation of state of solids under high pressures cannot be independent to the study on shock waves, which are based on the theory and experimental technology of shock waves. Furthermore, the characteristics of the high-pressure state equation of materials are implied and thus reflected in the propagation characteristics of shock waves. So one can invert quantitatively the high-pressure equation of state by studying the propagation characteristics of shock waves; mathematically, it belongs to the second kind of inverse problem. Therefore, it is no exaggeration to say that the dynamic experimental study of the solid high-pressure state equation is based on the shock wave theory.

It is necessary to introduce the basic theory of shock waves first.

4.1 Basic theory of shock waves

Most solid materials, under high pressures, are more and more difficult to compress with volume compression. That is, the tangential volumetric modulus (bulk modulus) K_t defined by the tangent of the pressure-specific volume curve (P—V) increases with the increase of pressure, as

Dynamics of Materials
ISBN: 978-0-12-817321-3
https://doi.org/10.1016/B978-0-12-817321-3.00004-8

shown qualitatively in Fig. 2.1 in Chapter 2. Since the propagation velocity C_B of a volumetric deformation wave perturbation depends on the bulk modulus K_t,

$$C_B = (K_t/\rho_0)^{1/2}$$

The propagation velocity of the disturbance increases with the increase of pressure in the loading process. This means that during a loading process, the propagation velocity of the high-amplitude disturbance is larger than that of the low-amplitude disturbance in front of it. As a result, the leading edge of the wave profile becomes steeper and steeper during the propagation of these disturbances, resulting in a shock wave finally (See Fig. 4.1). Of course, the propagation of expansion disturbances will generate weak discontinuous rarefaction waves.

Similar to the theory of gas dynamics, the shock wave front in a solid must also meet the following conservation of mass, momentum, and energy, i.e., the so-called **shock jump conditions**, or the **Rankine–Hugoniot** relationship (in short, R–H relationship).

In a material coordinate system (Lagrange coordinates), consider a one-dimensional strain plane shock wave propagating with the **material wave velocity** $\mathscr{D} = \frac{dX}{dt}$ in the X-axis direction. Here X denotes the Lagrange location of the wave front at the moment t. Stand on the wave front to observe a physical quantity $\psi(X,t)$, the total rate of $\psi(X,t)$ with respect to time, that is, the wave derivative is:

$$\frac{d\psi}{dt} = \frac{\partial\psi}{\partial t} + \mathscr{D}\frac{\partial\psi}{\partial X} \tag{4.1}$$

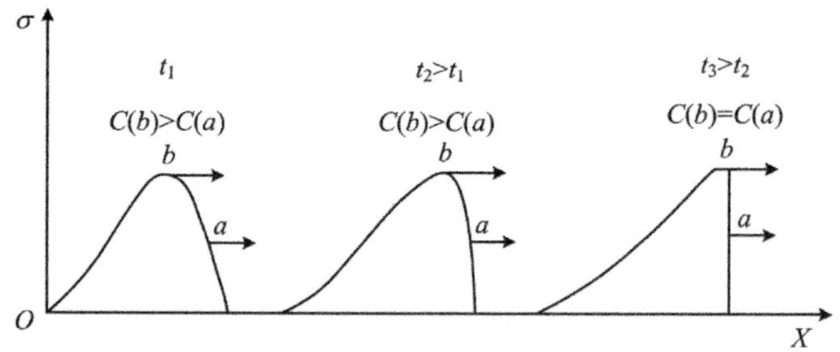

Figure 4.1 A shock wave generated in solids under high pressures.

Denote the values of ψ just before and after the wave front by ψ^+ and ψ^-, respectively, and the difference between them as $[\psi]$

$$[\psi] = \psi^- - \psi^+ \tag{4.2}$$

Obviously, if ψ is continuous across the wave front $[\psi] = 0$, otherwise $[\psi] \neq 0$. The quantity $[\psi]$ therefore represents the jump of ψ across the wave front. Taking the wave derivative of ψ^+ and ψ^- by Eq. (4.1), the difference of them is

$$\frac{d}{dt}[\psi] = \left[\frac{\partial \psi}{\partial t}\right] + \mathscr{D}\left[\frac{\partial \psi}{\partial X}\right] \tag{4.3}$$

This is the famous **Maxwell theorem**.

Now concretize ψ to the particle displacement $u(X, t)$. According to the displacement continuity condition, the displacement must be continuous across a wave front, namely $[u] = 0$, and its first derivatives $\frac{\partial u}{\partial t}(= v)$ and $\frac{\partial u}{\partial X}(= \varepsilon)$ are not equal across the shock wave; by Eq. (4.3), we have:

$$[v] = -\mathscr{D}[\varepsilon] \tag{4.4}$$

This is the **mass conservation condition across the shock wave front**.

Then the compatibility conditions of the related quantities across the wave front should be considered based on the law of dynamics.

Suppose the shock wave front is located at AB at time t (Fig. 4.2); in a time dt, it moves at location $A'B'$; the propagating distance is $dX = Ddt$; and by the momentum conservation of the region $ABA'B'$:

$$\left(\sigma^+ - \sigma^-\right)A_0 dt = \rho_0 A_0 dX\left(v^- - v^+\right)$$

which can be simplified as

$$[\sigma] = -\rho_0 \mathscr{D}[v] \tag{4.5}$$

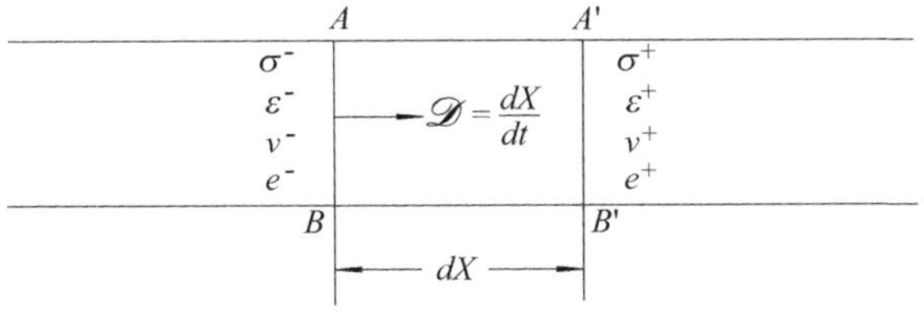

Figure 4.2 The state changes of an infinitesimal element *ABA'B'* in an impact process.

This is the **momentum conservation condition across the shock wave front**.

Now let us discuss the energy conservation across a shock wave front. Consider the energy change of the infinitesimal region *ABA'B'* in Fig. 4.2 in duration dt; if e is the internal energy in unit mass, then the energy conservation of this region is

$$\left(\sigma^{+}v^{+}-\sigma^{-}v^{-}\right)A_0dt=\left[e^{-}-e^{+}\right]\rho_0A_0dX+\frac{1}{2}\left\{\left(v^{-}\right)^2-\left(v^{+}\right)^2\right\}\rho_0A_0dX$$

This means that the work done by stress σ is converted into two parts: internal energy and kinetic energy. After simplification, we have

$$[\sigma v]=-\rho_0\mathscr{D}[e]-\frac{1}{2}\rho_0\mathscr{D}\left[v^2\right] \tag{4.6}$$

This is the **energy conservation condition across the shock wave front**.

In the discussion above and below, it is assumed that both stress and strain are positive in tension, and particle velocity is positive in the X-axis.

For solids under high pressure, according to the hydrodynamics approximation, the terms related to the distortion can be neglected. This is equivalent to replace the σ by $(-p)$ and the ε by $(V_0-V)/V_0$ in the above three equations. Moreover, in consistence with the traditional notations used in the discussion of shock waves in solids under high pressure, we change the notation for the particle velocity by u and for the internal energy by E, and consequently, the above equations are reduced to

$$[u]=-\rho_0\mathscr{D}[V] \tag{4.7}$$

$$[P]=\rho_0\mathscr{D}[u] \tag{4.8}$$

$$[E]=\frac{[Pu]}{\rho_0\mathscr{D}}-\frac{1}{2}\left[u^2\right]=-\frac{1}{2}\left(P^{-}+P^{+}\right)\ [V] \tag{4.9}$$

This is the **shock jumping condition** in the Lagrange form of one-dimensional strain shock wave in solids under high pressures.

Similarly, the same problem can be discussed in the spatial coordinates (Euler coordinates), where the plane shock wave propagates along the spatial coordinate x-axis with the spatial velocity D (See Fig. 4.3). Note that the

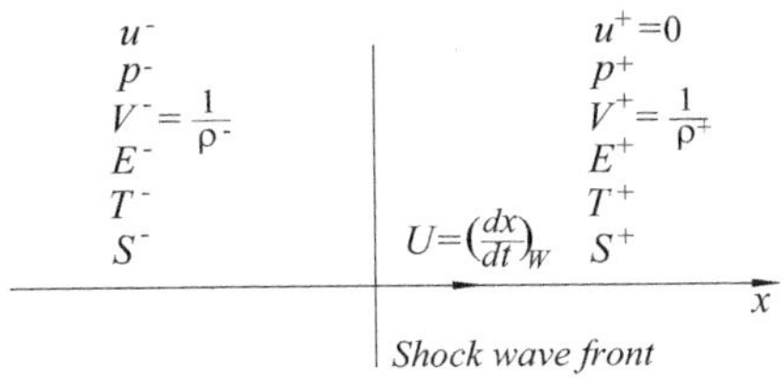

Figure 4.3 Shock wave front in spatial coordinates.

particle velocity relative to the wave front is (u-D). According to the condition of conservation of mass, we have:

$$\rho^-\left(u^- - D\right) = \rho^+\left(u^+ - D\right)$$

or

$$[\rho u] = D[\rho] \tag{4.10}$$

According to conservation of momentum, we have:

$$\rho^- u^-\left(u^- - D\right) - \rho^+ u^+\left(u^+ - D\right) = P^+ - P^-$$

By using the condition of mass conservation (4.10), the above equation can be simplified as

$$[P] = \rho^+\left(D - u^+\right)\ [u] \tag{4.11}$$

According to conservation of energy, we have:

$$\rho^-\left\{\frac{1}{2}\left(u^-\right)^2 + E^-\right\}\left(u^- - D\right) - \rho^+\left\{\frac{1}{2}\left(u^+\right)^2 + E^+\right\}\left(u^+ - D\right)$$
$$= p^+u^+ - p^-u^-$$

By using the conditions of mass conservation and momentum conservation, the above equation can be simplified as

$$[E] = -\frac{1}{2}\left(p^- + p^+\right)[V] \tag{4.12}$$

Eqs. (4.10)–(4.12) are the **shock jump condition** in the **Euler form** of one-dimensional strain plane shock wave.

It should be pointed out that the above two forms of shock jumping conditions are derived for the propagation of right traveling waves, and for left traveling waves, the sign in front of the wave velocity is required to change.

If the reference coordinate system is chosen to move along with the particle in front of the shock wave, the **relative spatial wave velocity** U of the shock wave (relative to the particle in front of the shock wave) is often introduced. Suppose the velocity of the particle in the front of a shock wave is u^+, thus the relation between the U and the Euler wave velocity D and the Lagrange wave velocity $\mathscr{D}$ is:

$$D = U + u^+ = \frac{V^+}{V_0} \cdot \mathscr{D} + u^+ \tag{4.13a}$$

Here $V^+(=1/\rho_0)$ is the specific volume in the front of a shock wave. Obviously, it is not difficult to derive from Eqs. (4.7)−(4.12), the material wave velocity $\mathscr{D}$, the relative spatial wave velocity U, and the absolute spatial wave velocity D of a shock wave can be expressed, respectively, as follows:

$$\mathscr{D} = V_0\sqrt{-\frac{[P]}{[V]}}, \tag{4.13b}$$

$$U = V^+\sqrt{-\frac{[P]}{[V]}}, \tag{4.13c}$$

$$D = u^+ + V^+\sqrt{-\frac{[P]}{[V]}} \tag{4.13d}$$

If the state in front of a shock wave is an uncompressed natural state, $V^+ = V_0$, then $U = \mathscr{D}$; and if the state in front of the shock wave is further at a rest state, i.e., $u^+ = 0$, then $D = U = \mathscr{D}$. In this case, the three kinds of wave velocities are consistent.

In this way, the three conservation conditions for the shock jump condition, together with an internal energy-type equation of state (Eq. 2.63), to characterize the material properties, there are totally four equations. In the given initial condition, namely the initial states before shock loading are given, the four equations contain five unknown quantities to characterize the final states after shock loading: P, V (or ρ), u, E, and $\mathscr{D}$ (Lagrange expression) or D (Euler expression). Once any of them is given by the boundary conditions, the remaining four quantities can be determined by the four equations. Therefore, for a certain material (that is, its internal energy-type state equation of solids under high pressure is known), under the given initial conditions and boundary conditions, the problem of a plane shock wave propagation is solvable definitely.

If no concrete boundary conditions are specified, then for a certain initial equilibrium state, the shock jump conditions, together with the internal energy-type state equation, provide the relations between any two of the five unknown quantities. There can be 10 possible forms and the three forms of them, P—V, P—u, $\mathscr{D}$—u, are commonly used. All of 10 relationships between shock wave quantities are called **shock adiabatic relationships**, or called **shock adiabatic curves**, or **Hugoniot curves** when they are described geometrically as graphs.

Because the shock jump process across the shock wave front is an irreversible nonequilibrium process, so the shock adiabatic curve actually represents only the connecting curve from the initial state of equilibrium (called the center point of the Hugoniot curve) to the possible final equilibrium states through a shock jump, which does not represent the locus of the successive states that the material experienced in the process. It is different from the relationship between the state quantities in an isothermal process (isothermal curve) and the relationship between the state quantities in an isentropic adiabatic process (isentropic adiabatic curve); the latter two are process curves. Furthermore, since the Hugoniot curve is for a certain initial equilibrium state (the center point), and all points on the Hugoniot curve are terminal points relative to that initial point, so the Hugoniot curves for different initial states are different.

The following three forms of Hugoniot curves, the P—V (or P—ρ) curves, the P—u curves, and the U—u curves, will be discussed in detail, respectively, as follows.

4.1.1 The P—V Hugoniot curves

Fig. 4.4 shows the Hugoniot curve in the P—V form. The chord AB that connects the final point B and the initial point A on the line has a special

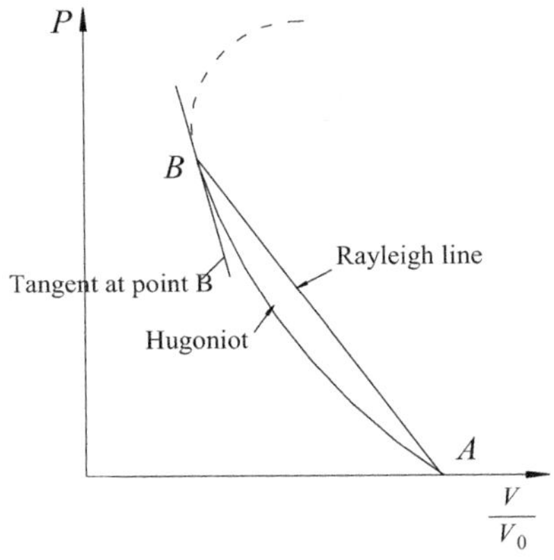

Figure 4.4 The P—V Hugoniot curve and Rayleigh line.

meaning, called the Rayleigh line (chord). By Eq. (4.13b), the material wave velocity $\mathscr{D}$ of a shock wave can be determined by the slope $= -\frac{[P]}{[V]}$ of the Rayleigh line. For a stable shock wave, each part of the wave front travels at the same wave velocity, which means that the trajectory of the state points experienced by the shock wave is just the Rayleigh line. In other words, the Rayleigh line is the process line of the shock jump. Of course, all points on the Rayleigh line are at nonequilibrium states except that the initial and final points are at thermodynamic equilibrium states.

A shock jump is a process having irreversible entropy increase. It can be proved that along the P–V Hugoniot curve, the entropy S increases with the increase of pressure P. As a matter of fact, taking the differential of Eq. (4.9) and denoting the initial state subscript 0, the final state without the subscript, we obtained:

$$dE = \frac{1}{2}(V_0 - V)dP - \frac{1}{2}(P + P_0)dV$$

This represents the differential relationship which should be satisfied between the two neighboring possible final states on a P–V Hugoniot curve. On the other hand, since any final state is at thermodynamic equilibrium, the first and second laws of thermodynamics should also be satisfied, namely:

$$dE = TdS - PdV \tag{4.14}$$

Eliminating dE from these two equations, we have:

$$\begin{aligned} TdS &= \frac{1}{2}(V_0 - V)dP + \frac{1}{2}(P - P_0)dV \\ &= \frac{1}{2}\left\{1 - \frac{\dfrac{P - P_0}{V_0 - V}}{\left(-\dfrac{dP}{dV}\right)}\right\}(V_0 - V)dP \end{aligned} \tag{4.15}$$

Referencing to Fig. 4.4, it is known that the $(P-P_0)/(V_0-V)$ is the slope of the Rayleigh line, while the $(-dP/dV)$ is the tangent slope of the P–V Hugoniot curve at the final point B. Since for normal materials, the P–V Hugoniot curve is concave upward, meaning that:

$$\frac{\left(\dfrac{p - p_0}{V_0 - V}\right)}{\left(-\dfrac{dp}{dV}\right)} < 1$$

Therefore, Eq. (4.15) shows that along the Hugoniot curve, the entropy increases with the increase of pressure. i.e.,

$$\frac{dS}{dP} > 0 \tag{4.16}$$

Let us examine some relations between the $P-V$ Hugoniot curve, the $P-V$ isentropic adiabatic curve, and the $P-V$ isothermal curve to understand their differences and interrelationships. If a total differential is taken along the $P-V$ Hugoniot curve, the relationship between the slope dP/dV of the $P-V$ Hugoniot curve and the slope $(\partial P/\partial V)_S$ of the isentropic $p-V$ can be established.

$$\frac{dP}{dV} = \left(\frac{\partial P}{\partial V}\right)_S + \left(\frac{\partial P}{\partial S}\right)_V \frac{dS}{dV} \tag{4.17}$$

where the second term on the right, $\left(\frac{\partial P}{\partial S}\right)_V$ is known as the stress function φ_s, and according to Eqs. (2.60), (2.59) and (2.54), we have.

$$\varphi_S = \left(\frac{\partial P}{\partial S}\right)_V = k_S \alpha_S = \frac{k_T \alpha_T T}{C_V} > 0$$

Noting again $\frac{dS}{dP} > 0$ along the Hugoniot curve (Eq. 4.16), this indicates that $\frac{dS}{dV}$ and $\frac{dP}{dV}$ have the same sign and are negative during a compressive process ($V < V_0$), so that

$$\frac{dP}{dV} < \left(\frac{\partial P}{\partial V}\right)_S$$

Or it can be rewritten as

$$-\left(\frac{dp}{dV}\right) > -\left(\frac{\partial P}{\partial V}\right)_S$$

This means on the $P-V$ graph (see Fig. 4.5), the Hugoniot curve AB is above the isentropic curve S_1 passing through the initial point A and below the isentropic curve S_2 passing through the final point B. The area below the $P-V$ isentropic curve represents the recoverable internal energy change during the isentropic process in the graph (see Eq. (4.15). Therefore, the area enclosed between the Rayleigh line AB and the expansion isentropic curve BC (the shaded part in Fig. 4.5) represents just the irreversible energy dissipation during the shock jump process. It corresponds to the irreversible entropy increase given by Eq. (4.16).

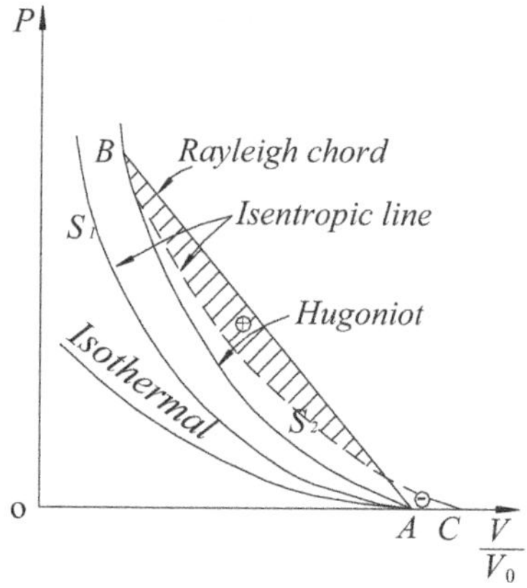

Figure 4.5 Isentropic curves, isotherm curve, and Hugoniot line.

As for the relation of the slope of P–V isentropic curve $\left(\frac{\partial P}{\partial V}\right)_S$ and the slope of P–V isothermal curve $\left(\frac{\partial P}{\partial V}\right)_T$, by differentiating $P=P(V,T)$ with respect to V under isentropic condition, we have:

$$\left(\frac{\partial P}{\partial V}\right)_S = \left(\frac{\partial P}{\partial V}\right)_T + \left(\frac{\partial P}{\partial T}\right)_V \left(\frac{\partial T}{\partial V}\right)_S$$

where the second term on the right, $\left(\frac{\partial P}{\partial T}\right)_V$, is the temperature stress function φ_T (Eq. 2.38), and the $\left(\frac{\partial T}{\partial V}\right)_S$ is the negative value of the stress function ($-\phi_S$) (Eq. 2.31), so there is

$$\left(\frac{\partial P}{\partial T}\right)_V \left(\frac{\partial T}{\partial V}\right)_S = -\phi_T \phi_S = -\frac{(K_T \alpha_T)^2 T}{C_V} < 0$$

Thus we have:

$$\left(\frac{\partial P}{\partial V}\right)_S < \left(\frac{\partial P}{\partial V}\right)_T$$

Or rewrite as:

$$-\left(\frac{\partial P}{\partial V}\right)_S > -\left(\frac{\partial P}{\partial V}\right)_T$$

This means on the $P-V$ graph (see Fig. 4.5), the isentropic curve is above the isothermal curve.

At the initial point A, ($V = V_0$), it is known that $\left.\frac{dS}{dV}\right|_A = 0$ from Eq. (4.15), and after substituting it into (4.17), we have

$$\left.\frac{dP}{dV}\right|_A = \left.\left(\frac{\partial P}{\partial V}\right)_S\right|_A \tag{4.18}$$

That is, at the initial point A, the slope of Hugoniot curve is equal to that of isentropic curve. Not only that, if we take further derivative of Eq. (4.17) and after some calculations, it is not difficult to obtain the following relations.

$$\left.\frac{d^2P}{dV^2}\right|_A = \left.\left(\frac{\partial^2 P}{\partial V^2}\right)_S\right|_A \tag{4.19a}$$

$$\left.\frac{d^3P}{dV^3}\right|_A \neq \left.\left(\frac{\partial^3 P}{\partial V^3}\right)_S\right|_A \tag{4.19b}$$

That is, at the initial state point A, the $P-V$ Hugoniot curve and the $P-V$ isentropic curve have the same slope and curvature until their third-order derivatives that make them different. This shows that near the initial state point, the $P-V$ isentropic curve is very close to the $P-V$ Hugoniot curve. Therefore, when the pressure P is not too high, we can approximately replace the $P-V$ Hugoniot curve with the $P-V$ isentropic curve.

4.1.2 The $P-u$ Hugoniot curves

The $P-u$ Hugoniot curve reflects the relationship between pressure P and particle velocity u. Because on the interaction interface of shock waves, both the pressure P and the particle velocity u are required to remain continuous, the Hugoniot curve in the $P-u$ form is most convenient for handling such issues as the interaction of two shock waves, as well as the reflection and transmission of shock waves in different media, etc. This will be further appreciated in the discussion that follows.

4.1.3 The $D-u$ Hugoniot curves

From the point of view of the experimental methods and test technologies for the research on shock waves, the parameters with velocity dimensions, such as the particle velocity u and the shock wave velocity D or $\mathscr{D}$, are generally easy to be determined directly. Because they can all come down to the measurement of the distance and the corresponding time interval, which is relatively easy to be achieved by the current testing technology, while the pressure, the specific volume and the temperature, etc. under the dynamic conditions are relatively difficult to be measured directly. Therefore, in the experimental determination of the Hugoniot curve and in the study of the high-pressure equation of state of materials by means of the determination of the Hugoniot curve, The ***D—u* Hugoniot curve** is most commonly used.

A large number of tests carried out on numerous of materials show that in a fairly wide range of test pressures without shock phase translation, the U—u Hugoniot curve is usually expressed in a simple linear relationship as shown in Fig. 4.6 and by Eq. (4.20).

$$U = a_0 + su \tag{4.20}$$

where a_0 and s are material constants. For some materials, such as iron, the U—u form of the Hugoniot line, which is experimentally measured, deviates from the linear relationship and requires the addition of a quadratic term of u, that is,

$$U = a_0 + su + qu^2 \tag{4.21}$$

where q is also a material constant.

The relationship between D and u can be also discussed from the P—V Hugoniot curve.

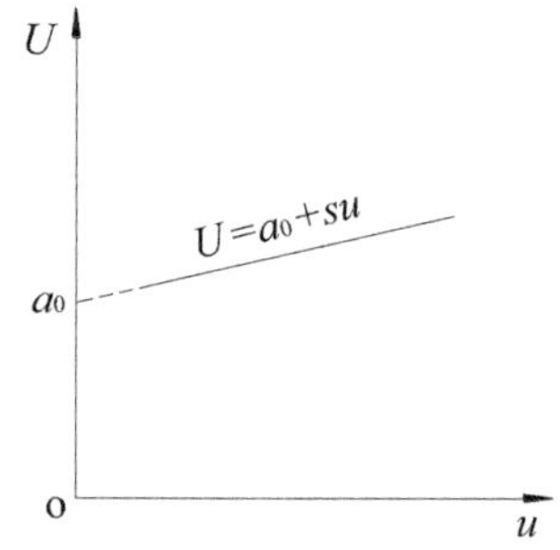

Figure 4.6 The Hugoniot curve in D—u form.

For the $P-V$ Hugoniot curve $P_H(V)$, by the Taylor series expansion at the V_0, we have:

$$P_H(V) = P_H(V_0) + \left.\frac{dP_H}{dV}\right|_{V_0}(V-V_0) + \left.\frac{1}{2}\frac{d^2P_H}{dV^2}\right|_{V_0}(V-V_0)^2 + \left.\frac{1}{6}\frac{d^3P_H}{dV^3}\right|_{V_0}(V-V_0)^3 + O\left(|V-V_0|^4\right) \tag{4.22}$$

where the subscript H denotes the quantities at the final point on the Hugoniot curve. The above equation can be rewritten as follows:

$$V_0^2\frac{P_H(V)-P_H(V_0)}{V_0-V} = V_0^2\left(\frac{dP_H}{dV}\right)_{V_0}\left[1-\frac{1}{2}\left(\frac{d^2P_H/dV^2}{dP_H/dV}\right)_{V_0}(V_0-V) + \frac{1}{6}\left(\frac{d^3P_H/dV^3}{dP_H/dV}\right)_{V_0}(V_0-V)^2\right] \tag{4.23}$$

where the minor quantities higher than the fourth order of $(V-V_0)$ have been omitted. On the other hand, from Eqs. (4.10) and (4.11), we have:

$$D-u_0 = V_0\sqrt{\frac{P-P_0}{V_0-V}} \tag{4.24}$$

$$u-u_0 = (V_0-V)\sqrt{\frac{P-P_0}{V_0-V}} \tag{4.25}$$

where P and P_0 are corresponding to $P_H(V)$ and $P_H(V_0)$, respectively. Eq. (4.24) is actually the result of applying Eq. (4.13d) to the Hugoniot curve. By using Eq. (4.18) showing that the slope of the Hugoniot curve is equal to the slope of the isentropic curve at the initial state V_0, we know that the sound velocity at the initial state c_0 can be determined by the following equation:

$$c_0^2 = \left(\frac{\partial P}{\partial \rho}\right)_{V_0} = \left(V^2\left(-\frac{\partial P}{\partial V}\right)_S\right)_{V_0} = V_0^2\left(-\frac{dP_H}{dV}\right)_{V_0} \tag{4.26}$$

Substituting the above results (from Eqs. 4.24—4.26) into Eq. (4.23), we have:

$$(D-u_0)^2 = c_0^2\left[1-\frac{V_0}{2}\left(\frac{d^2P_H/dV^2}{dP_H/dV}\right)_{V_0}\left(\frac{u-u_0}{D-u_0}\right)+\frac{V_0^2}{6}\left(\frac{d^3P_H/dV^3}{dP_H/dV}\right)_{V_0}\left(\frac{u-u_0}{D-u_0}\right)^2\right]$$

or

$$D-u_0 = c_0(1+A)^{1/2} \tag{4.27a}$$

where

$$A = -\frac{V_0}{2}\left(\frac{d^2P_H/dV^2}{dP_H/dV}\right)_{V_0}\left(\frac{u-u_0}{D-u_0}\right)+\frac{V_0^2}{6}\left(\frac{d^3P_H/dV^3}{dP_H/dV}\right)_{V_0}\left(\frac{u-u_0}{D-u_0}\right)^2 \tag{4.27b}$$

Let us do binomial expansion for Eq. (4.27).

$$D-u_0 \approx c_0\left(1+\frac{A}{2}-\frac{A^2}{8}\right)$$

Substitute the A of (4.27b) into the above equation and remain the quadratic terms of $\left(\frac{u-u_0}{D-u_0}\right)$, we have:

$$D-u_0 = c_0+\left[-\frac{V_0}{4}\left(\frac{d^2P_H/dV^2}{dP_H/dV}\right)_{V_0}\frac{c_0}{D-u_0}\right](u-u_0)+$$

$$\left\{\frac{1}{2c_0}\left[\frac{V_0^2}{6}\left(\frac{d^3P_H/dV^3}{dP_H/dV}\right)_{V_0}-\frac{V_0^2}{16}\left(\frac{d^2P_H/dV^2}{dP_H/dV}\right)_{V_0}^2\right]\left(\frac{c_0}{D-u_0}\right)^2\right\}(u-u_0)^2 \tag{4.28}$$

From the above equation, it is very difficult to directly solve the equation $D-u_0 = f(u-u_0)$ in explicit function form, which requires solving a cubic algebraic equation of $(D-u_0)$. The iterative solution method is used below. Take the first-order approximation:

$$D-u_0 \approx c_0$$

Substituting it into the right side of Eq. (4.28), we can get a second-order approximation.

$$D - u_0 \approx c_0 + \lambda_0(u - u_0) + \lambda'_0(u - u_0)^2 \tag{4.29}$$

where

$$\lambda_0 = -\frac{V_0}{4}\left(\frac{d^2P_H/dv^2}{dP_H/dv}\right)_{V_0}$$

$$\lambda'_0 = \frac{1}{2C_0}\left[\frac{V_0^2}{6}\left(\frac{d^3P_H/dv^3}{dP_H/dv}\right)_{V_0} - \lambda_0^2\right]$$

In view of the complexity of Eq. (4.29), only the first two terms on the right side are taken, namely

$$D - u_0 \approx c_0 + \lambda_0(u - u_0)$$

Substituting it into the right side of Eq. (3.28) as the second-order approximation, we can get the third-order approximation.

$$D - u_0 = c_0 + \lambda(u - u_0) + \lambda'(u - u_0)^2 \tag{4.30}$$

where

$$\lambda = \lambda_0$$

$$\lambda' = \lambda'_0 - \frac{\lambda_0^2}{c_0} = \frac{1}{2c_0}\left[\left(\frac{d^3p_H/dv^3}{dp_H/dv}\right)_{V_0} - 3\lambda_0^2\right]$$

Eq. (4.30) is a very useful equation, which establishes the relationship between the shock wave velocity D and the particle velocity u behind wave front and is confirmed by the experimental results.

Experiments show that the coefficient λ' of the quadratic term of $u-u_0$ is usually very small, so the linear relation can be directly used within a fairly wide range of pressure, that is,

$$D - u_0 = c_0 + \lambda(u - u_0) \tag{4.31}$$

If the state in front of the shock wave is in the natural state of stationary ($u_0 = 0$) and uncompressed ($V^+ = V_0$), then according to Eq. (4.13a), the above equation can be further simplified as:

$$D = \mathscr{D} = U = c_0 + \lambda u \tag{4.32}$$

For some solid materials, the linear relation (4.31) can be applied to 1–2 Mbar (1 Mbar = 0.1 TPa), or even higher pressures. As suggested by

Prieto and Renero (1970), the upper limit of pressure applicable to the above linear relationship is approximately 4 Mbar for pure elements, 2 Mbar for alloy, 0.4 Mbar for inorganic compounds, and 0.2 Mbar for organic compounds (Prieto and Renero, 1970).

According to Eq. (4.13a), the **relative space wave velocity** U on the left side of Eq. (4.31) is thus directly reduced to the linear relation (4.20). This shows that the physical meaning of a_0 in Eqn. (4.20) is the acoustic velocity c_0. In addition, the slope λ in this linear relationship, or the s in Eq. (4.20), is directly related to the Grüneisen coefficient γ_s. If the shock initial state temperature T_0 is the absolute temperature zero, the expression can be directly compared with the equation of γ_s, and the relationship is as follows:

$$\lambda = \frac{1}{2}(\gamma_s)_{0K} + \frac{2}{3} \tag{4.33}$$

Table 4.1 lists the values of c_0, λ of some materials, and the values calculated by using thermodynamic γ_{th}. The results show that the consistency between the two is satisfactory.

When the pressure P is higher, the linear relation will be significantly different from the actual situation. Eq. (4.30) must be used in this case. If $u_0 = 0$, this equation can be simplified to

$$U = D = c_0 + \lambda u + \lambda' u^2 \tag{4.34}$$

This is exactly Eq. (4.21).

4.2 Interaction, reflection, and transmission of shock waves in solids under high pressures

In the study of the equation of state of solids under high pressures, it is inevitable to deal with the incident, reflection, and transmission of shock waves and the interactions of shock waves. The basic principles of this research area must be mastered in advance.

The general principle to deal with shock wave interactions is that the requirement of continuity of the pressure P and the particle velocity u at the interfaces of wave interactions should be satisfied. In addition, the following two points should be noted, when it is supposed that the incident shock wave is a compression wave:

(1) when a reflected wave is a shock wave loaded with further compression, the final state point of the reflected shock wave shall fall on the Hugoniot curve with a new initial state point, i.e., a new central point,

Table 4.1 The values of c_0, λ experimentally measured and calculated by using sound wave formulas and thermodynamic γ_{th} for some materials (Jing, 1999).

Class	Experimental data[a]				Computed data[b]	
Parameters / Materials	Density (g/cm^3)	Pressure range (kbar)	Sound velocity c_0 (cm/μs)	λ	Sound velocity c_0 (cm/μs)	λ
Molybdenum	10.20	254−1633	0.516	1.24	0.522	1.12
Tantalum	16.46	272−547	0.337	1.16	0.350	1.21
Tungsten	19.17	394−2074	0.400	1.27	0.417	1.14
Aluminum	2.79	20−4930	0.525	1.39	0.520	1.42
Cobalt	8.82	244−1603	0.475	1.33	0.455	1.27
Nickel	8.86	235−9560	0.465	1.45	0.461	1.27
Copper	8.90	216−9550	0.396	1.50	0.387	1.31
Palladium	11.95	263−372	0.379	1.92	0.393	1.45
Silver	10.49	216−4010	0.324	1.59	0.307	1.53
Platinum	21.4	295−586	0.367	1.41	0.351	1.60
Gold	19.24	590−5130	0.308	1.56	0.297	1.85
Lead	11.34	390−7300	0.203	1.58	0.196	1.70

[a]The values of c_0 and λ in the table are obtained by fitting Eq. (4.32) of experimental data.
[b]The values are calculated from the sound velocity formula $C_0 = k_T/\rho_0$ and Eq. (4.33), where the values of k_T is from Table 2.1 in this book.

which is corresponding to the state in front of the reflected shock wave, instead of falling on the Hugoniot curve with the initial point corresponding to the initial state of the incident shock wave.

(2) when a reflected wave is a rerelease wave unloaded by expansion, its state is determined by the unloading isentropic curve, that is, when the rerelease wave passes through, all the states experienced by the medium fall on the isentropic expansion curve starting from the final state point of the incident shock wave.

Of course, as long as the equation of state is known, then the Hugoniot curve and the isentropic curve passing through any point are determined. In fact, if the Grüneisen coefficient of the material $\gamma(V)$ is known, it is not difficult to determine a Hugoniot curve or an isentropic curve with a new central point which is an arbitrary point on the known Hugoniot curve, as shown in Fig. 4.7. The concrete method is as follows.

Assuming that the point 0 denotes the initial state of the incident shock wave, the point 1 denotes the final state of the incident shock wave, and also the initial point of the reflected shock wave (Fig. 4.7A). For any point 2 on the Hugoniot curve of the reflection shock wave, the energy conservation condition across the shock wave front, Eq. (4.9), should be satisfied, that is,

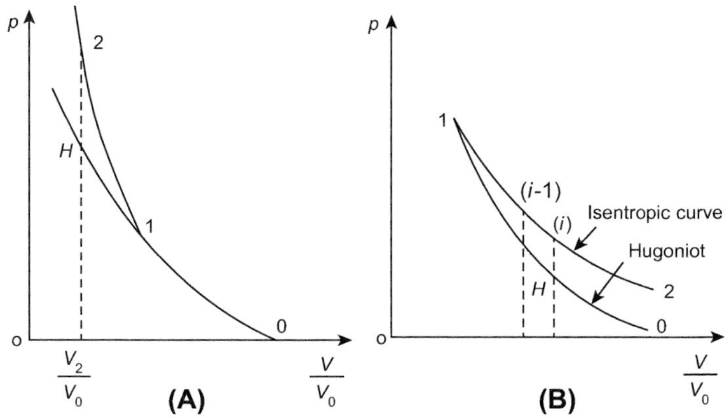

Figure 4.7 Determining (A) the loading Hugoniot curve and (B) the unloading isentropic curve from a known Hugoniot curve.

$$E_2 = E_1 + \frac{1}{2}(P_2 + P_1)(V_1 - V_2) \tag{4.35a}$$

At the same time, since the point 2 on the Hugoniot curve of the reflection shock wave and the point H on the original Hugoniot curve have the same specific volume V_2, they should satisfy the Grüneisen equation Eq. (2.67b), thus

$$E_2 = E_H + \frac{P_2 - P_H}{\left(\frac{\Upsilon}{V}\right)_2} \tag{4.35b}$$

where the subscripts *2* and H denote the corresponding quantities at the point 2 and H, respectively. Eliminating E_2 from the above two equations, we obtain:

$$E_1 - E_H = \frac{P_2 - P_u}{\left(\frac{\Upsilon}{V}\right)_i} - \frac{1}{2}(P_2 + P_1)(V_1 - V_2) \tag{4.36}$$

On the other side, both the point 1 and H are the points on the original Hugoniot curve with the point *0* being the central point. According to Eq. (4.9), they should satisfy:

$$E_1 - E_0 = \frac{1}{2}(P_1 + P_0)(V_0 - V_1)$$

$$E_H - E_0 = \frac{1}{2}(P_H + P_0)(V_0 - V_2)$$

Eliminating E_0 from the above two equations, we obtain:

$$E_1 - E_H = \frac{1}{2}(P_1 + P_0)(V_0 - V_1) - \frac{1}{2}(P_H + P_0)(V_0 - V_2) \qquad (4.37)$$

Then eliminating (E_1-E_H) from Eq. (4.36) and Eq. (4.37), and suppose $P_0 = 0$, we obtain:

$$P_2 = \frac{P_H - \left(\frac{\gamma}{V}\right)_2 \frac{(P_H - P_1)(V_0 - V_2)}{2}}{1 - \left(\frac{\gamma}{V}\right)_2 \frac{V_1 - V_2}{2}} \qquad (4.38)$$

This is the required $P-V$ Hugoniot curve with the point 1 as the central point for analyzing the reflected shock waves.

As to the isentropic curve passing through the point 1, it can be determined in a such way (Fig. 4.7B); since along an isentropic curve, there must be $dS = 0$, then from the thermodynamics law (Eq. 4.14), there should be:

$$dE = -PdV$$

or expressed in a difference form, we have:

$$E_i = E_{i-1} - \frac{1}{2}(P_i - P_{i-1})\Delta V \qquad (4.39)$$

On the other side, both the point i on the isentropic curve and the point H on the original Hugoniot curve have the identical specific volume V_i; they should satisfy the Grüneisen equation, Eq. (2.67b), so there is:

$$E_i = E_H + \frac{P_i - P_H}{\left(\frac{\gamma}{V}\right)_i} \qquad (4.40)$$

Eliminating E_i from the above two equations, we obtain:

$$P_i = \frac{P_H - \left(\frac{\gamma}{v}\right)_i \left[P_{i-1} \cdot \frac{\Delta v}{2} + E_H - E_{i-1}\right]}{1 + \left(\frac{\gamma}{v}\right)_i \cdot \frac{\Delta v}{2}} \qquad (4.41)$$

Therefore, the states at the point i can be determined from the states at the point (i-1) by a numerical method, and other points can also be determined point by point in the same way. Thus, the expansion isentropic curve $1-2$ is determined.

After the Hugoniot curves of reflected shock waves and the expansion isentropic curves of reflected unloading waves were determined, it is easy to deal with the interaction, reflection, and transmission of shock waves. As mentioned previously, it is convenient to analyze this sort of problems by using the P—u curves. In the following, we illuminate this by analyzing the reflection and transmission of shock waves at the interface between two materials.

Firstly, we discuss a problem that a plane shock wave S_i with pressure P_1 propagates from a material A with lower wave impedance into a material B with higher wave impedance (normal incidence), as shown in Fig.4.8A. Assume that both materials were undisturbed initially, which corresponds to the point 0 in the P—u plane. According to the momentum conservation condition on shock wave front, Eq. (4.8), it is known that the slope of the Rayleigh line connecting the initial state point 0 and the final state point 1 on the P—u Hugoniot curve of material A exactly equals to the shock wave impedance $(\rho_0 D)_A$ of material A. Since the impedances of those two materials are different, the reflection and the transmission simultaneously occur when an incident shock wave arrives at the interface of two materials. Since the shock-wave impedance $(\rho_0 D)_B$ of material B is higher than that $(\rho_0 D)_A$ of material A, from the previous discussion on the reflection of stress waves at the interface of different materials, it is clear that the reflected wave S_r is a shock wave that further loads the material A from the state 1 to the state 2, and the transmitted wave S_t is also a shock wave that compresses the material B from the undisturbed state 0 to the state 2. According to the continuity condition for P and u at the interface, respectively, and noticing the sign of D in the R—H relationship for leftward waves should be changed to negative, then the point 2 is the intersection of the positive Hugoniot curve of material B, whose central point is the point 0, and the negative Hugoniot curve of material A, whose central point is the point 1. Obviously, the intensity of transmitted wave S_t is higher than that of incident wave S_i. For instance, when a shock wave with pressure amplitude of 24 GPa propagates from the aluminum with lower shock wave impedance to the iron with higher shock wave impedance, the intensity of transmitted shock wave is about 34 GPa.

Now, we discuss the problem that a plane shock wave S_i with pressure P_1 rightward propagates from a material A with higher wave impedance into a material B with lower wave impedance, as shown in Fig. 4.8B. Similar to the

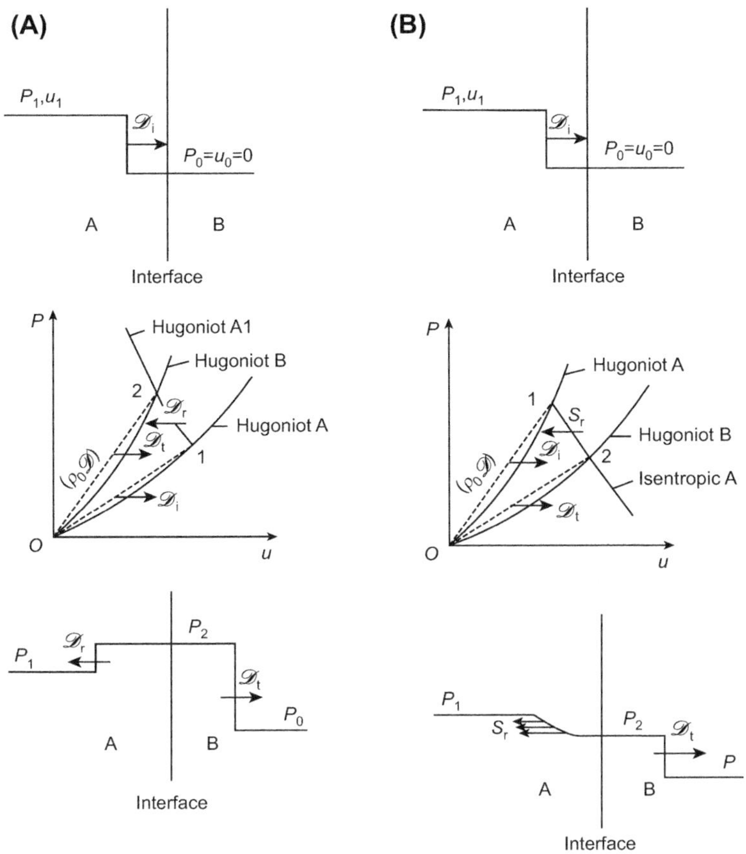

Figure 4.8 Reflection and transmission of a shock wave; (A) Incident into a high-wave impedance material from a low-wave impedance material and (B) Incident into a low-wave impedance material from a high-wave impedance material.

previous discussion, it is not difficult to conclude that unloading reflection will occur at the interface between the two materials, namely, the reflected wave *Rr* is an expansion rarefaction wave that unloads the material *A* from the state *1* to the state *2* and the transmitted wave is a shock wave that loads the material *B* from the undisturbed state *0* to the state *2*. The point *2* is the intersection point of the negative isentropic curve of material *A* passing through the point 1 and the positive Hugoniot curve of material *B*, whose center point is the point 0. Obviously, the intensity of transmitted wave S_t is lower than that of incident wave S_i. For instance, when a shock wave with pressure amplitude of 24 GPa propagates from the aluminum with high

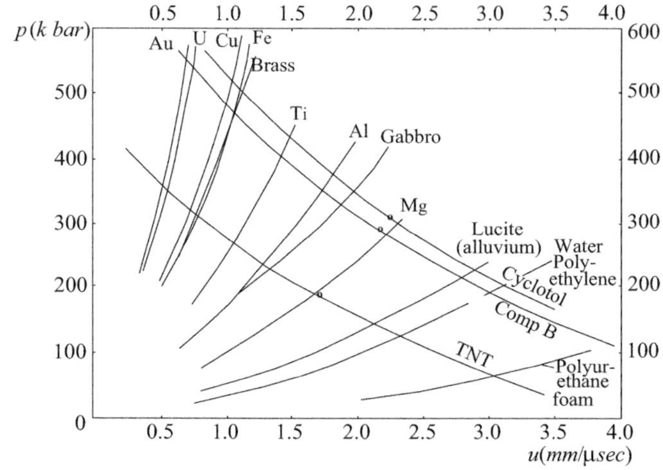

Figure 4.9 The positive Hugoniot curves for several materials, and the negative Hugoniot curves and negative isentropic curves for several explosives.

shock-wave impedance to the polyethylene with low shock-wave impedance, the intensity of transmitted shock wave is about 9.5 GPa.

Fig. 4.9 schematically shows the positive Hugoniot curves for several common materials, as well as three negative curves for three typical explosives. Note the dark point on each negative curve, which denotes the *C–J* (Chapman–Jouguet) detonation state point of each explosive. The dark point is the initial state point for the Hugoniot curve of the explosive (the part above the dark point of the negative curve) and also for the isentropic curve of the explosive (the part below the dark point of the negative curve). The intersection points of those positive curves for the relevant materials and those negative curves for the relevant explosives determine the wave intensity at the interfaces when the detonation wave of the corresponding explosive is normally incident into the relevant material. If the intersection point is located at a position higher than the *C–J* point (namely, the pressure of the intersection point is higher than the *C–J* detonation pressure), the shock wave is reflected into the detonation product; otherwise the rarefaction wave is reflected.

The loading reflection of the shock wave at the rigid wall and the unloading reflection at the free surface can be regarded as the special examples when the shock-wave impedance of the transmission material is ∞ and 0, respectively. However, due to the difference between the Hugoniot curve and the isentropic curve as well as the difference between the Hugoniot curves with different initial state points, unlike in elastic waves, the pressure amplitude of wave reflected at a rigid wall is no longer the twice of that of incident wave, and the particle velocity of wave reflected at a free surface is no longer the twice of that of incident wave.

For nonporous solid materials, especially metals, when the shock pressure is not very high (e.g., in the order of 10 GPa for metals), the differences between the Hugoniot curve and isentropic curve can often be neglected. This case is equivalent to the situation of weak shock waves, in which the entropy increase due to a shock jump can be neglected. In such case, both the Hugoniot curve of the reflected shock wave and the isentropic curve of the reflected rarefaction wave can approximately take the mirror image of the incident shock wave of the Hugoniot curve with respect to the perpendicular line *ab* passing through the final state point (point *1*) of the incident shock wave, as shown in Fig. 4.10. Then the problem can be simplified, and consequently the results related to the reflection of shock waves either at the rigid wall or at the free surface are consistent with that related to elastic waves.

It should be noticed that, for elastic waves, the wave impedance $\rho_0 C_0$ is constant; while for shock waves, the shock-wave impedance $\rho_0 D$ increases with pressure. Thereby, the relative value of wave impedance for two materials also changes with pressure. Even such a special situation may occur that under a certain critical pressure P_K, when $P < P_K$, the shock-wave impedance of material *A* is higher than that of material *B*; while when $P > P_K$, the shock-wave impedance of material *A* is lower than that of material *B*, as shown in Fig. 4.11, where the critical pressure P_K is the intersection point of two Hugoniot curves of the two materials.

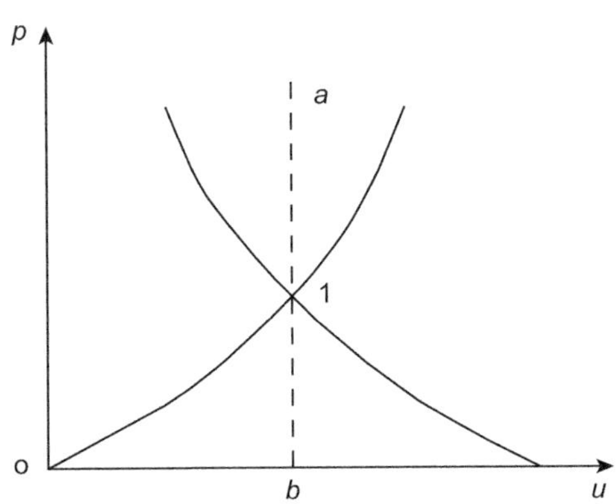

Figure 4.10 Approximate treatment of weak shock-wave reflection.

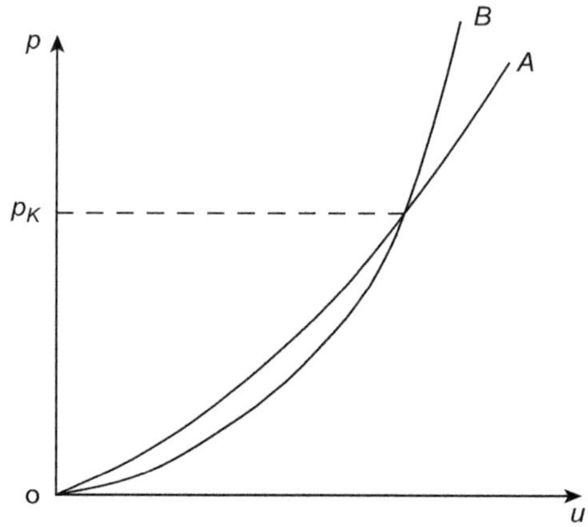

Figure 4.11 Wave impedance change of shock wave.

To give an example, let us analyze the explosive expansion process of a liquid-dispersed explosive device for (like a cloud bomb). For convenience, the problem is approximately analyzed in the following by replacing spherical waves with plane waves (Duvall, 1971), and the Lagrange coordinate system will be used here. Fig. 4.12 schemes the explosive device. The explosive region E (the center sphere) and the liquid to be dispersed B are separated by the inner spherical shell A; all of them are encased by the outer spherical shell C.

Assume that $(\rho_0\mathscr{D})_E > (\rho_0\mathscr{D})_A > (\rho_0\mathscr{D})_B < (\rho_0\mathscr{D})_C$. Firstly, as shown in Fig. 4.13, when the detonation wave whose state is denoted by the point 1 arrives at the shell A, an unloading reflection occurs, whose state corresponds to the point 2. Then, when the transmitted shock wave in A arrives at the interface of A and B, an unloading reflection occurs again,

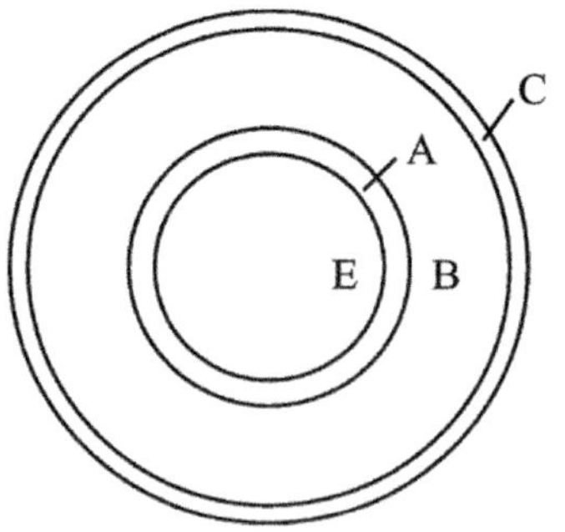

Figure 4.12 Schematic of the explosive device for liquid dispensing.

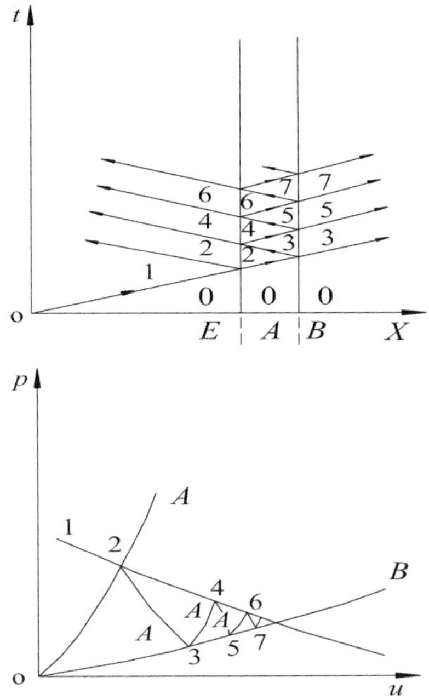

Figure 4.13 Wave propagation between the *A* and the liquid *B*.

whose state corresponds to the point 3. However, when the reflected wave returns to the interface of A and E, a loading reflection occurs, whose state corresponds to the point 4. On the analogy of this, there are a series of leftward rarefaction waves reflected back and a series of rightward shock waves reflected forth in the inner shell A, respectively, corresponding to the negative isentropic curves 2—3, 4—5 ..., and the positive Hugoniot curves 3—4, 5—6 .... Simultaneously there are a series of rightward waves transmitted into the liquid B, corresponding to the positive Hugoniot curves 0—3, 3—5, 5—7 ... and so on.

Secondly, as shown in Fig. 4.14, when the transmitted shock wave in liquid B arrives at the outer shell C, a loading reflection occurs, corresponding to the negative Hugoniot curve 1—2. Then, when the transmitted shock wave in C arrives at its free surface, an unloading reflection occurs, corresponding to the negative isentropic curve 2—3. This indicates that the expanding particle velocity of the outer shell increases, accelerating the outer shell outward. However, when the reflected wave returns to the interface of B and C, a rightward shock wave is reflected, corresponding to the positive Hugoniot curve 3—4. On the analogy of this, there are a series of leftward

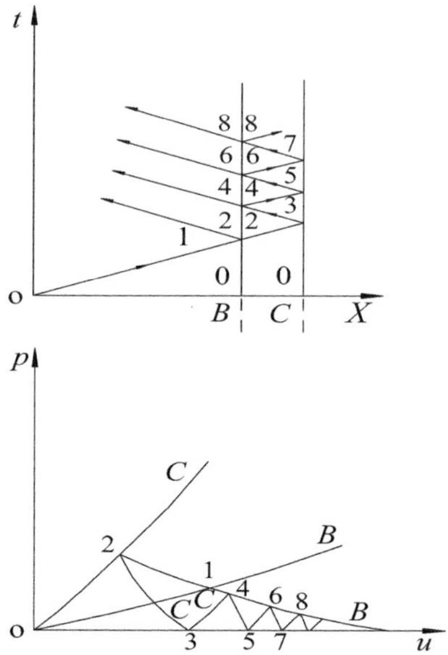

Figure 4.14 Wave propagation between the liquid *B* and the outer shell *C*.

rarefaction waves reflected back and a series of rightward shock waves reflected forth in the outer shell C, respectively, corresponding to the negative isentropic curves 4—5, 6—7 ... and the positive Hugoniot curves 5—6, 7—8 .... With each reflection at the free surface, the expanding velocity of outer shell increases once again, corresponding to the state 3, 5, 7..., and so on. As a result, the outer shell successively and acceleratory expands outward until the break of outer shell, and the liquid is dispersed as fog-like droplets of a 1000 times the volume of the original liquid.

4.3 High-pressure technique for shock waves

In order to carry out the experimental research of the equation of state of solids under high pressures, it is necessary to have the experimental device and dynamic test technique which can produce different high pressures. The dynamic test technique has been elaborated in other courses. The high-pressure experimental technique is mainly introduced in the following.

There are two kinds of techniques for producing high pressure in the laboratory: the static high-pressure technology and the dynamic high-pressure technology. It is very difficult to achieve high pressure with static high-pressure technology, and the maximum pressure that can be reached at

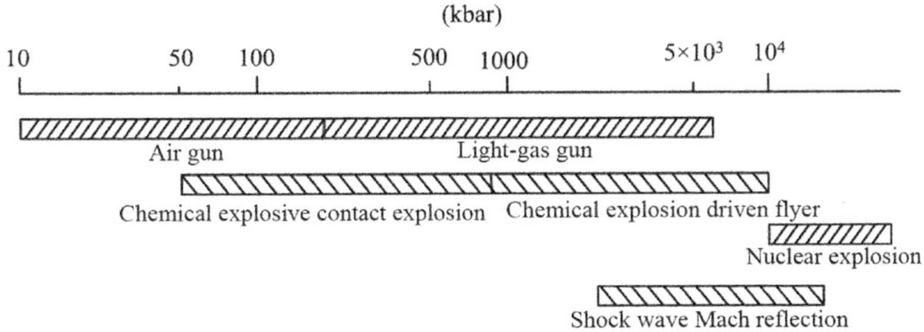

Figure 4.15 Pressure range covered by various shock wave and high-pressure methods (Jing, 1999).

present is only 10^2 GPa. Dynamic high-pressure technology can be easily obtained in various grades of high pressure (see Fig. 4.15), so this technology has been widely used.

There are two basic requirements for the high-pressure device used in the study of the equation of state of solids under high pressures by means of dynamic high-pressure technology or impact compression technology. First, the pressure could be adjustable and should be in a certain wide pressure range, so that the shock adiabatic curve could be experimentally measured from low pressures to high pressures. Second, the shock wave generated in a one-dimensional strain specimen should have a certain degree of planarity with respect to the cross-section of specimen to satisfy the requirement of one-dimensional strain experiments because the analysis of one-dimensional strain compression data is theoretically relatively simple.

At present, the dynamic high-pressure technologies, which are technically mature and widely used, mainly include chemical explosion high-pressure technology and high-pressure technology by gas gun. And the studies under further higher pressures mainly rely on the technology of nuclear explosion high pressure. At the same time, other nonnuclear explosive high-pressure technologies such as high-pressure laser generation (Trainor et al., 1978), electric gun (Steinberg et al., 1978), orbital gun (Hawke et al., 1981), and ion beam targeting (Sweeney et al., 1981) have also made great progress in recent years. The widely used chemical explosion and air gun will be mainly discussed in the following.

Chemical explosive has high chemical reaction—releasing energy and fast reaction speed, so it becomes one of the most commonly used high-pressure shock-wave energy. High pressure from several GPa to several dozen GPa can be generated by chemical explosion technology. In the study on measurement of the state equation, the chemical explosion high-pressure device

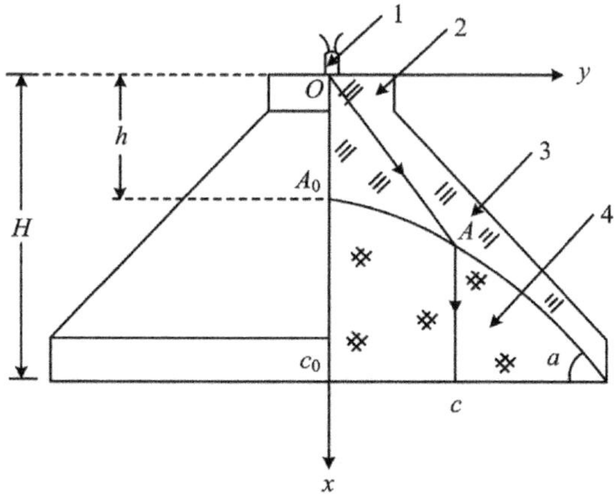

Figure 4.16 Schematic diagram of dynamite plane wave lens.

used is mainly a planar wave generator, whose function is to adjust the scattered detonation wave from point detonation into planar detonation wave in order to conduct shock compression on the sample under one-dimensional strain condition. The shape and internal structure of the planar wave generator are shown in Fig. 4.16 (Jing, 1999). According to its design principle, it can be called a planar wave lens or a detonation wave waveform adjustment controller. The base diameters of common planar wave generators are 100 mm or 200 mm and can be 300 mm in individual special case. The waveform quality of the plane shock wave obtained by the planar wave generator is very high, and the measured results show that the degree of planarity of the waveform produced by the planar wave generator having base diameter of 200 mm is not more than 1 mm.

One of the simplest methods in the chemical explosion high-pressure technique is to put the sample in direct contact with the explosive, as shown in Fig. 4.17 (Jing, 1999). The stable detonation wave (in the state of C—J detonation) generated by the planar wave generator will reflect and transmit at the interface between the explosive and the sample (substrate). According to the discussion in above section, whether the reflected wave is a shock wave or a rerelease wave depends on the material of the sample (substrate). The reflected wave may be a shock wave when the impact-wave impedance of the sample is higher than that of the explosive, or a rerelease wave when the shock-wave impedance of the sample is lower than that of the explosive; accordingly, the intensity of the transmission shock wave can be further enhanced or weakened. In general, the shock wave impedance of dense materials or high-density materials is relatively high, while that of porous

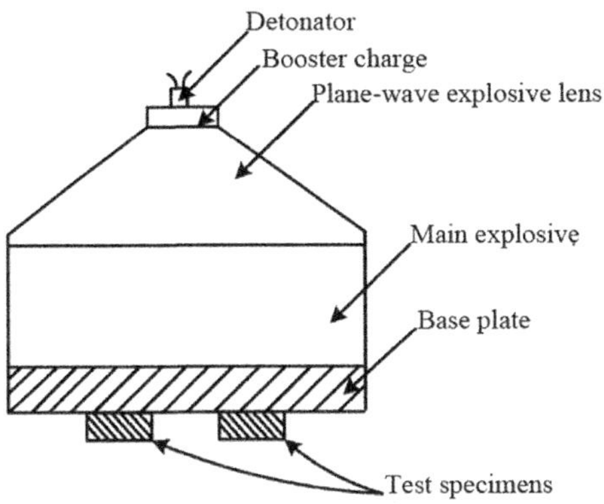

Figure 4.17 Schematic diagram of contact explosive device.

materials or low-density materials is relatively low. Aluminum, iron, and tungsten, respectively, represent typical low-impact wave impedance materials, medium-impact wave impedance materials, and high-impact wave impedance materials. The shock pressure generated by contact explosion between them and the four commonly used explosives is shown in Table 4.2 (Jing, 1999), respectively. It can be seen that the maximum shock pressure generated by the contact explosion method does not exceed 80 GPa.

In order to obtain higher shock pressure, the flyer pressurization technique can be used, and the structure of this device is shown in Fig. 4.18 (Jing, 1999). Its basic principles are: a plane-wave generator is used to accelerate a thin piece of flyer plate, which is driven by the detonation product, has fully absorbed the energy provided by the detonation product through a suitable distance, and hits the stationary target (the specimen) at a high speed. Thus a high compression shock wave is generated in the target to compress the target sample. The shock wave waveform (amplitude, width, etc.) in the target are mainly determined by the thickness, velocity, and physical properties of the flyer and the target material.

Table 4.2 Impact pressure during contact explosion for several typical materials (Jing, 1999) (kbar $\approx$ 0.1 GPa).

Sample material	**TNT**	**RDX**	$\frac{\text{RDX}}{\text{TNT}}\left(\frac{60}{40}\right)$	**HMX**
Aluminum	260	380	330	420
Iron	310	490	430	550
Tungsten	380	670	570	760

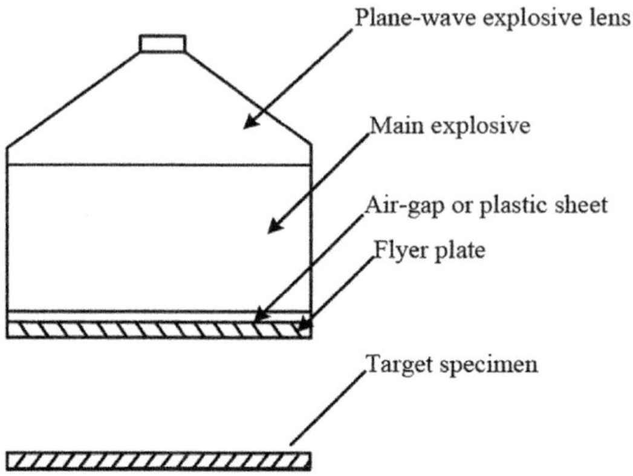

Figure 4.18 Schematic diagram of flying plate supercharger.

The mechanism of continuous acceleration of the flyer during flight is shown in Fig. 4.19. When the detonation wave hits the flyer, the flyer reaches the impact state 1. The shock wave in the flyer is reflected as rerelease wave on the free surface (front surface), so that the free surface of the flyer reaches the state 2. When this rerelease wave reaches the back surface of the

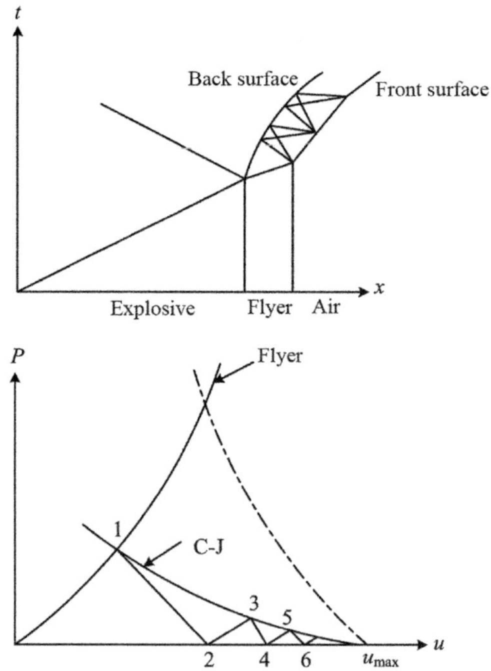

Figure 4.19 Schematic diagram of flyer pressurization principle.

flyer, the compression wave will be reflected to make the flyer reach the state 3. When the compression wave reaches the free surface of the flyer, it will be reflected as rerelease wave, which will accelerate the flyer further, reaching the state 4, and so on, until the flyer reaches its ultimate speed, u_{max}. At this point, the pressure inside the flyer drops to zero.

Fig. 4.19 expresses an ideal situation for an infinite thin slice. The actual situation is that the flyer has a certain thickness, the roundtrip propagation of waves in the flyer takes a certain time, therefore the velocity of the front and back surface of the flyer is different (ref. Fig. 4.20). But it is worth pointing out that when the flyer approaches its ultimate speed, the pressure inside the flyer is close to zero, the density of the flyer is close to the initial density, and the energy of the flyer is mainly kinetic energy. Therefore, we can treat the flyer material as an incompressible model (rigid body). So, according to Newton's second law, the velocity u $(= dx/dt)$ of the flyer satisfies the following equation of motion.

$$M\frac{du}{dt} = Sp \tag{4.42a}$$

where M is the mass of the flyer, S is the cross-section area of the flyer, and p is the detonation pressure on the back interface of the flyer.

According to detonation theory, detonation pressure depends on detonation wave velocity D_Z, sound velocity C_Z, and initial charge density ρ_0:

$$p = \frac{16}{27}\rho_0\frac{C_Z^3}{D_Z} \tag{4.42b}$$

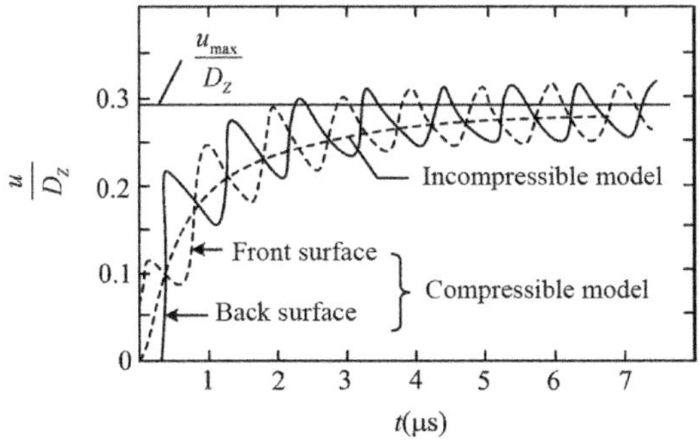

Figure 4.20 Speed–time curves of flyer plate.

Substituting the above eqn. into Eq. (4.42a), we have:

$$\frac{du}{dt} = \frac{\eta_Z C_Z^3}{lD_Z} \tag{4.42c}$$

where l is the length of charge, $\eta_Z = \frac{16m}{27M}$, and m is the mass of charge. Further suppose that

(1) The shock wave reflected by the detonation wave at the back interface of the flyer is a weak shock wave.

(2) The adiabatic exponential γ of the state equation of detonation product is approximated to 3.

Then the characteristic line equation describing the wave propagation in detonation product expressed in space coordinates can be expressed in linear equation as follows:

$$x = (u + C_Z)t \tag{4.42d}$$

Differentiate the above equation, we have:

$$\frac{dx}{dt} = u + C_Z + t\frac{du}{dt} + t\frac{dC_Z}{dt} \tag{4.42e}$$

This equation satisfies both characteristic line and boundary condition. Using $\frac{dx}{dt} = u$ at the boundary, by substituting it to Eqn.(4.42e), and then together with Eq. (4.42c) eliminating $\frac{du}{dt}$, we have:

$$\frac{dC_Z}{dt} + \frac{C_Z}{t} + \frac{\eta_Z C_Z^3}{lD_Z} = 0 \tag{4.42f}$$

So we can solve for C_Z and get:

$$C_Z = \frac{l}{t}\theta \tag{4.42g}$$

$$\theta = \left[1 - 2\eta_Z\left(1 - \frac{l}{D_Z t}\right)\right]^{-\frac{1}{2}} \tag{4.42h}$$

On the other hand, the boundary equation $\frac{dx}{dt} = u$ and the characteristic line equation (Eq. (4.42d)) are solved simultaneously, we have:

$$\frac{dx}{dt} = u = \frac{x - C_Z}{t}$$

Substituting (4.42g) into the above equation, and integrating it, we have:

$$x = D_Z t\left(1 + \frac{\theta - 1}{\eta_Z \theta}\right) \tag{4.42i}$$

The above equation expresses the law how the motion distance of the flyer changes with time. Again from $u = \frac{dx}{dt}$, finally one can obtain the relations of flyer velocity u changing with the time, that is,

$$u = D_Z\left(1 + \frac{\theta - 1}{\eta_Z \theta} - \frac{l\theta}{Dt}\right) \quad (4.42j)$$

In Fig. 4.20, the change curve of flyer velocity calculated by the incompressible model is represented by a dotted line. As can be seen from the figure, the results obtained by this model are quite different from the actual situation (compressible model) in the early stage of flyer motion, but not in the late stage. Therefore, we can use incompressible model to estimate the motion law of the flyer when its velocity is near its ultimate speed.

In Eq. (4.42j), taking $t \rightarrow \infty$, we get the ultimate speed of the flyer, u_{max}, which is equal to

$$u_{max} = D_Z\left(1 + \frac{1}{\eta_Z} - \sqrt{\frac{2}{\eta_Z} + \frac{1}{\eta_Z^2}}\right) \quad (4.43)$$

At this point, the energy density of the flyer per unit volume is $\rho_0 u_{max}^2$. When a flyer hits a target, the pressure p on the target is roughly:

$$p \propto \rho_0 u_{max}^2 \quad (4.44)$$

This shows that in order to increase the impact pressure in of the target, it is necessary to increase the flying flyer speed, or select a higher density of flyer material. In general, in the case of the same explosive, for the material with medium-shock wave impedance, the impact pressure generated at the target by the flying plate technique can be about three times higher than that of the contact explosion.

The chemical explosive high-pressure device also has its own inherent shortcomings. First, the pressure amplitude is difficult to control precisely; Second, the pressure range is not broad enough, especially for the low-pressure part, the low pressure usually we have is around 20 GPa, even if some supplementary measures are taken, it can only reduce to around 5 GPa. Third, the disadvantages caused by explosive explosion.

In order to overcome the above shortcomings, since the 1960s, the pneumatic gun has been widely used to drive flyer, and satisfactory results have been obtained. Compared with the chemical explosion high-pressure device, the pneumatic gun has the advantages of projectile (flyer) speed being able to control, smooth flight, good data repeatability, and high measurement accuracy. In a pressurized gas gun, the impact pressure of a one-stage gas gun can be extended to the low-pressure range which cannot be reached by the chemical explosive device, while the two-stage light gun can increase the impact pressure to the high-pressure range which can be compared with the chemical explosive device.

The one-stage gas gun is divided into piston type and diaphragm type according to different opening modes, and the specific structures are shown in Figs. 4.21 and 4.22, respectively (Jin, 1999). The diameter of a one-stage gas gun is usually 2.5−6 nches (about 63.5−152 mm), and the larger the gun diameter is, the larger and thicker the sample can be and the longer the observation time can be. The projectile is composed of a sabot and a flyer.

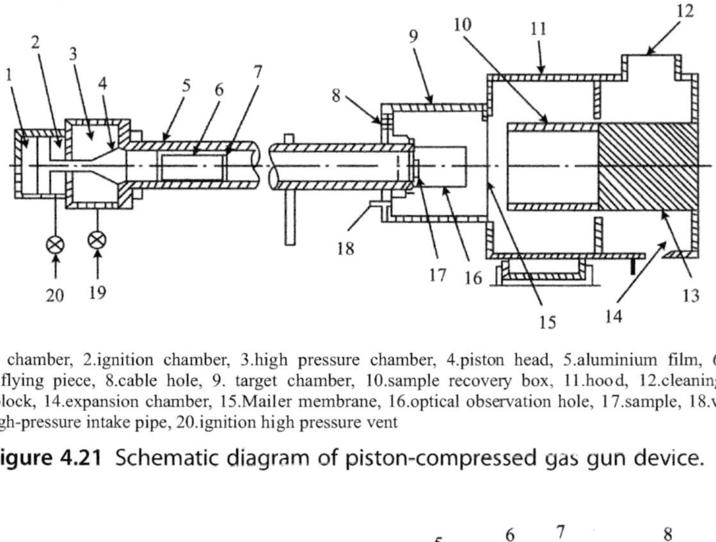

1.rear chamber, 2.ignition chamber, 3.high pressure chamber, 4.piston head, 5.aluminium film, 6.bullet bracket, 7.flying piece, 8.cable hole, 9. target chamber, 10.sample recovery box, 11.hood, 12.cleaning hole, 13.heavy block, 14.expansion chamber, 15.Mailer membrane, 16.optical observation hole, 17.sample, 18.vacuum tube, 19.high-pressure intake pipe, 20.ignition high pressure vent

Figure 4.21 Schematic diagram of piston-compressed gas gun device.

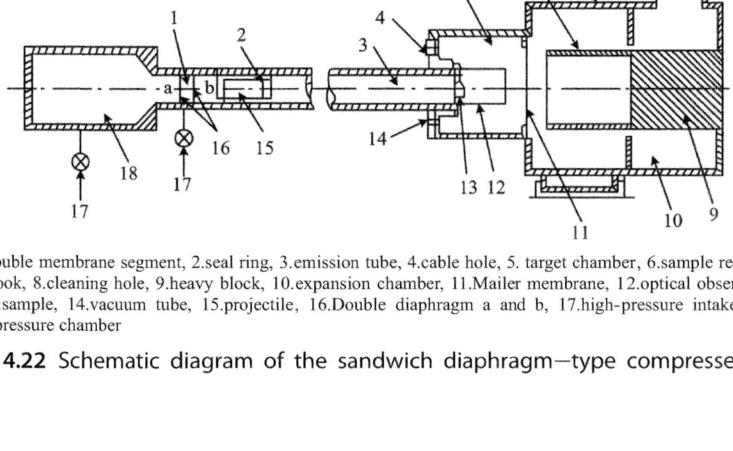

1.double membrane segment, 2.seal ring, 3.emission tube, 4.cable hole, 5. target chamber, 6.sample recovery box, 7.hook, 8.cleaning hole, 9.heavy block, 10.expansion chamber, 11.Mailer membrane, 12.optical observation hole, 13.sample, 14.vacuum tube, 15.projectile, 16.Double diaphragm a and b, 17.high-pressure intake pipe, 18.high pressure chamber

Figure 4.22 Schematic diagram of the sandwich diaphragm−type compressed gas gun.

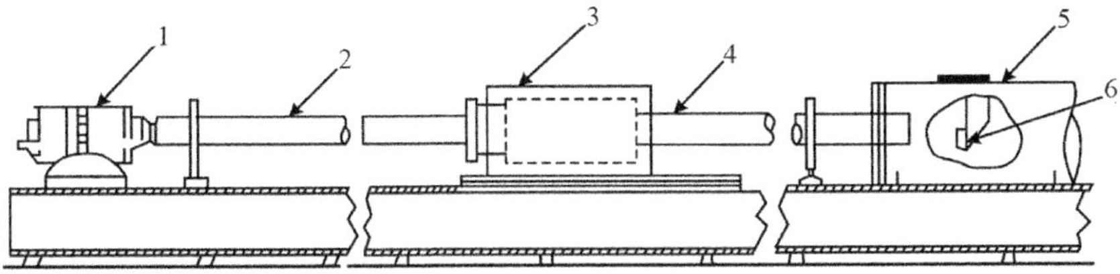

1.double membrane segment, 2. seal ring, 3. emission tube, 4. cable hole, 5. target chamber, 6. sample recovery box

Figure 4.23 Schematic diagram of a two-stage light gas.

In order to increase the velocity of the projectile, the weight of the projectile should be reduced as much as possible; in addition, by increasing the pressure of the working gas, increasing the length of the gun tube, and using low-density working gas. In order to eliminate the effect of gas on the acceleration of projectile and the collision of flyer with the target plate, the gun barrel and target chamber must be vacuumed. The maximum speed which a one-stage gas gun can achieve is about 1500 m/s.

There is no essential difference between a two-stage light gas gun (as shown in Fig. 4.23 (Jin, 1999)) and a one-stage gas gun; they are all driven by compressed gas to move the projectile. The difference between them is that the former consists of a powder chamber and a pump tube to form a high-pressure coupler, which replaces the high-pressure vessel of the latter. Because the pressure of the pump tube is in 1–2 orders of magnitude higher than that of the high-pressure vessel, the projectile velocity is much higher than that of the one-stage gas gun. Table 4.3 shows the gun diameter and projectile velocity of several typical high-pressure gas guns.

The unevenness and inclination of the pressure pulse wave front generated by the gas gun in the target depend on the flatness of the surfaces of

Table 4.3 The gun diameter and projectile velocity of several typical high-pressure gas gun (Jin, 1991).

Type of gun	Gun diameter (mm)	Projectile velocity (m/s) Minimum value	Projectile velocity (m/s) Maximum value
Single-stage gas gun	62.5	~100	1500
	101	~100	~1500
	152	~100	~600
Two-stage light gas gun	30	2000	8200
	50.8	1400	6500
	69	1000	4000

both the projectile and the target and the inclination of the projectile hitting the target, respectively. In order to obtain a satisfactory plane shock wave, the collision surface requires precision polishing, the flatness is about 0.5 μm, and the collision angle cannot be more than 0.03 degrees.

4.4 Measurement principle of shock adiabatic curve

The shock jump condition, together with the internal energy state equation, consists of four equations, including five unknown parameters *P*, *V*, *E*, *u*, and *D*; the relationship between any two parameters is called the shock adiabatic curve. The three most commonly used are $P-V$(or $P-\rho$), $P-u$, and $D-u$. It should be emphasized that since all forms of shock adiabatic curve should satisfy the three conservation relationships on the shock wave front and the internal energy state equations, different forms of shock adiabatic curve are not independent, but can be converted to each other.

At present, the measurement of shock adiabatic curve is mainly based on the flyer technology driven by the chemical explosion plane wave generator or the pressurized gas gun. From the point of view of experimental measurement technology, time difference Δt and distance difference ΔS are easy to realize high-precision measurement, so the measurement of the quotient of the two, namely the measurement of the velocity-type quantity, such as shock wave velocity D (or $\mathscr{D}$) and particle velocity u etc. is more mature and accurate. Let us start with the $D-u$ shock adiabatic curve.

4.4.1 Shock adiabatic curve in the *D*—*u* form

Eq. (4.30) gives that the $D-u$ Hugoniot curve satisfies generally the following quadratic equation:

$$D - u_0 = c_0 + \lambda(u - u_0) + \lambda'(u - u_0)^2$$

When $\lambda' \approx 0$(applicable to pressure range of 0.1—0.2T Pa), the above equation can be simplified to the following linear relation (Eq. 4.31):

$$D - u_0 = c_0 + \lambda(u - u_0)$$

Thus it can be seen that the problem comes down to the experiment to determine the material parameters c_0, λ, and λ'. Obviously, by measuring a series of experimental values corresponding to D and u one by one, it is not difficult in principle to determine c_0, λ, and λ' by fitting these data by Eq. (4.30).

In D and u measurements, D is usually measured directly, but u is not, which is usually obtained indirectly by means of the interaction relation of shock wave propagation when the flyer impinges on the specimen. Depending on the method used to determine the value of u, the current methods of measurement can be divided into three categories: **The flyer impact method (block method), contrast method (impedance-matching method), and free surface velocity method**. They are inseparable from the Hugoniot curve in the P—u form. The u value is determined by analyzing the incident, reflection, and transmission of shock waves in the P—u plane, so what we actually measured is the P—u type of shock adiabatic curve.

4.4.2 Shock adiabatic curve in the P—u form

Here, we discuss the block method (flying piece impact method), contrast method (impedance-matching method), and free surface velocity method of the Hugoniot line measurement based on the P—u form.

1. Flyer impact method

The principle of velocity measurement by the **flyer impact method** is shown in Fig. 4.24, where K_1 is a pair of probes used to measure the flyer velocity W, and the spacing ΔS_W is known; K_2 is a pair of probes used to measure the material wave velocity D of the shock wave in the target (sample), and the spacing ΔS_D is known. With the movement of the flyer and the propagation of the shock wave in the target, the probe at each measuring point is connected successively, and the signal source connected with the probe sends out a time signal to the recording system. If the time

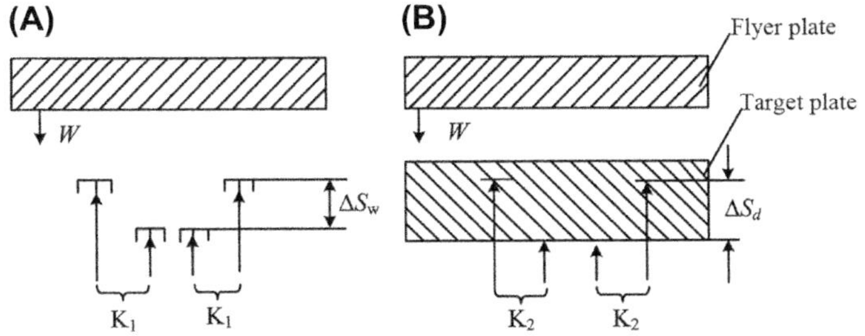

Figure 4.24 Schematic diagram of the block method-measuring device. (A) Measuring the flyer velocity and (B) Measuring the velocity of the shock wave in the target.

difference of sweep spacing ΔS_W measured by a pair of K_1 probes is Δt_W, the velocity W of the flyer is:

$$W = \Delta S_W / \Delta t_W$$

If the time difference between the shock wave sweeping space ΔS_d in a target measured by a pair of K_2 probes is Δt_d, the shock wave velocity is:

$$\mathscr{D} = \Delta S_\text{d} / \Delta t_d$$

How to determine the particle velocity behind the shock wave in the target by the flyer velocity W depends on the matching of the shock wave impedance $\rho_0\mathscr{D}$ of the flyer material and the target material and is inseparable from the analysis of shock adiabatic curve in the form of P—u.

When the flyer and the target plate are made of the same material, it can be seen from the analysis of the shock wave interaction given in Section 4.2 that the particle velocity u_T behind the shock wave in the target plate falls on the positive Hugoniot curve with the central point of ($P = 0$, $u = 0$). And the particle velocity u_F behind the shock wave in the flyer falls on the negative Hugoniot curve with the central point of ($P = 0$, $u = W$). Since the flyer material is the same as the target material, the two Hugoniot curves form the mirror symmetry, that is, the so called "symmetrical collision". And the particle velocity behind the shock wave front $u = u_T = u_F$ is determined by the intersection point of the two Hugoniot curves, equal to half the velocity at which the flyer impacts the target, that is,

$$u = \frac{W}{2}$$

When a pair of K_1 probes and a pair of K_2 probes are used at the same time to measure the results when the flyer impacts the target at a series of different speeds, both a series of data ($\mathscr{D}_i$,u_i) ($i = 1,2, \ldots$) and a series of data (P_i,u_i) ($i = 1,2, \ldots$) can be obtained. From those, we can determine the shock adiabatic curve in the D—u form and the shock adiabatic curve in the P—u form for the target material.

When the flyer and the target plate are made of different materials, besides measuring W and $\mathscr{D}$, the shock adiabatic curve in the P—u form of the flyer material must also be predicted. Based on the analysis of shock wave interaction given in Sec., the u of target material can be easily determined by graphic method, as shown in Fig. 4.25. In the figure, curve 1 is the predetermined positive shock adiabatic curve of the flyer material and curve 1′ is the P—u shock adiabatic curve of flyer material flying at the speed of W. After the flyer colliding with the target, a shock wave whose propagation

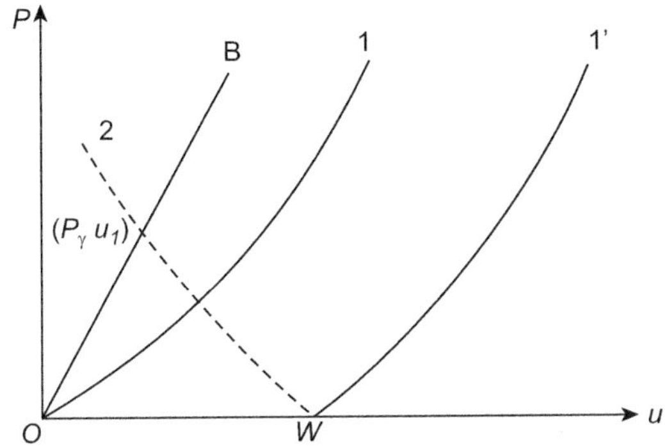

Figure 4.25 Determination of the particle velocity of target by the flyer impact method.

direction is opposite to the original flying direction is formed in the flyer, so the shock adiabatic curve should be the mirror image of curve 1′, that is the negative shock adiabatic curve 2 in the P—u form. On the other hand, the final state of the shock compression that the target material may reach must fall on the Rayleigh line OB with the slope of the shock wave impedance $(\rho_0\mathscr{D})_T$. From the condition that both the pressure and particle velocity on the flyer—target interface should remain continuous, it can be seen that the intersection (P_i, u_i) of the Rayleigh line OB with the curve 2 describes the shock pressure P_i and the particle velocity u_i behind the shock wave in the target plate after the collision. By a series of collisions at different speeds, a series of data $\mathscr{D}_i$ can be obtained and a series of the corresponding (P_i, u_i) ($i = 1,2,3, \ldots$) can be obtained. From this, the shock adiabatic curve in the $\mathscr{D}$—u form of the target material and the shock adiabatic curve in the P—u form of the target material can be determined.

2. The impedance matching method (the contrast method)

The measuring device of the impedance-matching method (the contrast method) is shown in Fig. 4.26, where the probe K_1 is used to measure the shock wave velocity $\mathscr{D}_A$ in the standard material, which is the incident wave velocity from the standard material A to the measured material B. The probe K_2 is used to measure the shock wave velocity $\mathscr{D}_B$ of the measured material B, which is the transmission wave velocity. The type of the reflected wave depends on the shock wave impedance of A and B materials could be either the shock wave or the rerelease wave.

The particle velocity u behind the shock wave front of the measured material can also be calculated by graphical method from the measurement results of the contrast method. The solution process is as shown in Fig. 4.27 (assuming the density of the measured material is known). The curve $P(u)$ in

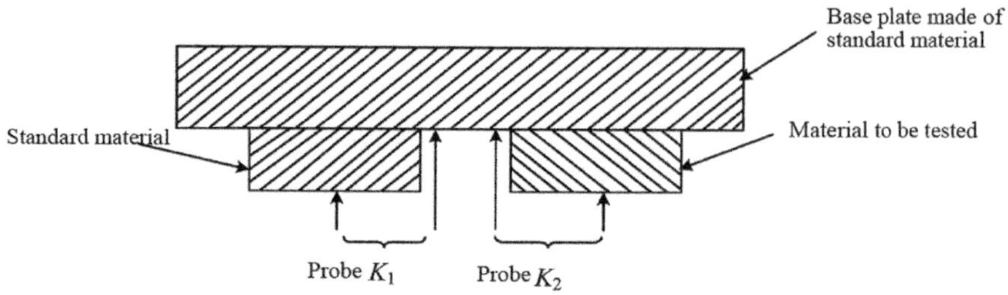

Figure 4.26 Determination of particle velocity of target by the contrast method.

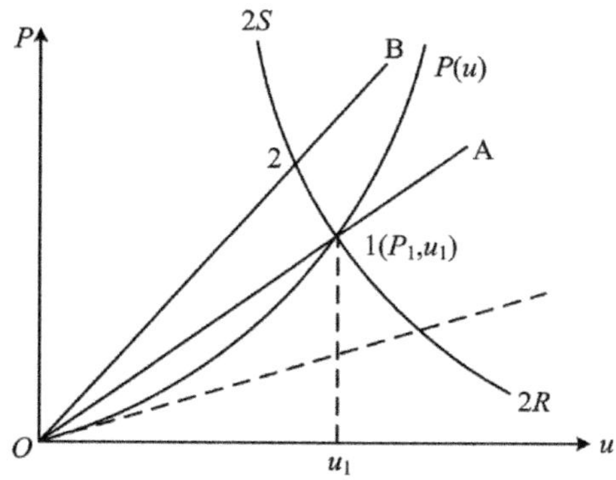

Figure 4.27 Determination of the particle velocity of the measured material by the contrast method.

the figure is the P—u shock adiabatic curve of the known standard material A. According to the measured incident wave velocity $\mathscr{D}_A$, the state point 1 of the incident wave can be determined. According to the measured transmission wave velocity $\mathscr{D}_B$, the Rayleigh line R_B on the shock adiabatic curve of the measured material can be determined. Obviously, the state point 2 behind the transmission wave should fall on this Rayleigh line R_B. If the shock wave impedance $(\rho_0\mathscr{D})_B$ of the measured material is greater than the shock wave impedance $(\rho_0\mathscr{D})_A$ of the standard material, then the Rayleigh line R_B is above the $P(u)$ curve (the line OB in Fig. 4.27). In this case, the reflected wave is a shock wave. The state behind the wave shall fall on the negative shock adiabatic curve with the state point 1 as the center point (the curve 1-2S in Fig 4.27). The intersection point (P_2,u_2) of the curve 1-2S with the Rayleigh line R_B gives the impact pressure and the particle velocity in the target after the collision, according to the continuous condition of the pressure and the particle velocity at the interface. If the shock wave impedance $(\rho_0\mathscr{D})_B$ of the measured material is smaller than the impact wave impedance $(\rho_0 \mathscr{D})_A$ of the standard material, then the Rayleigh line R_B is below the *P(u)* curve (the dotted line on Fig. 4.27). In this case, the reflected wave is a rerelease wave. The state behind the wave shall fall on the isentropic curve S_{A_1}, starting with the state point 1 (the curve 1-2R in Fig 4.27). Similarly, the intersection point of the S_{A_1} with the Rayleigh line R_B gives the impact pressure and particle velocity in the target after the collision. By a series of collisions at different velocities, a series of data $\mathscr{D}_{Bi}$ can be obtained, and a series of corresponding $(P_i, u_i)_B$ $(i = 1,2,3, \ldots)$ can be further obtained. From this, the shock adiabatic curve in the $D-u$ form and the shock adiabatic curve in the $P-u$ form of the material B can be determined.

In the process of graphical method, the approximate drawing of the shock adiabatic curve 1-2S or the isentropic curve 1-2R is as follows: a vertical line is made through the point 1, then the mirror inversion of the *P(u)* curve is made by using the vertical line as the symmetric axis. The mirror inversion above the point 1 is the shock adiabatic curve 1-2S and the mirror inversion below the point 1 is the isentropic curve 1-2R. The closer the two points (1, 2) is, the higher the accuracy of the above method is.

3. The free surface velocity method

The measuring device of **the free surface velocity method** is schematically shown in Fig. 4.28. In the figure, probe K_1 is used to measure the shock wave velocity $\mathscr{D}$ in the sample, K_2 is used to measure the flight velocity u_{fs} of the free surface of the sample, and then the particle velocity u is calculated by u_{fs}. So the free surface velocity method is named.

When the particle velocity u behind the incident shock wave is calculated from the free surface particle velocity u_{fs}, the conservation relation on the wave front of the reflected rerelease wave should be used. These relationships can easily be obtained from the conservation relation on the shock wave front given at the beginning of this chapter just by replacing the strong discontinuities with the weak discontinuities. For example, Eqs. (4.7), (4.8), and (4.13) can be transformed into following equations, respectively.

$$du = -\rho_0 C_S dV, \tag{4.7'}$$

$$du = \frac{dP}{\rho_0 C_S}, \tag{4.8'}$$

$$C_S = \sqrt{-\left(\frac{\partial P}{\partial V}\right)_S}, \tag{4.13b'}$$

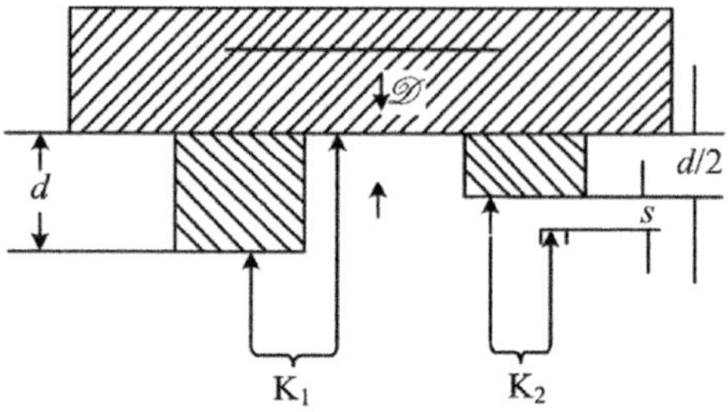

Figure 4.28 Schematic diagram of the measuring device for the free surface velocity method.

The C_S in the equation is the isentropic sound wave velocity determined by the slope of the $P-V$ isentropic curve of the rerelease wave. Thus, the relationship between the particle velocity u behind the shock wave front and the free surface velocity u_{fs} is (see Fig. 4.29A):

$$\begin{aligned} u_{fs} &= u + \int_{p_1=p_H}^{p_2=0} \frac{dp}{\rho_0 C_S} \\ &= u - \int_{V_1(p_1=p_H)}^{V_2(p_2=0)} \rho_0 C_S dV \\ &= u + u_r \end{aligned} \tag{4.45}$$

where u_r is the additional particle velocity caused by the center rerelease wave. The main problem of converting u_{fs} to u is how to solve for u_r, which comes down to how to determine the $P-V$ isentropic curve of the target material.

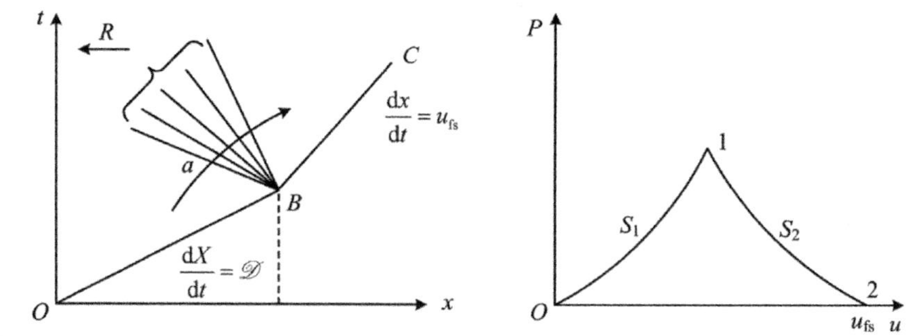

Figure 4.29 Determination of the particle velocity of the measured material by the free surface velocity method.

If the intensity of the incident shock wave is weak, that is, it can be approximately treated as the incident weak shock wave ignoring entropy increase; in this case, the shock adiabatic curve can be approximately replaced by the isentropic curve S_1 (see Fig. 4.29B), and the rerelease process of the free surface can be represented by the isentropic curve S_2, which is the mirror image inversion of S_1 through state point 1. Obviously, similar to the reflection of elastic waves on a free surface, there is approximately:

$$u_r \approx u$$

or

$$u \approx \frac{1}{2} u_{fs}$$

That is to say, in the case of a weak shock wave, the particle velocity is approximately equal to half the free surface velocity. This result is generally called the law of velocity multiplication of free surface for weak shock waves.

If the intensity of the incident shock wave is high, the law of free surface velocity multiplication is no longer applicable. Qualitatively (see Fig. 4.29A), at this time, the u_r is greater than u, or the particle velocity u is less than half the free surface velocity. In this case, u can be calculated iteratively from u_{fs}. The specific steps are: the first approximation $u^{(1)} = u_{f_s}/2$ of u is calculated by the law of free-surface velocity multiplication, and the corresponding shock adiabatic curve is obtained by $u_i^{(1)}$ and D_i, then the isentropic curve was obtained, and the corresponding free-surface velocity $\left|u_{fs}^{(n)} - u_{fs}\right| < \varepsilon$ is calculated by Eq. (4.45), and $u_{fs}^{(1)}/2$ is used as the second-order approximation of particle velocity $u^{(2)}$. Repeat the above steps until the difference between the $u_{fs}^{(n)}$ and the measured value u_{fs} is less than the measurement error ε_m, $\left|u_{fs}^{(n)} - u_{fs}\right| < \varepsilon_m$. Finally, the $u^{(n)}$, which corresponds to the $u_{fs}^{(n)}$, is the particle velocity value to be obtained. Experience shows that the convergence rate of the iterative process is very fast, and generally three or four times is enough.

Because the iterative process is relatively complicated, the block method or the contrast method should be used as far as possible when the incident shock wave is strong. The free-surface velocity method is mostly used in the case of weak shock waves.

4.4.3 Analytical expression of shock adiabatic curve in the P—V form

The shock adiabatic curve in the P—V form can be derived from the shock adiabatic curve in the D-u form.

After this, introduce the volumetric compression strain η as defined below:

$$\eta = 1 - \frac{V}{V_0} = 1 - \frac{\rho_0}{\rho}, \tag{4.46}$$

the mass conservation condition of shock jump in space coordinates, $[\rho u] = D[\rho]$ (Eq. 4.10), can be rewritten in the following simple form:

$$\eta = \frac{u - u_0}{D - u_0}.$$

We can establish an analytical expression of the shock adiabatic curve in the P—V form based on the linear relation (4.31) of the shock adiabatic curve in the D—u form. As a matter of fact, by solving Eq. (4.31) and the above equation simultaneously, we have:

$$u - u_0 = \frac{c_0 \eta}{1 - \lambda\eta} \tag{4.47a}$$

$$D - u_0 = \frac{c_0}{1 - \lambda\eta} \tag{4.47b}$$

Substituting the above two equations into the momentum conservation condition $[P] = \rho^+(D-u^+)\ [u]$, (Eq. 4.11) of shock jump in space coordinates, and setting $P_0 = 0$, an analytical expression of the shock adiabatic curve in the P—V form is obtained:

$$p_H = \rho_0 c_0^2 \frac{\eta}{(1-\lambda\eta)^2} \tag{4.48a}$$

It has been pointed out that linear relation (4.31) generally applies only to 1—2 Mbar (0.1—0.2 TPa), and at higher pressures, Eq. (4.30) with quadratic terms should be used. At this point, by solving simultaneously Eqs. (4.10), (4.11), and (4.30), after a more complex operation similar to that of Eq. (4.47a), a more general analytical expression for the Hugoniot curve in the P—V form can be obtained.

$$P_H = \rho_0 c_0^2 \frac{\eta}{(1-\lambda\eta)^2 - 2c_0\lambda'\eta^2} \cdot \left[1 + \frac{c_0^2\lambda'^2\eta}{(1+\lambda\eta)^2 - 2(c_0\lambda'\eta^2)} + \cdots\right] \tag{4.48b}$$

Obviously, when $\lambda' = 0$, the above equation is reduced to (4.48a). So the problem also boils down to how to determine the three characteristic constants c_0, λ, λ' characterizing the material properties. In other words, the problem of the experimental determination of the $P-V$ Hugoniot curve is then transformed into the problem of the experimental determination of the $D-u$ Hugoniot curve. Once the material parameters of the $D-u$ Hugoniot curve expressed by Eq. (4.30) or (4.31) are determined by experiments, the $P-V$ Hugoniot curve expressed by Eq. (4.48) is also determined.

With regard to the weak shock wave, recall that we have pointed out in the proof of Eq. (4.19), since the shock adiabatic curve in the $P-V$ form and the isentropic curve in the $P-V$ form have the same slope and curvature at the initial point, therefore, when the pressure P is not too high, we can replace the shock adiabatic curve in the $P-V$ form with the isentropic curve in the $P-V$ form approximately. In other words, for low-pressure weak shock wave that can ignore entropy increase, the Hugoniot curve in the $P-V$ form can be approximated by the isentropic curve in the $P-V$ form, which is the Murnaghan equation (Eq. (2.17b)) given in Chapter 2.

$$P = \frac{k_0}{n}\left[\left(\frac{V_0}{V}\right)^n - 1\right]$$

where k_0 is the zero-pressure isentropic bulk modulus, while n is the coefficient of the first derivative of isentropic bulk modulus with respect to pressure.

In addition to the direct measurement by the above impact tests, material parameters c_0 and λ in the $P-V$ shock adiabatic relation, or material parameters k_0 and n in the Murnaghan equation, can also be indirectly calculated or roughly estimated by other experimental parameters. The relevant parameters that can be used are: (1) the relative thermodynamic parameters of materials under normal conditions, such as γ_0, (2) static pressure measurement data, and (3) ultrasonic measurement data.

Let us start with the relevant thermodynamic parameters.

The relation between the Grüneisen coefficient γ and other thermodynamic parameters α_T, V, C_V has been given by Eq. (2.65) in Chapter 2:

$$\gamma = V\left(\frac{\partial P}{\partial E}\right)_V = V\cdot\frac{\phi_T}{C_V} = \frac{k_T\alpha_T V}{C_V}$$

The above eqn. shows that by using the isothermal bulk modulus k_T, isobaric thermal expansion coefficient α_T, isometric specific heat C_V, and specific volume V measured by general experiments, the Grüneisen parameter $\gamma(V_0)$ in normal state can be obtained.

On the other hand, the general relation between the Grüneisen coefficient $\gamma(V)$ and cold pressure $P_K(V)$ was given in Eq. (3.105) in Chapter 3.

$$\gamma(V) = -\frac{(2-\alpha)}{3} - \frac{V}{2}\frac{d^2\left(P_K(V)V^{2a/3}\right)/dV^2}{d\left(P_K(V)V^{2a/3}\right)/dV}$$

where $\alpha = 0, 1, 2$ corresponds to the Slate equation, Dugdale–MacDonald equation, and free volume equation, respectively. Substituting the P–V analytical equation (Eq. 4.48a) based on the D–u linear relation and the Murnaghan Eq. (2.17b) into the above equation, respectively, and using the characteristics of two-order tangent of the shock adiabatic curve in the P–V form and the isentropic curve in the P–V form at the central point, finally, the following two relations are obtained at $V = V_0$:

$$\lambda = \frac{1}{2}\gamma(V_0) + \frac{\alpha}{4} + \frac{1}{3} \tag{4.49}$$

$$n = 2\gamma(V_0) + \alpha + \frac{1}{3} \tag{4.50a}$$

If Eq. (4.49) is substituted into the above equation, we have:

$$n = 4\lambda - 1 \tag{4.50b}$$

The above equations give the relationship between $\gamma(V_0)$ and λ, n. Once $\gamma(V_0)$ is calculated by using thermodynamic parameters in normal states, the values of λ and n can be estimated. However, the values of λ and n are closely related to the values of α and generally tend to take the values of the Dugdale–MacDonald equation, that is, $\alpha = 1$.

With respect to c_0 and k_0, there are the following relationships:

According to the definition of initial point sound velocity (Eq. 4.26) and the definition of isentropic bulk modulus k_S (Eq. 2.43), we have:

$$c_0 = \sqrt{\left(\frac{\partial P}{\partial \rho}\right)_{V_0}} = \sqrt{\left[V^2\left(-\frac{\partial P}{\partial V}\right)_S\right]_{V_0}} = \sqrt{\frac{k_S(V_0)}{\rho}} \tag{4.51}$$

Here the k_S (V_0) is the k_0 in the Murnaghan equation (Eq. 2.17b). On the other hand, by Eqs. (2.62b) and (2.65) in Chapter 2, we have:

$$\frac{k_S}{k_T} = 1 + \frac{k_T\alpha_T^2 TV}{C_V}$$

$$\gamma = V \cdot \frac{\phi_T}{C_V} = \frac{k_T \alpha_T V}{C_V}$$

Then, the following relations between the isentropic bulk modulus k_S and the isothermal volume modulus k_T can be obtained.

$$k_S = k_T\left(1 + \frac{k_T \alpha_T v}{C_V}\alpha_T T\right) = k_T(1 + \alpha_T \gamma T) \quad (4.52a)$$

Therefore, by using the thermodynamic parameters α_T and γ, the isentropic bulk modulus k_S can be calculated by the isotherm bulk modulus k_T, and c_0 is determined. Under low pressures and room temperature, if $\alpha_T \gamma T << 1$, we have approximately:

$$k_S(V_0) \approx k_T(V_0) \quad (4.52b)$$

The above equation shows that we can determine or estimate c_0, λ and k_0, n, respectively, by using the thermodynamic experimental data.

Let us look at the relationship between the relevant static measurement data and the material parameters of the Murnaghan equation.

It is assumed that the static-pressure measurement data have been given, and the relevant material parameters a and b are determined according to the empirical Eq. (2.11) proposed by Bridgman, that is,

$$\frac{V_0 - V}{V_0} = -\Delta = aP - bP^2$$

The corresponding volumetric modulus $K_T(P)$, as shown in Eqs. (2.13) and (2.14), is

$$K_T(P) = \frac{1}{a - 2bP} = \frac{1}{a\left(1 - \frac{2b}{a}P\right)} \approx \frac{1}{a}\left(1 + \frac{2b}{a}P\right)$$

Here the subscript T is particularly added, indicating that it is the isotherm bulk modulus to distinguish it from the isentropic bulk modulus in the Murnaghan equation. The right side of the approximate equal sign is the approximate expression of K_T (P) obtained when $2bP/a \ll 1$, which expands into a power series and ignores the small quantity of the higher order.

It should also be noted that the K_T in Eq. (2.13) is the material isotherm modulus defined in the Lagrange coordinates. The isentropic bulk modulus k_S in the Murnaghan equation is defined in the Euler coordinates.

Therefore, if we want to relate the material parameters of the two, we first have to convert K_T into the space isotherm bulk modulus k_T defined in the Euler coordinates. That is

$$k_T = \frac{V}{V_o} K_T = \frac{\rho_0}{\rho} K_T = (1+\Delta) K_T \tag{4.53}$$

Substitute Eqs. (2.11) and (2.13) into the above equation, we have:

$$k_T(P) = \frac{1 - aP + bP^2}{a - 2bP}. \tag{4.54a}$$

After finishing and ignoring the small quantity above the second order, we have:

$$k_T(P) \approx \frac{1}{a}\left(1 + \frac{2b}{a} P\right) \tag{4.54b}$$

Substitute the above equation into Eq. (4.52a), we have:

$$k_0 = \frac{(1 + \alpha_T \gamma T)}{a}, \quad n = \left(\frac{2b}{a^2} - 1\right)(1 + \alpha_T \gamma T). \tag{4.55a}$$

This shows that when the material parameters k_0 and n in the Murnaghan isentropic equation are directly obtained by the material parameters a and b of the Bridgman isothermal equation, other thermodynamic parameters α_T and γ are also needed to be known. Only when $\alpha_T \gamma T << 1$ so that can be negligible, we have:

$$k_0 \approx \frac{1}{a}, \quad n \approx \left(\frac{2b}{a^2} - 1\right). \tag{4.55b}$$

This is equivalent to almost ignoring the difference between the isothermal bulk modulus k_T and the isentropic bulk modulus k_S.

Finally, let us discuss how to use the results of ultrasonic measurement to determine the above four parameters c_0, λ, k_0, and n.

The data measured by ultrasonic technology are usually the longitudinal wave sound velocity c_{l0} and the transverse wave sound velocity c_{t0} in normal states, and the longitudinal wave sound velocity c_l and the transverse wave sound velocity c_t under high pressures (below several kbar). The c_l and c_t under high pressures are linear to the pressure:

$$\begin{aligned} c_l &= c_{l0} + AP \\ c_t &= c_{t0} + BP \end{aligned} \tag{4.56}$$

where A and B are constants obtained by fitting experimental data. In addition, it is known by the definition of the spatial sound velocities:

$$c^2 = \frac{k_S}{\rho} \tag{4.57a}$$

$$c_l^2 = \frac{k_S + \frac{4}{3}g_S}{\rho} \tag{4.57b}$$

$$c_t^2 = \frac{g_S}{\rho}, \tag{4.57c}$$

where k_S and g_S in the equations are the isentropic bulk modulus and the isentropic shear modulus in the Euler space. So we have:

$$c^2 = c_l^2 - \frac{4}{3}c_t^2 \tag{4.58}$$

From this point, we have:

$$c_0 = c_{l0}^2 - \frac{4}{3}c_{t0}^2 \tag{4.59a}$$

$$k_0 = k_{S0} = \rho_0\left(c_{l0}^2 - \frac{4}{3}c_{t0}^2\right) \tag{4.59b}$$

$$n = \left(\frac{dk_S}{dP}\right)_0 = 2\rho_0\left(Ac_{l0} - \frac{4}{3}Bc_{t0}\right) + 1 \tag{4.59c}$$

Using Eq. (4.50b), thus we have:

$$\lambda = \frac{1}{2}\rho_0\left(Ac_{l0} - \frac{4}{3}Bc_{t0}\right) + \frac{1}{2} \tag{4.60}$$

It should be noted that the wave velocities in Eqs. (4.56)−(4.60) refer to the spatial wave velocity in the Euler coordinates. If the velocity measured is the material wave velocity in the Lagrange coordinates, then the corresponding material volume modulus Ks and so on can be derived in similar way of the above derivation, then according to Eq. (4.53), it can be converted into the space bulk modulus k_S and so on.

Of the above three methods for predicting c_0, λ, or k_0, n, the first method, namely, the results estimated by thermodynamic γ_{th} under normal conditions are in the best agreement with the experimental data of shock wave.

4.5 Determination of equation of state of solids under high pressures

First of all, it is emphasized again that do not confuse the shock adiabatic curve with the equation of state of solids under high pressures The shock adiabatic curve represents only the trajectory of the equilibrium final state point that a certain equilibrium initial state (i.e., the central point of the Hugoniot curve) may reach through the shock jump, rather than the equation of state of solids under high pressure itself. As far as the relationship between the shock adiabatic curve and the equation of state of the solid at high pressures is concerned, two different kinds of problems are actually involved: one is to find the shock adiabatic curve for a given equation of state, and the other is to find the equation of state for a given shock adiabatic curve. Much of what has been discussed previously belongs to the first category, which is a matter of complete certainty. The second kind of problem is an uncertain one, which can be solved according to the limited experimental results only by making some assumptions about the specific form of the equation of state. This section discusses how to use shock wave tests to determine the equation of state of solids under high pressures and belongs to the second category of problems.

Considering that the Grüneisen equation of state is a suitable model for the shock adiabatic curve measured under general laboratory conditions, so this section will use this model as the starting point for determining the equation of state under high pressures.

Eq. (2.67b) gives the general form of the Grüneisen equation of state:

$$P = P_k(V) + \frac{\gamma(V)}{V}[E - E_k(V)]$$

For different materials, the cold pressure $P_K(V)$, cold energy $E_K(V)$, and Grüneisen coefficients $\gamma(V)$ in the equation have different specific forms. In order to determine these three, we need to find three independent equations which are related to these three without other extra unknown variables.

The first equation is the Grüneisen state equation (Eq. (2.67b)) applied to the shock adiabatic curve.

$$P_H = P_k(V) + \frac{\gamma(V)}{V}[E_H - E_k(V)] \tag{4.61}$$

The P_H and E_H in the equation are measured on the shock adiabatic curve. The shock adiabatic curve in the D–u form measured by the impact compression test shows a linear relation for most materials, as shown in Eq. (4.31), and the corresponding P_H (Eq. 4.48a) and E_H are respectively:

$$p_H = \frac{\rho_0 C_0^2 \eta}{(1-\lambda\eta)^2} = \frac{C_0^2 \eta}{V_0(1-\lambda\eta)^2}$$

$$E_H = \frac{1}{2} P_H (V_0 - V) = \frac{1}{2} \frac{C_0^2 \eta^2}{(1-\lambda\eta)^2} \tag{4.62}$$

where η is given by Eq. (4.46):

$$\eta = 1 - \frac{V}{V_0}$$

On the other hand, according to Eq. (3.105) in chapter 3, the general relationship between the Grüneisen coefficient $\gamma(V)$ and cold pressure $P_K(V)$ is:

$$\gamma(V) = -\frac{(2-\alpha)}{3} - \frac{V}{2} \frac{d^2\left(P_K(V) V^{2a/3}\right)/dV^2}{d\left(P_K(V) V^{2a/3}\right)/dV}$$

where $\alpha = 0, 1, 2$ corresponds to the Slater $\gamma_s(V)$, Dugdale–MacDonald $\gamma_{DM}(V)$, and free volume $\gamma_f(V)$, respectively, thus the above equation provides the required second equation.

According to Eq. (2.68), the third equation will be

$$P_k(V) = -\frac{dE_k(V)}{dV}$$

Integrating the above equation along the isothermal process of $T = 0$, we have:

$$E_k(V) = -\int_{V_{0K}}^{V} P_k(V) dV - E_0 \tag{4.63}$$

where V_{0K} is the specific volume of the material at zero temperature and zero pressure, and E_0 is the integral constant. In order to calculate the integral constant, we take the internal energy in normal state as zero, that is,

$$E(T, V)\Big|_{V_0(P_0 = 0,\, T = T_0)}^{T_0 = 293\text{K}} = 0 \tag{4.64}$$

In addition, according to the definition of isometric specific heat C_V, we can obtain by integrating Eq. (2.55):

$$E(T,V) = \int_0^T C_V(T,V)dT + E_k(V)$$

Substitute Eqs. (4.63) and (4.64) into the above equation, we have:

$$E_0 = \int_0^{T_0} C_V(T,V_0)dT - \int_{V_{ok}}^{V_0} P_k(V)dV \tag{4.65}$$

Substituting the above equation back into Eq. (4.63), we have:

$$\begin{aligned} E_k &= -\int_0^{T_0} C_V(T,V_0)dT - \int_{V_0}^{V} P_k(V)dV \\ &= E_0^* - \int_{V_0}^{V} P_k(V)dV \end{aligned} \tag{4.66}$$

where $E_0^* = \int_0^{T_0} C_V(T,V_0)dT$ according to Eq. (3.85) in chapter 3, which gives the relationship of C_V as the function of Debye characteristic temperature θ_D, then E_0^* can be calculated according to the following equation.

$$E_0^* = 3NK_B \int_0^{T_0} \left\{ D\left(\frac{\theta_D}{T}\right) - \frac{\theta_D}{T} D'\left(\frac{\theta_D}{T}\right) \right\} dT \tag{4.67}$$

where K_B is Boltzmann constant, n is the number of atoms per unit mass, $D\left(\frac{\theta_D}{T}\right)$ is the Debye function and according to Eq. (3.86), the specific form of Debye function is:

$$D\left(\frac{\theta_D}{T}\right) = \frac{3}{\left(\frac{\theta_D}{T}\right)^3} \int_0^{\frac{\theta_D}{T}} \frac{x^3}{e^x - 1} dx$$

Its first derivative $D'\left(\frac{\theta_D}{T}\right)$ is:

$$D'\left(\frac{\theta_D}{T}\right) = \frac{3}{e^{\left(\frac{\theta_D}{T}\right)} - 1} - \frac{3}{\left(\frac{\theta_D}{T}\right)} D\left(\frac{\theta_D}{T}\right) \tag{4.68}$$

Substituting Eqs. (3.86) and (4.68) into Eq. (4.67) and integrating, after finishing it, we have:

$$E_0^* = 3NK_BT_0D\left(\frac{\theta_D}{T}\right) \tag{4.69}$$

The above three equations, (4.61), (3.105), and (4.66), have constituted the closed equations to determine $P_k(V)$, $E_k(V)$, and $\gamma(V)$.

A second-order integral–differential equation for $P_k(V)$ can be obtained by substituting Eqs. (4.61) and (4.66) into Eq. (3.105).

$$\frac{V(P_H - P_k(V))}{E_H + E_0^* + \int_{V_0}^{V} P_k(V)dV} = -\frac{(2-n)}{3} - \frac{V}{2}\cdot\frac{\left(P_k(V)V^{2n/3}\right)''}{\left(P_k(V)^{2n/3}\right)'} \tag{4.70}$$

where $n = 0, 1, 2$ corresponds to the Slate equation, Dugdale–MacDonald equation, and free volume equation, respectively. Because Eq. (4.70) is nonlinear, it is difficult to find its analytical solution. Thus, it is usually solved by using numerical calculation method. Furthermore, to solve Eq. (4.70), the initial values $P_k(V_0)$ and $P_k'(V_0)$ are needed to be determined. To do this, substituting Eq. (4.66) into (4.61), we have:

$$P_k(V) = P_H - \frac{\gamma(V)}{V}\left[E_H + E_0^* + \int_{V_0}^{V} P_k(V)dV\right] \tag{4.71}$$

Using $P_H(V_0) = 0$, $E_H(V_0) = 0$, from the above equation, we have:

$$P_k(V_0) = -\frac{\gamma(V_0)}{V_0}E_0^* \tag{4.72}$$

$$P'_k(V_0) = P'_H(V_0) + \frac{\gamma(V_0)}{V_0^2}E_0^*(1+\gamma(V_0)) - \frac{E_0^*}{V_0}\gamma'(V_0) \tag{4.73}$$

Obviously, in order to determine the values of $P_k(V_0)$ and $P'_k(V_0)$, we need to know the values of $\gamma(V_0)$ and $\gamma'(V_0)$.

In order to find $\gamma(V_0)$ and $\gamma'(V_0)$, we approximately consider that Eq. (3.105) is also true along the isentropic curve whose central point is $(V = V_0, P = 0)$, that is, the $\gamma(V)$ is approximately equal to:

$$\gamma(V) = -\frac{(2-n)}{3} - \frac{V}{2}\,\frac{\left(P_S(V)V^{2n/3}\right)''}{\left(P_S(V)V^{2n/3}\right)'} \tag{4.74}$$

thus

$$\gamma(V_0) = -\frac{(2+n)}{3} - \frac{V_0}{2}\frac{P''_S(V_0)}{P'_S(V_0)} \tag{4.75}$$

$$\gamma'(V_0) = -\frac{V_0}{2}\frac{P''_S(V_0) + (2n+1)V_0^{-1}P''_S(V_0) + \frac{2}{3}n(2n-1)V_0^{-2}P''_S(V_0)}{P'_S(V_0)} + \frac{V_0}{2}\left(\frac{P''_S(V_0)}{P'_S(V_0)} + \frac{4n}{3V_0}\right)^2 \tag{4.76}$$

When using the equation of Dugdale–MacDonald, $n = 1$, we have:

$$\gamma_{DM}(V_0) = -1 - \frac{V_0}{2}\frac{P'_S(V_0)}{P'_S(V_0)} \tag{4.77}$$

$$\gamma'_{DM}(V_0) = -\frac{V_0}{2}\frac{P''_S(V_0)}{P'_S(V_0)} + \left(\frac{V_0}{2}\frac{P''_S(V_0)}{P'_S(V_0)}\right)^2 - \frac{1}{6}\frac{P''_S(V_0)}{P'_S(V_0)} + \frac{5}{9V_0} \tag{4.78}$$

When using the equation of Slate, $n = 0$, we have:

$$\gamma_s(V_0) = -\frac{2}{3} - \frac{V_0}{2}\frac{P'_S(V_0)}{P'_s(V_0)} \tag{4.79}$$

$$\gamma'_s(V_0) = -\frac{1}{2}\frac{V_0P''_S(V_0) + P''_S(V_0)}{P'_S(V_0)} + \frac{V_0}{2}\left(\frac{P''_S(V_0)}{P'_S(V_0)}\right)^2 \tag{4.80}$$

The relationships between the derivatives at the center point of the isentropic curve and the derivatives at the center point of the shock adiabatic curve, as shown in Eqs. (4.18) and (4.19), are as follows:

$$\begin{aligned} p'_s(V_0) &= P'_H(V_0) \\ p''_s(V_0) &= P''_H(V_0) \end{aligned} \tag{4.81}$$

$$p'''_s(V_0) = P'''_H(V_0) + \frac{\gamma(V_0)}{2V_0}P''_H(V_0)$$

And the values of $P'_H(V_0)$, $P''_H(V_0)$, and $P'''_H(V_0)$ can be given directly by the experimental curve. If the experimental curve of the shock adiabatic curve in the $D-u$ form is linear (Eq. 4.32), then there is:

$$P'_H(V_0) = -\left(\frac{c_0}{V_0}\right)^2$$

$$P''_H(V_0) = 4\left(\frac{c_0^2\lambda}{V_0^3}\right) \tag{4.82}$$

$$P'''_H(V_0) = -18\left(\frac{c_0^2\lambda^2}{V_0^4}\right)$$

Substituting Eqs. (4.81) and (4.82) into (4.77–4.80), we have:

$$\gamma_{DM}(V_0) = 2\lambda - 1 \tag{4.83}$$

$$\gamma'_{DM}(V_0) = \frac{1}{V_0}\left(\lambda^2 - \frac{\lambda}{3} + \frac{5}{9}\right) \tag{4.84}$$

$$\gamma_s(V_0) = 2\lambda - \frac{2}{3} \tag{4.85}$$

$$\gamma'_s(V_0) = \frac{1}{V_0}\left(\lambda^2 + \frac{4}{3}\lambda\right) \tag{4.86}$$

Once $\gamma(V_0)$ and $\gamma'(V_0)$ are found, as mentioned previously, the corresponding Grüneisen equation of state can be obtained.

4.6 Shock phase transition

The effects of material phase transition on shock adiabatic curve have not been discussed in the previous sections.

Materials under different pressures and temperatures have different aggregation states, such as solid state, liquid state, gaseous state, and so on. People use "phase" to express the "appearance" of materials in different forms of solid, liquid, and gas. In a broad sense, phase refers to the homogeneous part of a material system with the same physical and chemical properties, which is separated from other parts by a certain interface. With the change of external pressure and temperature, the phase transition, reflecting as the change of inner microstructure and the corresponding properties of the crystal, will occur in solids.

Since shock waves in solids are usually generated under high pressure, irreversible entropy increase in the process of shock jump will cause a sharp rise in temperature. Under such high temperature and high-pressure conditions, some materials may undergo phase transition. The possibility of phase transition caused by shock wave was first proposed by H. Schardin in 1941 (Schardin, 1941). In 1956, Bancroft et al. (1956) first discovered the abnormal phenomenon caused by phase transition of iron at 13 GPa pressure in shock wave experiments, which corresponds to the transformation from phase α (body-centered cubic crystal) to phase ε (closely packed hexagonal crystal). Since then, the study of phase transition by shock wave experiment has been widely used.

Phase transition is a discontinuous quantitative change process related to temperature T and pressure P. Gibbs free energy $G(T, P)$, as a function of T and P, is the most widely used thermodynamic potential function in the study of phase transition thermodynamics, and the Gibbs free energy per unit mass is defined as chemical potential μ (P, T). During the phase transition process, the chemical potential μ (P, T), temperature T, and pressure P between the two phases remain equal, that is,

$$T_1 = T_2 = T, P_1 = P_2 = P, \mu_1 = \mu_2$$

Reviewing the relation between thermodynamic potential functions in Chapter 2, Eq. (2.26), the chemical potential μ (P, T) in the equation can be expressed as:

$$\mu_1 = e_1 - s_1 T + v_1 P, \mu_2 = e_2 - s_2 T + v_2 P$$

where e, s, v are internal energy, entropy, and specific volume of unit mass, respectively. In (μ, T, P) three-dimensional space, the chemical potential $\mu_1 = \mu_1(P,T)$ and $\mu_2 = \mu_2(P,T)$ represent two surfaces, respectively. The fact that μ, T, P are each equal, respectively, in phase transition means that the phase transition process occurs either on the intersection of the two surfaces or on the tangent of the two surfaces. The first case is usually defined as a first-order phase transition and the second case as the second-order phase transition. The phase transition process is going to be a decrease in the chemical potential.

In the first-order phase transition process, the chemical potential between phases is equal, but their first-order partial derivatives are not equal (there is a jump), and the thermodynamic expression for such a transition is:

$$\mu_1 = \mu_2$$

$$\frac{\partial \mu_1}{\partial P} \neq \frac{\partial \mu_2}{\partial P}, \quad \text{namely } v_1 \neq v_2,$$

$$\frac{\partial \mu_1}{\partial T} \neq \frac{\partial \mu_2}{\partial T}, \quad \text{namely } s_1 \neq s_2,$$

That is, in the first-order phase transition, the volume jump and the entropy jump will occur before and after the phase transition. Because there is volume jump in this process, it is easy to observe in shock wave experiment.

According to $\Delta\mu = \mu_1 - \mu_2 = 0$, we have: After finishing, the slope of p—t phase line in the first-order phase transition can be obtained.

$$\frac{\partial(\Delta\mu)}{\partial T} dT + \frac{\partial(\Delta\mu)}{\partial P} dP = 0$$

After finishing, the slope of $P-T$ phase line in the first-order phase transition can be obtained.

$$\frac{dP}{dT} = \frac{s_2 - s_1}{v_2 - v_1} \tag{4.87a}$$

This is the famous Clausius—Clapeyron equation. Since the phase transition is carried out under isobaric pressure, according to the definition (Eq. 2.39) of the thermodynamic potential function enthalpy H in Chapter 2, the entropy jump in the above equation can be rewritten as the enthalpy jump.

$$\frac{dP}{dT} = \frac{s_2 - s_1}{v_2 - v_1} = \frac{H_2 - H_1}{T(v_2 - v_1)} \tag{4.87b}$$

The enthalpy jump $\Delta H = H_2 - H_1$ in the equation is the heat absorbed or released in the phase transition process, known as latent heat of phase transition.

In the second-order phase transition, not only the chemical potentials are equal, but the first-order partial derivatives are each equal, respectively, but the second-order partial derivatives are not each equal. The thermodynamic expressions for this kind of phase transition are:

$$\mu_1 = \mu_2$$

$$v_1 = v_2$$

$$s_1 = s_2$$

Referring to the relevant Eqs. (2.48)−(2.50) about the Gibbs free energy *G(T, P)* of thermodynamic potential function in Chapter 2, we have:

$$\frac{\partial^2 \mu_1}{\partial P^2} \neq \frac{\partial^2 \mu_2}{\partial P^2}, \text{ that is } \beta_1 \neq \beta_2$$

$$\frac{\partial^2 \mu_1}{\partial P \partial T} \neq \frac{\partial^2 \mu_2}{\partial P \partial T}, \text{ that is } (\alpha_T)_1 \neq (\alpha_T)_2$$

$$\frac{\partial^2 \mu_1}{\partial T^2} \neq \frac{\partial^2 \mu_2}{\partial T^2}, \text{ that is } (c_p)_1 \neq (c_p)_2$$

where compression coefficient β is the reciprocal of the isothermal bulk modulus k_T, $\beta = \frac{1}{k_T}$. It can be seen that there is no volume jump and entropy jump in the second-order phase transition, but there are jumps in some thermodynamic parameters such as compression coefficient β, thermal expansion coefficient α_T, and isobaric specific heat C_p. The jump of compression coefficient can also be observed directly in shock wave experiment.

Because there is no volume jump and entropy jump,

$$\Delta v = v_2 - v_1 = \frac{\partial \mu_2}{\partial P} - \frac{\partial \mu_1}{\partial P} = 0$$

$$\Delta s = s_2 - s_1 = \frac{\partial \mu_2}{\partial T} - \frac{\partial \mu_1}{\partial T} = 0$$

we have:

$$\frac{\partial(\Delta v)}{\partial T} dT + \frac{\partial(\Delta v)}{\partial P} dP = 0$$

$$\frac{\partial(\Delta s)}{\partial T} dT + \frac{\partial(\Delta s)}{\partial P} dP = 0$$

After finishing, we have:

$$\frac{dP}{dT} = \frac{(\alpha_T)_2 - (\alpha_T)_1}{\beta_2 - \beta_1}$$

$$\frac{dP}{dT} = \frac{(c_p)_2 - (c_p)_1}{\left[(\alpha_T)_2 - (\alpha_T)_1\right] Tv}$$

These two equations are called the Ehrenfest equations.

McQueen et al. (1967) have made a detailed analysis of the first-order phase transition characteristics under shock wave, which can be described by the slope of $P-T$ phase curve.

If the slope $\frac{dP}{dT} > 0$, that is, $\frac{\Delta H}{\Delta v} > 0$ and the $\left(\frac{dP}{dT}\right)_H$ along the shock adiabatic curve is smaller than the slope of the phase curve, then the phase transition must occur in the direction from the high-density phase (that is, the low-temperature phase) to the low-density phase (that is, the high-temperature phase).

Shock melting and shock vaporization are typical examples of such phase transitions. The effect of such phase transition on the shock adiabatic curve is shown in Fig. 4.31A, where the phase curve is represented by solid line, the shock adiabatic curve is represented by dotted line, and the initial phase (region I) of the shock adiabatic curve is high-density phase (low-temperature phase). Since the shock adiabatic curves before and after the phase transition present a relatively smooth transition between the two phases (from region I to region II) on the $P-V$ diagram and the $D-u$ diagram, such kind of phase transition is not easy to observe in the shock wave experiment; in other words, it is difficult in shock wave experiments to determine whether this kind of shock phase transition occurs.

If $\frac{dP}{dT} > 0$, but the $\left(\frac{dP}{dT}\right)_H$ along the shock adiabatic curve is greater than the slope of the phase curve, then the phase transition must occur in the direction from the low-density phase (that is, high-temperature phase), to the high-density phase (that is, low-temperature phase). It usually corresponds to a solid–solid phase transition. The effect of such phase transition on the shock adiabatic curve is shown in Fig. 4.30B. This kind of phase transition is easy to be observed in shock wave experiments since the shock adiabatic curves before and after phase transition have obvious inflection between the two phases on the $P-V$ and $D-u$ graphs.

(Note that in this diagram and in Fig. 4.31, the $H-V$ lines were drawn under constant pressure; in the $D-u$ diagram, the $D-u$ shock adiabatic curves (represented by solid line) were determined by experiments).

Another first-order phase transition corresponds to the slope of $P-T$ phase curve being less than 0, $\frac{dP}{dT} < 0$, that is, $\frac{\Delta H}{\Delta v} < 0$, in this case, the $\left(\frac{dP}{dT}\right)_H$ along the shock adiabatic curve is positive and must be greater than the slope of the phase curve. Therefore, the phase transition can only occur in the direction from the low-temperature phase (and thus the low-density phase) to the high-temperature phase (and thus the high-density phase). Abnormal melting, such as observed in niobium and silicon, and crystalline phase transitions, such as observed in iron and zirconium, are

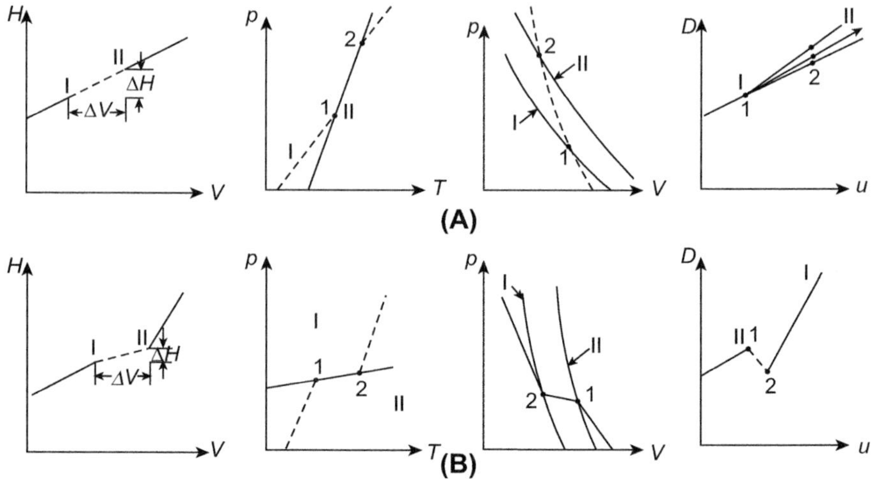

Figure 4.30 Effect of phase transition on shock adiabatic curves when $dP/dT > 0$.

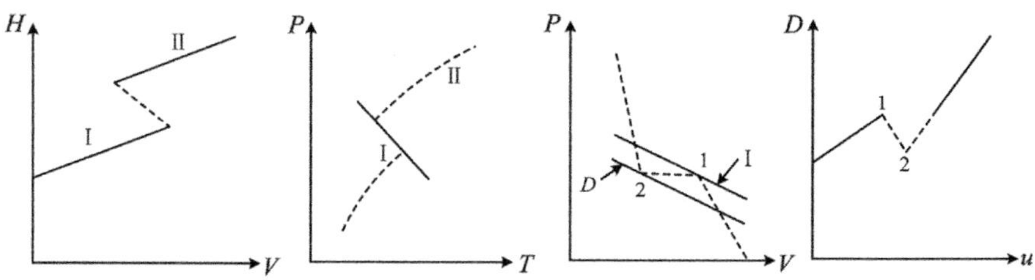

Figure 4.31 Effect of phase transition on shock adiabatic curves when $dP/dT < 0$.

belonged to such cases. The effect of such phase transition on the shock adiabatic curve is shown in Fig. 4.31. Obviously, this kind of phase transition is also easier to observe in shock wave experiments.

In principle, a similar discussion on the second-order phase transition can be made, which is not described here.

Finally, it should be noted that the shock phase transition is a very rapid process, which can be completed in microseconds or even nanoseconds. The experimental results show that the shock phase transition and static phase transition have exactly the same result.

PART TWO

Distortion law of the dynamic constitutive relationship of materials

As mentioned in the first chapter, the constitutive relationship of materials is a mathematical expression of the mechanical behavior of materials. The strain rate–dependent dynamic constitutive relationship of materials is usually expressed as the functional relationship between stress tensor σ_{ij}, strain tensor ε_{ij}, strain rate tensor $\dot{\varepsilon}_{ij}$, and temperature T.

Because stress tensor σ_{ij} can be decomposed into the sum of the spherical tensor of stress σ_m and the deviatoric tensor of stress S_{ij}, strain tensor ε_{ij} can be decomposed into the sum of the spherical tensor of strain e_m and the deviatoric tensor of strain e_{ij}, and strain rate tensor $\dot{\varepsilon}_{ij}$ can be decomposed into the sum of the spherical tensor of strain rate $\dot{e}_m$ and the deviatoric tensor of strain rate $\dot{e}_{ij}$:

$$\sigma_{ij} = \sigma_m \delta_{ij} + S_{ij}, \quad \varepsilon_{ij} = e_m \delta_{ij} + e_{ij}, \quad \dot{\varepsilon}_{ij} = \dot{e}_m \delta_{ij} + \dot{e}_{ij} \tag{II.1}$$

then the strain rate–dependent constitutive relationship can usually also be expressed in two parts: (1) the spherical part, which describes the change in volume (the volumetric deformation law in scalar form), and (2) the deviatoric part, which describes the change in shape (the law of distortion in functional form):

$$\text{Volumetric deformation law:} \quad \sigma_m = f_v(e_m, \dot{e}_m, T), \tag{II.2}$$

$$\text{Distortion law:} \quad S_{ij} = \overline{\Phi}\left(e_{ij}, \dot{e}_{ij}, T\right), \tag{II.3}$$

Here, the assumption is that the volumetric deformation law and the law of distortion are decoupled and unrelated to each other.

In the case of high pressures that allow distortion to be ignored and under the assumption that plastic volumetric deformation does not occur under the influence of spherical stress (isoaxial stress) and that the strain-rate effect of volumetric deformation is absent (no bulk viscosity), the solid is approximately an ideal compressible fluid. Then, the volumetric law is reduced to a nonlinear elastic, volumetric deformation relationship. From a thermodynamic perspective, the relationship between mechanical quantities at this time is unrelated to the path but is only a function of state. This relationship is often referred to as the equation of state for solids under high pressures. This has been discussed in the first chapter.

This Part 2 focuses on the dynamic constitutive distortion law of materials. The core difference between this law and the quasistatic constitutive distortion law lies in the **strain rate effect**. In recent decades, in-depth macroscopic, microscopic, and experimental research on the strain rate effect has made considerable progress. This progress reflects the urgent needs and drivers of a wide range of practical problems in different fields. In this part, the fifth chapter provides a discussion on the basic problems of macromechanics; the sixth chapter presents a discussion on the microphysical basis from the perspective of dislocation dynamics; and the seventh chapter gives a discussion on dynamic experimental studies based on stress wave theory.

CHAPTER FIVE

Dynamic distortion law of materials: macroscopic representation

5.1 Experimental phenomena characterizing the mechanical behavior of materials under high-speed deformation

The aforementioned fluid dynamics model is not suitable to describe the flow stress that characterizes material shape change when it is not negligible compared with the isoaxial pressure that characterizes volume change. Thus, the dynamic distortion law of materials must be simultaneously examined. A large number of experiments have shown that dynamic distortion mainly involves the strain-rate effect and its associated temperature effect.

5.1.1 Strain-rate effect

First, several representative experimental phenomena that concern the mechanical properties of materials under high-speed deformation are observed below.

The historic impact experiment on steel wire under drop-weight impact performed by Hopkinson, J (1872) and his son Hopkinson, B (1905) (see Fig. 5.1) provided three important and surprising results: (1) The height H of the falling weight is the main factor that controls impact fracture of the wire. That is, impact fracture mainly depends on impact velocity but has nothing to do with the mass of falling weight. (2) The impact fracture is not located at impact end A but is located at suspension consolidation end B. (3) The measured dynamic yield strength is approximately twice the static yield strength. In this experiment, the steel wire was subjected once to a stress 50% greater than the static yield strength to 100 μs without apparent yielding (yield lag). The first two results are mainly attributed to the inertial effect (that is, stress wave propagation) of the steel wire, and the third result is completely attributed to the strain-rate effect of the material's mechanical properties.

Dynamics of Materials
ISBN: 978-0-12-817321-3
https://doi.org/10.1016/B978-0-12-817321-3.00005-X

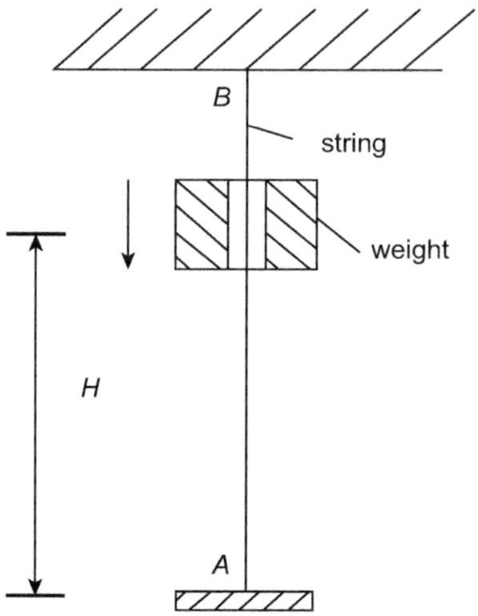

Figure 5.1 Schematic of Hopkinson and son experiment.

Hopkinson and son experiment and a series of subsequent experiments showed that the mechanical properties of materials are related to strain rate. At the same time, the inertial effect of the experimental system should also be considered in the analysis of the dynamic mechanical properties of materials under impact conditions, that is, the stress wave theory should be used to analyze and explain the experimental phenomenon. In impact experiments, the stress wave effect (inertia effect) and the material strain-rate effect are often coupled to each other; it is important to distinguish the inertial effect in the experimental system and the strain-rate effect of the material itself.

The strain-rate effect of materials is firstly manifested as that the yield strength of the material increases as strain rate increases, as well as the yield lag. This phenomenon has been observed in the Hopkinson and son experiments as described above. Davies and Hunter (1963) further demonstrated this phenomenon through a series of experiments on various metal and polymer materials by using the split Hopkinson pressure bar technology (see Chapter 7 for details). As shown in Table 5.1, the dynamic yield strengths of low-carbon steels, soft irons (Armco irons), and high-purity molybdenum (750 MPa, 560 MPa, and 1120 MPa) are 2.6 times, 3.6 times, and 3.3 times higher, respectively, than their static yield strengths.

Similarly, the flow stress of the material increases as strain rate increases, that is, the dynamic stress—strain curve in the stress—strain plot moves up with the increase of strain rate. The typical experimental results for metallic

Table 5.1 Dynamic and static yield strengths of several metallic materials (Davies and Hunter, 1963).

Material	Static yield strength/Y_s (MPa)	Dynamic yield strength/Y_d (MPa)	Ratio (Y_d/Y_s)
Low-carbon steel	290	750	2.6
Soft iron	155	560	3.6
High-purity molybdenum	340	>1120	>3.3

materials with different crystal structures, such as mild steel (body-centered cubic lattice), industrial pure aluminum (face-centered cubic lattice), and titanium alloy TB-2 (close-packed hexagonal) are shown in Figs. 5.2–5.4. This phenomenon is similar to the strain hardening discussed in the context of plastic mechanics, called strain-rate hardening. Of course, an anomaly

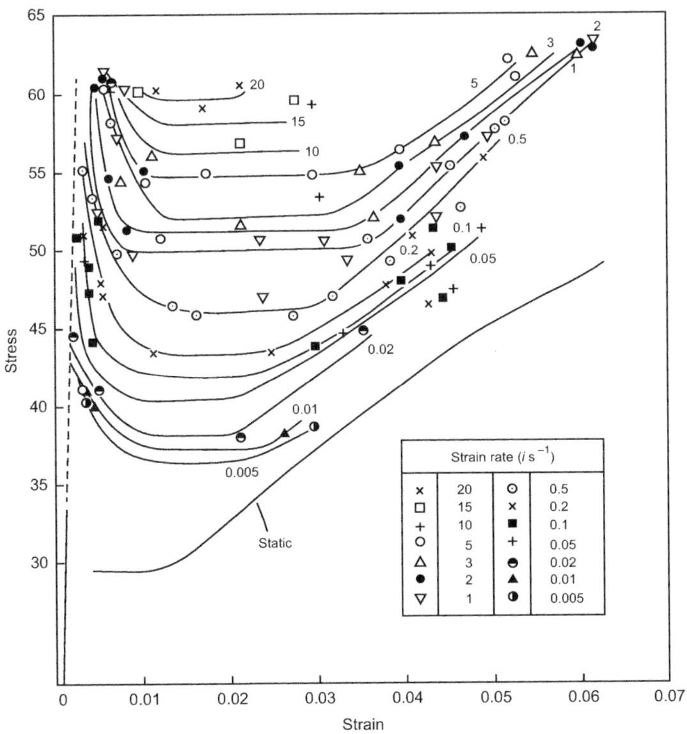

Figure 5.2 Stress–strain curves of mild steel under different uniaxial compressive strain rates. *From Marsh, K J., Campbell, J D., 1963. The effect of strain rate on the post-yield flow of mild steel. J. Mech. Phys. Solids. 11 (1), 49–63, Fig. 6, p.54. Reprinted with permission of the publisher.*

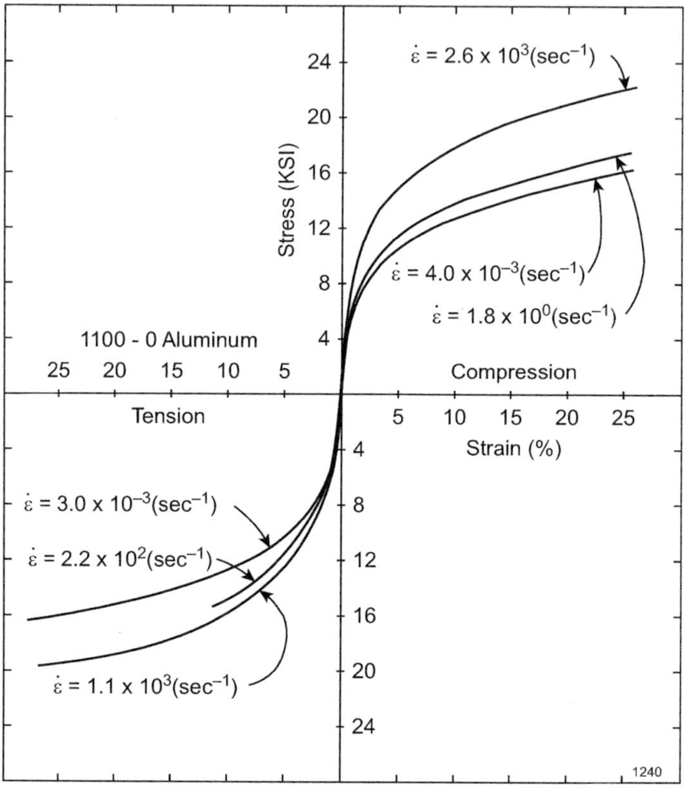

Figure 5.3 Stress—strain curves of industrial pure aluminum (1100-0 Al) under different uniaxial compressive strain rates. *From Lindholm, U S., Yeakley, L M., 1968. High strain-rate testing: tension and compression, Exp. Mech. 8 (1), 1—9, Fig. 8, p.7. Reprinted with permission of the publisher.*

may exist. For example, the strain rate of the material may not be sensitive or may be negatively sensitive to strain rate (see Chapter 6).

In general, materials with different microscopic crystal structures exhibit different strain-rate sensitivity. For example, an HCP-type metal or alloy (see Fig. 5.4) is not as strain-rate sensitive as an FCC-type (see Fig. 5.3). However, a BCC-type (see Fig. 5.2) is highly sensitive to strain rate.

However, abnormal phenomena may exist. For example, although pure aluminum shows high strain-rate sensitivity (Fig. 5.3), the alloying-enhanced aluminum alloy 6061-T6 is insensitive to strain rate (Maiden and Green, 1966), as shown in Fig. 5.5. This apparent insensitivity to strain rate is often the result of a combination of multiple factors, which will be

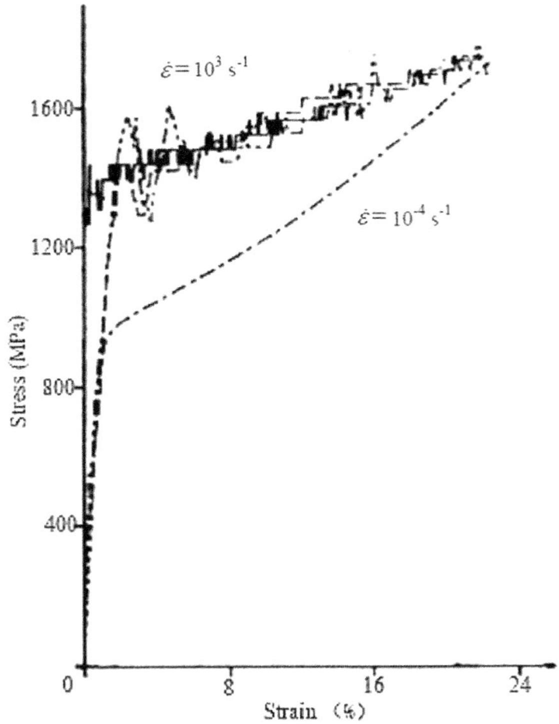

Figure 5.4 Stress—strain curves of titanium alloy TB-2 under different uniaxial compressive strain rates. *From Wang, L., Hu, S., 1985. The dynamic stress-strain relation of Ti-alloy TB-2 under high strain rates. Explos. Shock Waves 5 (1), 9—15.*

further discussed in the next chapter on the basis of dislocation dynamics and other mechanisms.

Historically, in reference to the large number of experimental data on the one-dimensional stress of metallic materials, the dependency of flow stress on strain rate is usually subdivided into two types: the following power function law

$$\frac{\sigma}{\sigma_0} = \left(\frac{\dot{\varepsilon}}{\dot{\varepsilon}_0}\right)^n \tag{5.1a}$$

or

$$\frac{\sigma}{\sigma_0} = 1 + \left(\frac{\dot{\varepsilon}}{\dot{\varepsilon}_0}\right)^n \tag{5.1b}$$

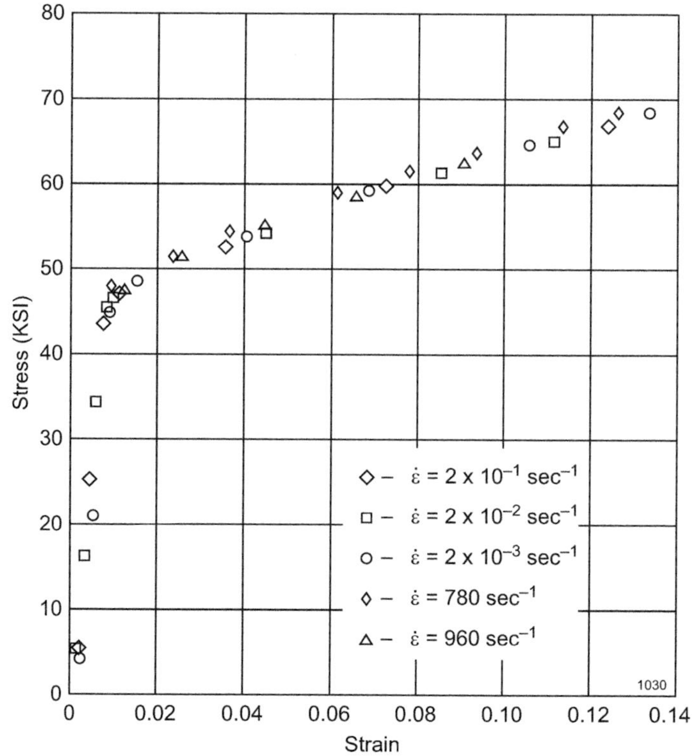

Figure 5.5 Stress—strain curves of aluminum alloy 6061-T6 under different uniaxial compressive strain rates. *From Lindholm, U S., Yeakley, L M., 1968. High strain-rate testing: tension and compression, Exp. Mech. 8 (1), 1—9, Fig. 10, p.8. Reprinted with permission of the publisher.*

and the logarithm law

$$\frac{\sigma}{\sigma_0} = 1 + \lambda \ln\left(\frac{\dot{\varepsilon}}{\dot{\varepsilon}_0}\right) \tag{5.2}$$

where σ_0 is flow stress under the quasistatic test ($\dot{\varepsilon}=\dot{\varepsilon}_0$). Obviously n and λ

$$n = \frac{\partial \ln \sigma}{\partial \ln \dot{\varepsilon}} \tag{5.3}$$

$$\lambda = \frac{\partial \sigma}{\partial \ln \dot{\varepsilon}} \tag{5.4}$$

represent the strain-rate sensitivity coefficient for the power function law and the logarithm law, respectively. Eqs. (5.1a) and (5.2) are linear in double

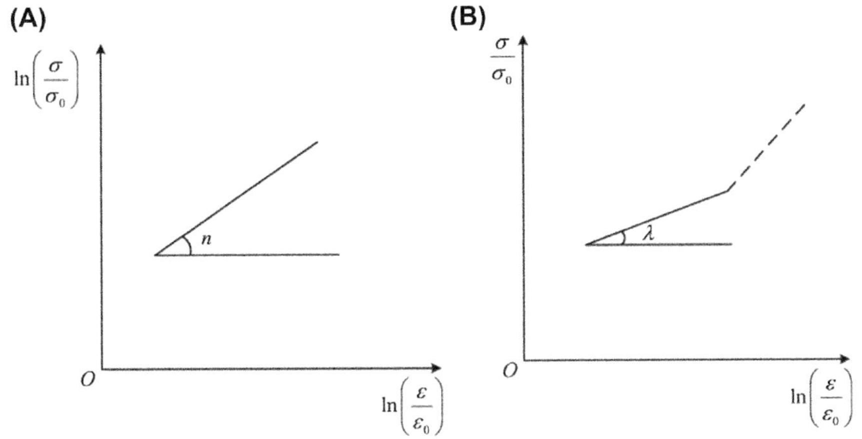

Figure 5.6 Two types of strain-rate correlation: (A) power function law and (B) logarithm law.

logarithmic and semilogarithmic coordinates, respectively (Fig. 5.6). The latter indicates that the flow stress will change significantly only when there is a strain rate change in the order of magnitude.

Note that $(\sigma - \sigma_0)$ is the difference between dynamic stress and static stress and is often referred to as **overstress** or **extra stress**. $(\sigma - \sigma_0)/\sigma_0$ is nondimensionalized overstress. Thus, Eqs. (5.1b) and (5.2) indicate that overstress is a function of strain rate. The results of this experimental study can be traced back to the work of Lüdwik (1909).

In fact, Eqs. (5.1) and (5.2) are applicable to different materials and different strain-rate ranges, respectively.

For example, according to the data of stress—strain curves provided by Lindholm and Yeakley (1968) for industrial pure aluminum (1100-0 Al) under different uniaxial compressive strain rates (Fig. 5.3), if replot the same data in the $\sigma - \log\dot{\varepsilon}$ coordinates for different given strain, there appear two straight lines with different slopes in different strain-rate ranges (Fig. 5.7). This behavior indicates that the strain-rate sensitivity coefficient λ in high strain-rate range (10^2—10^3 s^{-1}) is higher than that in quasistatic low strain-rate range (10^{-3}—10^1 s^{-1}).

Follansbee and Kocks (1988) experimentally studied high-purity oxygen-free copper (OFHC) and found that when strain $\varepsilon = 0.15$, the relationship between flow stress and logarithm of the strain rate is as shown in Fig. 5.8. As can be seen, it satisfies the linear relation in semilogarithmic coordinates (Eq. 5.2) in the wide strain-rate range of $10^{-4} - 10^3$ s^{-1}.

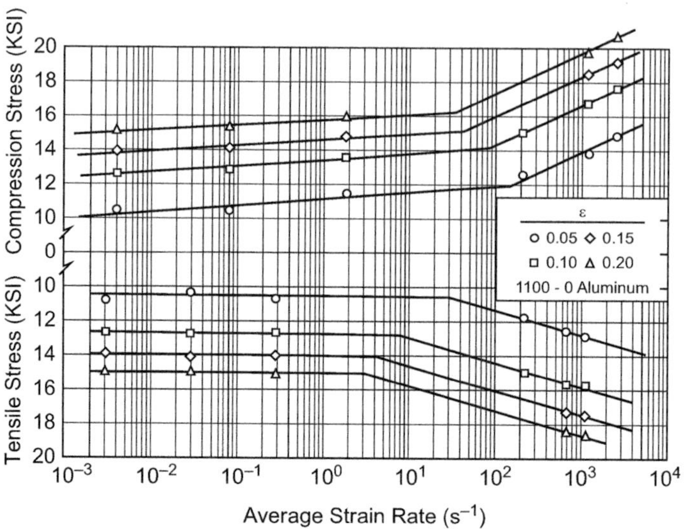

Figure 5.7 $\sigma - \log\dot{\varepsilon}$ curve of industrial pure aluminum (1100-0 Al) under different strains. *From Lindholm, U S., Yeakley, L M., 1968. High strain-rate testing: tension and compression, Exp. Mech. 8 (1), 1–9, Fig. 9, p.8. Reprinted with permission of the publisher.*

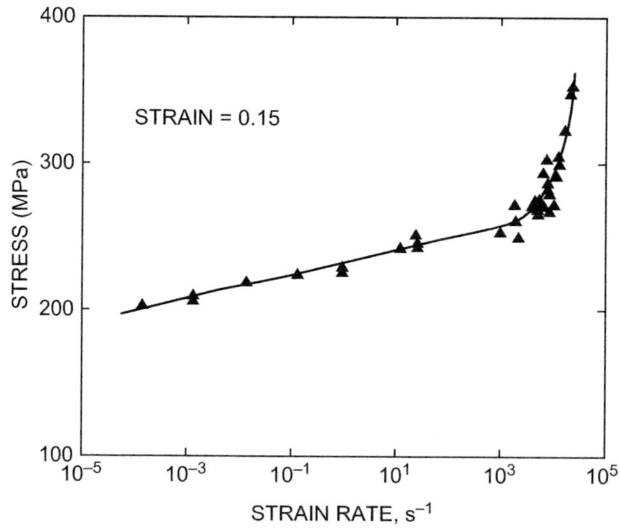

Figure 5.8 $\sigma - \log\dot{\varepsilon}$ curve for oxygen-free copper at $\varepsilon = 0.15$. *From Follansbee and Kocks, 1988, Fig. 1, p.82. Reprinted with permission of the publisher.*

However, for the vast majority of metallic materials, the $\sigma - \log\dot{\varepsilon}$ curve changes sharply when the strain rate $\dot{\varepsilon} \geq 10^4\ s^{-1}$ and flow stress increase sharply, as shown in Fig. 5.8. The inflection of this curve indicates that the plastic flow mechanism of the material has essentially changed. It is generally believed that the plastic flow mechanism of the material has changed substantially; the thermal activation mechanism of dislocation movement, which will be discussed further in the next chapter, has been replaced by a new linear viscosity mechanism to control plastic flow, as shown in Fig. 5.9 (Campbell and Ferguson, 1970). σ is linearly related with $\dot{\varepsilon}$ itself:

$$\frac{\partial \sigma}{\partial \dot{\varepsilon}} = \eta \tag{5.5}$$

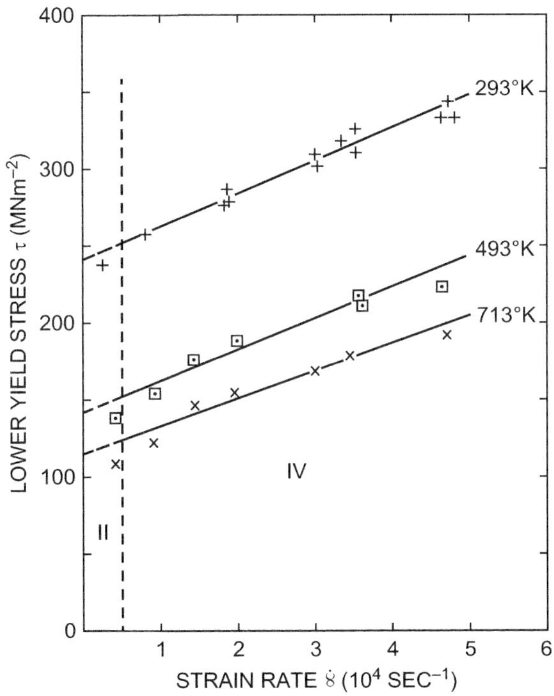

Figure 5.9 Strain-rate sensitivity of mild steel under a very high strain rate. *From Campbell, J D., Ferguson, W G., 1970. The temperature and strain rate dependence of shear strength of mild steel. Philos. Mag. 21 63–82, Fig. 11, p.77. Reprinted with permission of the publisher.*

Table 5.2 Measured values ($kNsm^{-2}$) of η for several metallic materials (Lindholm, 1974).

Material	Aluminum	Copper	Zinc	Brass	Soft steel
Polycrystalline	2.1[b]	3.6[b]	–	5.5[b]	2.1[a]
	1.4				2.8[b]
Single crystal	1.2[c]	10.8[e]	0.5[d]		

[a](Campbell and Ferguson, 1970).
[b](Dowling et al., 1970).
[c](Ferguson et al., 1967b).
[d](Ferguson et al., 1967a).
[e](Greenman et al., 1967).

where the constant η is the viscosity coefficient. The measured values of several metallic materials (unit $kNsm^{-2}$) are shown in Table 5.2 (Lindholm, 1974).

The effect of strain rate on the strength limit σ_b is basically similar to that of strain rate on the flow stress; that is, the strength limit as the critical point of flow stress increases with the increase of strain rate and presents the delayed fracture phenomenon. At the same time, the material failure forms change, and most materials will experience **impact embrittlement**, but also have abnormal **impact toughening**. The former refers to the transformation of material failure from ductile rupture to brittle fracture, showing that the elongation decreases with the increase of strain rate. Molybdenum is taken as an example to illustrate the impact embrittlement; in the static tensile test, the elongation of molybdenum is $\delta = 40\%$. When the strain rate is equal to $2\,s^{-1}$, the elongation δ of molybdenum tends to 0 (Fig. 5.10) (Lindholm, 1974). By contrast, for the alloy of molybdenum and the rare earth element rhenium, 50Mo–50Re, its elongation δ and strength limit simultaneously increase with strain rate (Fig. 5.11) (Xu et al., 2008).

The Part 3 below will provide an in-depth discussion of the dynamic failure of materials, which will not be discussed in detail here.

5.1.2 Combined effects of strain rate and temperature and rate–temperature equivalence

The temperature-softening effect of materials, namely the flow stress decreases as temperature increases, is well known before the strain-rate effect. Here, this phenomenon is illustrated by using the experimental results for the quasistatic yield stress of iron varying with temperature as presented

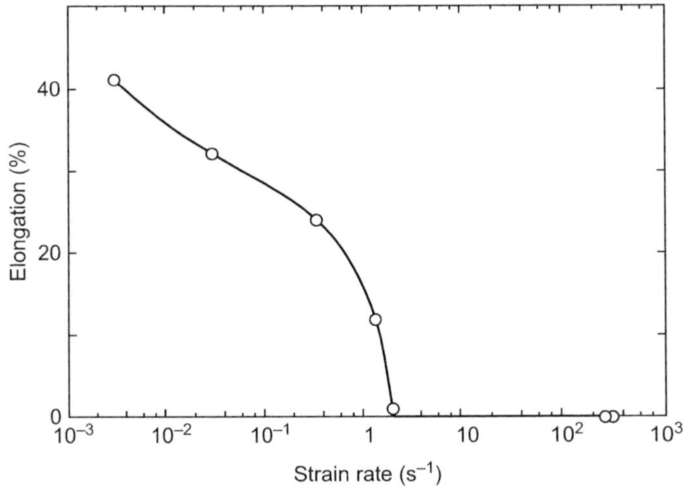

Figure 5.10 Relationship between the tensile elongation and strain rate of molybdenum. *From Lindholm, 1974, Fig. 11, p.16. Reprinted with permission of the publisher.*

in Fig. 5.12 (Vohringer, 1989); the fitting curve can be expressed by the following formula:

$$\sigma = \sigma_r \left(1 - \left(\frac{T - T_r}{T_m - T_r} \right)^m \right) \tag{5.6}$$

where T_m is the melting point temperature, T_r is the reference temperature, σ_r is the yield stress at the reference temperature, and m characterizes the temperature-softening characteristics.

The combined effect of strain rate and temperature under impact loading has received increased attention. Lindholm's results for aluminum under different strain rates and temperatures for different given strain are shown in Fig. 5.13 (Lindholm, 1968). These results show that the linear relationship in $\sigma - \log\dot{\varepsilon}$ coordinates is well satisfied at different given temperatures (Fig. 5.2).

The experimental results of Campbell and Ferguson (1970) for the low yield stress of mild steel under different strain rates and temperatures are shown in Fig. 5.14. They found that the yield stress of mild steel under the different combined effects of strain rate—temperature can be partitioned in accordance with different strain-rate sensitivities (different viscoplastic mechanisms), where zone I is a low strain rate—high temperature zone that is less sensitive to strain rate; zone II is a high strain rate—low temperature zone where the strain-rate sensitivity coefficient λ obeys the linear

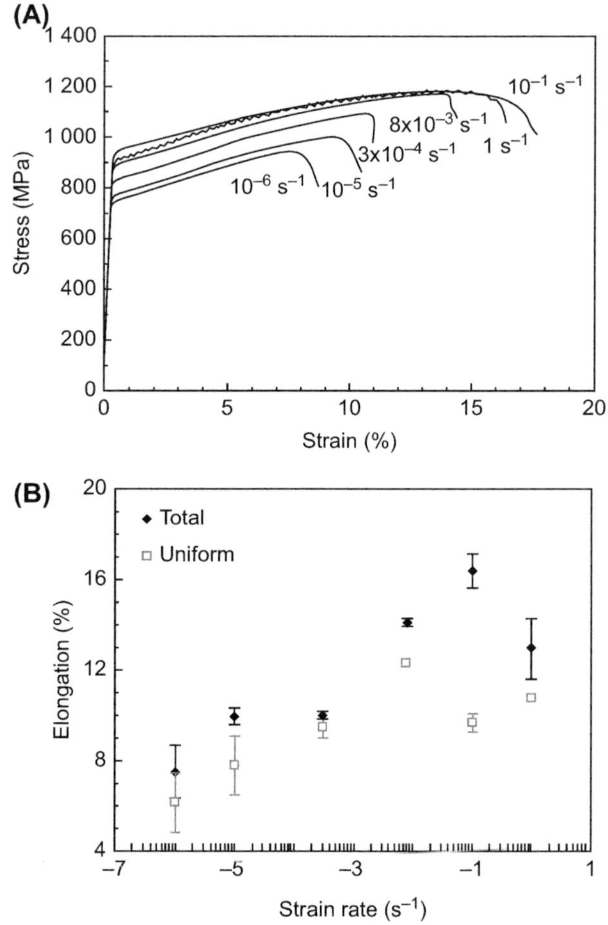

Figure 5.11 (A) Stress–strain curve and (B) tensile elongation of 50Mo–50Re molybdenum–rhenium alloys under different strain rates at room temperature. *From Xu, J., Leonhardt, T., Farrell, J., Effgen, M., Zhai, T., 2008. Anomalous strain-rate effect on plasticity of a Mo-Re alloy at room temperature. Mater. Sci. Eng. A479 76–82, Fig. 2 on p.77 and Fig. 4 on p.78. Reprinted with permission of the publisher.*

relationship (Eq. 5.2) in $\sigma - \log\dot{\varepsilon}$ coordinates; and zone III is the region of high strain rate where strain-rate sensitivity rises rapidly and the strain-rate sensitivity coefficient obeys the linear relationship (Eq. 5.5) in $\sigma - \dot{\varepsilon}$ coordinates.

A large number of experimental observation shows that, in general, the strain-rate effect and temperature effect on the mechanical properties of materials have very similar or closely related intrinsic relationships, manifesting

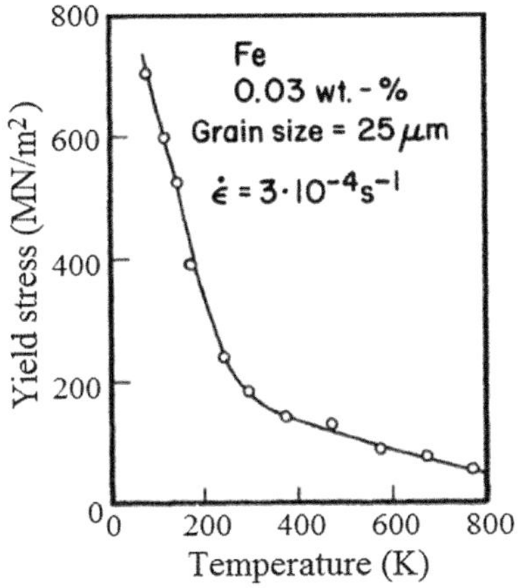

Figure 5.12 Changes in the quasistatic yield stress of iron with temperature. *From Vohringer, O., in Deformation Behavior of Metallic Materials, ed. C.Y. Chiem, International Summer School on Dynamic Behavior of Materials, ENSM. Nantes Sept. 11–15, 1989, p.7.*

as that the consequences of decreasing environmental temperature (low-temperature hardening/embrittlement) are often equivalent to the consequences of increasing strain rate (strain-rate hardening/embrittlement). This equivalent phenomenon is called **strain rate–temperature equivalence** or **rate–temperature equivalence.** Given that strain rate and time t are reciprocal, this equivalence can also manifest as **time–temperature equivalence.** It can be described by introducing a strain rate $\dot{\varepsilon}$ and temperature T-combined parameters, for example T^* as suggested by Lindholm (1974).

$$T^* = T\ln\left(\frac{\dot{\varepsilon}_0}{\dot{\varepsilon}_p}\right) \approx T\ln\left(\frac{\dot{\varepsilon}_0}{\dot{\varepsilon}}\right) \tag{5.7}$$

where $\dot{\varepsilon}_p$ is the plastic strain rate (which is simplified to the total strain rate $\dot{\varepsilon}$ when small elastic deformation is negligible) and $\dot{\varepsilon}_0$ is the reference strain rate. Other researchers have proposed other similar parameters, such as the **velocity-modified temperature** proposed by McGregor and Fisher (1946), $T_V = T\left(1 - A\ln\frac{\dot{\varepsilon}_p}{\dot{\varepsilon}_0}\right)$. These rate–temperature equivalent parameters are equivalent to each other and can be traced back to the Zener–Holloman parameter $Z = \dot{\varepsilon}\exp\left(\frac{U}{kT}\right)$ proposed by Zener and Hollomon (1944) which is based on dislocation dynamics (to be discussed in

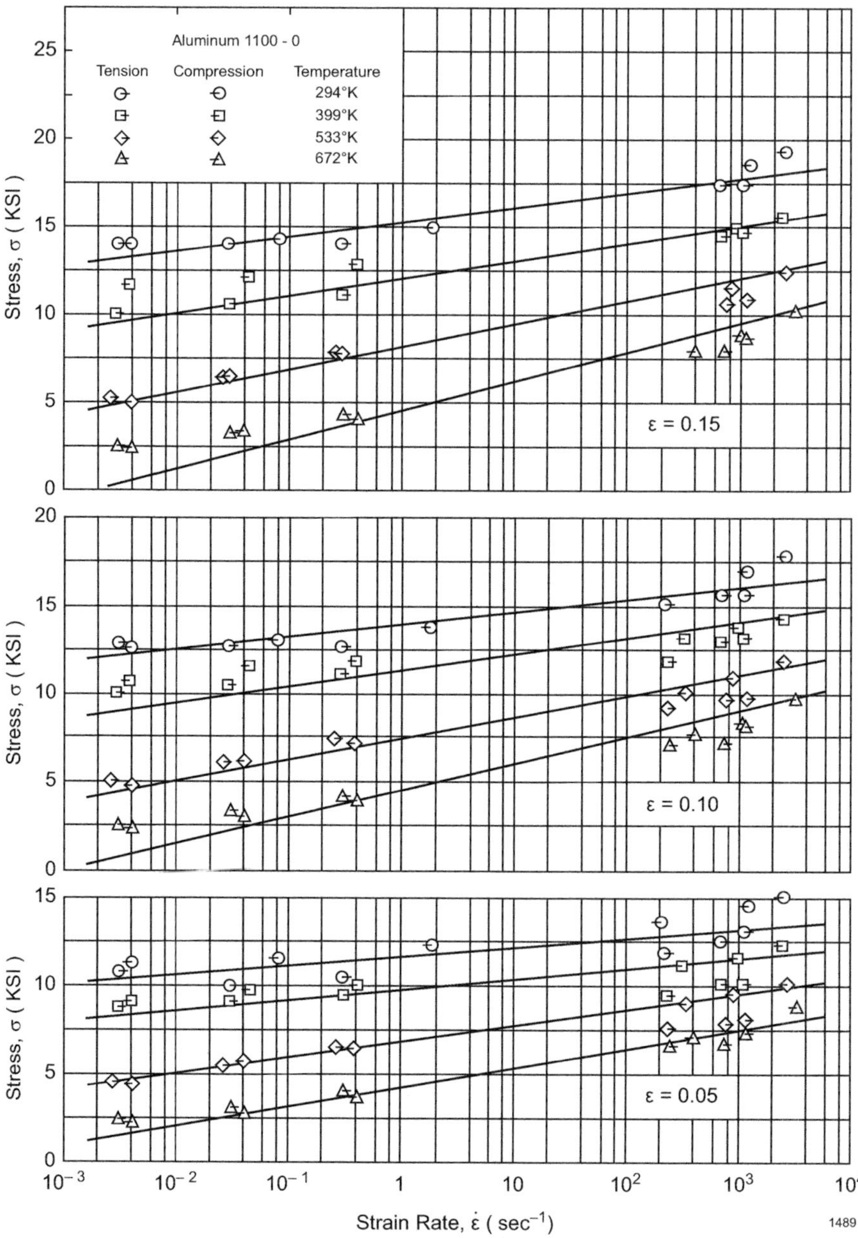

Figure 5.13 Changes in aluminum flow stress ($\varepsilon = 0.15$) with strain rate and temperature. *From Lindholm, U S., 1968. Some experiments in dynamic plasticity under combined stress. In Mechanical Behavior of Materials under Dynamic Loads, Springer Verlag, New York, 77–95., Fig. 5 on p.88. Reprinted with permission of the publisher.*

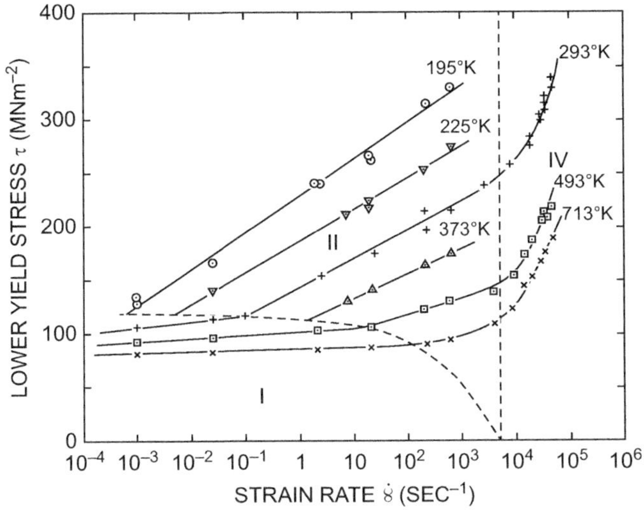

Figure 5.14 Change in the yield stress of mild steel with strain rate and temperature. *From Campbell, J D., Ferguson, W G., 1970. The temperature and strain rate dependence of shear strength of mild steel, Philos. Mag. 21 63–82, Fig. 5, p.68. Reprinted with permission of the publisher.*

the next chapter) and all of those can be referred to as Zener–Holloman type rate–temperature equivalent parameter.

After introducing the rate–temperature equivalent parameter T^* shown in Eq. (5.7), the distortion law that simultaneously accounts for the strain-rate effect and temperature effect can be normalized as illustrated in Fig. 5.15 (Lindholm, 1974). That is, as T^* decreases (meaning that the temperature decreases or the strain rate increases), the stress–strain curve moves toward the direction of stress increase (reflecting the hardening effect) and transforms from tough (plastic) rupture to brittle fracture.

Fig. 5.15 divides the rate–temperature dependent constitutive distortion properties of the material into three regions: the **elastic/viscoelastic (E) region**, the **stable plastic (SP) region, and the unstable plastic (UP) region.** The E region develops before plastic yielding. Although the elastic response in this region is generally strain rate-dependent (viscoelastic), it is often assumed to be the ideal elasticity that is independent of strain rate and is bounded on the left side by the limit of transient response such that strain rate approaches infinity. The SP region characterizes strain hardening $\left(\frac{d\sigma}{d\varepsilon} > 0\right)$. The UP region characterizes strain softening

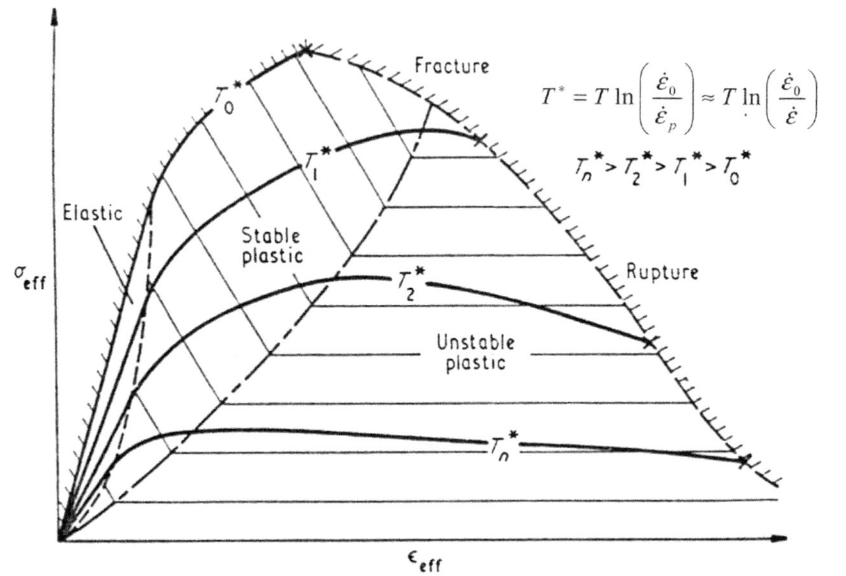

Figure 5.15 Schematic of the thermoviscoplastic constitutive behavior of metallic materials accounting for rate–temperature equivalence. *From Lindholm, 1974, Fig. 1, p.4. Reprinted with permission of the publisher.*

$\left(\frac{d\sigma}{d\varepsilon} < 0\right)$ and is bounded on the right side by the failure limit. The boundary that bounds these three regions is determined by the corresponding critical yield criterion, constitutive instability criterion $\left(\frac{d\sigma}{d\varepsilon} = 0\right)$, and failure criterion, respectively. As can be seen from Fig. 5.15 that these criteria are generally the functions of strain rate and temperature. In other words, when accounting for the rate–temperature effect, the dynamic yield criterion, dynamic constitutive instability criterion, and dynamic failure criterion cannot be described by the "univariate criterion" such as the criterion of critical stress type or the criterion of critical strain type, commonly used in quasistatic mechanics, and must be replaced with the corresponding "multivariate criteria".

With regard to dynamic failure, Fig. 5.15 illustrates the phenomenon of impact embrittlement and low-temperature embrittlement, that is, as strain rate increases or temperature decreases, the failure mode will transform from ductile failure (indicated by the term **rupture** in the figure and generally referring to failure after plastic deformation) to brittle failure (indicated by the term **fracture** in the figure and generally referring to fracture without plastic deformation). In correspondence to the concept of toughening–embrittlement transition temperature, an equivalent toughening–embrittlement transition strain rate that corresponds to the boundary point between rupture and fracture exists on the failure limit line on the right side of the figure.

In order to generalize the rate–temperature-related dynamic distortion relation under one-dimensional stress to the general three-dimensional stress state, similar to the von Mises yield condition in classical plasticity, Fig. 5.15 introduces the effective stress σ_{eff} based on the second invariant J_2 ($=S_{ij}S_{ij}/2$) of the deviatoric tensor of stress and the corresponding effective strain ε_{eff} based on the second invariant K_2 ($=e_{ij}e_{ij}/2$) of the deviatoric tensor of strain and the effective strain rate $\dot{\varepsilon}_{eff}$ based on strain-rate deviator D_2 ($=\dot{e}_{ij}\dot{e}_{ij}/2$) of the second invariant.

$$\sigma_{eff} = \frac{\sqrt{2}}{2}\left[(\sigma_1 - \sigma_2)^2 + (\sigma_2 - \sigma_3)^2 + (\sigma_3 - \sigma_1)^2\right]^{1/2} \tag{5.8a}$$

$$\varepsilon_{eff} = \frac{\sqrt{2}}{3}\left[(\varepsilon_1 - \varepsilon_2)^2 + (\varepsilon_2 - \varepsilon_3)^2 + (\varepsilon_3 - \varepsilon_1)^2\right]^{1/2} \tag{5.8b}$$

$$\dot{\varepsilon}_{eff} = \frac{\sqrt{2}}{3}\left[(\dot{\varepsilon}_1 - \dot{\varepsilon}_2)^2 + (\dot{\varepsilon}_2 - \dot{\varepsilon}_3)^2 + (\dot{\varepsilon}_3 - \dot{\varepsilon}_1)^2\right]^{1/2} \tag{5.8c}$$

In this case, the $\dot{\varepsilon}_p$ or $\dot{\varepsilon}$ in the rate–temperature equivalent parameter T^* (Eq. 5.7) should be understood as the effective strain rate defined in Eq. (5.8c).

Whether the effective stress σ_{eff} and effective strain ε_{eff} introduced by the von Mises yield condition in quasistatic classical plastic mechanics can be generalized to high strain-rate conditions is an issue of widespread concern and requires further in-depth study. This assumption is supported by relevant experimental results for homogeneously isotropic materials. For example, Randall and Campbell (1974) carried out experiments on thin-walled tubes for EN2E steel subjected to pull-twisted composite loads at three different strain rates. The results of the low yield stress are shown in Fig. 5.16, indicating that the Mises criterion continued to hold with the increase of strain rates (Randall and Campbell, 1974).

As another example, the experimental results for the yield stress $\sigma_{1\%}$ by using the tension–torsion composite Hopkinson bar under a wide strain-rate range for the German mold steel 40CrMnMo7 are shown in Fig. 5.17 (Meyer, 2004), where $\sigma_{1\%}$ are the yield stresses defined with the residual plastic strain of 1%. This figure also shows that the Mises criterion continues to hold as the strain rate increases.

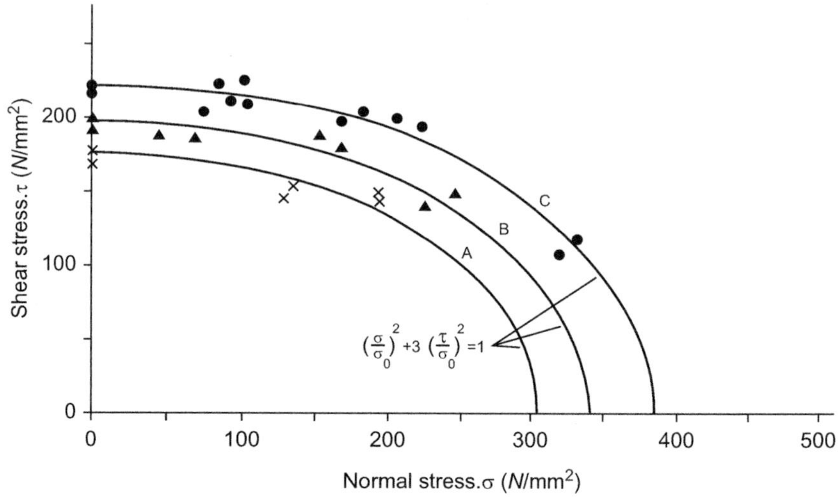

Figure 5.16 Effect of strain rate on the yield stress of EN2E steel. *From Lindholm, 1974, Fig. 4, p.11. Reprinted with permission of the publisher.*

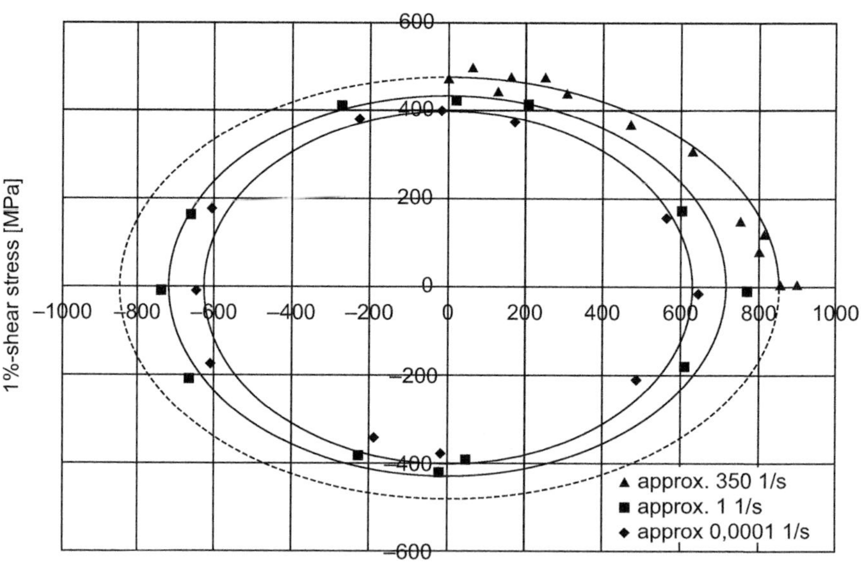

Figure 5.17 Effect of strain rate $\sigma_{1\%}$ on yield stress of 40CrMnMo7 steel. *From Meyer, L W., 2004. Material behaviour at high strain rates, Proc. 1st Int. Conf. on High Speed Forming- CHSF. Dortmund, 45–56, Fig. 15, p.54. Reprinted with permission of the publisher.*

5.1.3 Effect of strain-rate history

In addition to the strain-rate effect, the dynamic mechanical response of materials generally depends on the history of strain rate. This dependence is called the effect of strain-rate history and is illustrated in Fig. 5.18, where the *AD* and *BC* curves correspond to the stress–strain curves at low strain rate $\dot{\varepsilon}_1$ and high strain rate $\dot{\varepsilon}_2(>\dot{\varepsilon}_1)$, respectively.

As shown in Fig. 5.18A, if the material is insensitive to strain-rate history and only the strain-rate effect exists, then when the material, at a given strain ε_T, changes from the strain-rate $\dot{\varepsilon}_1$ state at the point *A* on the curve *AD* to the strain rate $\dot{\varepsilon}_2$ ($>\dot{\varepsilon}_1$) at the point *B* on the curve *BC*, the stress–strain relation of the material will immediately shift to the new $\dot{\varepsilon}_2$ curve *BC*, that is, it will jump from *A* point to *B* point, and the corresponding stress increment is $\varDelta\sigma = \sigma_B - \sigma_A$. However, if the material is sensitive to strain-rate history, as shown in Fig. 5.18B, the original strain rate $\dot{\varepsilon}_1$ retains its historical influence, and the stress–strain relationship of the material will gradually approach the stress–strain curve at the new strain rate $\dot{\varepsilon}_2$. That is, it will jump from point *A* to point *B*', and the corresponding stress increment is $\Delta\sigma_s = \sigma_{B'} - \sigma_A$. The other part of the stress increment $\Delta\sigma_h = \sigma_B - \sigma_{B'}$ is gradually obtained after this jump. $\Delta\sigma_h$ is regarded as a measure of the effect of strain-rate history and characterizes the strain-rate sensitivity of strain hardening that corresponds to microscopic structural states (Klepaczko, 1975). This kind of "memory" of the material with regard to the original strain-rate history will gradually fade over time, is known as the fading memory effect, and is vividly illustrated by the experimental results for aluminum, as shown in

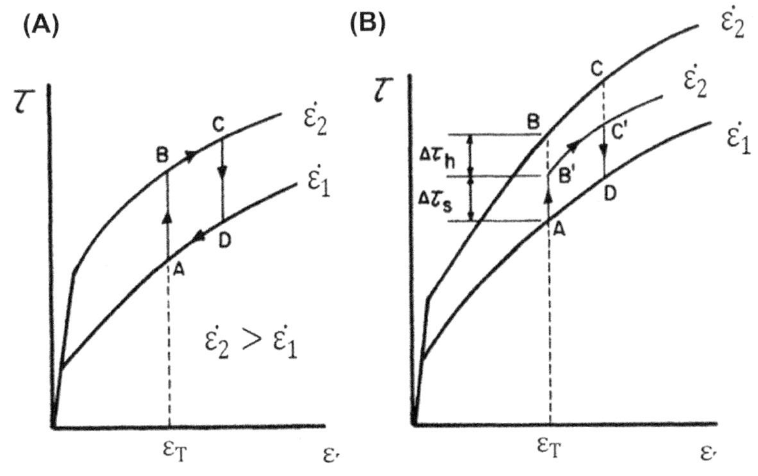

Figure 5.18 Schematic of (A) insensitive and (B) sensitive to the strain-rate history.

Fig. 5.19 (Klepaczko, 1968). Fig. 5.19A shows experimental results obtained under high strain rates after deformation at low strain rates. Fig. 5.19B shows experimental results obtained under low strain rates after deformation at high strain rates.

Some typical results of strain-rate jump experiments performed with a split Hopkinson torsion bar further indicate that the strain-rate history effect varies from material to material (Eleiche and Campbell, 1976). Copper is sensitive to strain-rate history, as shown in Fig. 5.20A. By contrast, titanium

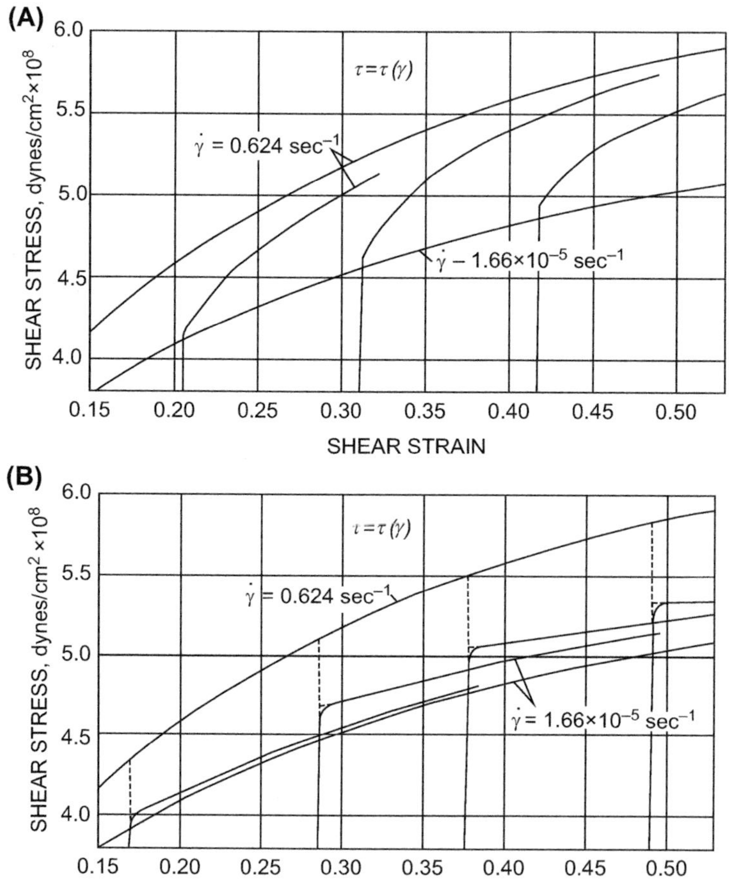

Figure 5.19 Sensitivity of aluminum to strain-rate history. *From Klepaczko, J., 1968. Strain rate history effects for polycrystalline aluminum and theory of intersections. J. Mech. Phys. Solids. 16 255–266, Fig. 2 on p.257 and Fig. 3 on p.258. Reprinted with permission of the publisher.*

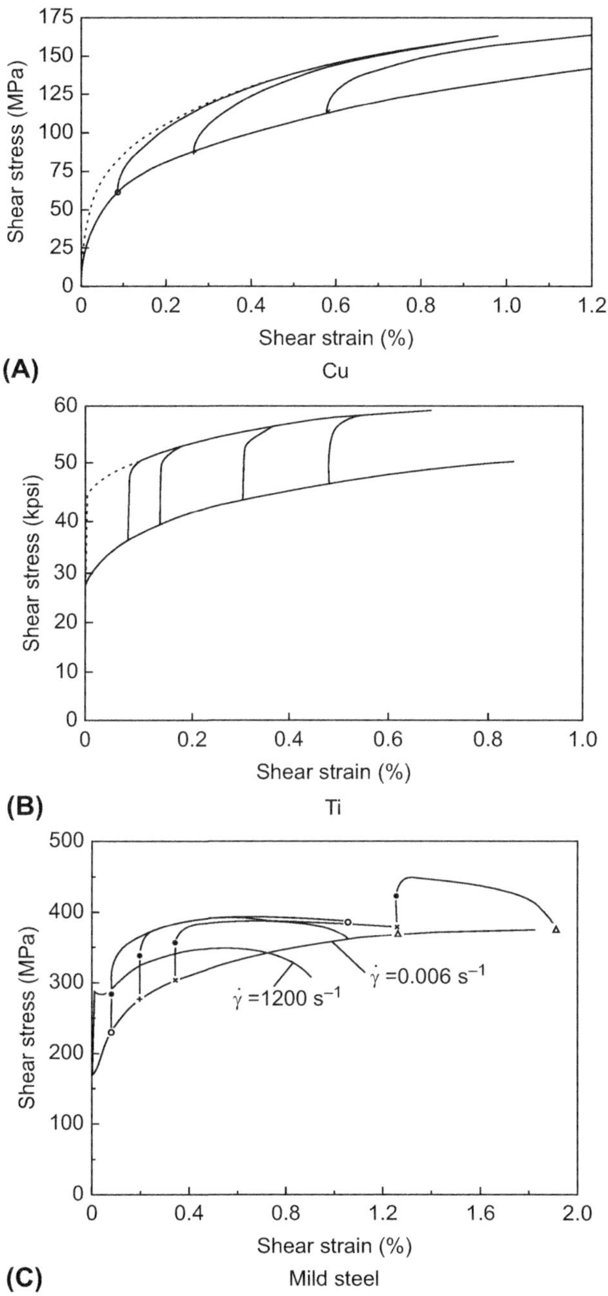

Figure 5.20 Sensitivity of different materials to strain-rate history. *From Eleiche., A M, Campbell., J D, 1976. The Influence of Strain-Rate History and Temperature on the Shear Strength of Copper, Titanium and Mild Steel. (Tech. Rep. AFML-TR-76-90). Air Force Materials Laboratory, Wright-Patterson Air Force Base, Ohio, USA.*

is not sensitive to strain-rate history, as shown in Fig. 5.20B, although is highly sensitive to strain rate. Mild steel presents the special phenomenon of overshooting when jumping from the low strain-rate curve to the high strain rate. This phenomenon is illustrated in Fig. 5.20C. When the strain rate jumps from low strain rate to high strain rate, the stress overshoots the stress that corresponds to the high strain-rate curve and then gradually approaches the high strain-rate curve.

Once the effect of strain-rate history is considered, the description of the material dynamic constitutive distortion law becomes increasingly complex. This complexity is a problem that needs further study (see Eq. (6.49) and related discussion in Chapter 6). At present, as will be described in the following section, most of dynamic constitutive material distortion laws used in practical engineering in fact ignore the effect of strain-rate history.

5.2 Viscoplastic constitutive equations (phenomenological models)

The strain-rate effect is usually ignored in quasistatic solid mechanics. Therefore, taking the one-dimensional form as an example, the constitutive equation can be generally expressed as:

$$f(\sigma, \varepsilon) = 0 \tag{5.9}$$

In material dynamics, when considering strain-rate effects, the above formula should be rewritten as:

$$f(\sigma, \varepsilon, \dot{\varepsilon}) = 0 \tag{5.10a}$$

From the perspective of macromechanics, the role of strain rate in rate-dependent plastic rheological behavior is similar to the viscosity of viscous fluids. This type of equation is commonly referred to as the viscoplastic constitutive equation of materials. If the temperature effect is further considered, then

$$f(\sigma, \varepsilon, \dot{\varepsilon}, T) = 0 \tag{5.10b}$$

This equation is often called the thermoviscoplastic constitutive equation. Or, by introducing a rate—temperature equivalent parameter, such as T^* defined by Eq. (5.7), then

$$f(\sigma, \varepsilon, T^*) = 0 \tag{5.10c}$$

The next question comes down to how to determine the specific functional form. This section mainly provides a discussion from the perspective of macromechanics, including empirical formulas, such as the Cowper–Symonds (C–S) equation and the Johnson–Cook (J–C) equation; the Sokolovsky–Malvern–Perzyna (SMP) model, which is based on the initial yield surface and overstress concepts; and the Bodner–Parton (B–P) model, which is based on the concept of no yielding surface. All of those are referred to as the phenomenological viscoplastic constitutive equation. In addition, some constitutive equations derived from the perspective of material microstructure and dislocation dynamics will be described in the next chapter.

5.2.1 Cowper–Symonds equation

Cowper and Symonds (1957) proposed the following rate-dependent constitutive equation based on a large amount of experimental data for the low yield stress of metallic materials under different strain rates.

$$\dot{\varepsilon} = D\left(\frac{\sigma_0'}{\sigma_0} - 1\right)^q \tag{5.11a}$$

The above equation can be rewritten as:

$$\frac{\sigma_0'}{\sigma_0} = 1 + \left(\frac{\dot{\varepsilon}}{D}\right)^{1/q} \tag{5.11b}$$

where σ_0 is quasistatic flow stress, σ_0' is flow stress when the strain rate is $\dot{\varepsilon}$, and D and q are material constants. In contrast to Eq. (5.1b) in the previous section, D and q correspond to $\dot{\varepsilon}_0$ and $\frac{1}{n}$ in Eq. (5.1b), respectively, and the physical meaning of n is the strain-rate sensitivity coefficient defined by Eq. (5.3). Table 5.3 lists the D and q experimental values of several metallic materials:

For mild steel, when $D = 40.4$ and $q = 5$, the comparison between Eq. (5.11) and the experimental data collected by Symonds over 30 years is shown in Fig. 5.21 (Symonds, 1967). Considering that the microscopic grain sizes and heat treatment conditions of mild steel studied in different laboratories are different, and the test systems are different, the data in the figure are

Table 5.3 Experimental values of D and q for several metallic materials.

Material	D (s^{-1})	q
Soft steel (Cowper and Symonds, 1957)	40.4	5
Aluminum alloy (Bodner and Symonds, 1962)	6500	4
α-titanium (Symonds and chon, 1974)	120	9
304 stainless steel (Forrestal and Sagartz, 1978)	100	10

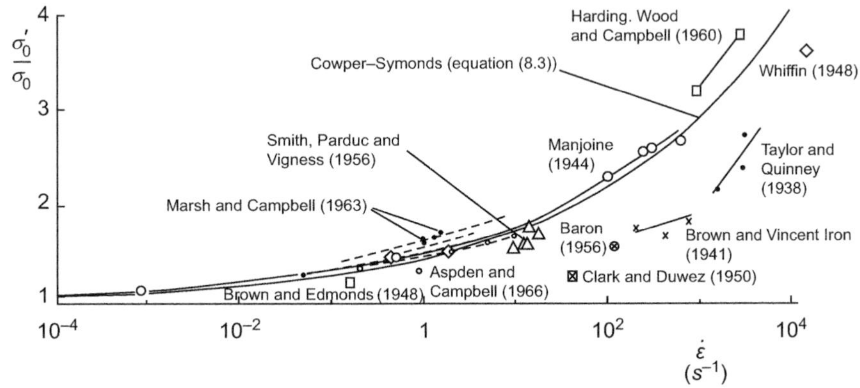

Figure 5.21 Comparison of the results obtained with the Cowper–Symonds equation and the experimental results of lower yield stress under different strain rates for mild steel (Symonds, 1967).

inevitably some scattered. Even so, the Cowper–Symonds equation is supported by so many experimental data in general, which is one of the important reasons why it is still widely adopted in engineering applications.

If $\dot{\varepsilon} = D(= \dot{\varepsilon}_0)$, then we have $\sigma_0' = 2\sigma_0$ regardless the value of q and strain, which is equivalent to the dynamic rigid-plastic model.

In the original form of the C–S formula (Eq. 5.11), the strain-hardening effect of the material is in fact not reflected. To account for strain-hardening effects, Eq. (5.11) can be rewritten as (Malvern, 1951):

$$\dot{\varepsilon} = D\left(\frac{\sigma_0' - \sigma(\varepsilon)}{\sigma_0}\right)^q \tag{5.11c}$$

or as (Symonds, 1965)

$$\dot{\varepsilon} = D\left(\frac{\sigma_0'}{\sigma(\varepsilon)} - 1\right)^q \tag{5.11d}$$

where $\sigma(\varepsilon)$ is a quasistatic uniaxial stress–strain curve taking account of strain hardening, and $(\sigma_0' - \sigma(\varepsilon))$ is the overstress that was emphasized in the discussion of (5.1b) and (5.2) in the previous section. In such case, the C–S equation can be also classified as an empirical equation on the basis of overstress.

The original form of the C–S equation (Eq. 5.11) is based on the experimental results in one-dimensional stress state. To generalize the equation to the general three-dimensional stress state, one-dimensional stress, strain, and strain rate in the equation need only be replaced by the effective stress, effective strain, and effective strain rate defined in Eq. (5.8).

The corresponding viscoplastic constitutive equation in tensor form can be easily written on the basis of Eq. (5.11). In fact, according to the Levy–Mises incremental theory in classical plastic mechanics (Wang Ren et al., 1992), the increment of the plastic deviatoric strain tensor de_{ij}^p is proportional to the deviatoric stress tensor S_{ij}.

$$de_{ij}^p = d\lambda S_{ij} \tag{5.12a}$$

where $d\lambda$ is a nonnegative scale factor. Square both sides of the above equation after calculation; note the definitions of effective stress σ_{eff} (Eq. 5.8a) and the effective plastic deviatoric tensor of strain de_{eff}^p (change $\dot{\boldsymbol{\varepsilon}}$ in Eq. (5.8c) to de^p), then $d\lambda$ can be determined as:

$$d\lambda = \frac{3}{2}\frac{de_{eff}^p}{\sigma_{eff}}$$

Substituting the above equation into Eq. (5.12a) yields

$$de_{ij}^p = \frac{3}{2}\frac{de_{eff}^p S_{ij}}{\sigma_{eff}} \tag{5.12b}$$

When volumetric plastic deformation is negligible, distinguishing between the plastic deviatoric tensor of strain rate $\dot{e}_{ij}^p$ and the plastic full tensor of strain rate $\dot{\boldsymbol{\varepsilon}}_{ij}^p$ is unnecessary. Thus, the above equation in the incremental form can be rewritten as the following equation in terms of rate:

$$\begin{aligned}
\dot{\boldsymbol{\varepsilon}}_{ij}^p &= \left(\frac{D_2^p}{J_2}\right)^{1/2} S_{ij} = \frac{3}{2}\frac{\dot{\boldsymbol{\varepsilon}}_{eff}^p}{\sigma_{eff}} S_{ij} \\
D_2^p &= \frac{1}{2}\dot{\boldsymbol{\varepsilon}}_{ij}^p\dot{\boldsymbol{\varepsilon}}_{ij}^p = \frac{3}{4}\left(\dot{\boldsymbol{\varepsilon}}_{eff}^p\right)^2 = \frac{1}{6}\left[\left(\dot{\boldsymbol{\varepsilon}}_1^p - \dot{\boldsymbol{\varepsilon}}_2^p\right)^2 + \left(\dot{\boldsymbol{\varepsilon}}_2^p - \dot{\boldsymbol{\varepsilon}}_3^p\right)^2 + \left(\dot{\boldsymbol{\varepsilon}}_3^p - \dot{\boldsymbol{\varepsilon}}_1^p\right)^2\right] \\
J_2 &= \frac{1}{2}S_{ij}S_{ij} = \frac{1}{3}(\sigma_{eff})^2 = \frac{1}{6}\left[(\sigma_1 - \sigma_2)^2 + (\sigma_2 - \sigma_3)^2 + (\sigma_3 - \sigma_1)^2\right]
\end{aligned} \tag{5.12c}$$

Expressing the C–S equation (Eq. 5.11a) in terms of the effective stress, effective strain, and effective strain rate, then substituting it into the above equation and ignoring the elastic strain rate yields

$$\dot{\boldsymbol{\varepsilon}}_{ij} = \frac{3}{2}\frac{D}{\sigma_{eff}}\left(\frac{\sigma_{eff}}{\sigma_0} - 1\right)^q S_{ij} \tag{5.12d}$$

This is the tensor form of the strain rate-dependent viscoplastic constitutive equation based on the C–S equation. Note that Eqs. (5.12a) – (5.12c) are not dependent on whether a yield surface or static stress—strain curve exists. By contrast, Eq. (5.12d) incorporated with the C—S equation takes that there exists a yield surface or static stress—strain curve as the premise and is thus considered as the overstress-type viscoplastic constitutive equation.

The C—S equation has been widely used in the structural impact mechanics of metallic structures (Jones, 2011), including car collisions, ship collisions, ship—bridge collisions (Wang et al., 2008; Chen et al., 2013), and other practical engineering.

5.2.2 Johnson—Cook (J—C) equation

It is known that the quasistatic stress—strain relationship of the material taking account of strain hardening is often described by the exponential hardening term as below:

$$\sigma = \sigma_0 + B\varepsilon^n \tag{5.13}$$

where σ_0 is the yield stress, both the coefficient B in front of the exponential term and the strain-hardening coefficient n characterize strain hardening. It is also known that the dependence of flow stress on temperature T can be described by Eq. (5.6) as described above. It is moreover known that the dependency of flow stress on strain rate$\dot{\varepsilon}$, when strain rate is not too high, can be described by Eq. (5.2). On the basis of the above comprehensive considerations, Johnson and Cook (1983) combined the above factors and proposed the following equation to describe the dynamic constitutive relationship of materials (Johnson et al., 1983):

$$\sigma = (\sigma_0 + B\varepsilon^n)\left(1 + C\ln\frac{\dot{\varepsilon}}{\dot{\varepsilon}_0}\right)\left(1 - \left(\frac{T - T_r}{T_m - T_r}\right)^m\right) \tag{5.14}$$

where the reference strain rate is $\dot{\varepsilon}_0$ (normally taken as 1 s^{-1} or as the quasistatic strain rate), the reference temperature T_r (usually the temperature at which the yield stress σ_0 is measured), and the melting point temperature T_m all can be determined in advance. Therefore, there are five constitutive constants to be experimentally determined, including σ_0, B, n, C (i.e., the strain-rate sensitivity coefficient in Eq. 5.2), and temperature-softening coefficient m. According to a series of experimental results, Johnson and Cook provided the relevant data of 12 metal materials (Johnson and Cook, 1983; Johnson et al., 1983), as shown in Table 5.4, which greatly facilitated the engineering applications.

Table 5.4 Johnson–Cook constitutive constants for different materials ($\dot{\varepsilon} = 1\ s^{-1}$) (Johnson and Cook, 1983).

	Description				Constitutive constants for $\sigma = [\sigma_0 + B\varepsilon^n][1 + C\ln\varepsilon^*][1 - T^*m]$				
Material	**Hardness (Rockwell)**	**Density (kg/m^3)**	**Specific heat (J/kg K)**	**Melting temperature (K)**	**σ_0 (MPa)**	**B (MPa)**	**n**	**C**	**m**
OFHC copper	F-30	8960	383	1356	90	292	0.31	0.025	1.09
Cartridge brass	F-67	8520	385	1189	112	505	0.42	0.009	1.68
Nickel 200	F-79	8900	446	1726	163	648	0.33	0.006	1.44
Armco iron	F-72	7890	452	1811	175	380	0.32	0.060	0.55
Carpenter electrical iron	F-83	7890	452	1811	290	339	0.40	0.055	0.55
1006 steel	F-94	7890	452	1811	350	275	0.36	0.022	1.00
2024-T351 aluminum	B-75	2770	875	775	265	426	0.34	0.015	1.00
7039 aluminum	B-76	2770	875	877	337	343	0.41	0.010	1.00
4340 steel	C-30	7830	477	1793	792	510	0.26	0.014	1.03
S-7 tool steel	C-50	7750	477	1763	1539	477	0.18	0.012	1.00
Tungsten alloy (0.07Ni, 0.03Fe)	C-47	17,000	134	1723	1506	177	0.12	0.016	1.00
Depleted uranium-0.75% Ti	C-45	18,600	117	1473	1079	1120	0.25	0.007	1.00

The J–C equation is an empirical formula that is based on one-dimensional stress experimental data. When generalized to the general three-dimensional stress state, the Mises criterion is assumed to hold continuously under high strain rates. That is, what needs is only to replace the one-dimensional stress, strain, and strain rate in the equation by the effective stress, effective strain, and effective strain rate defined by Eq. (5.8).

Given that the J–C equation is an empirical formula that is fitted on the basis of experimental data, it is not difficult to understand that the corresponding empirical fitting formula will change depending on the material. Therefore, since the formulation of the J–C equation, different modifications or improvements to Eq. (5.14) have been proposed. For example, other functional forms may be used to describe the strain-hardening term in the equation or other functional forms may be used to describe the strain-rate-hardening term in the equation. They can all be classified as the Johnson–Cook type equation.

Among them, it is particularly worth mentioning that for brittle materials, such as ceramics, rocks, and concretes, considering that the dynamic mechanical behavior of such materials is very sensitive to hydrostatic pressure p (triaxial isoaxial pressure or spherical pressure) and damage D, but not to temperature, Holmquist et al. (1993) suggested the following **Holmquist–Johnson–Cook Equation** (HJC Equation):

$$\sigma^* = \left[A(1-D) + Bp^{*N}\right](1 + C\ln\dot{\varepsilon}^*) \tag{5.15a}$$

To nondimensionalize the equation, each quantity in the formula is normalized, i.e., $\sigma^* = \frac{\sigma_{eff}}{f_c}$ is the normalized effective stress ($\leq S_{max}$), f_c is the quasistatic uniaxial compressive strength, and S_{max} is the maximum normalized effective stress that the material can achieve. Moreover, $p^* = \frac{p}{f_c}$ is the normalized isoaxial pressure, and $\dot{\varepsilon}^* = \frac{\dot{\varepsilon}_{eff}}{\dot{\varepsilon}_0}$ is the normalized effective strain rate. The reference strain rate $\dot{\varepsilon}_0$ is often taken as $1\ \mathrm{s}^{-1}$. A is the normalized noninvasive cohesive strength for materials without damage, B is the pressure-hardening coefficient, N is the pressure-hardening index, C is the strain-rate sensitivity coefficient, and D ($0 \leq D \leq 1.0$) is a macroscopic scalar that characterizes damage, where $D = 0$ corresponds to no damage and $D = 1$ corresponds to destruction. The damage evolution dD in the HJC model is described by the ratio of the total plastic strain increment and the total failure strain, the former is the sum of the effective plastic strain increment $d\varepsilon_p$ and the volumetric plastic strain increment $d\mu_p$, while the latter is the sum of the effective plastic failure strain ε_f and the volumetric plastic failure strain μ_f. The following damage evolution equation is thus obtained:

$$\frac{dD}{dt} = \frac{1}{(\varepsilon_f + \mu_f)} \frac{d(\varepsilon_p + \mu_p)}{dt} \tag{5.15b}$$

The total failure strain ($\varepsilon_f + \mu_f$), as shown by the following equation, is the function of normalized isoaxial pressure p^* and normalized maximum tensile stress $T^*\left(=\frac{T}{f_c}\right)$, where T is the tensile strength of the material.

$$\varepsilon_p + \mu_p = D_1\left(p^* + T^*\right)^{D_2} \geq \varepsilon_{f\min} \tag{5.15c}$$

where D_1 and D_2 are the material constants that characterize damage evolution, and ε_{fmin} is the minimum plastic failure strain, which is used to control the failure induced by the tensile stress wave, which is a kind of brittle failure. The dynamic evolution of damage will be further discussed in Part 3.

The significance of Eq. (5.15) is not only to extend the J—C equation derived from metallic materials to brittle materials and to take into account of the influence of hydrostatic pressure p (spherical triaxial pressure) and damage D, but also to be not limited to the assumption that the volumetric law is decoupled from the distortion law. That is, on the one hand, the influence of the triaxial isoaxial pressure (spherical stress tensor) p is considered in the dynamic distortion relationship of materials, and on the other hand, the volumetric deformation is no longer limited to the elastic deformation in the category of reversible thermodynamics, but the existence of irreversible plastic volumetric deformation is considered.

Similar to the derivation of the C—S equation in tensor form (Eq. 5.12d), rewriting the J—C tensor equation as the form that $\dot{\varepsilon}^p_{eff}$ be a function of other mechanical quantities, and substituting it into the $\dot{\varepsilon}^p_{eff}$ in (Eq. 5.12c), the viscoplastic constitutive equation based on the J—C equation in tensor form can be derived.

The J—C equation is widely used to given that it simultaneously accounts for the strain-hardening effects, strain-rate—hardening effects, and temperature-softening effects of materials; has a simple form that facilitates engineering applications; and provides material constants for formula application based on a large number of experiments. However, the use of the multiplicative form of these effects to describe the dynamic distortion behavior of materials is not supported by theory. For example, the multiplicative form of these effects is inconsistent with the mechanism of dislocation dynamics, including the aforementioned rate–temperature equivalence, as discussed in the next chapter. For another example, when the strain rate is higher than $10^4\ s^{-1}$, the linear strain-rate sensitivity $\sigma - \log\dot{\varepsilon}$ described by Eq. (5.2) no longer holds. Thus, the J–C equation established on the basis of Eq. (5.2) will introduce large errors. The applicable conditions and limitations of this formula must be mastered when it is used.

5.2.3 Sokolovsky–Malvern–Perzyna equation

Although Eqs. (5.1b) and (5.2), which are based on extensive experimental data, are expressed in different functional forms, they have one point in common, that is, the strain rate is considered as the function of overstress, i.e., the function of the difference between the dynamic flow stress σ and the quasistatic flow stress σ_0, $(\sigma - \sigma_0)$ or the function of the dimensionless overstress $\frac{\sigma-\sigma_0}{\sigma_0}$. Strictly speaking, the strain rate here should refer to the plastic strain rate $\dot{\varepsilon}_p$ that corresponds to plastic flow stress and ignores the elastic strain rate.

In the one-dimensional form, the total strain rate $\dot{\varepsilon}$ is assumed to consist of an elastic part (set as rate-independent transient response) $\dot{\varepsilon}^e$ and a nonelastic part (rate-dependent nontransient response, viscoplastic response) $\dot{\varepsilon}^p$.

$$\dot{\varepsilon} = \dot{\varepsilon}^e + \dot{\varepsilon}^p \tag{5.16}$$

where the elastic part $\dot{\varepsilon}^e$ stands for rate-independent transient response, and the nonelastic part $\dot{\varepsilon}^p$ is taken as a function of overstress and represents the plasticviscous coupling response. The Sokolovsky–Malvern–Perzyna elastic–viscoplastic model was developed on the basis of the above assumptions.

Sokolovsky (Соколовский, 1948) first proposed the following elastic–viscoplastic constitutive relations on the basis of a perfect plastic body:

$$\dot{\varepsilon} = \frac{\dot{\sigma}}{E} + Sign\sigma \cdot g\left[\frac{|\sigma|}{Y_0} - 1\right] \tag{5.17}$$

That is, $\dot{\varepsilon}^p$ is considered as a function $g\left[\frac{|\sigma|}{Y_0} - 1\right]$ of the overstress defined by the difference between the dynamic stress σ and the static yield limit Y_0 of perfect plastic body. Eq. (5.17) shows that at different constant strain rates, the stress–strain curves of the viscoplastic deformation part are a set of parallel lines in the $\sigma - \varepsilon$ plane (Fig. 5.22A). Function g is generally a nonlinear function.

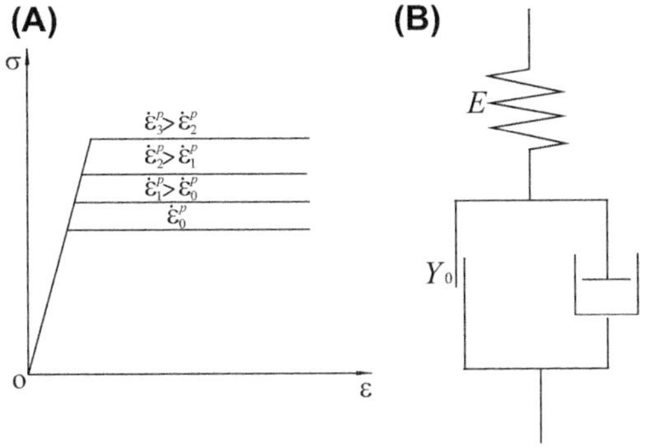

Figure 5.22 (A) Elasto–viscoplastic stress–strain curves and (B) corresponding rheological model represented by Eq. (5.17).

When g is a linear function of overstress, Eq. (5.17) can be written as follows:

$$\dot{\varepsilon} = \frac{\dot{\sigma}}{E} + \frac{|\sigma| - Y_0}{\eta} = \frac{\dot{\sigma}}{E} + Sign\sigma \cdot \gamma^* \frac{|\sigma| - Y_0}{Y_0} \tag{5.18}$$

The $\frac{|\sigma|-Y_0}{\eta}$ term characterizing the $\dot{\varepsilon}_p$ is similar to that in the Newtonian viscosity law for viscous fluids, only that the stress in the Newtonian viscosity law is replaced with overstress. If $\dot{\varepsilon}_p$ is characterized by dimensionless overstress, then as shown by the second equal sign of the above formula, $\gamma^*\left(=\frac{Y_0}{\eta}\right)$ becomes a material constant characterizing the material viscosity. In terms of rheological models, if the perfect plastic body is described by a sliding plate with constant friction Y_0, then a viscoplastic element consists of a sliding plate element and a viscous dashpot element in parallel, called the Bingham model. Thus, Eq. (5.17) can be expressed as a three-element elasto–viscoplastic model that consists of a spring element and a viscoplastic element in series, as shown in Fig. 5.22B.

Malvern (1951) proposed that strain hardening should be taken into account, so the quasistatic stress–strain curve $\sigma_0(\varepsilon)$ taking consideration of strain hardening is used instead of Y_0 in Eq. (5.17). The following elasto–viscoplastic constitutive equation was provided:

$$\dot{\varepsilon} = \frac{\dot{\sigma}}{E} + Sign\sigma \cdot g\left[\frac{\sigma}{\sigma_0(\varepsilon)} - 1\right] \tag{5.19}$$

This equation corresponds to that in the model shown in Fig. 5.22B, wherein the friction force Y is set as a function of plastic strain ε_p,

$Y = Y(\varepsilon_p)$. At different constant strain rates, Eq. (5.19) is such a family of stress–strain curves in $\sigma - \varepsilon$ plane, of which the viscoplastic deformation curves are equidistant from the static stress–strain curve $\sigma = \sigma_0(\varepsilon)$, respectively, as shown in Fig. 5.23.

According to Sokolovsky–Malvern model, the viscoplastic deformation occurs only when the overstress is greater than zero, and when the overstress is less than or equal to zero, only the elastic deformation occurs, which can be uniformly expressed as the following constitutive equation:

$$\left.\begin{aligned} \dot{\varepsilon} &= \frac{\dot{\sigma}}{E} + \left\langle g\left[\frac{\sigma}{\sigma_0(\varepsilon)} - 1\right]\right\rangle \\ \langle g\rangle &= \begin{cases} 0, & for\ \dfrac{\sigma}{\sigma_0(\varepsilon)} - 1 \le 0, \\ g, & for\ \dfrac{\sigma}{\sigma_0(\varepsilon)} - 1 > 0, \end{cases} \end{aligned}\right\} \tag{5.20a}$$

Or similar to Eq. (5.18), the viscosity coefficient γ^* is introduced. That is, set $g(\xi) = \gamma^*\phi(\xi)$.

Then, Eq. (5.20a) can be written as:

$$\begin{aligned} \dot{\varepsilon} &= \frac{\dot{\sigma}}{E} + \gamma^* < \phi(F) > \\ < \phi(F) > &= \begin{cases} 0, & F \le 0 \\ \phi(F), & F > 0 \end{cases} \end{aligned} \tag{5.20b}$$

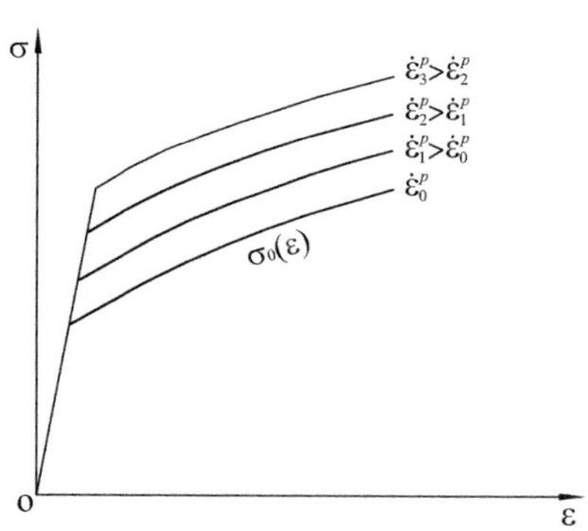

Figure 5.23 Elasto–viscoplastic constitutive model represented by Eq. (5.19).

$$F = \frac{\sigma}{\sigma_0(\varepsilon)} - 1$$

Note that whether the Sokolovsky—Malvern elasto—viscoplastic body is loaded into the viscoplastic zone from the elastic zone or unloaded from the viscoplastic zone back to the elastic zone is completely determined by whether the overstress is greater than zero. As long as the overstress exceeds zero, regardless of whether the stress increases $\left(\frac{\partial\sigma}{\partial t} > 0\right)$ or decreases $\left(\frac{\partial\sigma}{\partial t} < 0\right)$ over time, it follows the same constitutive relationship. Viscoplastic deformation may continue to develop even if stress decreases. For example, as shown in Fig. 5.24A, when the stress increases linearly with time (*OAB*) and then decreases linearly (*BCD*), the $\sigma - \varepsilon$ curve calculated in accordance with Eq. (5.20) is shown in Fig. 5.24B. As can be seen in this figure, after stress has decreased from *B* point, viscoplastic deformation continues to develop because overstress continues to exceed zero. The viscoplastic state is transferred to the elastic unloaded state only when the stress is unloaded to *C* point where overstress is 0. The *C* point does not correspond to the point where the stress begins to fall and does not correspond to the point where the strain drops. In these aspects, elasto—viscoplastic materials are completely different from elastoplastic materials.

Eq. (5.20) given above remains limited to the one-dimensional stress problem. Perzyna (1966) further extended the overstress model to the three-dimensional stress state and provided the following general expression in the tensor form, called the Sokolovsky—Malvern—Perzyna (SMP) elasto—viscoplastic constitutive equation:

$$\begin{aligned} \dot{\varepsilon}_{ij} &= \dot{\varepsilon}_{ij}^{e} + \dot{\varepsilon}_{ij}^{p} \\ &= \frac{1}{2\mu}\dot{S}_{ij} + \frac{1}{3K}\dot{\sigma}_{ij}\delta_{ij} + \gamma < \phi(F) > \frac{\partial f}{\partial \sigma_{ij}} \end{aligned} \tag{5.21}$$

$$\langle \phi(F) \rangle = \begin{cases} 0, F \leq 0 \\ \phi(F), F > 0 \end{cases}$$

$$F = \frac{f\left(\sigma_{ij}, \dot{\varepsilon}_{ij}^{p}\right)}{k(w_p)} - 1$$

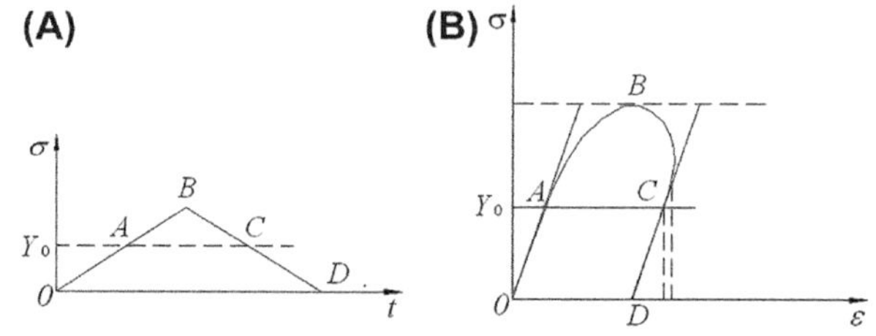

Figure 5.24 (A) Linear loading—unloading of stress over time. (B) Stress—strain curve based on the elastic—viscoplastic material according to Eq. (5.20).

where F is the plastic flow function that measures the difference (overstress) between the instantaneous stress state and the static flow stress $k(w_p)$ that accounts for hardening, and w_p is the plastic work. In fact, the case that F equals to zero corresponds to the static loading process ($\dot{\varepsilon}_{ij}^p = 0$). At this time, the loading function can be written as $f\left(\sigma_{ij}, \dot{\varepsilon}_{ij}^p\right) = f_s = k(w_p)$. The case that F greater than 0 corresponds to the dynamic loading process. At this time, the load function $f\left(\sigma_{ij}, \dot{\varepsilon}_{ij}^p\right)$ can be written as f_d. Substituting the results into Eq. (5.21) yields

$$F = \frac{f_d - f_s}{f_s}$$

In fact, the SMP equation (Eq. 5.21) implies two basic assumptions: first, the total strain-rate tensor is the sum of the elastic strain-rate tensor and the viscoplastic strain-rate tensor (rather than the tensor product); second, the rate-independent strain-rate tensor (instantaneous strain-rate tensor) only contains the elastic strain-rate tensor. However, as Cristescu (1967) pointed out, the instantaneous deformation of the material is not only has instantaneously elastic deformation but also instantaneously plastic deformation, as shown in Fig. 5.25. Therefore, he proposed the following general quasilinear constitutive equation:

$$\frac{\partial \varepsilon}{\partial t} = \left[\frac{1}{E} + \phi(\sigma, \varepsilon)\right] \frac{\partial \sigma}{\partial t} + \psi(\sigma, \varepsilon) \tag{5.22}$$

After integration, the following is obtained.

$$\begin{aligned} \varepsilon &= \frac{\sigma}{E} + \int_0^{\sigma} \phi(\sigma, \varepsilon) d\sigma + \int_0^{t} \psi(\sigma, \varepsilon) dt \\ &= \varepsilon^E + \varepsilon^{IP} + \varepsilon^{VP} \end{aligned} \tag{5.23}$$

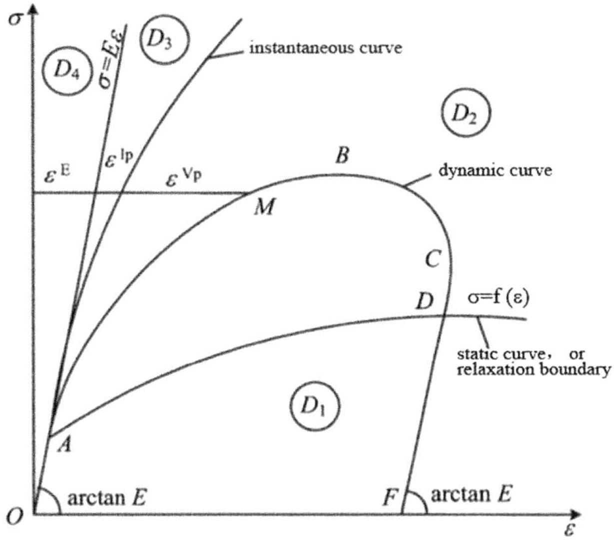

Figure 5.25 Schematic of overstressed viscoplastic stress—strain curves taking account of instantaneous plastic response.

where ε^E, ε^{IP}, and ε^{VP} represent the instantaneous elastic strain, instantaneous plastic strain, and viscoplastic strain in the total strain, respectively, as shown in Fig. 5.25. ε^E and ε^{IP} together form the instantaneous curve. By taking the elastic line $\sigma = E\varepsilon$, instantaneous curve, dynamic curve, and static curve (or relaxation boundary) $\sigma = f(\varepsilon)$ as the boundaries, the four zones can be divided in $\sigma - \varepsilon$ coordinates: below the static curve $\sigma = f(\varepsilon)$ is the D_1 zone that characterizes the static response, between the static curve and the transient curve is the D_2 zone that characterizes the overstressed viscoplastic response, between the instantaneous curve and the elastic line is the D_3 zone that characterizes the instantaneous plastic response, and the left of the elastic line is the D_4 zone that characterizes the instantaneous elastic response.

If the material does not possess any instantaneous plasticity, that is $\phi(\sigma,\varepsilon) = 0$, Eq. (5.22) is simplified to the Sokolovsky—Malvern equation.

5.2.4 Bodner—Parton equation

The above discussions are all the viscoplastic equations based on overstress, which is determined by the difference between the dynamic stress—strain curve $\sigma = \sigma_d(\varepsilon)$ and the static stress—strain curve $\sigma = \sigma_0(\varepsilon)$. If you ask, what is the strain rate under which the static stress—strain curve $\sigma = \sigma_0(\varepsilon)$ be measured? There have at least two possible understandings: first, from the perspective of practical application, it is often understood as that the stress—strain curve is measured under quasistatic experimental conditions (e.g., the strain rate of 10^{-4} s^{-1}). Following this logic, then for the stress—strain curve measured under the strain rate less than the quasistatic experimental strain rate of 10^{-4} s^{-1} (e.g., the creep strain rate of 10^{-8} s^{-1}), overstress will be negative. Of course, the overstress theory can be limited to only the strain-rate range above the quasistatic value. This is what the engineering community is actually doing. Nevertheless, some shortcomings exist in this theoretical framework. Second, from the theoretical logic, $\sigma = \sigma_0(\varepsilon)$ can be understood as some kind of limit curve, such as a relaxation boundary. This will correspond to the limit where the strain rate tends to zero. In other words, as long as the strain rate exceeds zero, viscoplastic flow will exist. So it means that we are essentially taking $\sigma = \sigma_0(\varepsilon)$ as a zero curve where overstress is zero.

From the perspective of dislocation dynamics that will be discussed in the next chapter, viscoelastic flow exists as long as the strain rate is not zero. On the basis of this perspective, Bodner and Partom (1975) proposed a viscoplastic constitutive equation with strain hardening but without the yield surface (Bodner, 1968, 2000; Bodner and Partom, 1975).

Similar to Eq. (5.16), the total strain is assumed to be divisible into two parts of reversible elasticity and irreversible inelasticity:

$$\dot{\varepsilon}_{ij} = \dot{\varepsilon}^e_{ij} + \dot{\varepsilon}^p_{ij} \tag{5.24a}$$

where the elastic strain rate and stress rate are subject to Hooke's law, that is:

$$\dot{\varepsilon}^e_{ij} = \frac{\dot{\sigma}_{ij}}{2\mu_e} - \frac{\lambda_e \dot{\sigma}_{kk}\delta_{ij}}{[2\mu_e(3\lambda_e + 2\mu_e)]} \tag{5.24b}$$

where λ_e and μ_e are Lame elastic constants; the inelastic strain rate is subject to the classical Levy—Mises flow rule (Eq. 5.12a), or directly from the form shown in Eq. (5.12c).

$$\begin{gathered}
\dot{\varepsilon}^p_{ij} = \left(\frac{D^p_2}{J_2}\right)^{1/2} S_{ij} = \frac{3}{2}\frac{\dot{\varepsilon}^p_{eff}}{\sigma_{eff}} S_{ij} \\
D^p_2 = \frac{1}{2}\dot{\varepsilon}^p_{ij}\dot{\varepsilon}^p_{ij} = \frac{3}{4}\left(\dot{\varepsilon}^p_{eff}\right)^2 = \frac{1}{6}\left[\left(\dot{\varepsilon}^p_1 - \dot{\varepsilon}^p_2\right)^2 + \left(\dot{\varepsilon}^p_2 - \dot{\varepsilon}^p_3\right)^2 + \left(\dot{\varepsilon}^p_3 - \dot{\varepsilon}^p_1\right)^2\right] \\
J_2 = \frac{1}{2}S_{ij}S_{ij} = \frac{1}{3}(\sigma_{eff})^2 = \frac{1}{6}\left[(\sigma_1 - \sigma_2)^2 + (\sigma_2 - \sigma_3)^2 + (\sigma_3 - \sigma_1)^2\right]
\end{gathered} \tag{5.24c}$$

Bodner—Partom gave up the traditional view of the existence of yield condition or static stress—strain curve $\sigma = \sigma_0(\varepsilon)$, assumed that D_2^p is a function of J_2, $D_2^p = f(J_2)$, that is, under the assumption that $\dot{\varepsilon}_{eff}^p$ is a function of σ_{eff}, $\dot{\varepsilon}_{eff}^p = f(\sigma_{eff})$. This is equivalent to not having to assume that there exists a yield surface in the stress space (including the subsequent yield surface that accounts for the strain hardening) as well as the associated loading and unloading conditions. Therefore, for any given stress deviator S_{ij}, a corresponding inelastic strain rate $\dot{\varepsilon}_{ij}^p$ exists but no yield limit and overstress exist. Free from the view of overstress, the Bodner—Partom (B—P) equation can be applied to a wide range of strain rates from creep (e.g., strain rate of $10^{-8}\ \mathrm{s}^{-1}$) up to high-speed flow/deformation (e.g., strain rate of $10^6\ \mathrm{s}^{-1}$).

Note that Eq. (5.24) is applicable to both loading and unloading, while in the theory based on the existence of yield surface, only elastic unloading occurs when the overstress is zero.

Obviously, the specific characteristics of the viscoplastic flow/deformation of materials will be determined by the specific function form of $D_2^p = f(J_2)$ or $\dot{\varepsilon}_{eff}^p = f(\sigma_{eff})$. Bodner—Partom suggested a generic function as follows:

$$D_2^P = D_0^2 \exp\left[-\left(\frac{Z^2}{\sigma_{eff}^2}\right)^n\right] \tag{5.25a}$$

where D_0 is a scalar factor that characterizes the limit value of D_2^p under high stress, and Z is a scalar parameter with a stress dimension that depends on loading history to characterize the hardening behavior, and Z is a scalar parameter that characterizes the strain-rate sensitivity.

Substituting (5.25a) into (5.24c) yields

$$\dot{\varepsilon}_{ij}^p = D_0 \exp\left[-\frac{1}{2}\left(\frac{Z^2}{\sigma_{eff}^2}\right)^n\right]\frac{\sqrt{3}S_{ij}}{\sigma_{eff}} \tag{5.25b}$$

Under the conditions of one-dimensional uniaxial stress σ_{11} and simple shear τ_{12}, the above equation is simplified, respectively, as:

$$\dot{\varepsilon}_{11}^p = \frac{2}{\sqrt{3}}\left(\frac{\sigma_{11}}{|\sigma_{11}|}\right) D_0 \exp\left[-\frac{1}{2}\left(\frac{Z}{\sigma_{11}}\right)^{2n}\right] \tag{5.25c}$$

$$\dot{\varepsilon}^p_{12} = \dot{\gamma}^p_{12} = D_0\left(\frac{\tau_{12}}{|\tau_{12}|}\right)\exp\left[-\frac{1}{2}\left(\frac{Z}{\sqrt{3}\tau_{12}}\right)^{2n}\right] \tag{5.25d}$$

where $\dot{\gamma}^p_{12}$ and τ_{12} are the engineering shear strain rate and the engineering shear stress, respectively.

For Eq. (5.25a) under one-dimensional uniaxial stress condition, the plot of the dimensionless inelastic strain rate $\frac{\sqrt{3}}{2}\frac{\dot{\varepsilon}^p_{11}}{D_0}$ versus the dimensionless stress $\frac{\sigma_{11}}{Z}$ is shown in Fig. 5.26, and the plot of the dimensionless stress $\frac{\sigma_{11}}{Z}$ versus the logarithm of the dimensionless inelastic strain rate $\log\left(\frac{\sqrt{3}}{2}\frac{\dot{\varepsilon}^p_{11}}{D_0}\right)$ is shown in Fig. 5.27. As can be seen from Fig. 5.26, the function in the form of Eq. (5.25) has the ability to describe different ranges of viscoplastic flow/deformation including incubation, rapid growth, and saturation. As can be seen too from Figs. 5.26 and 5.27 that n can reflect and control strain-rate sensitivity well and can also affect stress level. As can be further seen from Fig. 5.27 that Eq. (5.25) can reflect the linear strain-rate sensitivity in $\sigma - \log\dot{\varepsilon}$ coordinates as well as a rapid increase in strain-rate sensitivity under higher strain rates. Although there is no explicit temperature term in Eq. (5.25), both Z and n imply dependence on temperature. These are the main advantages provided by the B–P equation (Eq. (5.25)).

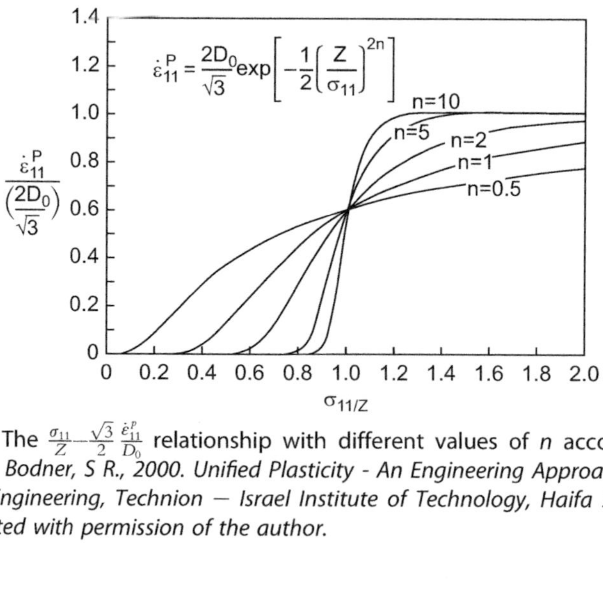

Figure 5.26 The $\frac{\sigma_{11}}{Z}$–$\frac{\sqrt{3}}{2}\frac{\dot{\varepsilon}^p_{11}}{D_0}$ relationship with different values of n according to Eq. (5.25c). *From Bodner, S R., 2000. Unified Plasticity - An Engineering Approach, Faculty of Mechanical Engineering, Technion – Israel Institute of Technology, Haifa 32000. Israel., Fig. 1. Reprinted with permission of the author.*

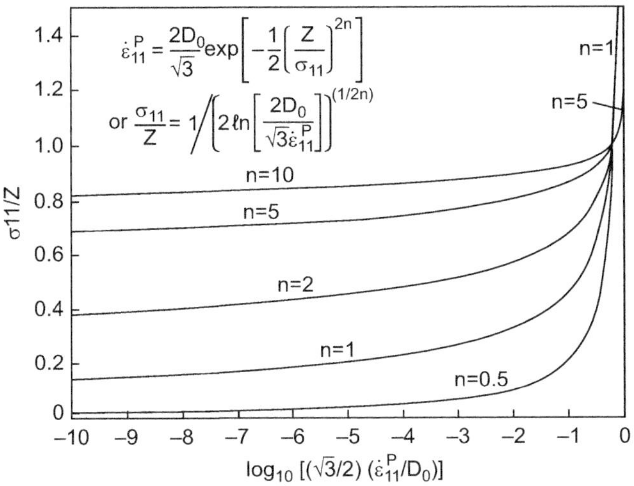

Figure 5.27 The $\frac{\sigma_{11}}{Z}$–$\log\left(\frac{\sqrt{3}}{2}\frac{\dot{\varepsilon}_{11}^{p}}{D_0}\right)$ relationship with different values of *n* according to Eq. (5.25c). *(From Bodner, S R., 2000. Unified Plasticity - An Engineering Approach, Faculty of Mechanical Engineering, Technion – Israel Institute of Technology, Haifa 32000. Israel., Fig. 2. Reprinted with permission of the author).*

Notably, the B–P model not only can reflect the linear strain-rate sensitivity in $\sigma - \log\dot{\varepsilon}$ coordinates but also the transition of strain-rate sensitivity at higher strain rates. Tong Wei, Clifton, and Huang Shihui (1992) obtained the experimental results under shear strain rates $\dot{\gamma}$ up to $10^6\ s^{-1}$ by using pressure–shear plate collision experiments. The distribution of experimental points in $\tau - \log\dot{\gamma}$ coordinates for the given shear strain of $\gamma = 0.20$ is shown in Fig. 5.28. Bodner and Rubin (1994) used a B–P model that accounts for the strain-rate effects of the strain hardening, and their numerical simulation results (dotted line in Fig. 5.28) were in good agreement with these experimental points, which successfully reflect the linear rise of the $\tau - \log\dot{\gamma}$ curve from the strain rate of $10^4\ s^{-1}$.

The "no yield surface" view of the B–P equation originates from the study of dislocation dynamics. In fact, some other forms of viscoplastic constitutive equations have been proposed on the basis of theory and experiments of dislocation dynamics, which will be discussed in the next chapter.

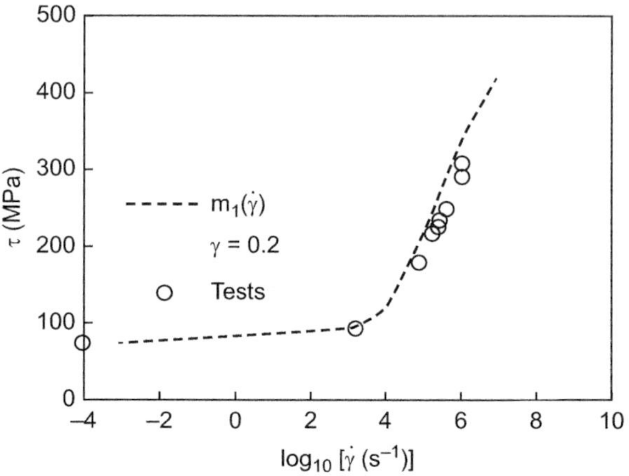

Figure 5.28 Comparison of results of numerical simulation by the B—P model (- - -) and experimental data (○). *From Bodner, S R., 2000. Unified Plasticity - An Engineering Approach, Faculty of Mechanical Engineering, Technion — Israel Institute of Technology, Haifa 32000. Israel., Fig. 5. Reprinted with permission of the author.*

5.3 Nonlinear viscoelastic constitutive equations under high strain rates

So far, the content discussed is mainly about metal materials. The strain-rate sensitivity of the elastic deformation of metal materials can often be disregarded, unlike that of plastic deformation. Therefore, it is often assumed that the elastic deformation part follows Hooke's law regardless of strain-rate effect, as shown in Formula (5.16) or (5.24b). In fact, even for metal materials, the internal friction phenomenon in elastic deformation stage, the physical dissipation in the elastic wave propagation, as well as the heating and fracture under elastic fatigue load and so on, all indicate that without exception, the real metal materials actually have viscoelastic properties.

The main reason why people are interested in the study of viscoelastic constitutive relations is that polymer materials, such as rubber and plastics, exhibit obvious strain-rate effect even at room temperature and quasistatic load. The total global production of polymer materials has been equal to that of iron and steel in terms of volume and has been listed as one of the three major engineering materials in the world together with metal, and ceramic materials, which further promotes the study of viscoelastic constitutive relations.

However, with regard to the study of the viscoelastic constitutive relation of polymer materials, considerable attention has been devoted to linear viscoelastic behavior under quasistatic and low strain rates (Yang Tingqing et al., 2004). Investigating nonlinear viscoelastic behavior has been an important research direction in recent decades (Lockett, 1972). Research on dynamic nonlinear viscoelastic behavior under high strain rates of impact load only emerged after the development of the split Hopkinson pressure bar technique. In this section, the nonlinear viscoelastic constitutive relation of polymer materials at high strain rates is discussed.

Researchers have extensively used the research approaches of dynamic constitutive relation of metal materials to conduct experimental research on polymer materials; thus, the dynamic behavior of polymer materials at high strain rates is often included into the aforementioned viscoplastic category. One of the reasons is that experimental observation only focuses on the loading characteristics of materials but ignores the unloading characteristics. In fact, it is difficult to accurately distinguish the constitutive characteristics of materials only by the loading stress—strain curves, as shown in Fig. 5.29. The loading stress—strain curves of *a*, *b*, and *c* in Fig. 5.29 are the same (an initial linear section then a nonlinear section), but it is not sufficient to determine its constitutive type without examining its unloading characteristics. When the unloading curve coincides with the loading curve (case *a*), the constitutive type is nonlinear elastic. When the elastic unloading is carried out according to the same slope as the initial linear segment of loading (case *c*), the constitutive type is nonlinear elasto—plastic. When the unloading and loading curves form a hysteresis curve (case *b*), the constitutive type is nonlinear viscoelastic. Only a comprehensive study of the entire loading—unloading process can help distinguish these constitutive types.

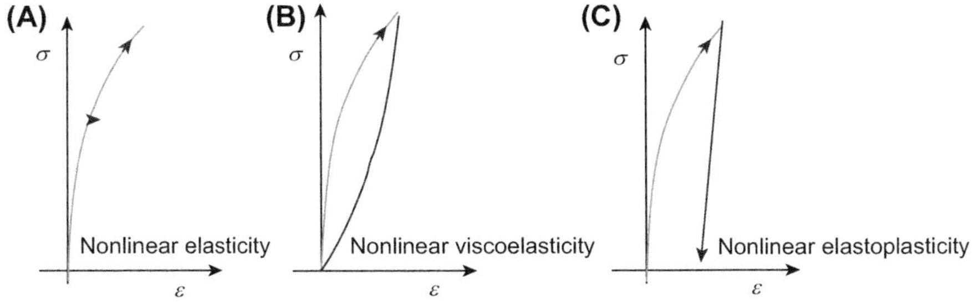

Figure 5.29 The loading stress—strain curves of different materials may be the same but may differ according to the unloading curves.

5.3.1 Nonlinear viscoelastic constitutive equation (Zhu–Wang–Tang equation)

It is from the viewpoint of attaching importance to both loading and unloading characteristics; Zhu Zhaoxiang, Wang Lili, and their coworkers conducted a series of experimental studies on several typical engineering plastics (e.g., epoxy resin, polymethyl methacrylate (PMMA), polycarbonate (PC), polyamide (PA or nylon), acrylonitrile-butadiene-styrene (ABS), polybutylene terephthalate (PBT), PP/PA polymer blends, and fiber-reinforced polymer composites) since the 1980s (Wang and Yang, 1992; Wang et al., 2000; Kobayashi and Wang Lili, 2001). Figs. 5.30 (Tang et al., 1981) and 5.31 (Yang Li-Ming et al., 1986) show the results for a thermosetting plastics and a thermoplastic plastics, respectively.

These experimental results indicate that up to a large strain range of approximately 7%, the experimental materials exhibit nonlinear viscoelastic responses with hysteresis curves. If the experimental data for a given strain are plotted in $\sigma - \log\dot{\varepsilon}$ coordinates, e.g., obtaining Fig. 5.32 from the data in Fig. 5.31, it is represented as a family of broken lines with different slopes of straight lines in semilogarithmic coordinates, which means that the experimental material would exhibit different strain-rate sensitivities at the

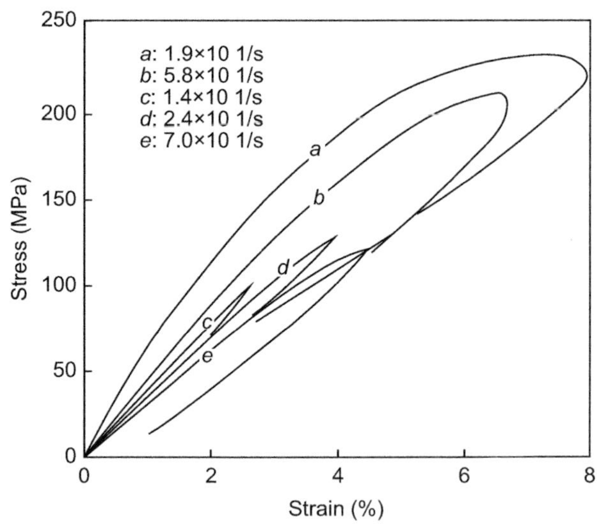

Figure 5.30 Experimental results of epoxy resin under strain rates of 10^{-4}–10^{3} s^{-1}. *From Tang, Z., Tian, Lan-qiao, Chu, Chao-hsiang, Wang, Li-li, 1981. Mechanical behavior of epoxy resin under high strain rates, Second National Conference of Explosive Mechanics. Chinese Society of Theoretical and Applied Mechanics, Yangzhou, (in Chinese).*

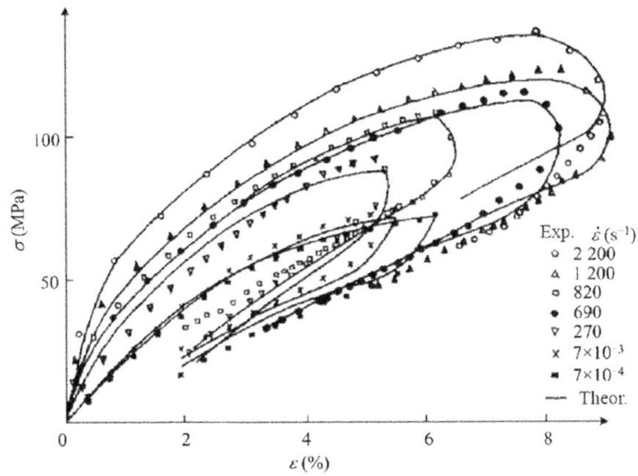

Figure 5.31 Experimental results of polycarbonate (PC) under strain rates of 10^{-4}–10^{3} s^{-1}. *Yang, Li-Ming, Chu, Chao-Hsiang, Wang, Li-Li, 1986. Effects of short-glass-fiber reinforcement on nonlinear viscoelastic behavior of polycarbonate. Explos. Shock Waves. 6 (1) 1–9 (in Chinese)*

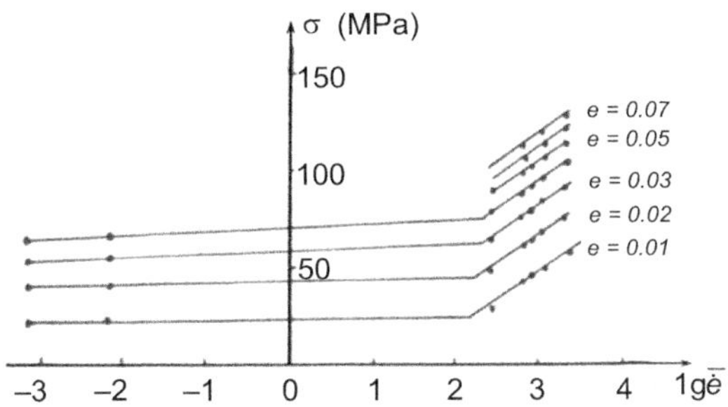

Figure 5.32 Experimental results of polycarbonate resin PC in the $\sigma - \log\dot{\varepsilon}$ coordinate system. *Yang, Li-Ming, Chu, Chao-Hsiang, Wang, Li-Li, 1986. Effects of short-glass-fiber reinforcement on nonlinear viscoelastic behavior of polycarbonate. Explos. Shock Waves. 6 (1) 1–9 (in Chinese)*

quasistatic strain rates and at the impact high strain rates and becomes more sensitive under high strain rates.

According to viscoelasticity theory (Wang and Yang, 1992; Tingqing et al., 2004) the strain-rate sensitivity of viscoelastic materials depends mainly on viscosity coefficient η or the corresponding relaxation time $\theta = \eta/E$ (where E is the elastic constant). Fig. 5.32 suggests that from the quasistatic strain rate to the impact high strain rate, there are two dominant relaxation times, namely, θ_1 and θ_2, representing the low strain-rate response and high strain-rate response of viscoelastic materials, respectively.

Thus, Zhu Zhaoxiang, Wang Lili, and Tang Zhiping (Tang et al., 1981) first proposed the following nonlinear viscoelastic constitutive relation (called Zhu—Wang—Tang equation or briefly ZWT equation):

$$\sigma = f_e(\varepsilon) + \sigma_1(\varepsilon, \dot{\varepsilon}) + \sigma_2(\varepsilon, \dot{\varepsilon})$$
$$= f_e(\varepsilon) + E_1 \int_0^t \dot{\varepsilon} \exp\left(-\frac{t-\tau}{\theta_1}\right) d\tau + E_2 \int_0^t \dot{\varepsilon} \exp\left(-\frac{t-\tau}{\theta_2}\right) d\tau \tag{5.26a}$$

$$f_e(\varepsilon) = E_0\varepsilon + \alpha\varepsilon^2 + \beta\varepsilon^3 \tag{5.26b}$$

Alternatively, to avoid the possibility that Eq. (5.26b) may lead to false "elastic strain softening" as the strain increase, f_e (ε) can be presented as the following exponential function (Zhou Feng-hua et al., 1992).

$$f_e(\varepsilon) = \sigma_m \left[1 - \exp\left(-\sum_{i=1}^{n} \frac{(m\varepsilon)^i}{i}\right)\right] \tag{5.26c}$$

Formula (5.26a) can also be equivalently presented as the following differential form.

$$\frac{\partial \sigma}{\partial t} = \frac{df_e(\varepsilon)}{d\varepsilon}\frac{\partial \varepsilon}{\partial t} + \frac{\partial \sigma_1}{\partial t} + \frac{\partial \sigma_2}{\partial t} = \left[\frac{df_e(\varepsilon)}{d\varepsilon} + E_1 + E_2\right]\frac{\partial \varepsilon}{\partial t} - \frac{\sigma_1}{\theta_1} - \frac{\sigma_2}{\theta_2} \tag{5.26d}$$

Fig. 5.33 presents the rheological model of Eq. (5.26). The model is composed of a nonlinear spring, a Maxwell Unit I characterizing the viscoelastic response under low strain rates, and a Maxwell Unit II characterizing the viscoelastic response under high strain rates in parallel, corresponding to

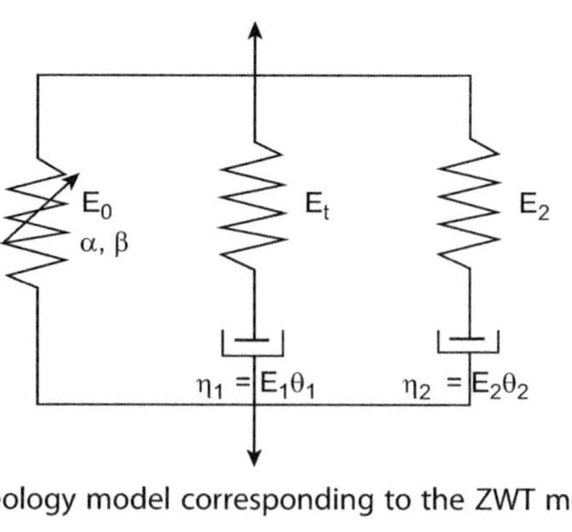

Figure 5.33 Rheology model corresponding to the ZWT model (Eq. 5.26).

Eq. (5.26a) that the total stress is the sum of f_e (ε), σ_1, and σ_2. f_e (ε) describes the nonlinear elastic equilibrium response (corresponding to the nonlinear spring in Fig. 5.32); when it is described by Eq. (5.26b), E_0, α, and β refer to the corresponding elastic constants; when it is described by Eq. (5.26c), σ_{m}, m, and the positive integer n are material parameters. Their physical meanings are as follows: σ_m denotes the asymptotic maximum of $f_e(\varepsilon)$ when $\varepsilon \rightarrow \infty$, m is the ratio of E_0 and σ_m, and the positive integer n is a material parameter characterizing the initial linearity of f_e (ε). The first integral item on the left of the equal sign in Eq. (5.26a) describes the viscoelastic response under low strain rates (corresponding to Maxwell Unit I in Fig. 5.33); E_1 and θ_1 refer to the elastic constant and relaxation time of the corresponding Maxwell unit, respectively. The next integral item describes the viscoelastic response under high strain rates (corresponding to Maxwell Unit II in Fig. 5.33); E_2 and θ_2 refer to the elastic constant and relaxation time of the corresponding Maxwell unit, respectively.

Table 5.5 shows the material parameters in the ZWT equation obtained from the experiment for several typical engineering plastics (Wang and Yang, 1992).

The fitting curve (solid line) of the ZWT equation in Fig. 5.31 coincides with the experimental points in the range of strain rates across seven orders of magnitude, not only for the loading curves but also for the unloading curves. These results indicate that the ZWT equation is suitable for characterizing the nonlinear viscoelastic behavior of polymers under a wide strain-rate range from the quasistatic strain rates up to the impact high strain rates.

To understand how relaxation time θ_j characterizes the strain-rate dependence (viscosity characteristic) of viscoelastic materials, we examine an arbitrary Maxwell element. Its constitutive equation in differential form is:

$$\dot{\varepsilon} = \frac{\dot{\sigma}_j}{E_j} + \frac{\sigma_j}{\eta_j} = \frac{\dot{\sigma}_j}{E_j} + \frac{\sigma_j}{E_j \theta_j}$$

Under a constant strain rate ($\dot{\varepsilon}$ = const, $\varepsilon = \dot{\varepsilon} t$),

$$\dot{\varepsilon} = \frac{\dot{\sigma}_j}{E_j} + \frac{\sigma_j}{\eta_j} = \frac{\dot{\sigma}_j}{E_j} + \frac{\sigma_j}{E_j \theta_j}$$

When $\dot{\varepsilon}$ approaches infinity, σ_j reaches its maximum value, namely, transient response $\sigma_I = E_j \varepsilon$; when $\dot{\varepsilon}$ approaches zero, σ_j reaches its minimum value, namely, equilibrium response $\sigma_E = 0$. If a dimensionless stress relaxation response is introduced, as defined below, then:

$$\overline{\sigma}_j = \frac{\sigma_j}{\sigma_{\max} - \sigma_{\min}} = \frac{\sigma_j}{\sigma_I - \sigma_E} = \frac{\sigma_j}{E_j \varepsilon}$$

It can be modified as:

$$\overline{\sigma}_j = \frac{\theta_j \dot{\varepsilon}}{\varepsilon}\left\{1 - \exp\left(-\frac{\varepsilon}{\theta_j \dot{\varepsilon}}\right)\right\} = \frac{\theta_j}{t}\left\{1 - \exp\left(-\frac{t}{\theta_j}\right)\right\}$$

If $\overline{\sigma}_j = 0.995$ and $\overline{\sigma}_j = 0.005$ are approximately regarded as the beginning and the end of a relaxation process, respectively, then the "effective influence domain" (abbreviated as EID) can be determined (Chu et al., 1985). If the EID is in terms of time, then:

$$10^{-2} \le \frac{t}{\theta_j} \le 10^{2.3}$$

If the EID is in terms of strain rate, then (assuming $\varepsilon = 1$):

$$10^2 \ge \frac{\dot{\varepsilon}}{\left(\frac{\varepsilon}{\theta_j}\right)} \ge 10^{-2.3}$$

It means that the EID of any relaxation time, regardless of being in terms of time or strain rate, is approximately 4.5 orders of magnitude.

For a viscoelastic material consisting three Maxwell elements in parallel, namely, *i*, *j*, and *k*, with the corresponding relaxation times of $\theta_i = 10^2$, $\theta_j = 10^{-5}$, and $\theta_k = 10^{-10}$ s^{-1}, respectively, the stress relaxation response σ as a function of $\dot{\varepsilon}$ is shown in Fig. 5.34, showing each Maxwell element has its own EID approximately 4.5 orders of magnitude. It can be seen that the EID of Maxwell element *j* with $\theta_j = 10^{-5}$ s^{-1} (solid line in Fig. 5.34) is exactly within the range of strain rate $10^{0.7}$–10^5 s^{-1} that we care about in the impact experiment. In this strain-rate range, Maxwell element *k* with $\theta_k = 10^{-10}$ s^{-1} (dotted line in Fig. 5.34) has relaxed to its equilibrium state (the viscoelastic response is zero), and Maxwell element *i* with $\theta_i = 10^2$ s^{-1} does not have sufficient time to relax, only has not yet relaxed transient response ($\sigma_i = E_i \varepsilon$).

Given that the EID of any relaxation time is 4.5 orders of magnitude, it can be imagined that the viscoelastic model with two relaxation times in parallel connection (such as θ_{Low} and θ_{High} to dominate quasistatic and impact responses, respectively) is sufficient to characterize the viscoelastic response

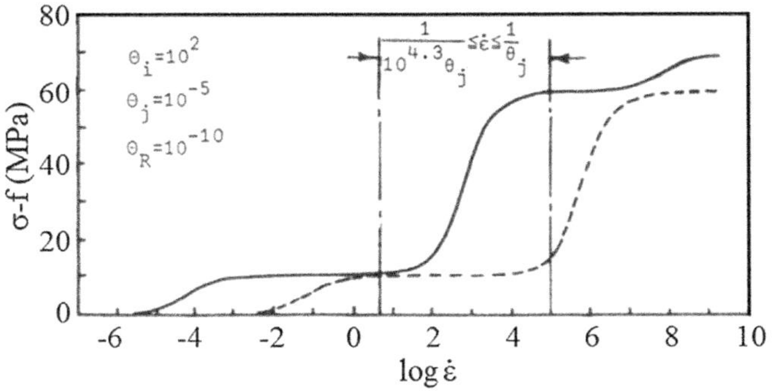

Figure 5.34 The effective influence domain (EID) for relaxation time θ_j. *From Chu, C., Wang, L., Xu, D., 1985. A nonlinear thermo-viscoelastic constitutive equation for thermoset plastics at high strain-rates. Proceedings of the International Conference on Nonlinear Mechanics (Oct. 28-31, 1985, Shanghai, China). Ed. Wei-Zang, C. Science Press, Beijing, 92–97.*

in the strain-rate range of about nine orders of magnitude (from quasistatic to impact dynamics). This means that in this case, it is no longer necessary to pursue a viscoelastic model consisting of N (>2) relaxation time.

In connection with the previous analysis and Table 5.5, it is particularly necessary to emphasize the following characteristics of the ZWT equation (Eq. 5.26).

(a) The constitutive nonlinearity only originates from purely elastic responses, and all viscoelastic responses or rate (time)-related responses are essentially linear. This constitutive nonlinearity is a type of weak nonlinearity, which may be called rate-independent nonlinearity. If materials of this type (satisfying Formula (5.26)) are called ZWT materials, then it is not difficult to generalize the classical linear viscoelasticity theory to the rate-dependent response of ZWT materials.

Table 5.5 ZWT parameters of epoxy, polymethyl methacrylate (PMMA), and polycarbonate (PC).

	Epoxy	PMMA-1	PMMA-2	PMMA-3	PC
ρ_o (kg/m^3)	1.20×10^3	1.19×10^3	1.19×10^3	1.19×10^3	1.20×10^3
E_o (GPa)	1.96	2.04	2.19	2.95	2.20
$\alpha\alpha$ (GPa)	4.12	4.17	4.55	10.9	23
β (GPa)	−181	−233	−199	−96.4	−52
E_1 (GPa)	1.47	0.897	0.949	0.832	0.10
θ_1 (s)	157	15.3	13.8	7.33	470
E_2 (GPa)	3.43	3.07	3.98	5.24	0.73
θ_2 (μs)	8.57	95.4	67.4	40.5	140

(b) The experiments on typical engineering plastics (Table 5.5) show that the ratio of α/E_0 refers to 1—10 orders of magnitude, and the ratio of β/E_0 refers to $10-10^2$ orders of magnitude. Thus, if $\varepsilon > 0.01$, nonlinearity should be considered; on the contrary, if $\varepsilon < 0.01$, nonlinearity can be approximately neglected.

(c) The experiments (Table 5.5) also show that θ_1 is usually $10-10^2$ s orders of magnitude (the unit *i* in Fig. 5.33), and θ_2 is usually $10^{-4}-10^{-6}$ s orders of magnitude (the unit *j* in Fig. 5.34). Therefore, θ_1 is responsible for the low strain-rate response and θ_2 is responsible for the high strain-rate response. Since θ_1 is four to six orders of magnitude higher than θ_2, and since the EID of each relaxation time accounts for approximately 4.5 orders of magnitude, θ_1 and θ_2 will play their roles in their own effective influence area.

(d) Consequently, under the quasistatic loading condition with time scales of $1-10^2$ s, the high-frequency Maxwell element with the relaxation time $\theta_2 = 10^{-4}-10^{-6}$ s has completely relaxed at the beginning of the quasistatic loading. Therefore, Formula (5.26) is simplified as:

$$\sigma = f_e(\varepsilon) + E_1 \int_0^t \dot{\varepsilon}(t)\exp\left(-\frac{t-\tau}{\theta_1}\right)d\tau \tag{5.27a}$$

(e) On the contrary, under the impact loading condition with time scales of $1-10^2$ μs, the low-frequency Maxwell element with the relaxation time $\theta_1 = 10-10^2$ s does not have sufficient time to relax until the end of the loading. At this moment, the low-frequency Maxwell element is simplified as a simple spring with an elastic constant of E_1, and Formula (5.26) is simplified as:

$$\begin{aligned}\sigma &= f_e(\varepsilon) + E_1\varepsilon + E_2 \int_0^t \dot{\varepsilon}(t)\exp\left(-\frac{t-\tau}{\theta_2}\right)d\tau \\ &= \sigma_e(\varepsilon) + E_2 \int_0^t \dot{\varepsilon}(t)\exp\left(-\frac{t-\tau}{\theta_2}\right)d\tau\end{aligned} \tag{5.27b}$$

This result indicates that the propagation characteristics of nonlinear viscoelastic waves of polymers under impact loading are governed by Formula (5.27b). The constitutive parameters in the equation must be determined at a specific strain rate because their respective material constitutive parameters have different physical meanings and are applicable to different loading conditions.

It is worth noting that, although Eqs. (5.27a) and (5.27b) are formally equivalent to the three-element model, but their respective constitutive parameters have different physical meanings and are applicable to different loading conditions, so the constitutive parameters in the formulas must be determined at a specific strain rate. Specifically, $\sigma_e(\varepsilon)$ in Eq. (5.27b) includes not only the elastic equilibrium response but also the elastic response of the low-strain-rate Maxwell element. Therefore, this formula should not be simply regarded as an ordinary three-element model but should be viewed as a simplified case of the ZWT model under given conditions.

Theoretically, Eq. (5.26) can be derived on the basis of viscoelastic constitutive theory of modern continuum mechanics (rational mechanics) (Zhu Zhao-xiang, 2015), e.g., either from Coleman—Noll's finite linear viscoelastic theory (Coleman and Noll, 1960, 1961) or from Green—Rivlin's multiple integral constitutive theory (Green and Rivlin, 1957). In fact, the one-dimensional form of Green—Rivlin's multiple integral is shown as:

$$\sigma(t)=\int_{-\infty}^{t}\varphi_1(t-\tau_1)\dot{\varepsilon}(\tau_1)d\tau_1+\int\int_{-\infty}^{t}\varphi_2(t-\tau_1,t-\tau_2)\dot{\varepsilon}(\tau_1)\dot{\varepsilon}(\tau_2)d\tau_1 d\tau_2$$
$$+\int\int\int_{-\infty}^{t}\varphi_3(t-\tau_1,t-\tau_2,t-\tau_3)\dot{\varepsilon}(\tau_1)\dot{\varepsilon}(\tau_2)\dot{\varepsilon}(\tau_3)d\tau_1 d\tau_2 d\tau_3+\cdots \tag{5.28}$$

where Φ_i is the relaxation function, the first term of the single integral on the right of the equal sign represents the linear term that obeys the superposition principle; the second term of the double integral refers to the accumulation of the influence of the joint action of the strain increment at τ_1 with the strain increment at τ_2 on material behavior at current moment t; the triple integral refers to the accumulation of influence of strain increments at three time points of τ_1, τ_2, and τ_3 on material behavior at the current moment; and so forth. However, Eq. (5.28) is unsuitable for engineering application. Even if only the first three terms are used, at least 28 groups of different experiments will be required to determine the relaxation functions f_x (Lockett, 1972).

On the basis of the experimental results of the dynamic mechanical behavior of epoxy resin, polycarbonate, nylon, PMMA, ABS, PBT, and numerous types of solid polymer materials (Wang and Yang, 1992), we can reasonably assume that

$$\varphi_1(t-\tau_1)=E_o+E_1\exp\left(-\frac{t-\tau_1}{\theta_1}\right)+E_2\exp\left(-\frac{t-\tau_1}{\theta_2}\right)$$

$$\varphi_2(t-\tau_1, t-\tau_2) = \text{const.} = \alpha$$

$$\varphi_3(t-\tau_1, t-\tau_2, t-\tau_3) = \text{const.} = \beta$$

At this moment, Green—Rivlin's multiple integral (Eq. 5.28) is simplified as the ZWT equation (Eq. 5.26).

The ZWT equation is not only applicable to the polymer itself, but also applicable to composites with polymer matrix showing by the research combining the micromechanics theoretical analysis and the experimental research (Yang et al., 1986; 1993; 1994; Yang and Wang, 1991), and can even be applied to concrete materials (Chen and Wang, 1997, 2000; Wang et al., 2000).

Eq. (5.26) is the one-dimensional stress form of ZWT equation, which can be generalized to a three-dimensional stress state by using tensor description (Zhou and Liu, 1966; Yang and Shim, 2005; Feng Zhen-zhou et al., 2007). Yang and Shim (2005) and Feng Zhen-zhou et al. (2007) compiled the corresponding subprogram in a commercial code (e.g., LS-DYNA and ABAQUS) in the tensor form of the ZWT equation. ZWT equation is not only successfully used to describe the dynamic impact behavior of rubber, foam plastics, and biomaterials (Wang et al., 2000) but also to simulate the impact response of engineering structures such as mobile phone dropping (Yang and Shim, 2005) and bird strike aircraft windshield (Wang et al., 1992; Feng Zhen-zhou et al., 2007).

5.3.2 Nonlinear thermoviscoelastic constitutive equation and rate—temperature equivalence

The abovementioned nonlinear viscoelastic constitutive equation does not consider the effect of temperature yet. However, in the engineering application of polymer materials, a change in ambient temperature exerts a significant effect on mechanical properties. Many studies have been conducted under quasistatic or even low strain rates (e.g., creep and relaxation experiments) (Ward, 1983). In this section, we discuss the temperature effect of nonlinear viscoelastic materials at high strain rates and the related rate—temperature equivalence.

- **Experimental study on the temperature effect at high strain rates**

With the experimental technique of temperature-controlled split Hopkinson bar technique (SHBT), Fig. 5.35A shows the stress—strain curves at the high strain rate of $10^3\ \text{s}^{-1}$ of epoxy resin at different temperatures (−20°C−100°C) (Zhu Zhaoxiang et al., 1988). The curves at high strain rate ($10^3\ \text{s}^{-1}$) between the stress (for different given strains) versus the

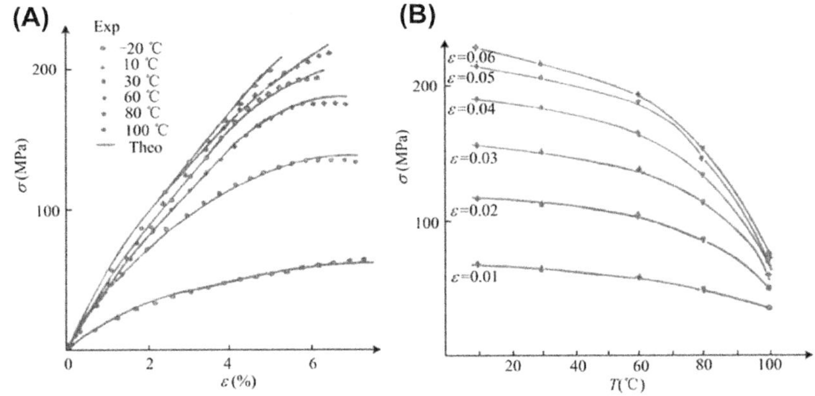

Figure 5.35 The experimental results for epoxy resin under high strain rate f_x: (A) stress–strain curves at different temperatures and (B) stress–temperature curves for different strain values. *From Zhu Zhao-xiang, Xu, Da-ben, Wang, Lli., 1988. Thermo-Viscoelastic constitutive equation and time-temperature equivalence of epoxy resin at high strain rates. J. Ningbo Univ. (Nat. Sci. Eng. Ed.). 1 (1) 58–68 (in Chinese).*

temperature can be replotted from the data of Fig. 5.35A, as shown in Fig. 5.35B, indicating an obvious temperature effect.

Thermoplastic materials have more sensitive temperature effect than thermosetting plastics (such as epoxy resin). Fig. 5.36 shows the stress–strain curves for polymethyl methacrylate (PMMA) at the high strain rate of $9 \times 10^2\ s^{-1}$ and at different temperatures (−60°C–100°C) (Shaoqiu

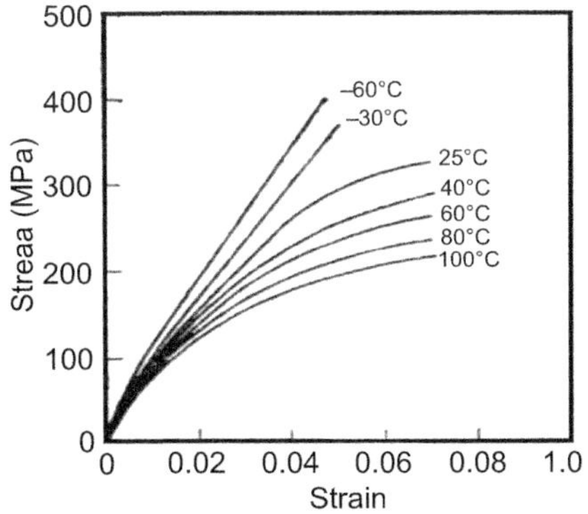

Figure 5.36 Stress–strain curves for polymethyl methacrylate under different temperature at high strain rate ($\dot{\varepsilon} = 9 \times 10^2\ s^{-1}$). *From Shi Shao-qiu, Gan, S., Wang, L., 1990. The thermo-viscoelastic mechanical behaviour of an aeronautical PMMA under impact loading. J. Ningbo Univ. (Nat. Sci. Eng. Ed.). 3 (2) 66–75 (in Chinese).*

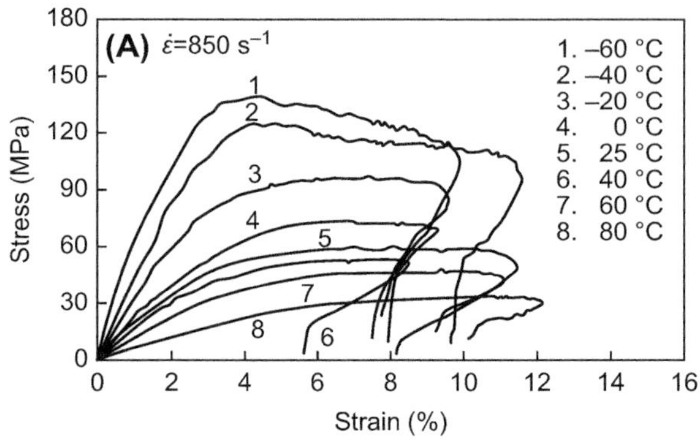

Figure 5.37 Stress—strain curves for PP/PA copolymer under different temperature at high strain rate (f_x). *From Shi Shao-qiu, Yu Bing, Wang L., 2007a. The thermo-viscoelastic constitutive equation of PP/PA blends and its rate temperature equivalency at high strain rates. Explos. Shock Waves. 27 (3) 210—216 (in Chinese); Shi Shao-qiu, Yu B., Wang L., 2007b. The thermo-viscoelastic constitutive equation of PP/PA blends and its rate temperature equivalency at high strain rates. Macromol. Symp. 247 28—34*

et al., 1990). Fig. 5.37 presents the stress—strain curves for PP/PA copolymers at a high strain rate of $8.5 \times 10^2\ s^{-1}$ and at different temperatures (−60°C—80°C) (Shaoqiu et al., 2007).

These results show that polymer materials are sensitive to not only strain rate but also temperature, showing that the stress—strain curves decrease with the increase of temperature (thermal softening effect). Thus, the temperature decrease and the strain rate increase show a certain equivalent effect.

- **Nonlinear thermoviscoelastic constitutive equation**

According to the previous experimental result, the material parameters in the ZWT equation (Eq. 5.26), namely, E_o, α, β, E_1, θ_1, E_2, and θ_2, should be functions of temperature T in general. Hence,

$$\sigma = f_e(T,\varepsilon) + E_1(T)\int_0^t \dot{\varepsilon}\exp\left(-\frac{t-\tau}{\theta_1(T)}\right)d\tau + E_2(T)\int_0^t \dot{\varepsilon}\exp\left(-\frac{t-\tau}{\theta_2(T)}\right)d\tau \tag{5.29a}$$

$$f_e(T,\varepsilon) = E_0(T)\varepsilon + \alpha(T)\varepsilon^2 + \beta(T)\varepsilon^3 \tag{5.29b}$$

This formula is the ZWT nonlinear thermoviscoelastic constitutive equation that considers both strain rate and temperature effects.

As shown by Eq. (5.27b), under impact loading conditions with time scales of $1-10^2$ μs, the low-frequency Maxwell element is simplified as a simple spring with an elastic constant E_1 due to not having sufficient time to relax. Thus, the previous formula is simplified as:

$$\sigma = \sigma_{eff}(T,\varepsilon) + E_2(T)\int_0^t \dot{\varepsilon}\exp\left(-\frac{t-\tau}{\theta_2(T)}\right)d\tau \tag{5.30a}$$

$$\begin{aligned}\sigma_{eff}(T,\varepsilon) &= [E_0(T)+E_1(T)]\varepsilon + \alpha(T)\varepsilon^2 + \beta(T)\varepsilon^3 \\ &= E_a(T)\varepsilon + \alpha(T)\varepsilon^2 + \beta(T)\varepsilon^3\end{aligned} \tag{5.30b}$$

where $E_a(T) = E_0(T) + E_1(T)$. The rate-independent nonlinear elastic equilibrium term $\sigma_{eff}(T,\varepsilon)$ can be moved from the right side of the equation to the left side; so that the right side contains only the viscoelastic term, namely, Eq. (5.30a) can be rewritten as:

$$\sigma_{over} = \sigma - \sigma_{eff}(T,\varepsilon) = E_2(T)\int_0^t \dot{\varepsilon}\exp\left(-\frac{t-\tau}{\theta_2(T)}\right)d\tau \tag{5.30c}$$

where $\sigma_{over} = \sigma(T,\varepsilon,\dot{\varepsilon}) - \sigma_{eff}(T,\varepsilon)$, similar to the overstress in visco-plastic theory, can be referred to the overstress in viscoelastic theory. Then the previous formula can also be described in a dimensionless form as:

$$\overline{\sigma} = \frac{\sigma - \sigma_{eff}(T,\varepsilon)}{E_2(T)} = \int_0^t \dot{\varepsilon}\exp\left(-\frac{t-\tau}{\theta_2(T)}\right)d\tau \tag{5.31}$$

where $\overline{\sigma} == \frac{\sigma-\sigma_{eff}(T,\varepsilon)}{E_2(T)}$ refers to dimensionless overstress.

The material parameters in the ZWT nonlinear thermoviscoelastic constitutive equation can be determined by fitting with experimental data. Tables 5.6–5.8 show the parameters of epoxy resin (Zhaoxiang et al., 1988), polymethyl methacrylate (PMMA) (Shaoqiu et al., 1990), and PP/PA copolymers (Shaoqiu et al., 2007), respectively.

According to the ZWT material parameters given in Table 5.6, the theoretical curves (solid lines) is plotted in Fig. 5.35A. It is shown that before the stress–strain curve becomes unstable ($\frac{d\sigma}{d\varepsilon} = 0$), the stress–strain curves under

Table 5.6 ZWT material parameters of epoxy resin at different temperatures, $\eta = E\theta$.

T (°C)	E_a (GPa)	α (GPa)	β (GPa)	E_2 (GPa)	θ_2 (μs)	η_2 (KPa ·s)
10	3.92	14.2	−411	5.42	3.32	1.80
30	3.91	8.87	−360	5.02	3.19	1.60
60	3.84	−1.93	−257	5.29	2.67	1.41
80	3.81	−27.8	2.6	5.67	2.01	1.14
100	1.41	−9.12	−3.0	5.23	1.99	1.04

Table 5.7 ZWT material parameters of PMMA at different temperatures.

T (°C)	E_a (GPa)	α (GPa)	β (GPa)	E_1 (GPa)	θ_1 (μs)	E_2 (GPa)	θ_2 (μs)
−60	4.16	−5.34	23.3	1.22	80.0	4.95	54.1
−30	3.66	−5.38	−65.3	1.12	7.89	4.91	49.0
25	2.95	−10.9	−96.4	0.832	7.33	5.24	40.5
40	2.66	−13.3	−95.6	0.776	5.26	5.02	38.9
60	2.39	−14.6	−80.1	0.649	4.43	4.98	34.8
80	2.05	−15.3	−65.3	0.564	2.38	4.98	33.9
100	1.60	−16.6	−23.0	0.537	0.89	5.14	31.9

Table 5.8 ZWT material parameters of PP/PA copolymer at different temperatures, $\eta = E\theta$.

T (°C)	E_0 (GPa)	α (GPa)	β (GPa)	E_1 (GPa)	θ_1 (s)	E_2 (GPa)	θ_2 (μs)	η (Pa ·s)
25	0.67	−3.32	−18.14	0.30	34.04	0.86	6.82	5872
40	0.62	−6.39	28.03	0.24	23.91	0.83	6.34	5276
60	0.61	−8.40	52.60	0.16	14.06	0.82	6.03	4913
80	0.45	−10.51	86.94	0.12	2.41	0.81	5.67	4567

the high strain rate (10^3 s^{-1}) at −20°C−100°C can be described by the ZWT nonlinear thermoviscoelastic constitutive equation satisfactorily.

- **Time−temperature and rate−temperature equivalence of nonlinear thermoviscoelastic responses at high strain rates**

Eq. (5.31) shows that the mechanical response of nonlinear viscoelastic materials at high strain rates is generally a function of both strain rate and temperature. The experimental results in Figs. 5.35−5.37 indicate that the temperature decrease is similar to the strain-rate increase.

Therefore, similar to the rate−temperature equivalence of metal materials discussed in Section 5.1.2, the study mechanical properties of polymer materials on the equivalence of strain rate f_x (or time t) and temperature T are also one of the important topics that scientists in both mechanics and chemistry fields pay attention to.

In the study of linear viscoelastic at low strain rates, the strain-rate dependence of mechanical response usually presents as time dependence, such as the decrease of stress with time under a given strain (stress relaxation) or the increase in strain with time under a given stress (creep) and so on. Such time dependence is essentially a manifestation of constitutive strain-rate dependence. In dimension, strain rate and time are reciprocal to each other. The Maxwell viscoelastic model with spring element and dashpot element in series was initially proposed to elucidate stress relaxation. The

Kelvin—Voigt viscoelastic model with spring element and dashpot element in parallel was proposed to elucidate the creep phenomenon.

In history, people initially paid attention to whether a certain transformation or equivalence relation exists between the time-dependence and temperature-dependence of mechanical responses under low strain rates, namely, time—temperature equivalence (Ward, 1983). That is, by changing the time scale (horizontal movement of a quantity $\log\alpha_T$ on time scale), the viscoelastic behavior at one temperature can be superimposed to or associated with the viscoelastic behavior at another temperature. $\alpha_T\,(T)$ is called a shift factor.

Introduce the reduce time ξ as a dimensionless time in terms of $\alpha_T\,(T)$.

$$\xi = \frac{t}{a(T)} \tag{5.32}$$

If the isothermal constitutive equation under known reference temperature T_0 is:

$$\sigma = \int_0^t \varphi(t-\tau)\,\dot{\varepsilon}\,d\tau \tag{5.33}$$

then, the constitutive equation under ambient environment T can be presented as:

$$\sigma = \int_0^\xi \varphi(\xi-\xi')\frac{\partial\varepsilon}{\partial\xi'}\,d\xi' \tag{5.34}$$

A material with these properties is called a thermo-rheologically simple material.

It is noted that $\theta_2(T)$ in the ZWT nonlinear thermoviscoelastic equation under high strain rates (Eq. 5.30) is the control factor of viscosity characteristics, and it can be seen from Tables 5.6—5.8 that $\theta_2(T)$ significantly decreases with the increase in temperature. We can define the shift factor $\alpha_T\,(T)$ in Eq. (5.32) as the following dimensionless relaxation time.

$$\alpha(T) = \frac{\theta_2(T)}{\theta_{2\mathrm{R}}(T)} \tag{5.35}$$

where $\theta_{2\mathrm{R}}(T)$ refers to $\theta_2(T_R)$ under reference temperature T_R. On the basis of Eqs. (5.32), (5.35) and (5.31), it can be written under any ambient environment T as (Chu et al., 1985, Zhu Zhao-xiang et al., 1988):

$$\overline{\sigma} = \frac{\sigma - \sigma_{eff}(T,\varepsilon)}{E_2(T)} = \int_0^\xi \frac{d\varepsilon}{d\xi'}\exp\left(-\frac{\xi-\xi'}{\theta_{2R}}\right)d\xi' \tag{5.36}$$

This means that if dimensionless overstress $\overline{\sigma}$ is used to describe the thermoviscoelastic response of materials under high strain rates, then it is only a function of reduce time ξ, and T and t no longer appear as two independent variables. In other words, provided that the history of $\boldsymbol{\varepsilon}$ (ξ) is the same, the $\overline{\sigma}$-ξ relations at any temperature will be superimposed on the same curve, namely, the so-called master curve.

If the dimensionless parameter Z_{VE} is introduced:

$$Z_{VE} = \dot{\boldsymbol{\varepsilon}}\ \theta_2(T) = \frac{d\boldsymbol{\varepsilon}}{d\xi}\theta_{2R} \tag{5.37}$$

Since in the experiment of constant temperature and constant strain rate, Z is a constant, Eq. (5.36) is changed to:

$$\frac{\overline{\sigma}}{Z_{VE}} = 1 - \exp\left(-\frac{\varepsilon}{Z_{VE}}\right) \tag{5.38}$$

The above formula describes the unified stress—strain curve with Z_{VE} as a single parameter, $\frac{\overline{\sigma}}{Z_{VE}} - \frac{\varepsilon}{Z_{VE}}$(Chu et al., 1985; Zhu Zhao-xiang et al., 1988), reflecting the time—temperature equivalence or rate—temperature equivalence. In fact, when the temperature increases, $\theta_2(T)$ decreases (see Tables 5.6—5.8), which means that Z_{VE} decreases when $\dot{\boldsymbol{\varepsilon}}$ is constant; this case is equivalent to the decrease in $\dot{\boldsymbol{\varepsilon}}$ when T is constant. For epoxy resin and PP/PA copolymer, all experimental data are drawn in $\frac{\overline{\sigma}}{Z_{VE}} - \frac{\varepsilon}{Z_{VE}}$ coordinates, as shown in Fig. 5.38 (Zhu Zhaoxiang et al., 1988) and Fig.5.39 (Shaoqiu et al., 2007), respectively. The theoretical curves predicted by Eq. (5.38) are provided in these figures too, respectively, for comparison. As can be seen that the high strain-rate curves at different temperatures originally shown in Figs. 5.35 and 5.37 can be approximately normalized to a master curve, respectively. Note that Eq. (5.38) is established under the condition of constant temperature and constant strain rate, while such condition is not strictly satisfied in any SHPB experiment due to e.g., adiabatic temperature increases with increasing strain and strain rate often changes in an experimental process, and so on. Thus, the experimental results showing that most data are distributed near the theoretical curve with a certain error is surely understandable. Even with a certain error, it is sufficient to reflect the time—temperature equivalence or rate—temperature equivalence under high strain rates.

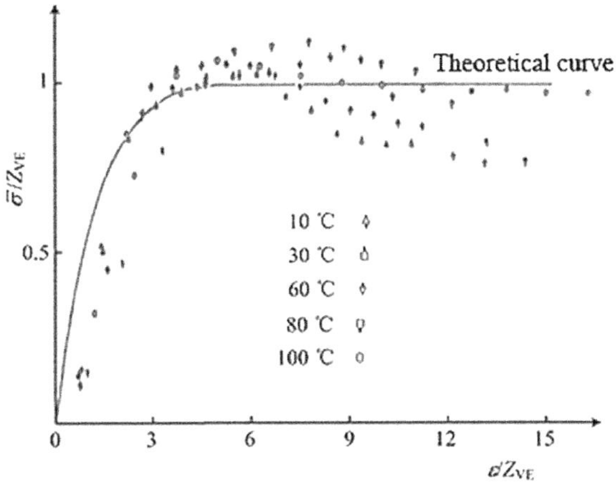

Figure 5.38 $\frac{\overline{\sigma}}{Z_{VE}} - \frac{\varepsilon}{Z_{VE}}$ curve of epoxy resin under different temperatures and high strain rate ($\dot{\varepsilon} = 9 \times 10^2$ s^{-1}). *From Zhu Zhao-xiang, Xu Da-ben, Wang L., 1988. Thermo-Viscoelastic constitutive equation and time-temperature equivalence of epoxy resin at high strain rates. J. Ningbo Univ. (Nat. Sci. Eng. Ed.). 1 (1), 58–68 (in Chinese).*

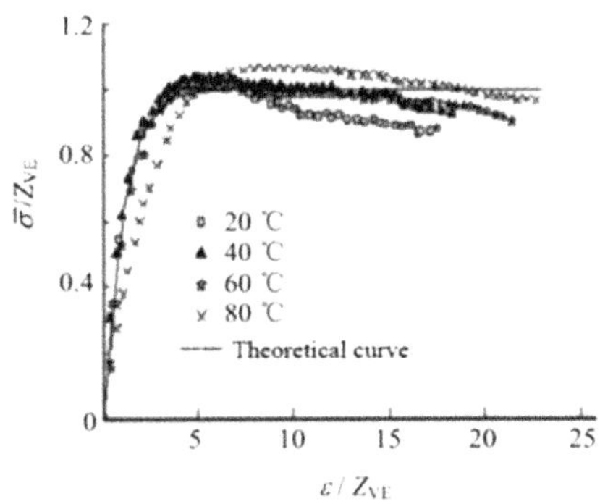

Figure 5.39 $\frac{\overline{\sigma}}{Z_{VE}} - \frac{\varepsilon}{Z_{VE}}$ curve of PP/PA copolymer under different temperatures and high strain rate ($\dot{\varepsilon} = 8.5 \times 10^2$ s^{-1}). *From Shi Shaoqiu et al. (2007).*

It is worthwhile pointing out that if the temperature dependence of relaxation time θ *(T)* follows the Arrhenius equation, then:

$$\theta(T) = \theta_0 \exp\left(-\frac{A}{kT}\right) \tag{5.39}$$

where k refers to the Boltzmann constant and A represents the activation energy. Substitute the above equation into Eq. (5.37), the following is obtained.

$$Z_{VE} = \dot{\varepsilon}\theta(T) = \theta_0\dot{\varepsilon}\exp\left(-\frac{A}{kT}\right) \tag{5.40}$$

Thus, the Zener–Hollomon parameter $Z = \dot{\varepsilon}\exp\left(\frac{U}{kT}\right)$, as discussed in Section 5.1.2, "The combined effects of strain rate and temperature and the rate–temperature equivalence", can be regarded as a special case of Z_{VE}. The rate–temperature equivalence parameters in metal materials are essentially interrelated with the rate–temperature equivalence parameters in the polymer materials discussed here.

5.4 Constitutive model under one-dimensional strain

So far, we have discussed the rate-independent volumetric deformation law of materials in the first chapter and the rate-dependent distortion law of materials in this chapter. When the pressure is so high that the distortion-resistant shear strength of materials can be ignored, it is only needed to consider the state equation of materials under high pressure, namely, the so-called hydrodynamic model can be adopted. However, when the pressure is not considerably high, the shear strength of materials cannot be ignored; the rate-independent volumetric deformation law and the rate-dependent distortion law of materials must be considered simultaneously. The combination of the two forms a complete constitutive relation.

For a large number of actual problems under explosion/shock loading, whether dealing with specific engineering problems or conducting experimental studies, the most basic and common three-dimensional stress state is the one-dimensional strain state. In this case, the transverse displacement, strain, and particle velocity are all zero, whereas the two transverse principal stresses are symmetric and equal.

$$u_Y = u_Z = \varepsilon_Y = \frac{\partial u_Y}{\partial Y} = \varepsilon_Z = \frac{\partial u_Z}{\partial Z} = v_Y = \frac{\partial u_Y}{\partial t} = v_Z = \frac{\partial u_Z}{\partial t} = 0, \sigma_Y = \sigma_Z \tag{5.41}$$

Therefore, axial stress σ_X can be decomposed into the sum of hydrostatic pressure P (spherical term) and maximum shear stress τ (distortional term).

$$\sigma_X = -P + S_{XX} = -p + \frac{2}{3}(\sigma_X - \sigma_Y) = -P + \frac{4}{3}\tau \tag{5.42}$$

where P increases with the increase of volumetric compression and there is no limit to its value, which can be as high as 10^5–10^6 MPa orders of magnitude in practice; τ increases with the increase of shear deformation of the medium and takes the shear strength of the material as the limit, which for most engineering materials is approximately 10^1–10^3 MPa orders of magnitude. Therefore, if the impact pressure is two orders higher than the shear strength of materials, namely, $\tau/P \leq .01$, then the τ term in the previous formula can be neglected, and approximately $\sigma_X \approx -p$. Only at this moment can the hydrodynamic model provide a good enough approximation. On the contrary, when the impact pressure is close or equal to the magnitude of shear strength of materials, the hydrodynamics approximation is no longer applicable. The constitutive relation composed of the volumetric deformation law and the distortional law must be fully considered.

In order to account for the influence of shear strength of solids, Zheng Zhe-min and Xie Bo-min (1965), Lee (1971), and Chou and Hopkins (1972) independently developed the hydroelastoplastic model.

The hydroelastoplastic model combines the equation of state of solids under high pressures, reflecting the nonlinear volumetric deformation law of materials (e.g., the internal energy–type state equation $P = P(E, V)$ such as Eq. (2.65)), with the nonlinear elastoplastic distortion law. Note that under one-dimensional strain conditions, the Mises criterion is in the same form as the Tresca criterion.

$$\sigma_X - \sigma_Y = \pm Y \tag{5.43}$$

then the hydroelastoplastic model can be expressed as:

$$\text{Volumetric deformation law: } P = P(E, V) \tag{5.44a}$$

$$\text{Distortion law: } \sigma_X - \sigma_Y = \begin{cases} 2G\varepsilon_X & \text{(elastic)} \\ \pm Y & \text{(plastic)} \end{cases} \tag{5.44b}$$

The elastic–plastic distortion law can also be expressed as the following differential form.

$$\frac{dS_{XX}}{d\varepsilon_X} = \begin{cases} \dfrac{4}{3}G, & |S_{XX}| < \dfrac{2}{3}Y, \\ \dfrac{4}{3}G_p, & |S_{XX}| = \dfrac{2}{3}Y. \end{cases} \tag{5.44c}$$

where G_p refers to the plastic shear modulus, which is generally the function of plastic work.

On the σ_X–σ_Y stress plane (Fig. 5.40), Eq. (5.43) is described as two parallel lines with a slope of 1, called yield locus. The elastic region is bounded by the upper and lower branches of yield loci. The two yield loci are parallel

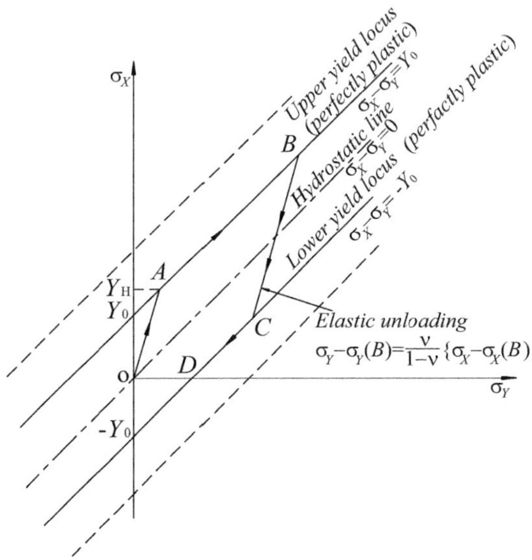

Figure 5.40 The upper and the lower yield locus on the $\sigma_X - \sigma_Y$ stress plane.

and symmetrical to the hydrostatic pressure line $\sigma_X = \sigma_Y\ (=\sigma_Z)$. This is the expression of the assumption that hydrostatic pressure has no effect on yield. For perfect plastic materials, $Y = Y_0$; then the yield locus is fixed. For isotropic hardening materials, Y is a function of plastic deformation or plastic work; with the increase of plastic deformation or plastic work, the upper and lower yield loci remain symmetrical to the hydrostatic pressure line and expand outward in parallel, as shown by the two dotted lines in Fig. 5.40.

Let us take another look at the axial stress—strain relation $\sigma_X - \varepsilon_X$ under the one-dimensional strain condition, as shown in Fig. 5.41.

In the elastic phase, in accordance with Hooke's law, Eq. (5.42) can be rewritten as:

$$\begin{aligned} \sigma_X &= -P + S_{XX} = K\Delta + 2Ge_{XX} = \left(K + \frac{4}{3}G\right)\varepsilon_X \\ &= (\lambda + 2\mu)\varepsilon_X = \frac{(1-\nu)E}{(1+\nu)(1-2\nu)}\varepsilon_X = E_L\varepsilon_X \end{aligned} \tag{5.45}$$

which corresponds to the OA section in the figure. In the formula, E_L is the lateral modulus of elasticity; λ and μ denote elastic Lame coefficients; and ν refers to Poisson's ratio, generally, $0 < \nu < 0.5$. The above formula shows that $E_L > E$, that is, the lateral modulus of elasticity of 1D strain is greater than the modulus of elasticity E of 1D stress. The ν of commonly used metal

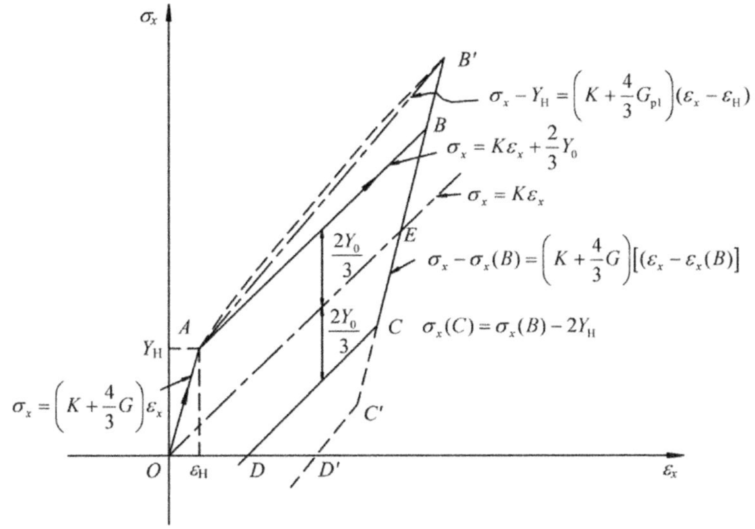

Figure 5.41 Axial elasto–plastic stress–strain relation under the one-dimensional strain condition.

materials is in the range of 1/4—1/3, and correspondingly, E_L/E is in the range of 1.2—1.5.

In the elastic phase, the relation between the transverse lateral stress σ_Y ($=\sigma_Z$) and the axial stress σ_X is as follows:

$$\sigma_Y = \sigma_Z = \frac{\nu}{1-\nu}\sigma_X = \frac{\lambda}{\lambda + 2\mu}\sigma_X \tag{5.46}$$

Substitute this relation into the yield condition (Eq. 5.43), thus, the initial yield limit Y_H of axial stress σ_X under the one-dimensional strain condition can be determined as:

$$Y_H = \frac{1-\nu}{1-2\nu}Y_0 = \frac{\lambda + 2\mu}{2\mu}Y_0 = \frac{K + 4G/3}{2G}Y_0 \tag{5.47}$$

where Y_H is the lateral yield limit or called Hugoniot elasticity limit, which corresponds to Point A in Fig. 5.41. Y_H is higher than the initial yield limit Y_0 under uniaxial stress. For example, if $\nu = 1/3$, then $Y_H = 2\ Y_0$.

In Fig. 5.41, OE refers to the volumetric deformation law. The bulk modulus under higher pressures is no longer a constant; then OE will be a curve. For perfect plastic materials, $Y = Y_0$; the plastic section of the $\sigma_X - \varepsilon_X$ curve maintains an equal distance ($2Y_0/3$) to the OE line, shown as the AB line in Fig. 5.41. For isotropic hardening materials, plastic shear modulus G_p is a function of plastic work W_p, G_p (W_p), which corresponds to the AB' line in Fig. 5.41. For linear hardening materials, $G_p = G_{p1} =$ constant. If unloading begins after the plastic loading at Point B (or Point B'), and is assumed to be elastic unloading, then unloading is carried out along *BC line* (or *B'C line*) parallel to OA line. When unloading to Point E, the medium only bears hydrostatic pressure. As σ_X continues to decrease, the shear stress corresponding to the distortion increases in the direction opposite to the sign of the original loading, that is, the elastic distortion occurs opposite to the direction of loading. Thus, the medium along the EC line is in elastic distortion loading in the opposite direction. In Point C, Eq. (5.43) satisfies the inverse yield condition $\sigma_X - \sigma_Y = -Y$; the material starts to enter reverse plastic deformation. Afterward, the so-called loading for σ_X is actually the reverse plastic loading along the CD line, that is, the reverse plastic loading along the lower yield locus.

For the hydroelastoplastic model discussed previously, people may ask this question: this chapter has been mainly discussing the distortion law related to the strain rate, so in the hydroelastoplastic model, where is the strain-rate—dependent distortion law reflected?

At first glance, the hydroelastoplastic model appears to be a strain rate-independent constitutive model. However, considering that the strain rate under impact load is many orders of magnitude higher than that under quasistatic load, the distortion law in the hydroelastoplastic model can be understood as follows: within a certain strain-rate range of the impact load, the material has a unique **dynamic stress—strain relation** in the average meaning, which is different from the quasistatic stress—strain relation; in this sense, the effect of strain rate has been taken into consideration. However, this strain-rate effect does not appear explicitly in the constitutive equation and is much more convenient for mathematical processing.

Of course, we can also replace Eq. (5.44b) with the various strain rate—dependent distortion laws discussed in this chapter to construct an explicit strain rate—dependent hydroviscoelastic—plastic constitutive model. The discussion on the hydroelastoplastic model in the present section is actually not only due to its wide application until now, but also because it provides a method for the construction of explicit strain rate—dependent hydroviscoelastic—plastic models.

Several researchers have also discovered from experimental studies that under high-explosionessure, the two most basic material characteristic parameters in the distortion law, namely, the elastic shear modulus G and the yield strength Y, which characterize the resistances to the elastic distortion and plastic distortion of materials, respectively, are actually dependent on pressure (spherical stress) and temperature, respectively; while in contrast, the strain-rate effect can be disregarded. At this moment, the study on the constitutive distortion law of solids can be simplified and concluded as the determination of $G = G(P, T)$ and $Y = Y(P, T)$, such as the SCG model proposed by Steinberg et al., (1980):

$$G = G_0\left[1 + \frac{G_0'}{G_0}\frac{P}{\eta^{1/3}} + \frac{G_T'}{G_0}(T - 300)\right] \tag{5.48a}$$

$$Y = Y_0[1 + \beta(\varepsilon + \varepsilon_0)]^n\left[1 + \frac{Y_P'}{Y_0}\frac{P}{\eta^{1/3}} + \frac{G_T'}{G_0}(T - 300)\right] \tag{5.48b}$$

where η $(=V_0/V)$ refers to the compression rate, G_P', G_P', and G_T' represent the partial derivatives of G, Y with respect to pressure P and temperature T, respectively; the subscript "0" indicates the initial state; β and n denote the working hardening parameters. Notably, the distortion law depends on the spherical stress at this moment. This means that the volumetric deformation law and the distortion law are no longer decoupled as assumed previously, but there is a certain coupling relationship. In fact, the study on the constitutive relation of the mutual coupling between the volumetric deformation law and the distortion law has become an important research direction.

CHAPTER SIX

Dynamic constitutive distortional law of materials—micro mechanism based on dislocation dynamics

The discussion in the Chapter 5 indicates that the core problem of what distinguishes the dynamic constitutive distortional law with the quasistatic one is the strain-rate effect. This core problem was discussed in the previous chapter from the view of macroscopic mechanics. Its microscopic physical foundation will be correspondingly discussed from the view of dislocation dynamics in this chapter.

The problem of dislocation was raised in 1930s; it is mainly due to that the difference in orders of magnitude exists between the theoretical magnitude of crystal plastic distortion (shear) strength and the actual magnitude. The theoretical magnitude mentioned here is calculated for the perfect crystal as discussed in the Chapter 3, while the actual magnitude is experimentally measured for the actual crystal, which contains various microdefects, e.g., point defects, dislocations, twins, microcracks, phase transformations, and so on.

From the microscopic view, the dislocation, i.e., the so-called one-dimensional defect of crystals, plays a dominated role for the viscoplastic distortional law, especially its strain-rate sensibility. Consequently, the dislocation dynamics plays a leading role for the dynamic strain-rate—dependent constitutive distortional law. Thus in the following, the discussion will be emphatically focused on the dislocation dynamics.

6.1 Theoretical shear strength

Recalling the discussions in Sections 3.2 and 3.4 in Chapter 3, if a plastic shear slip occurs along a certain lattice plane in a perfect crystal under the action of shear stress τ, the stress τ must be large enough to overcome the binding force between the related crystal particles.

Dynamics of Materials
ISBN: 978-0-12-817321-3
https://doi.org/10.1016/B978-0-12-817321-3.00006-1

Consider a plastic shear slip between two adjacent rows of crystal atoms under the action of shear stress τ, as shown in Fig. 6.1, where a denotes the row spacing and b denotes the atomic spacing. Since a crystal has periodic lattice structure, the distribution of crystal binding energy (potential energy) U and binding force τ $(=-\frac{\partial U}{\partial X})$ must be described by period functions. If the locations A, B, C ... in Fig. 6.1 represent balance positions, where the potential energy reaches the minimum (energy barrier valley) and the binding force is zero, then the binding force reaches its maximum when the slip distance equals $b/4$. Furthermore, when the slip distance equals $b/2$, the binding force reaches zero again, but the potential energy reaches its maximum (energy barrier peak), which is a substable state. After overcoming this maximum energy barrier, once the plastic slip distance reaches a magnitude of an atomic spacing b from a stable state, the potential energy decreases to the minimum again. Assume that the potential energy can be approximately expressed by a cosine function.

$$U = -2A\cos\frac{2\pi X}{b} \tag{6.1}$$

where X denotes the slip displacement and $2A$ denotes the amplitude, then the corresponding shear stress τ can be expressed as:

$$\tau = -\frac{\partial U}{\partial X} = \frac{4\pi A}{b}\sin\frac{2\pi X}{b} \tag{6.2a}$$

When the slip displacement $X << b$, Eq. (6.2a) can be simplified to:

$$\tau = -\frac{\partial U}{\partial X} = \frac{4\pi A}{b}\cdot\frac{2\pi X}{b} \tag{6.2b}$$

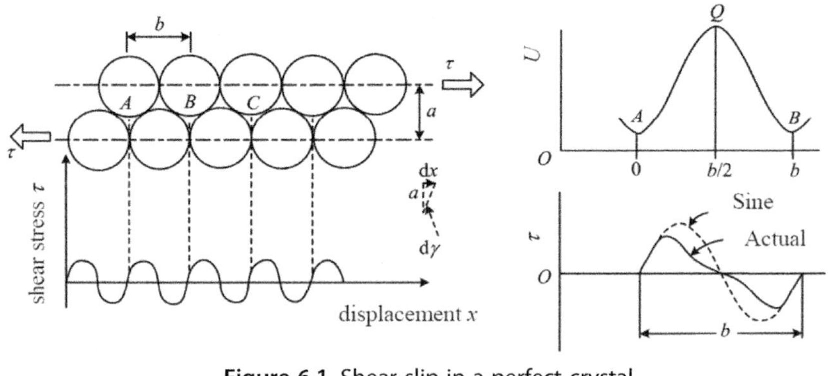

Figure 6.1 Shear slip in a perfect crystal.

Moreover, when the slip displacement $X<<a$, since the shear strain can be approximately expressed as $\gamma \approx \frac{X}{a}$, then according to the Hooke's law, we have:

$$\tau = G\gamma \approx G\frac{X}{a} \tag{6.3}$$

where G denotes the elastic shear modulus. Comparing Eqs. 6.2b and Eq. 6.3, we have the following approximate equation:

$$\frac{4\pi A}{b} = \frac{Gb}{2\pi a} \tag{6.4}$$

Substituting Eq. (6.4) into Eq. (6.2a), we have:

$$\tau = \frac{Gb}{2\pi a}\sin\frac{2\pi X}{b} \tag{6.5}$$

Obviously, the shear stress which can make the upper row of atoms produce a **whole plastic slip** relative to the lower row of atoms, namely the so-called theoretical shear strength (theoretical yield strength), is equal to the maximum of the above equation.

$$\tau_{\max} = \frac{Gb}{2\pi a} \tag{6.6a}$$

The b/a ratio is different for different crystal structures, e.g., the surface core cubic (FCC), the body-centered cubic (BCC), and the hexagonal close-packed (HCC), etc. as well as for different lattice slip planes. But, b and a are generally in the same order of magnitude, so approximately we have:

$$\tau_{\max} \approx \frac{G}{2\pi} \tag{6.6b}$$

Note that the results from Eqs. (6.1)−(6.6) are obtained under the assumption that the potential energy U can be approximately described by a simple cosine function. A more accurate analysis gives that the $\tau_{\max}$ is approximately equal to $G/30$.

The theoretical shear strength based on a **whole plastic slip** assumption, either calculated by $\tau_{\max} = G/2\pi$ or by $\tau_{\max} = G/30$, is very far from the actual experimental measurement. Clearly it is because that the ideal whole plastic slip assumption is not realistic. Comparisons between the theoretical shear strength calculated by $\tau_{\max} = G/2\pi$ and the experimentally measurements for different materials are listed in Table 6.1 (Tegart, 1966; Hertzberg, 1976). As can be seen, the magnitude difference between them is as large as

Table 6.1 Theoretical and experimental shear strength in various metals.

Metal	Theoretical value ($\tau_m = G/2\pi$) (GPa)	Measured value (MPa)
Silver, Ag	12.6	0.37
Aluminum, Al	11.3	0.78
Copper, Cu	19.6	0.49
Nickel, Ni	32	3.2–7.35
Iron, Fe	33.9	27.5
Molybdenum, Mo	54.1	71.6
Niobium, Nb	16.6	33.3
Cadmium, Cd	9.9	0.57
Magnesium, Mg (basal slip)	7	0.39
Magnesium, Mg (prism slip)	7	39.2
Titanium, Ti (prism slip)	16.9	13.7
Beryllium, Be (basal slip)	49.3	1.57
Beryllium, Be (prism slip)	49.3	52

in two to four orders of magnitude, i.e., their ratio is in the range of 10^2 to 10^4. This great difference initiates a series of research focused on the imperfections of actual crystals, particularly the dislocation.

All in all, such great difference between the theoretical and experimental values is mainly attributed to the imperfections in actual crystals. According to the dimensions of imperfection in crystal structures, they are classified into four kinds, cf. Figs. 6.2 and 6.3 (Hertzberg, 1976; Meyers, 1994): (1) Point (atomic or electronic) defects, of which the dimensions in all three directions are very small, thus are called zero-dimension defects, such as vacancy and interstitial atom, etc. (2) Line (or one-dimensional) defects, of which the dimensions in two directions are very small, such as dislocation, etc. (3) Interfacial (or two-dimensional) defects, of which the dimension in one direction is very small, such as grain boundary, twin boundary, phase boundary, and stacking fault, etc. (4) Volume (or three-dimensional) defects, of which the dimensions in all three directions are not small, such as void and inclusion, etc.

Dimensional ranges of different classes of defects are shown in Fig. 6.3.

The line defect, i.e., dislocation, plays the key role to slip deformation, so in the following, we will mainly discuss the dislocation and its influence on plastic slip deformation.

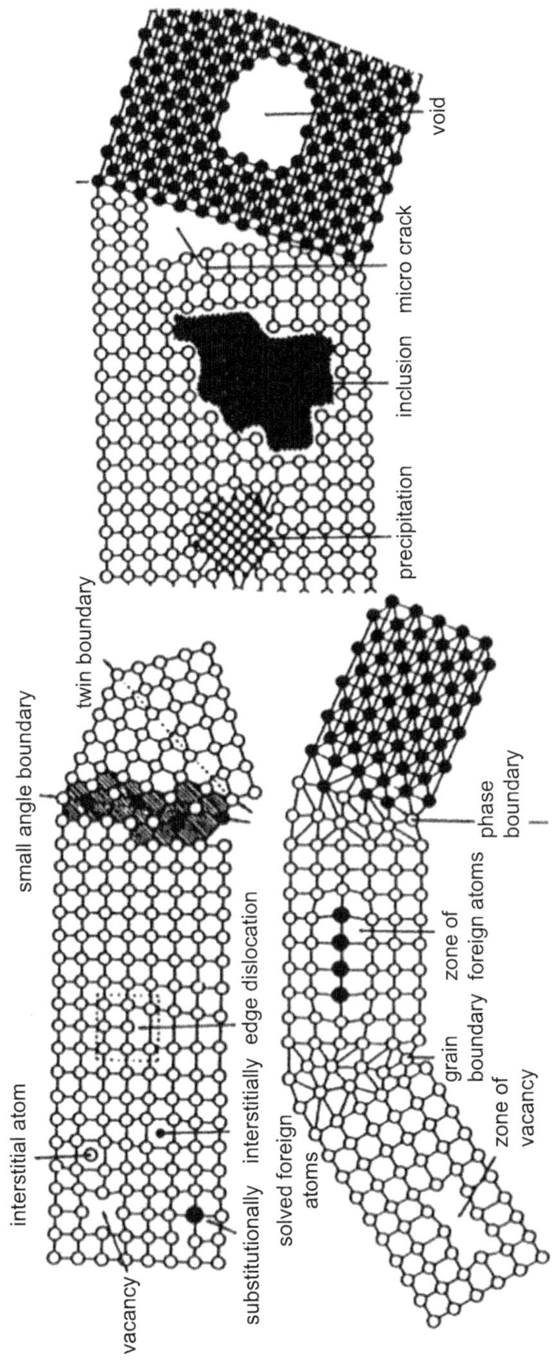

Figure 6.2 Different types of defects/obstacles in crystalline materials. *From Vohringer, O., 1989. Deformation behavior of metallic materials. In: Chiem, C.Y. (Ed.), Int. Summer School on Dynamic Behaviour of Materials for Europ. Engineers and Scientists, EMSN, Nantes, September 11-15, 1989, 7–50., Fig. 12, p.40.*

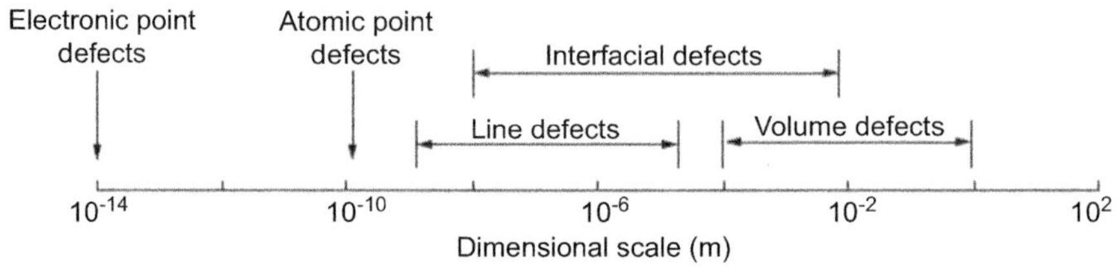

Figure 6.3 Dimensional ranges of different classes of defects. *From Meyers, M.A., Chawla, K.K., 2009. Mechanical Behavior of Materials. Cambridge University Press, Cambridge., Fig. 4.1, p.252. Reprinted with permission of the publisher.*

6.2 Basic knowledge of dislocations

6.2.1 Introduction of dislocation concept

Polanyi, M (1934), Orowan, E (1934), and Taylor, G 1 (1934), respectively, in 1934, almost simultaneously but independently, introduced the concept of dislocation.

The movement of dislocation under the action of shear stress τ is illustrated in Fig. 6.4. In Fig. 6.4A, a crystal has not been disturbed yet, and the dashed line indicates the slip plane where the upper half of the crystal will slip relative to the lower half. In Fig. 6.4B, the atomic bonds of the left outer plane in the upper half, relative to the lower half, have been broken under the action of shear stress τ, and thus this plane is pushed rightward to form a step. In Fig. 6.4C, an extra row of atoms appears in the upper half, which results in misplacement relative to the lower half; this extra row of atoms is called **dislocation**. After that, the dislocation continues to move rightward under the action of shear stress, as shown in Fig. 6.4D and C, and finally, as shown in Fig. 6.4F, the dislocation emerges on the right outer plane, and a slip with a surface step is completed.

Obviously, the shear stress required to move the dislocation step by step (the actual yield strength) is much less than the shear stress required making crystal a whole slip. This is the basic reason why a difference in orders of magnitude exists between the shear strength of perfect crystals and of actual crystals with internal defects.

The effect of dislocation on plastic slip can be qualitatively understood from the following two analogies.

It is often difficult to move a large rug as a whole in a large room. However, if create a drum on the one side of rug and then push this drum to the other side, as shown in Fig. 6.5, the rug displacement with a

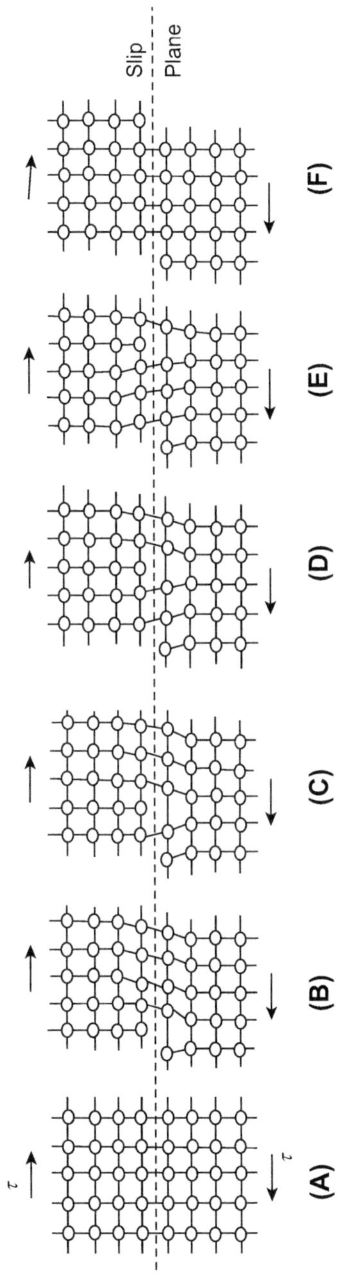

Figure 6.4 Slip formations due to dislocation movement.

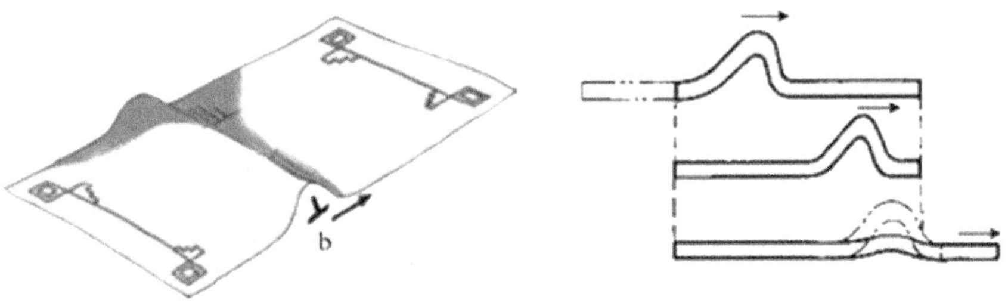

Figure 6.5 To move a rug by pushing its drum.

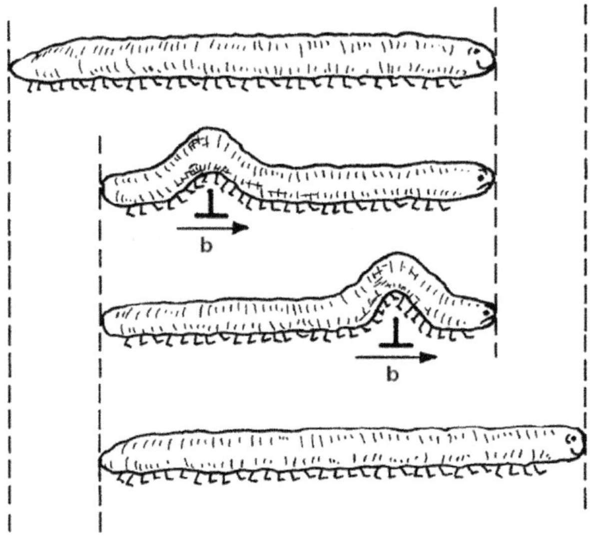

Figure 6.6 Caterpillar moves by means of propagation of bulge "dislocation".

drum width can be completed by using a smaller force. Here, the rug drum is equivalent to a dislocation.

Similarly, a caterpillar moves by means of forming a bulge "dislocation" on its tail, as shown in Fig. 6.6, by the way of the movement of bulge "dislocation" from the back to the front, step by step, to realize the caterpillar "displacement". People sometimes compared the dislocation movement in the crystal to a "caterpillar crawling".

6.2.2 Experimental observation of dislocations

"Does dislocation exist?" The dislocation, which had been in doubt and controversy for as long as 20 years (1935—55), was not recognized until the transmission electron microscopy (TEM) observed the dislocation and thus confirmed the existence of dislocations.

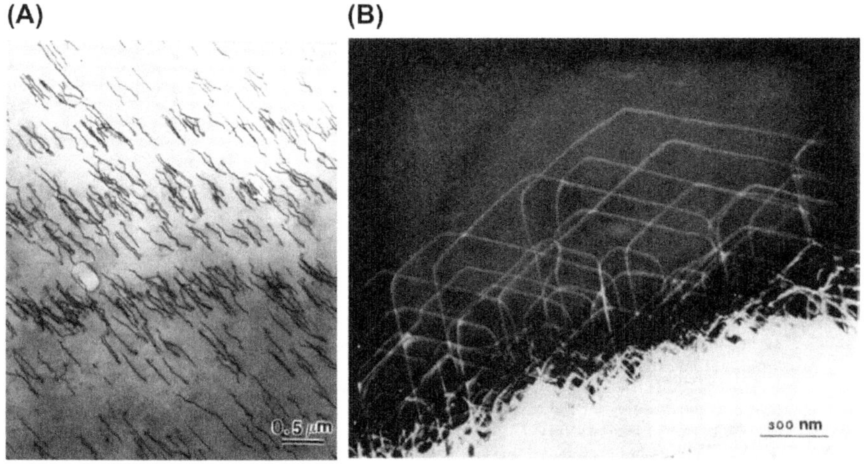

Figure 6.7 Dislocations observed by transmission electron microscopy (TEM), (A) Titanium (bright-field image), (B) Silicon (dark-field image). *From Meyers, M.A., Chawla, K.K., 2009. Mechanical Behavior of Materials. Cambridge University Press, Cambridge., Fig. 4.22, p.271. Reprinted with permission of the publisher.*

Fig. 6.7 shows typical dislocations observed by the TEM (Meyers and Chawla, 2009); in Fig. 6.7A, the dislocations in titanium appear as dark lines in bright-field image, while in Fig. 6.7B, the dislocations in silicon appear as white lines in dark-field image.

Fig. 6.8 presents a high-resolution (in the order of nm) electron micrograph of a dislocation in a thin PbTe foil (Messerschmidt, 2010), where the bright dots represent columns of atoms. As can be seen in the center part, an extra half-plane which is perpendicular to the image plane is inserted, namely the dislocation. Moreover, as shown in the image, the regular arrangement of atoms is disturbed by the dislocation, causing a far-reaching elastic strain field, visible as a bending of the atom planes.

6.2.3 Basic properties of dislocations

1. Classification of dislocation and the Burgers vector

Dislocations can be classified into edge dislocation, screw dislocation, and mixed dislocation; the latter is a hybrid of the former two basic forms of dislocation.

The dislocations shown in Figs. 6.4 and 6.8 are all the edge dislocation. In Fig. 6.9A, the lower end line EF of the extra half atomic plane EFGH represents the dislocation line; in other words, the dislocation line is the

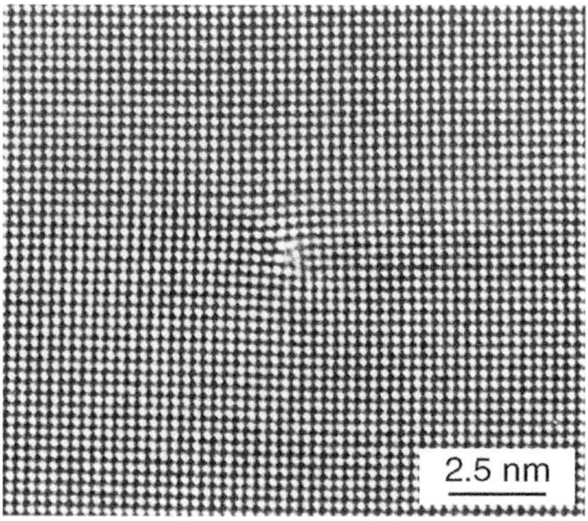

Figure 6.8 A dislocation in a thin PbTe foil observed by high-resolution TEM. *From Messerschmidt, U., 2010. Dislocation Dynamics during Plastic Deformation. Springer Science & Business Media., Fig. 1.3, p.7. Reprinted with permission of the publisher.*

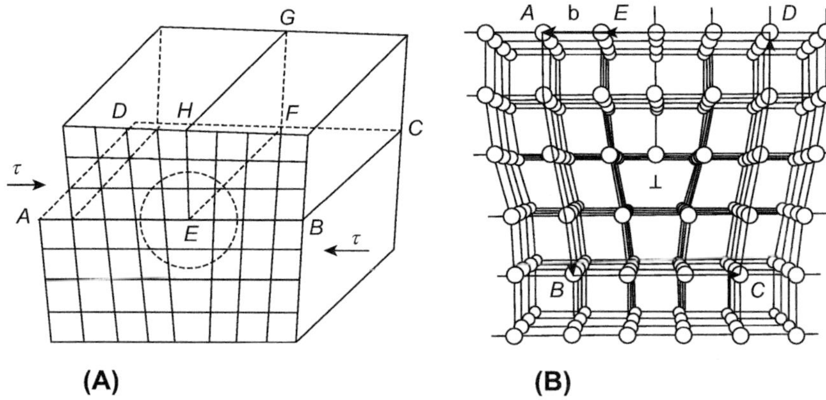

Figure 6.9 Edge dislocation.

intersection of the extra half atomic plane and the slip plane. The extra half atomic plane which appears on the upper half of crystal is called the positive edge dislocation and represented by the symbol ⊥; On the contrary, the extra half atomic plane which appears on the lower half of crystal is called the negative edge dislocation and represented by the symbol T. Due to the insertion of a half atomic plane in the crystal, the periodic arrangement of the crystal atoms is distorted and the stress–strain field changes. For a positive edge dislocation, the upper half is compressed, while the lower half is expanded.

The crystal slip caused by different dislocations is a vector with different directions and different magnitude. In order to quantitatively describe such characteristic vector, the line integral defined by the following equation is introduced, namely the line integral of displacements vector $\overrightarrow{u}$ along the closed curve C around the dislocation.

$$\overrightarrow{b} = \oint_C d\overrightarrow{u} \tag{6.7}$$

This integral was first proposed by Burgers (Burgers, 1939), so is called the Burgers vector, and the integral closed curve C is called the Burgers circuit. The Burgers vector $\overrightarrow{b}$ characterizes the intensity of dislocation. The specific method to determine the Burgers vector for edge dislocation is illustrated in Fig. 6.9B, namely, to form a Burgers circuit $ABCDFA$ from atoms to atoms around the dislocation line in the crystal lattice by step to step. The circuit starts from the point A and goes through the same lattice vector of both positive and negative, in both horizontal and vertical directions, then reaches the point F. The lattice vector from the point F to the start point A is the dislocation Burgers vector $\overrightarrow{b}$. For an edge dislocation, the Burgers vector $\overrightarrow{b}$ is perpendicular to the dislocation line vector $\overrightarrow{l}$, namely $\overrightarrow{b} \perp \overrightarrow{l}$.

Fig. 6.10 shows an edge dislocation and the lattice distortions around it in molybdenum (BCC structure) observed by using an atomic resolution TEM (Meyers and Chawla, 2009). Each dark spot represents one atom and the foil plane imaged is the lattice plane (100). The dislocation line is perpendicular to the foil plane. A Burgers circuit is drawn around the edge dislocation. The right-hand side of the picture shows a unit cell.

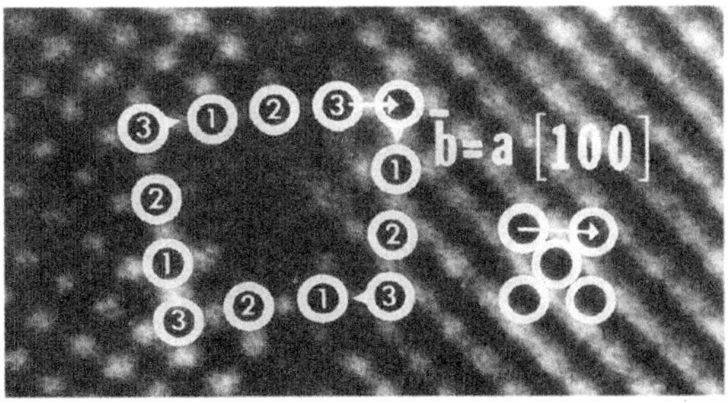

Figure 6.10 High-resolution TEM micrograph of edge dislocation in molybdenum with a Burgers circuit around it. *From Meyers, M.A., Chawla, K.K., 2009. Mechanical Behavior of Materials. Cambridge University Press, Cambridge., Fig. 4.24, p.272. Reprinted with permission of the publisher.*

Helical dislocation is caused by the relative antiplane displacement (off-plane displacement) of two parts of the sliding surface. The spiral dislocation is formed by the relative movement of the right and left parts of the crystal on one side of the dislocation line.

The **screw dislocation** is generated by the relative antiplane displacement (off-plane displacement) of the two parts on the two sides, respectively, of the dislocation line. For example, a screw dislocation is shown in Fig. 6.11A by the line *BC*, which is formed by a relative motion of the left part in the front of *BD* moving upward and the right part moving downward. Its Burgers circuit is in a spiral form (Fig. 6.11B), and thus is called the screw dislocation. For a screw dislocation, the Burgers vector $\overrightarrow{b}$ is parallel to the dislocation line vector $\overrightarrow{l}$, namely $\overrightarrow{b} \parallel \overrightarrow{l}$.

In general, the boundary of the slip area is not necessarily a straight line. For example, the boundary of the slip area *AC* shown in Fig. 6.12 is a curve, which is a more general dislocation line. The Burgers vector $\overrightarrow{b}$ at point *A* is parallel to the dislocation line vector $\overrightarrow{l}$, which is a pure screw dislocation. The Burgers vector $\overrightarrow{b}$ at point *C* is perpendicular to the dislocation line vector $\overrightarrow{l}$, which is a pure edge dislocation. In the middle part of the curve, the angle between the dislocation line vector $\overrightarrow{l}$ and the Burgers vector $\overrightarrow{b}$ varies between 0 and $\pi/2$, which is a **mixed dislocation**.

A dislocation line is a boundary between the slipped and unslipped portions in a crystal. In other words, the movement of the dislocation line on the slip plane is similar to the extension or propagation of a crack front, and the crystal behind the dislocation line has been already slipped, while the crystal in front of

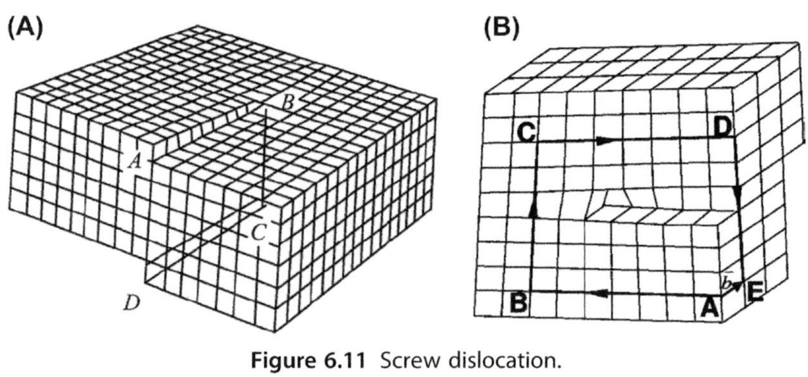

Figure 6.11 Screw dislocation.

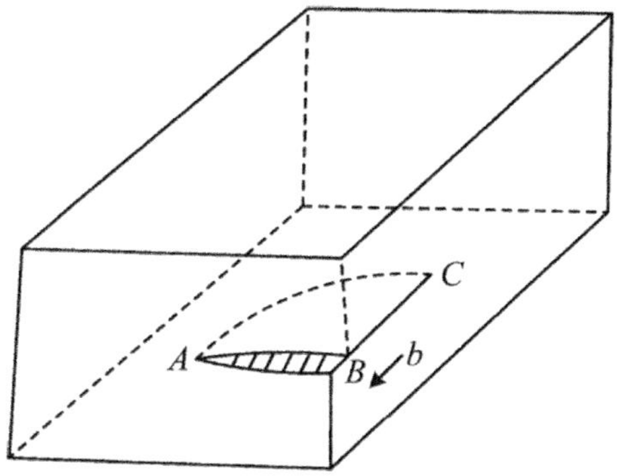

Figure 6.12 Mixed dislocation.

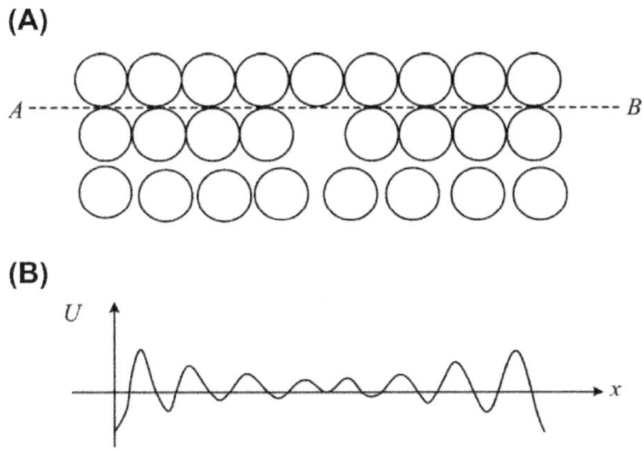

Figure 6.13 The change of lattice potential energy induced by edge dislocation.

the dislocation line has not been slipped yet. Comparing with *crack mechanics* (e.g., see discussions and Fig. 9.1 in Chapter 9), the edge dislocation is similar to the sliding cracker in-plane shear mode (Mode-II), while the screw dislocation is similar to the tearing crack or antiplane shear mode (Mode-III).

2. Peierls—Nabarro stress

Comparing with the whole slip of perfect crystal, the slip deformation of actual crystals becomes much easier due to the existence of dislocation. Consider an edge dislocation perpendicular to the plane of Fig. 6.13A where *AB* denotes the slip plane; above the *AB*, there are N+1 atoms and below the *AB*, there correspondingly are N atoms. The distance between two atoms is *b*. The distance between two rows of atoms is *a*.

In such a case, the potential energy U of the slip plane AB which is jointly influenced by the N+1 atoms above AB and the N atoms below AB is a combination of the following two parts:

$$U = -A\cos\left[2\pi\frac{X}{b}\left(\frac{N+1}{N+\frac{1}{2}}\right)\right] - A\cos\left[2\pi\frac{X}{b}\left(\frac{N}{N+\frac{1}{2}}\right)\right] \tag{6.8}$$

Atomic vibration frequencies of these two parts are very close, so the potential energy curve appears beat phenomenon as shown in Fig. 6.13B, resulting in the amplitude of potential energy curve near the dislocation center is reduced, namely the potential energy in the dislocation center is in a minimum valley. To make the dislocation move across the barrier shown in Fig. 6.13A, namely to move the dislocation a distance (either rightward or leftward) to the next potential energy valley, it requires to overcome the internal resistance of the lattice. When the atomic thermal fluctuations are not taken into account, this is a problem related to a pure mechanical analysis of the stress field in dislocation core zone.

Peierls (1940) and Nabarro (1947) adopted the elastic mechanical model of continuum to analyze this problem (called the Peierls—Nabarro model) and calculated the critical stress for initiating dislocation motion. Similar to Eq. (6.1) for analyzing the theoretical strength of perfect crystal, if assume that the periodic potential energy of crystal could be approximately expressed by a cosine function, the calculation shows that the critical shear stress required to overcome the peak resistance for dislocation motion is

$$\tau_{P-N} = \frac{2G}{1-\nu}\exp\left(-\frac{2\pi a}{b(1-\nu)}\right) \tag{6.9}$$

τ_{P-N} is called the Peierls—Nabarro stress (shortly P—N stress). For the crystal of $a = b$, if the Poison ratio $\nu = 0.3$, then $\tau_{P-N} = 10^{-4}G$. In more general case, the ratio of τ_{P-N} to the elastic shear module G is in the range of 10^{-4}—10^{-2}. As can be seen from Table 6.1, this value is far less than the theoretical shear strength of a perfect crystal, but is near the actual shear yield stress. Obviously, the τ_{P-N} calculated by the P—N model here is a simplified result, where the atomic thermal fluctuations are not taken into account. Thus, the τ_{P-N} actually characterizes the intrinsic friction stress or the internal friction stress of crystal lattices under low temperature 0 K.

3. Dislocation density

Another important physical quantity to characterize the microstructure of dislocation is the dislocation density ρ, which is defined as the total length of dislocation lines per unit volume. In an infinitesimal element $dX_1dX_2dX_3$ in the (X_1, X_2, X_3) rectangular coordinate system, let there be N linear dislocations parallel to the X_2 axis in the area of dX_1dX_3, then there are:

$$\rho = \frac{NdX_2}{dX_1dX_2dX_3} = \frac{N}{dX_1dX_3} \tag{6.10a}$$

For the strongly deformed materials, the dislocation density ρ could be in a range of $1-10^3$ km/m^3. As can be seen from the second equal sign of the above equation, the dislocation density can also be approximately measured as the number of emergence points per unit area (m^{-2}). However, because of the different geometries of dislocation arrangement, Eq. (6.10a) which gives an equivalent relation between the body density ρ_v and the surface density ρ_s, is not strictly true. According to the calculation by Schoeck (1962), when the dislocation distribution in a crystal is isotropic random chaotic, the body density ρ_v and the surface density ρ_s has the following relationship.

$$\rho_v = 2\rho_s \tag{6.10b}$$

Characteristic dislocation densities in some materials showing different deformed states are given in Table 6.2.

4. Dislocation multiplication

According to the dislocation motion mechanism shown in Fig. 6.4, when the dislocation takes off on the right outer plane of the crystal under shear stress, a slip of magnitude b is completed, and meanwhile the dislocation itself disappears. Accordingly, it could be perhaps imagined that the dislocation density will gradually decrease with the gradual increase of plastic deformation. However, the actual situation given in Table 6.2 shows that the well-annealed metal crystals after undergoing slightly plastic deformation (about 10%); the dislocation density will increase from 10^7-10^8 m^{-2} to $10^{10}-1^2$ m^{-2}, increasing by three to four orders of magnitude. It means that the dislocations under the action of shear stress, in addition to producing

Table 6.2 Characteristic dislocation densities in some materials.

Material	**ρ_s(m^{-2})**
Semiconductor single crystals	0 or very few
Well-annealed metal crystals	10^7-10^8
Slightly deformed crystals	$10^{10}-1^2$
Strongly deformed crystals	$10^{12}-1^5$

From Messerschmidt, U., 2010. Dislocation Dynamics during Plastic Deformation. Springer Science & Business Media., Table 1.2, P.8. Reprinted with permission of the publisher.

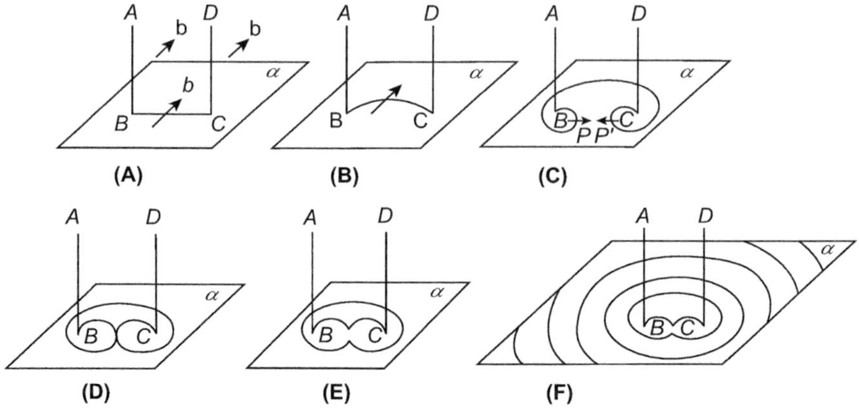

Figure 6.14 Schematics of dislocation multiplication by the Frank–Read mechanism, where *ABCD* shows the dislocation with Burgers vector *b*, but only the segment *BC* in the slip plane α is mobile; under the applied shear stress, the whole development situation of *BC* is schematically shown from (A) to (F).

plastic slips, must also have a mechanism for generating new dislocations continuously from the original dislocations namely the **dislocation multiplication mechanism**. Otherwise, it is difficult to explain why the dislocation density simultaneously and continuously increases when the plastic slips are continuously produced in a plastic deformation process.

The **Frank-Read source** is the most widely cited one of the dislocation multiplication mechanisms (Frank and Read, 1950), although it is not the only mechanism. Consider the dislocation ABCD with Burgers vector $\overrightarrow{b}$ shown in Fig. 6.14A. Notice that only the segment BC in the slip plane α is mobile; while segments AB and CD are not in the slip plane α do not move. Therefore, the segment BC is pinned at both ends B and C. Under the applied shear stress, the mobile segment BC becomes curve in the slip process, since its two ends are pinned and the stress applied to the dislocation is always perpendicular to the dislocation line. The whole development situation of BC is schematically shown in Fig. 6.14A–F. With the increase of stress, the curvature radius of the dislocation segment BC decreases until the BC becomes a semicircle, namely the curvature radius reaches its minimum. At this point, the stress reaches the maximum and the dislocation reaches a critical condition of instability. Since then, the semicircle of the BC is converted to a gyroscope, as shown in Fig. 6.14C, where the dislocation segments P and P' have opposite signs, so that they attract each other. When they approach and touch each other (Fig. 6.14D), they can annihilate to form a complete dislocation loop (Fig. 6.14E), and at the same time, the

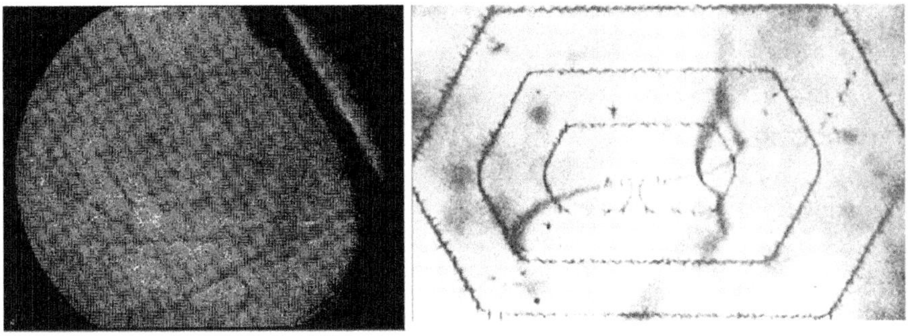

Figure 6.15 Photomicrograph showing the Frank–Read source in silicon crystal. *From Tyler, W.W., Dash, W.C., 1957. Dislocation Arrays in Germanium. Journal of Applied Physics 28 (11), 1221–1224., Fig. 7, p.1223. Reprinted with permission of the publisher.*

original source dislocation BC is restored. This procedure can operate repeatedly to generate a greater number of dislocation loops (Fig. 6.14F). However, as loops are formed, they establish a back stress, so that the critical shear stress τ_c required for generating new loops increases too. This reflects the effect that the plastic flow stress increases with the development of plastic deformation, namely the so-called strain-hardening or work-hardening effect. In terms of dislocation density ρ, the following Bailey–Hirsch relation is well-known, namely the plastic shear flow stress τ is proportional to the square root of the dislocation density ρ:

$$\tau \propto \rho^{1/2} \tag{6.11}$$

Some direct experimental observations confirm the existence of Frank–Read source in crystals. A typical example is that Dash (1956) experimentally observed the Frank–Read source and its multiplication of dislocation loops in the silicon single crystal by infrared light method, as shown in Fig. 6.15.

6.3 Dislocation dynamics

The above discussions show that the microscopic mechanism of macroscopic plastic deformation of actual materials can be summed up as the dislocation movement which overcomes all kinds of microenergy barrier, which is the dynamic response of dislocation in crystals under external loading. This is the research object of dislocation dynamics.

However, the foregoing discussion is carried out at the microscopic level of the lattice scale. How to extend the results at the microscopic level to the macroscopic mechanics level requires establishing a cross-scale equation linking microscopic–macroscopic scale. This is the key to any cross-scale research.

The equation which links the microscopic dislocation motion with the macroscopic plastic distortion was first proposed by Orowan (1940).

6.3.1 Orowan equation

As shown in Fig. 6.16, consider a macroscopic crystal with the macroscopic scale of $l \times l \times l$ under the action of shear stress τ. At the microscopic level, there are N parallel mobile edge dislocations with Burgers vector $\overrightarrow{b}$, which move on the parallel slip planes. Obviously, the magnitude of total displacement should be Nb, and since the mobile dislocation density ρ_m is N/l^2, according to Eq. (6.10a), the macroscopic plastic shear strain γ^P could be expressed as:

$$\gamma^p = \tan\theta = \frac{Nb}{l} = \frac{Nbl}{l^2} = \phi\rho_m bl \tag{6.12a}$$

where ϕ is the orientation factor (Schmid factor), since the N mobile dislocations are actually not all parallel as shown in Fig. 6.16.

Differential Eq. (6.12a) to time t, we have:

$$\dot{\gamma}^p = \phi\rho_m bv_d + \phi bl\dot{\rho}_m \tag{6.12b}$$

As can be seen, the plastic strain rate $\dot{\gamma}^p$ $(=d\gamma^p/dt)$ depends on the dislocation moving velocity v_d $(=dL/dt)$ and the rate of mobile dislocation density with respect to time t, $\dot{\rho}_m(= d\rho_m/dt)$.

Assume that the change of mobile dislocation density ρ_m with respect to t is slow so that it can be ignored, then we have:

$$\dot{\gamma}^p = \phi\rho_m bv_d \tag{6.12c}$$

This equation establishes the relationship between the macroscopic plastic strain rate $\dot{\gamma}^p$ and the microscopic parameters—the magnitude of Burgers vector $\overrightarrow{b}$ and the dislocation moving velocity v_d. Eq. (6.12) is called the **Orowan equation**.

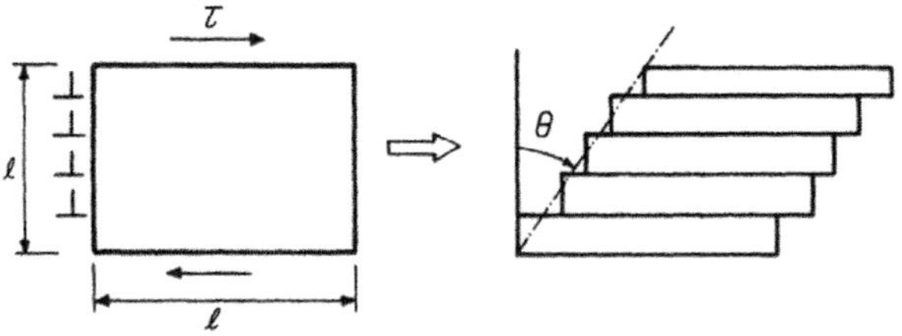

Figure 6.16 The macroscopic plastic shear strain γ^p(=tanθ) caused by the motion of a row of parallel dislocations.

Rice (1970) generalized the Orowan equation to a general three-dimensional situation in form of tensor, which for homogenous isotropic materials is expressed as:

$$\dot{\varepsilon}_{ij}^{p} = \frac{1}{V}\int_{0}^{L}\frac{1}{2}(n_i b_j + n_j b_i)v_d dl \tag{6.13}$$

where L is the total length of dislocation in volume V.

After having the relationship between the macroscopic plastic strain rate $\dot{\gamma}^p$ and the dislocation moving velocity v_d, the next step is we should further discuss the relationship between the dislocation moving velocity v_d and the applied shear stress τ (or equivalently the normal stress σ). Thereby, we can establish the relationship between the applied shear stress τ (or equivalently the normal stress σ) and the plastic strain rate $\dot{\gamma}^p$, that is called the rate-dependent viscoplastic distortional law.

6.3.2 Experimental study of dislocation velocity

Johnston and Gilman (1959) carried out the pioneering experiments to study how the dislocation velocity v_d changes with the shear stress τ. The results are drawn in the log-log coordinates, as shown in Fig. 6.17. As can be seen, with increasing the shear stress τ, the dislocation velocity v_d increases rapidly, increasing in several orders. After reaching a certain high velocity,

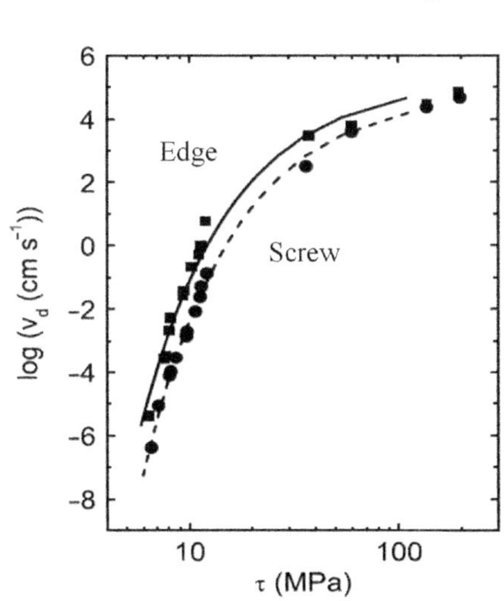

Figure 6.17 The relation between the applied stress and the dislocation velocity experimentally measured in LiF. *From Johnston, W.G., Gilman, J.J., 1959. Dislocation velocities, dislocation densities, and plastic flow in lithium fluoride crystals. J Appl Phys. 30 (2), 129–144., Fig. 5, p.132. Reprinted with permission of the publisher.*

the slope of the curve gradually decreases and is gradually close to the sound velocity. The velocity of edge dislocations is slightly higher than that of screw dislocations, while the trend is consistent.

According to Fig. 6.17, there exists approximately a linear relation in the log-log coordinates; thus, Eq. (6.12c) can be expressed as:

$$\dot{\gamma}^p \propto v_d \propto \tau^m \tag{6.14}$$

This provides to the empirical power function of viscoplastic constitutive relation $\frac{\sigma}{\sigma_0} = \left(\frac{\dot{\varepsilon}}{\dot{\varepsilon}_0}\right)^n$ (Eq. 5.1 in Chapter 5) a physical support on the basis of dislocation dynamics mechanism.

With increasing the v_d, the slop of $logv_d$—$\log\tau$ curve in Fig. 6.17 decreases. When the slope deceases to 1, it is equivalent to $m = 1$ in Eq. (6.14), namely $\dot{\gamma}^p \propto v_d \propto \tau$. This provides to the empirical linear viscous equation of viscoplastic constitutive relation $\frac{\partial \sigma}{\partial \dot{\varepsilon}} = \eta$ (Eq. 5.5 as well as the curve broken into two segments in Fig. 5.7) a physical support on the basis of dislocation dynamics mechanism.

Since the experiment of Johnston and Gilman on the dislocation velocity v_d, other researchers have conducted further experimental studies on the dislocation velocity of different materials, and the results are summarized in Fig. 6.18 (Meyers and Chawla, 1984). It can be seen that in different dislocation velocity range, different materials can be characterized by the linear relations $logv_d$—$\log\tau$ with different slopes. According to the magnitude of slope m, it can be divided into two categories: one is the situation at a relatively low dislocation velocity, the slope $m > 1$; another is the situation at a relatively high dislocation velocity, the slope $m = 1$. Studies found that the former situation corresponds to the mechanism of thermally activated dislocation motion, of which the details will be further discussed below. And the latter situation corresponds to (a) the interaction between dislocation and lattice thermal vibration and (b) the interaction between dislocation and electron cloud. The (a) situation is displayed as phonon viscosity or phonon drag mechanism (phonon refers to the propagation of sound waves in lattices, see Section 3.5 in Chapter 3). The (b) situation is displayed as electron viscosity mechanism. Such viscous effect, as a primary approximation, is assumed to follow the Newton viscosity theorem, so we have $\dot{\gamma}^p \propto v_d \propto \tau$. This is the equivalent of taking a dislocation motion as moving in a viscous solid, which is like a ship anchor being dragged in a viscous fluid, is subjected to viscous resistance. Thus, it is called the **dislocation drag** mechanism.

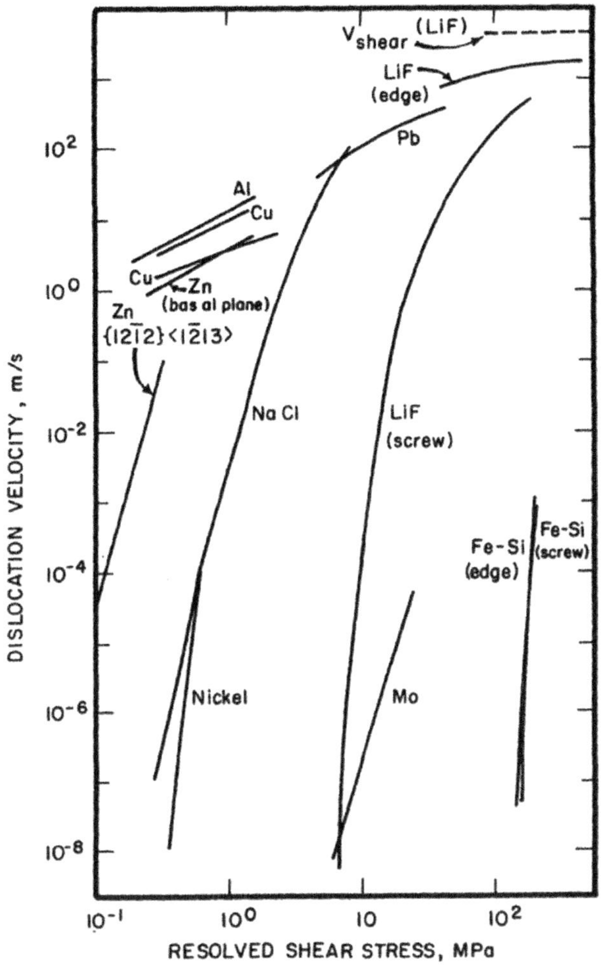

Figure 6.18 The relationship between the experimentally measured dislocation velocity and the applied shear stress for different materials. *From Meyers, M.A., Chawla, K.K., 1984. Mechanical Metallurgy, Principles and Applications. Prentice-Hall, Englewood Cliffs, NJ. Fig. 8.9, p.304. Reprinted with permission of the Author.*

6.3.3 Short-range barrier and long-range barrier

The applied stress required to microscopically overcoming all kinds of obstacles, or the corresponding barriers, is expressed macroscopically as the plastic flow stress τ_p. The barriers of dislocation motion could be small and narrow (in the Burgers vector order of magnitude) or large and wide. The former is called the **short-range barrier**, as the barrier shown in Fig. 6.13 during the discussion of P—N stress τ_{P-N}. The latter is called the **long-range barrier**,

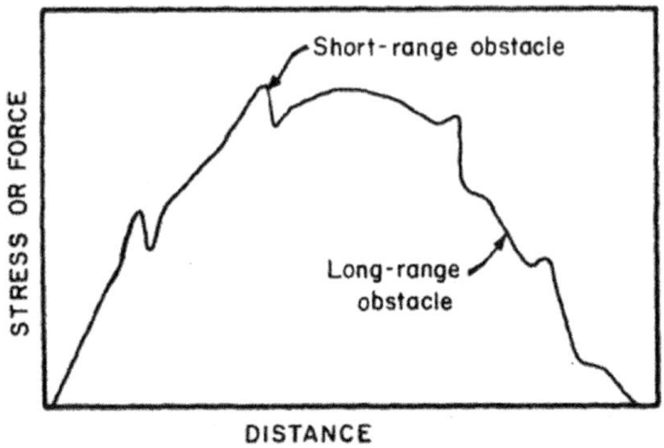

Figure 6.19 The barriers in the path of dislocation motion.

as those barriers corresponding to various obstacles shown in Fig. 6.2. Both the short-range barriers and long-range barriers in the path of dislocation motion are schematically shown in Fig. 6.19.

The energy required to overcome the short-range barrier is relatively small and is sensitive to strain rate and temperature. In fact, by Section 3.5 in Chapter 3, it is easy to understand that as the temperature rises, the amplitude of lattice atomic thermal vibration increases; the corresponding heat can help the dislocation to move across the short-range potential barrier. Such barrier is called thermally activated. On the contrary, the energy required to overcome the long-range barrier is relatively large; the heat energy of the lattice atomic thermal vibration is too small to help it so is insensitive to temperature and strain rate. Such barrier is called athermal or nonthermally activated.

Correspondingly, the shear stress required for plastic deformation can be divided into two parts: one required for overcoming athermal (long-range) barriers τ_G and one required for overcoming thermally activated (short-range) barriers $\tau^\star$.

$$\tau = \tau_G(G) + \tau^*(T, \dot{\gamma}) \tag{6.15a}$$

Or expressed in the form of normal stress σ, we have:

$$\sigma = \sigma_G(G) + \sigma^*(T, \dot{\varepsilon}) \tag{6.15b}$$

where τ_G is the nonthermally activated component, and the temperature dependence of τ_G is equivalent to the temperature dependence of the material elastic constants (usually represented by the shear modulus G), so it is usually expressed as $\tau_G(G)$. However, $\tau^\star$ is the thermally activated component and depends on temperature and strain rate, so it is usually expressed as $\tau^*(T, \dot{\gamma})$. Eq. (6.15) is schematically shown in Fig. 6.20. Since the temperature sensitivity of G is relatively weak, the temperature variation of $\tau_G(G)$ is also relatively weak, while $\tau^*(T, \dot{\gamma})$ varies significantly with the temperature and strain rate.

As an instance, Fig. 6.21 shows the variation of yield stress of iron and tantalum with temperature (Meyers and Chawla, 2009), from which the $\sigma_G(G)$ is estimated to be about 50 MPa, while the $\sigma^*(T, \dot{\gamma})$ is over 10^3 MPa at 0 K. With increasing temperature, since heat energy can help dislocation to move across the short-range barrier, thus the $\sigma^*(T, \dot{\gamma})$ decreases markedly with increasing temperature.

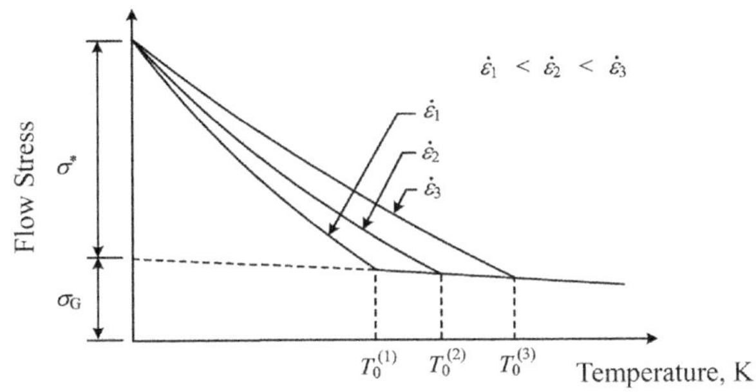

Figure 6.20 The nonthermal components and the thermally activated components of the flow stress σ.

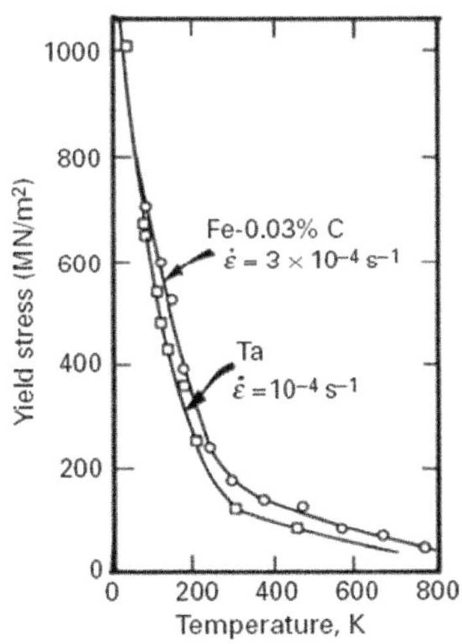

Figure 6.21 The yield stress of iron and tantalum varies markedly with temperature. *From Meyers, M.A., Chawla, K.K., 2009. Mechanical Behavior of Materials. Cambridge University Press, Cambridge., Fig. 4.62B, p.312. Reprinted with permission of the publisher.*

It is $\tau^*(T, \dot{\gamma})$, which has a thermal activation characteristic, and has the closest relationship with the macroscopic thermoplastic distortion law. Thus, the thermal activation mechanism will be further discussed in the following.

6.3.4 Thermally activated mechanism

As can be seen from the discussion in Section 3.5 of Chapter 3, with increasing temperature T, the amplitude of lattice atomic thermal vibration increases, and the corresponding heat energy can help dislocation to overcome the short-range barrier. In other words, the energy required for overcoming barrier is composed by two parts: the work done by shear stress τ and the thermally activated energy, as illustrated in Fig. 6.22. The barrier curves at $T_0(=0\text{ K})$, T_1, T_2, and T_3 ($T_0<T_1<T_2<T_3$) are schematically plotted in coordinates of shear stress τ versus distance x as shown in Fig. 6.22A. The total integral area under the barrier curve is equal to the potential energy required for overcoming this barrier, which corresponds to the Helmholtz free energy A (T, V) under the isothermal and isovolumetric condition in Chapter 2. At the absolute zero temperature $T_0 = 0$ K, thermal activation energy ΔU_0 is zero; the energy required for dislocation to overcome the barrier depends entirely on the work done by the P—N stress. As temperatures rise to T_1, T_2, T_3, respectively, the thermal activation energy provided by atomic thermal vibration, ΔU, increases accordingly, as shown by ΔU_1, ΔU_2, ΔU_3 with shadow lines in the figure, respectively, then the effective height of the barrier is reduced; therefore, the shear stress τ required for overcoming the potential barrier decreases to τ_1、τ_2、$\tau_3(=0)$, respcctively, as shown in Fig. 6.22B.

According to statistical mechanics, the probability P_B of a dislocation overcoming a potential barrier by thermal activation energy U is

$$P_B = \exp\left(-\frac{U}{kT}\right) \tag{6.16a}$$

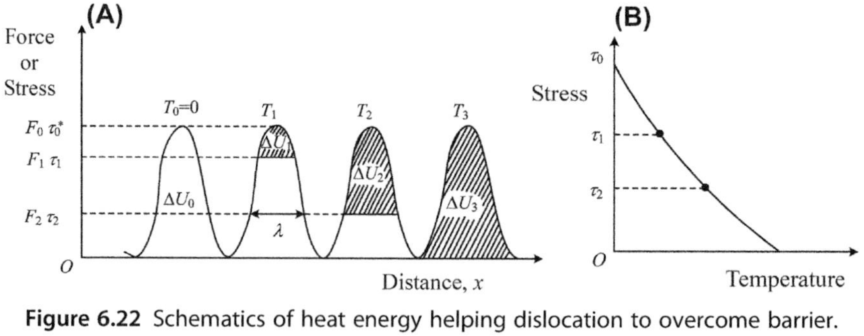

Figure 6.22 Schematics of heat energy helping dislocation to overcome barrier.

where k is the Boltzmann constant and U the thermal activation energy under isothermal and isobaric condition, which corresponds to the Gibbs free energy $G(T, P)$ in Chapter 2. On the other side, the probability P_B of a dislocation overcoming a potential barrier is also the ratio of the frequency f_1 of a dislocation successfully overcoming a potential barrier to the dislocation vibration frequency f_0; thus, we have:

$$f_1 = f_0 \exp\left(-\frac{U}{kT}\right) \tag{6.16b}$$

Notice that the dislocation velocity v_d is also the product of the frequency f_1 of a dislocation successfully overcoming a potential barrier and the average moving distance χ, then we have:

$$v_d = \chi f_1 = \chi f_0 \exp\left(-\frac{U}{kT}\right) = v_0 \exp\left(-\frac{U}{kT}\right) \tag{6.16c}$$

This is the **Arrhenius equation for dislocation velocity**. Substitute Eq. (6.16c) into the Orowan equation (Eq. 6.12), we obtain:

$$\dot{\gamma}^p = \phi\rho_m b v_d = \phi\rho_m b v_0 \exp\left(-\frac{U}{kT}\right) = \dot{\gamma}_0 \exp\left(-\frac{U}{kT}\right) \tag{6.17a}$$

where $\dot{\gamma}_0$ is the preexponential factor, a combination of the parameters before the exponential function. This is the **Arrhenius equation for plastic strain rate**. Considering that the thermal activation energy is a function of shear stress τ (or equivalently normal stress σ), the above equation can be rewritten as the below form:

$$U(\tau) = -kT \ln\frac{\dot{\gamma}^p}{\dot{\gamma}_0} = kT \ln\frac{\dot{\gamma}_0}{\dot{\gamma}^p} \tag{6.17b}$$

Or equivalently

$$U(\sigma) = -kT \ln\frac{\dot{\varepsilon}^p}{\dot{\varepsilon}_0} = kT \ln\frac{\dot{\varepsilon}_0}{\dot{\varepsilon}^p} \tag{6.17c}$$

This expression indicates that the activation energy increases with increasing the temperature, while decreases with increasing the strain rate. Correspondingly, the flow stress decreases with increasing the temperature, while increases with increasing the strain rate, as schematically shown in the previous Fig. 6.20.

Notice that the combination parameter of the strain rate $\dot{\varepsilon}$ and the temperature T appeared on the right-hand side of Eq. (6.17), $T \ln\frac{\dot{\varepsilon}^p}{\dot{\varepsilon}_0}$, which is what the rate–temperature equivalent parameter T^* defined by Eq. (5.7) in Chapter 5. Therefore Eq. (6.17) provides a dislocation dynamics basis to the description of rate—temperature equivalency of viscoplastic distortional law of materials.

Eq. (6.17) also indicates that there exists a viscoplastic flow as long as the strain rate is not zero. It is based on this view that Bodner and Partom (1975) proposed an elastic—viscoplastic constitutive equation without yield surface but taking account of strain hardening (see the related discussion on Eq. (5.25) in Chapter 5).

Eq. (6.17) actually presents the thermoplastic distortion law in an implicit function form.

$$\tau = \tau(\gamma, \dot{\gamma}^p, T) \tag{6.18a}$$

$$\sigma = \tau(\varepsilon, \dot{\varepsilon}^p, T) \tag{6.18b}$$

The next key is to determine the specific form of above implicit function, which we will further discuss in the next section.

6.4 Thermoviscoplastic constitutive equation based on dislocation dynamics

Eq. (6.17) shows that the specific function of a thermoviscoplastic distortional law depends on the function form $U(F,T)$ of how the activation energy U is related to the applied force F. Under the isothermal condition, the activation energy U can be calculated with the shape of barrier curve in the F-X coordinates, known as the **barrier shape**, as shown in Figs. 6.22 and 6.23A.

$$U(F, T) = \int_F^{F_0} X(F, T)dF \tag{6.19a}$$

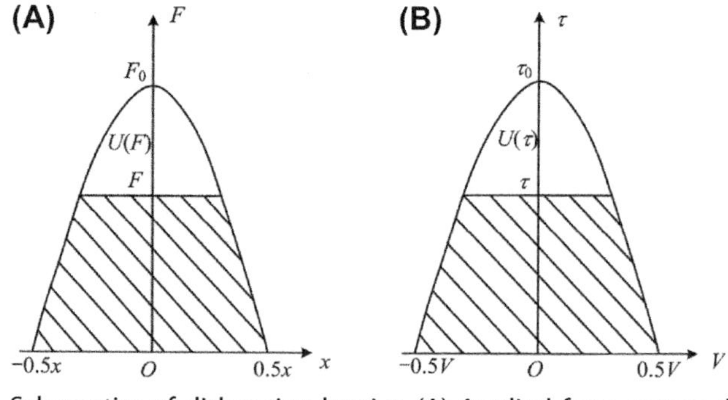

Figure 6.23 Schematics of dislocation barrier. (A) Applied force versus displacement; (B) Shear stress versus activation volume.

where X is the activation distance, namely the width of the activation barrier. The $U(F,T)$ can be equivalently expressed as $U(\tau,T)$, namely the shape of barrier curve in the τ-X coordinates under the isothermal condition, as shown in Fig. 6.23B.

$$\begin{aligned} U(\tau, T) &= \int_{\tau}^{\tau_0} V(\tau, T)d\tau \\ &= \int_{0}^{\tau_0} V(\tau, T)d\tau - \int_{0}^{\tau} V(\tau, T)d\tau \\ &= U^*(\tau_0, T) - U^*(\tau, T) \end{aligned} \tag{6.19b}$$

where U^* is defined as $U^* = \int_0^{\tau} V(\tau, T)d\tau$. For a dislocation with length l and Burgers vector $\vec{b}$, the product of the activation distance X and the dislocation length l is the area that the dislocation has swept in the activation, which is called the activation area. The product of the activation area and the magnitude of Burgers vector b is the activation volume V.

$$V = blX \tag{6.20}$$

For convenience, the following dimensionless parameters are introduced.

$$\overline{\tau} = \frac{\tau}{\tau_c}, \quad \overline{V} = \frac{V}{V^*}, \quad \overline{U} = \frac{U}{\tau_c V^*}, \quad \overline{T} = \frac{kT}{\tau_c V^*}, \quad \overline{\dot{\gamma}} = \frac{\dot{\gamma}^p}{\dot{\gamma}_0}, \tag{6.21}$$

where τ_c is the characteristic stress, which can be the maximum stress of potential barrier as shown in Fig. 6.23B (in such case $\overline{\tau} < 1$) or can be the long-range stress τ_G (in such case $\overline{\tau} > 1$), etc., and V^* is the activation volume at $\tau = 0$, $V^* = V(0,T)$. Therefore, Eqs. 6.19b and 6.17a can be rewritten in the dimensionless form as:

$$\overline{U}(\overline{\tau}) = \int_{\overline{\tau}}^{1} \overline{V}(\overline{\tau})d\overline{\tau} \tag{6.22}$$

$$\overline{\dot{\gamma}} = \exp\left(-\frac{\overline{U}(\overline{\tau})}{\overline{T}}\right) \tag{6.23}$$

Thus, it can be seen that the concrete function of the macroscopic thermoviscoplastic distortional law based on the thermal activation mechanism ultimately depends on the shape of dislocation potential barrier, namely the dimensionless function $\overline{U}(\overline{\tau})$ or $\overline{V}(\overline{\tau})$. The following discussion will point out that some of the currently used macroscopic thermoviscoplastic distortion laws, respectively, have its corresponding dislocation barrier shape (Davidson and Lindholm, 1974; Meyers, 1994).

However, regardless of the barrier shape, the following two conditions must be satisfied in order to be consistent with the definition of $V^{\star}(=V(0,T))$, and the requirement that $\overline{V}$ should be a decreasing function of $\overline{\tau}$:

$$\overline{V}=1 \ for \ \overline{\tau}=0, \tag{6.24}$$

$$\frac{d\overline{V}}{d\overline{\tau}}<0 \tag{6.25}$$

Starting from the barrier shape $\overline{V}(\overline{\tau})$, several representative thermo-viscoplastic distortional laws will be discussed below.

6.4.1 Rectangular potential Barrier—Seeger's model

The simplest barrier shape is a rectangle, as shown in Fig. 6.24. In such case, we have:

$$\overline{V}=1 \quad \overline{U}=1-\overline{\tau} \tag{6.26a}$$

It means that the thermal activation energy U is a linear function of shear stress τ. Substituting it into Eq. (6.22), we have:

$$\dot{\overline{\gamma}}=\exp\left(-\frac{1-\overline{\tau}}{\overline{T}}\right) \tag{6.26b}$$

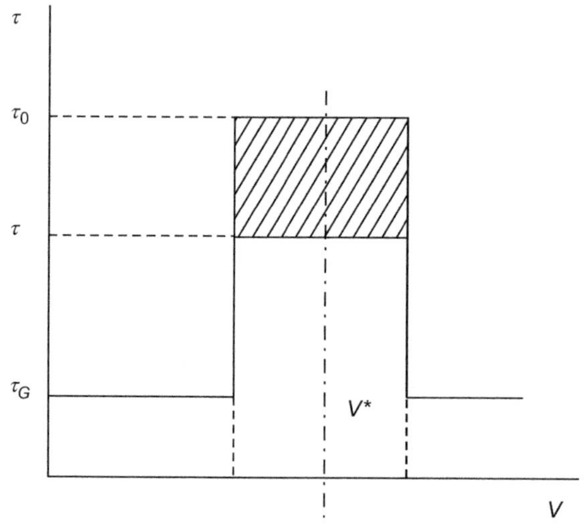

Figure 6.24 The schematic τ–V diagram for a rectangular dislocation barrier.

Or in the form of

$$\overline{\tau} = 1 + \overline{T}\ln\overline{\dot{\gamma}} \tag{6.26c}$$

Rewriting this dimensionless equation to the corresponding dimensional equation, and noticing that $\overline{U} = 1 = \frac{U_0}{\tau_c V*}$, where U_0 is the total potential barrier at $\tau = 0$, we have:

$$\tau = \frac{U_0}{V*} + \frac{kT}{V*}\ln\frac{\dot{\gamma}^p}{\dot{\gamma}_0} \tag{6.26d}$$

Please note that the τ in the discussion of thermal activation mechanism, including the τ in the above equation, is only in correspond to the short-range stress in the thermal activation process, i.e., is actually the $\tau^\star$ in Eq. (6.15a). If take the athermal long-range stress τ_G into account, the above equation should be more completely rewritten as:

$$\tau = \tau_G + \frac{U_0}{V^*} + \frac{kT}{V^*}\ln\frac{\dot{\gamma}^p}{\dot{\gamma}_0} \tag{6.26e}$$

The τ here is the total shear stress, including the athermal long-range stress τ_G and the thermal activation short-range stress $\tau^\star$; this is the well-known **Seeger's thermoviscoplastic distortional law** (Seeger, 1955). It provides to the empirical equation Eq. (5.2) in Chapter 5 a theoretical basis based on dislocation dynamics.

The experimental results by Lindholm (1968) for aluminum in a strain-rate range of seven orders of magnitude and in a temperature range of 294–672 K have been given in Fig. 5.13 in Chapter 5, where the solid lines are plotted in accordance with the Seeger's model (Eq. 6.26) and shown linear relationship in the $\sigma - \log\dot{\varepsilon}$ coordinates. Fig. 5.13 shows that the Seeger's model is in good agreement with the experimental data even under different temperature.

Recall the Johnson–Cook equation (Eq. 5.14) in Section 5.2.2 of Chapter 5, where the second term $\left(1 + C\ln\frac{\dot{\varepsilon}}{\dot{\varepsilon}_0}\right)$, which reflects the "strain-rate effect", is obviously in consistent with the Seeger's model based on the microscopic dislocation dynamics mechanism.

About Eq. (6.26e), it is worthwhile to further note the role played by the athermal long-range stress τ_G. A large number of experimental results of different researchers show that although the strain-rate sensitivity of aluminum is quite high, the strain-rate sensitivity of aluminum alloy is

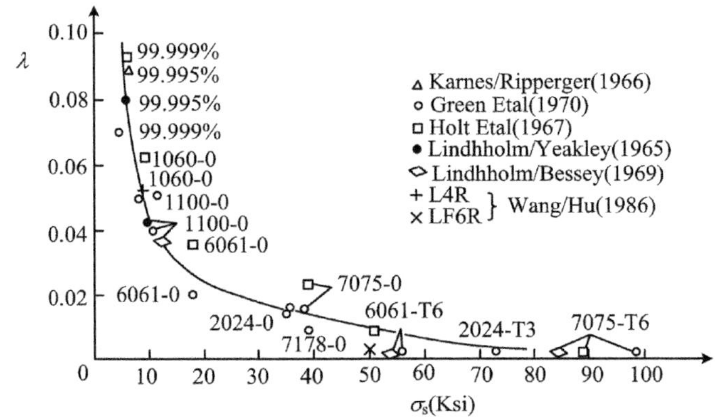

Figure 6.25 The strain-rate sensitive rate λ $\left(=\frac{\sigma_D-\sigma_S}{\sigma_S\,\log(\dot{\varepsilon}_D/\dot{\varepsilon}_S)}\right)$ for a given strain $\varepsilon = 6\%$ changes with the quasistatic yield stress σ_S for aluminum and aluminum alloys. *From Wang Lili and Hu Shisheng, 1986, Fig. 5, p.165.*

significantly reduced with increasing its quasistatic strength, as shown in Fig. 6.25 (where the strain-rate sensitive coefficient λ is defined as $\lambda = \frac{\sigma_D-\sigma_S}{\sigma_S\,\log(\dot{\varepsilon}_D/\dot{\varepsilon}_S)}$, in which the subscripts D and S denote dynamic and quasistatic, respectively). On the microscopic mechanism, this phenomenon is interpreted as that alloy elements through the precipitation hardening and solid-solution hardening mechanism mainly improve the long-range stress τ_G and increase the strength of aluminum alloy, but had no obvious effect on the thermal activation short-range stress $\tau^{\star}$ (Holt, etc. 1967; Wang Lili and Hu Shisheng, 1986).

6.4.2 Nonlinear potential barrier—Davidson–Lindholm model

The rectangular dislocation potential barrier corresponds to a linear relationship between the activation energy U and the stress τ, accordingly corresponds to a linear $\tau - \log\dot{\gamma}$ relation. However, a large of experimental investigation shows that this is often true only within a certain range of strain rate.

Other dislocation barrier shapes correspond to nonlinear U—τ relations. Davidson and Lindholm (1974) pointed out that a series of typical barrier shapes can be characterized by a single function.

$$\overline{V} = \left(1-\overline{\tau}^{Z}\right)^{1/Z} \tag{6.27}$$

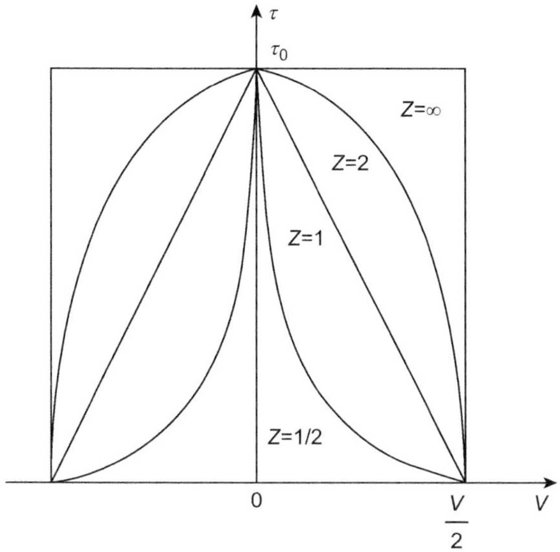

Figure 6.26 Several barrier shapes described by Eq. (6.27).

The above equation requires that $\overline{\tau} \le 1$ to satisfy $\overline{V}$ being real and $Z > 0$ to satisfy Eq. (6.25). When Z takes a different value, it corresponds to a different barrier shape, as shown in Fig. 6.26. In particular,

(1) when $Z = \infty$, $\overline{V} = 1$—corresponding to the Seeger's model;

(2) when $Z = 1$, $\overline{V} = (1 - \overline{\tau})$—corresponding to the triangle barrier proposed by Makin (1968);

(3) when $Z = 2$, $\overline{V} = \left(1 - \overline{\tau}^2\right)^{1/2}$—corresponding to the elliptic barrier;

(4) when $Z = 1/2$, $\overline{V} = \left(1 - \overline{\tau}^{1/2}\right)^2$—corresponding to the hypocycloidal barrier proposed by Fleisher (1962).

Once we have the function $\overline{V}(\overline{\tau})$ corresponding to different Z values, it is not difficult to find the nonlinear relation $\overline{U}(\overline{\tau})$, and then the corresponding thermoviscoplastic distortion law can be obtained by Eq. (6.22).

6.4.3 Nonlinear potential barrier—Kocks–Argon–Ashby model

Kocks et al. (1975) proposed the following nonlinear model which contains two parameters P and q $(0 < p \le 1, 1 \le q \le 2)$ and is obviously more general than the Davidson–Lindholm model (Eq. 6.27).

$$U = U_0\left[1 - \left(\frac{\sigma}{\sigma_0}\right)^p\right]^q \tag{6.28a}$$

The corresponding thermoviscoplastic constitutive equation is

$$\ln\left(\frac{\dot{\varepsilon}_0}{\dot{\varepsilon}}\right)=\frac{U_0}{kT}\left[1-\left(\frac{\sigma}{\sigma_0}\right)^p\right]^q \tag{6.28b}$$

For example, when $P=1$ and $q=2$, it corresponds to the following sinusoidal barrier:

$$U=U_0\left[1-\frac{\sigma}{\sigma_0}\right]^2 \tag{6.29}$$

Hoge and Mukherjee (1977) used this sinusoidal barrier to describe how the flow stress depends on the strain rate and temperature for tantalum.

6.4.4 Nonlinear potential barrier—spectrum of hyperbolic barriers

The Davidson–Lindholm model (Eq. 6.27 and Fig. 6.26) not only satisfies the restrictive conditions of Eqs. (6.24) and (6.25), but also implicitly satisfies the following conditions (except the rectangular barrier),

$$\overline{V}=0,\ for\ \overline{\tau}=1 \tag{6.30}$$

which means that there exists the extreme point.

Wang L.L. (1984) proposed the following hyperbolic shape barrier, which is not restricted by Eq. (6.30),

$$\overline{V}=(1+\overline{\tau})^{-m},\quad m\geq 0 \tag{6.31}$$

where $m\geq 0$ to satisfy the requirement of Eq. (6.25).

Different m values give different shapes of hyperbolic potential barrier, as shown in Fig. 6.27.

Eq. (6.31) has wide applicability and can be reduced to a series of known models by taking different m values.

(1) When $m=0$, $\overline{V}=1$—corresponding to the Seeger's model (ref. Eq. 6.26);

(2) When $m=1$, $\overline{V}=(1+\overline{\tau})^{-1}$, then:

$$\overline{U}=\ln 2-\ln(1+\overline{\tau}) \tag{6.32a}$$

$$\overline{T}\ln\overline{\dot{\gamma}}=\ln(1+\overline{\tau})-\ln 2 \tag{6.32b}$$

If the characteristic stress τ_c in the dimensionless stress $\overline{\tau}(=\tau/\tau_c)$ is taken as the long-range stress τ_G, and assume $\overline{\tau}>>1$, then we have:

$$\ln\overline{\dot{\gamma}}\propto\ln\overline{\tau} \tag{6.32c}$$

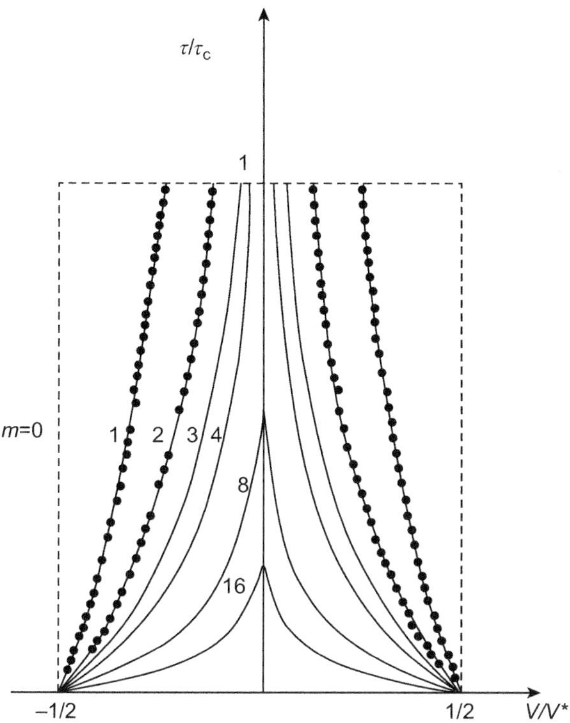

Figure 6.27 Several shapes of hyperbolic potential barrier described by Eq. (6.29).

This is consistent with the power function law (Eq. 5.1) in Chapter 5, and in the double log coordinates $\ln\overline{\tau} - \ln\overline{\dot{\gamma}}$, it is presented as a linear line, of which the linear slope characterizes the strain-rate sensitivity (Eq. 5.3). Therefore, for all the empirical formulas expressed as various power function, including the experimentally determined empirical formula $v_d \propto \tau^n$ describing the relation between the dislocation velocity v_d and stress τ, it provides a theoretical support based on the dislocation dynamics.

(3) When $m = 2$, $\overline{V} = (1+\overline{\tau})^{-2}$, then:

$$\overline{U} = (1+\overline{\tau})^{-1} - \frac{1}{2} \tag{6.33a}$$

$$\overline{T}\ln\overline{\dot{\gamma}} = \frac{1}{2} - (1+\overline{\tau})^{-1} \tag{6.33b}$$

Furthermore, when $\overline{\tau} >> 1$, we have:

$$\ln\overline{\dot{\gamma}} \propto -\frac{1}{\overline{\tau}} \tag{6.33c}$$

Recall the pioneering experiment results by Johnston and Gilman (1959) regarding the relation between the dislocation velocity v_d and shear stress τ for LiF, which was represented as a power function law in $lnv_d - ln\tau$ coordinates at that time, as shown in Fig. 6.17. Gilman (1960, 1965) later found that the same experimental results can be characterized by a more concise empirical formula.

$$\frac{v_d}{v_0} = \exp\left(-\frac{D}{\tau}\right) \tag{6.34a}$$

where v_0 is the velocity limit close to sound velocity and D the characteristic drag stress. This equation is also supported by other researchers' experimental results as shown in Fig. 6.28. The figure gives the experimental results for the dislocation velocity of the silicon iron by Stein and Low (1960), and the experimental results for sodium chloride by Gutmanas et al., (1963). As can be seen, all three fit well with Eq. (6.34a). Substituting Eq. (6.34a) into the Orowan equation, we have:

$$\dot{\gamma} \propto \exp\left(-\frac{D}{\tau}\right) \tag{6.34b}$$

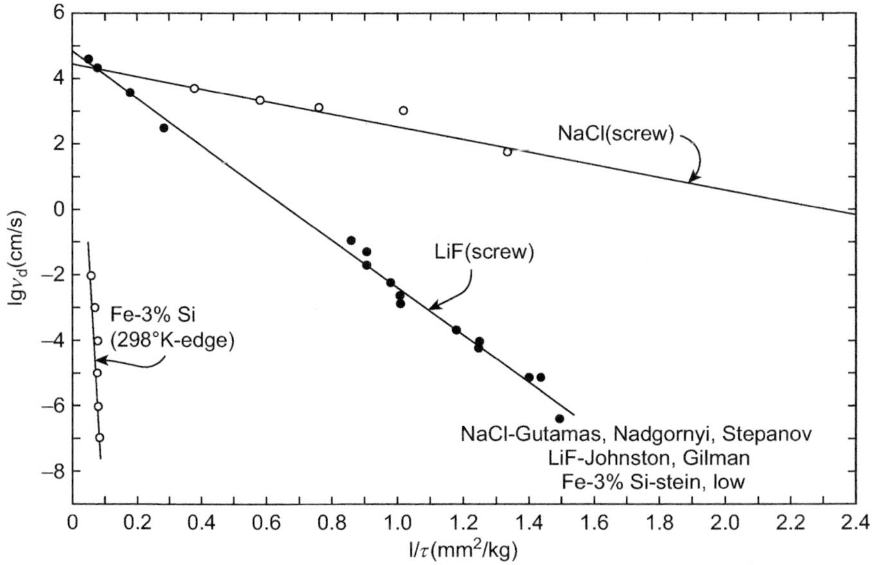

Figure 6.28 The measured dislocation velocity v_d is inversely proportional to shear stress τ, meeting the Gilman formula (Eq. 6.34). *From Gilman, J.J, 1965. Dislocation Mobility in Crystals, J Appl Phys. 36 (10), 3195–3206., Fig. 2, p.3196. Reprinted with permission of the publisher.*

Eq. (6.34) is called the **Gilman equation**. This empirical equation is obviously consistent with Eq. (6.33c), then becomes the special case during $m = 2$ of the hyperbolic barrier, and thus obtains the theoretical support from dislocation dynamics. Some past researchers challenged the Gilman empirical formula (Davidson and Lindholm, 1974) and criticized that Eq. (6.34) corresponds to an unbounded barrier, namely the requirement that the $U^\star$ and τ_0 should have finite bounds is not satisfied. However, from the view of hyperbolic barrier (ref. Fig. 6.27), in fact, this objection does not hold true.

(4) When m is assumed to be an arbitrary value, substituting $\overline{V} = (1 + \overline{\tau})^{-m}$ into Eq. (6.22) and by using Eq. (6.23), the following equation in general form can be obtained (but $m \neq 1$).

$$(1 + \overline{\tau})^{1-m} = 2^{1-m} + (1 - m)\overline{T}1n\bar{\dot{\gamma}} \tag{6.35a}$$

$$\bar{\dot{\gamma}} = \exp\left(\frac{(1 + \overline{\tau})^{1-m} - 2^{1-m}}{(1 - m)\overline{T}}\right) \tag{6.35b}$$

This is the general form of the thermoviscoplastic distortion law based on the hyperbolic thermal activation barrier.

For any value of $m > 0$, if $\overline{\tau} << 1$, Eq. (6.31) for hyperbolic barriers can be expanded into series.

$$\overline{V} = 1 - m\overline{\tau} + \frac{m(m+1)}{2!}\overline{\tau}^2 - \cdots \tag{6.36}$$

Obviously, neglecting the higher order terms, Eq. (6.36) reduces to the triangular-shape barrier proposed by Makin (1968), if $m = 1$; and it is consistent with the parabolic-shape barrier proposed by Mott (1956), if $m = 1/2$.

From what has been discussed above, a lot of existing barrier shapes can be regarded as special cases of hyperbolic-shape barrier.

6.4.4.1 Spectrum of hyperbolic barriers

Moreover, consider that a single thermal activation barrier generally plays a leading role in a certain range of strain rate and temperature, so in a broader range of strain rate and temperature multiple mechanisms may exist, which are characterized by different barrier shapes, respectively. In other words, there exist a spectrum of barriers (Makin, 1968; Wang, L.L., 1984); in fact, different mechanism plays leading role, respectively, under different condition. In this way, based on the hyperbolic barrier shown in Eq. (6.31), the corresponding barrier spectrum can be established.

From Eqs. 6.35a and Eq.35b, for the ith barrier, we have:

$$\overline{\tau}_i = \left(2^{1-m_i} + (1-m_i)\overline{T}_i 1n\dot{\overline{\gamma}}_i\right)^{\frac{1}{1-m_i}} - 1 \tag{6.37a}$$

$$\dot{\overline{\gamma}}_i = \exp\left(\frac{(1+\overline{\tau}_i)^{1-m_i} - 2^{1-m_i}}{(1-m_i)\overline{T}_i}\right) \tag{6.37b}$$

If for a given stress $\overline{\tau} = \overline{\tau}_i$ and temperature $\overline{T} = \overline{T}_i$, the total thermoviscoplastic strain rate $\dot{\overline{\gamma}}$ is the sum of each $\dot{\overline{\gamma}}_i$ multiplied by the corresponding weight function ψ_i, then we have:

$$\dot{\overline{\gamma}} = \sum_{i=1}^{n} \psi_i \dot{\overline{\gamma}}_i = \sum_{i=1}^{n} \psi_i \exp\left(\frac{(1+\tau)^{1-m_i} - 2^{1-m_i}}{(1-m_i)\overline{T}}\right) \tag{6.38a}$$

where the "strain-rate weight function" ψ_i is a function of $\overline{\gamma}$, $\dot{\overline{\gamma}}$, and T in general and satisfies

$$\sum_{i=1}^{n} \psi_i(\gamma, \dot{\gamma}, T) = 1 \tag{6.38b}$$

If for a given strain rate $\dot{\overline{\gamma}} = \dot{\overline{\gamma}}_i$ and temperature $\overline{T} = \overline{T}_i$, the total stress $\overline{\tau}$ is the sum of each $\overline{\tau}_i$ multiplied by the corresponding weight function ϕ_i, then we have:

$$\overline{\tau} = \sum_{i=1}^{n} \varphi_i \overline{\tau}_i = \sum_{i=1}^{n} \varphi_i \left\{ \left((2^{1-m_i} + (1-m_i)\overline{T}\ln\dot{\overline{\gamma}} \right)^{\frac{1}{1-m_i}} - 1 \right\} \tag{6.39a}$$

where the "stress weight function" ϕ_i is also a function of γ, $\dot{\gamma}$, and T in general and satisfies

$$\sum_{i=1}^{n} \phi_i(\gamma, \dot{\gamma}, T) = 1 \tag{6.39b}$$

Thermoviscoplastic distortional equations Eqs. (6.38) and (6.39) are, respectively, correspondent to the combination of barriers "in series" and "in parallel," similar to the generalized Maxwell model and the generalized Kelvin model in the viscoelastic theory.

Obviously, the role of each barrier in the different range of strain rate and temperature is characterized by the weight function ψ_i or ϕ_i. How to determine the weight function ψ_i and ϕ_i is another important problem.

6.4.4.2 Experiment verification of the hyperbolic barrier model

In order to experimentally verify the hyperbolic barrier model, the dimensionless Eq. (6.35) is expanded to the corresponding dimensional equation:

$$(\tau+\tau_0)^{1-m}=(2\tau_0)^{1-m}+\frac{(1-m)k}{\tau_0^m V^*}T\ln\dot{\gamma}+\frac{(m-1)k\ln\dot{\gamma}_0}{\tau_0^m V^*}T \tag{6.40a}$$

For convenience to fit with experimental data, it can be regarded as a linear equation in the following form:

$$\begin{aligned} &y=A+B_1x_1+B_2x_2\\ &y=(\tau+\tau_0)^{1-m},\quad x_1=T\ln\dot{\gamma},\quad x_2=T\\ &A=(2\tau_0)^{1-m},\quad B_1=\frac{(1-m)k}{\tau_0^m V^*},\quad B_2=\frac{(m-1)k\ln\dot{\gamma}_0}{\tau_0^m V^*}\end{aligned} \tag{6.40b}$$

It is shown that the stress τ is not only dependent on the rate—temperature coupled term (namely the so-called rate—temperature equivalent term) but also simultaneously dependent on the temperature T. For a certain $\dot{\gamma}$, by using the two-dimensional linear regression analysis, regression coefficients A, B_1, and B_2 can be determined from experimental data, thereby τ_0, $V^\star$, and $\dot{\gamma}_0$ can be calculated subsequently as:

$$\tau_0=\frac{1}{2}A^{\left(\frac{1}{1-m}\right)},\quad V^*=\frac{(1-m)k}{\tau_0^m B_1},\quad \dot{\gamma}_0=\exp(-B_2/B_1) \tag{6.40c}$$

Note that in real case, τ_0 is necessary to be determined through an iteration process, since y also depends on τ_0.

The experimental data of Lindholm (1968) for aluminum under a wide range of strain rate 10^{-3} to 10^3 s^{-1} and temperature 294 to 672°K (see Fig. 5.13 in Chapter 5) are used to examine the availability of Eq. (6.40), by assuming $m=0$, 1, 2, and 3, respectively. Note that the case of $m=0$ is equivalent to the Seeger's model. Each family of regression curves is the best least-squares fit through the whole Lindholm's data. It is found that the correlation coefficient has the highest value at $m=2$. In Fig. 6.29, the regression curves for $m=2$ are drawn by solid lines, while dashed lines in the same diagram are Lindholm's fitting lines corresponding to the Seeger's model. As can be seen, the nonlinear hyperbolic thermal activation barrier shows a better agreement with experimental data than the linear Seeger's barrier, especially near two ends of curves.

The experimental data of the lower yield stress of mild steel, for a range of temperature from 195 to 713 K and of strain rate from 10^{-3} s^{-1} to

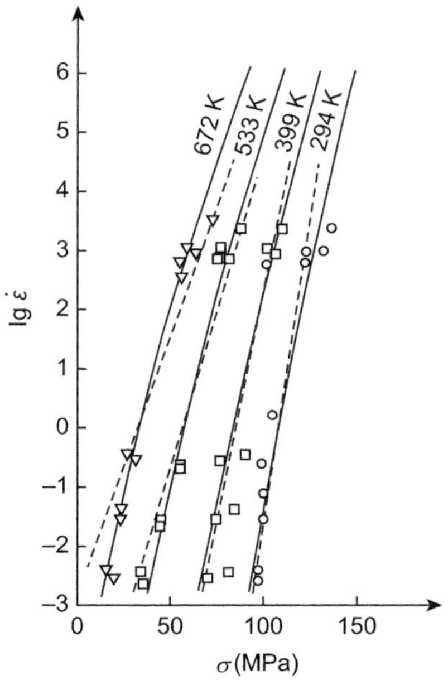

Figure 6.29 Comparisons between the Lindholm's experimental results for aluminum (for a given strain of 0.15) and the regression curves of hyperbolic barrier model for $m = 2$. *From Wang, L.L., 1984. A Thermo-Viscoplastic Constitutive Equation Based on Hyperbolic Shape Thermo-Activated Barriers. Trans ASME. J Eng Mat Tech. 106 331–336., Fig. 1, p.334. Reprinted with permission of the publisher.*

$4 \times 10^4\ s^{-1}$, have been given by Campbell and Ferguson (1970), as having shown in Fig. 5.14 in Chapter 5, and now are given again in Fig. 5.30B. According to the strain-rate sensitivity, these experiment results are divided into three regions: Zone I (low strain rate—high temperature zone) with less sensitivity to strain rate; Zone II (high strain rate—low temperature zone) that follows the linear relation in $\sigma - \log\dot{\varepsilon}$ coordinates (Seeger's model); and Zone IV (high strain rate zone) that follows the linear viscous relationship in $\sigma - \dot{\varepsilon}$ coordinates. For the experimental data of Zone I and Zone II, when the regression analysis of the least square method is carried out according to the hyperbolic barrier model, the correlation coefficient of m = 2 is also found to be the highest. The fitting result is shown in Fig. 6.30A by the solid line. As can be seen that the experimental data of Zone I and Zone II can be completely characterized by a single thermal activation mechanism with hyperbolic barrier (m = 2).

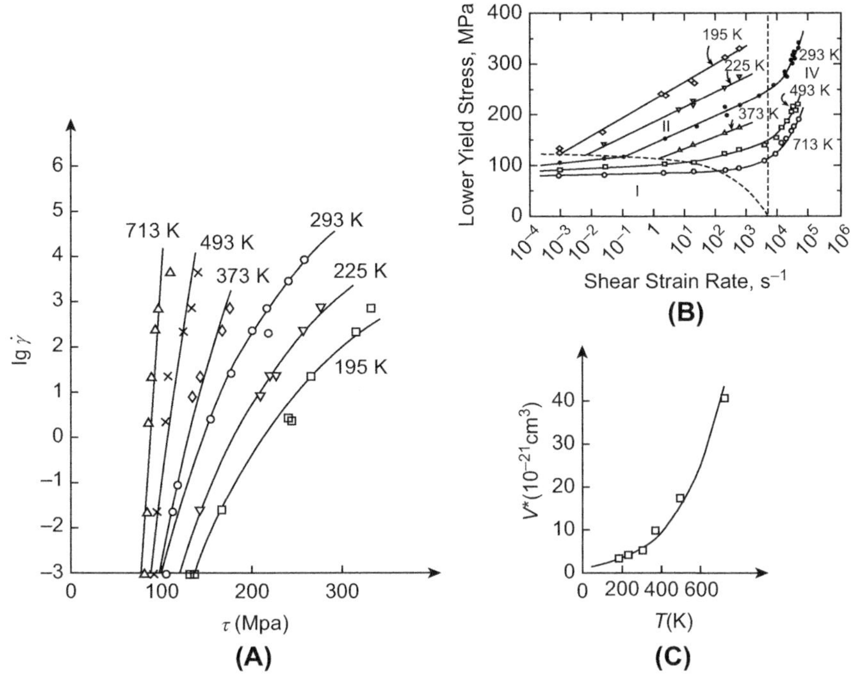

Figure 6.30 Comparisons between the experimental results by Campbell–Ferguson for mild steel and the regression curves by hyperbolic barrier model. *From Wang, L.L., 1984. A Thermo-Viscoplastic Constitutive Equation Based on Hyperbolic Shape Thermo-Activated Barriers. Trans ASME. J Eng Mat Tech. 106 331–336., Fig. 2, p.334 and Fig. 4, p.335. Reprinted with permission of the publisher.*

The above two famous experimental results, the former is for FCC metal and the latter is for BCC metal, consistently provided a strong support for the dislocation dynamics mechanism with the hyperbolic barrier model.

During the regression analysis of experimental results of Campbell and Ferguson, it is also found that the activation volume at $\tau = 0$, $V^\star$ is a function of temperature T, $V^\star = V(0,T) = V^\star(T)$, as shown in Fig. 6.30C, and can be expressed in a dimensionless form as:

$$\frac{V^*}{V_0} = (1+\alpha)^{T/T_0} \tag{6.41}$$

For convenience, take the characteristic temperature $T_0 = 1$ K, V_0 represents the activation volume at 0 K, and α represents the relative change of activation volume at 1 K. For the mild steel tested by Campbell and Ferguson, $V_0 = 1.419 \times 10^{21}$ cm^3 and $\alpha = 0.00{,}485$.

Substituting Eq. (6.41) into Eq. (6.35a), we have:

$$\left(1+\frac{\tau}{\tau_0}\right)^{1-m} = (2)^{1-m} + \frac{(m-1)kT}{\tau_0 V_0 (1+\alpha)^T} \ln\left(\frac{\dot{\gamma}_0}{\dot{\gamma}}\right) \tag{6.42}$$

Recall that in the discussion on the rate–temperature equivalency (ref. Section 5.1.2 in Chapter 5), a parameter $T^{\star}$ has been introduced, which is defined as:

$$T^* = T \ln\left(\frac{\dot{\varepsilon}_0}{\dot{\varepsilon}_p}\right) \approx T \ln\left(\frac{\dot{\varepsilon}_0}{\dot{\varepsilon}}\right), \tag{5.7}$$

Taking account of the $V^{\star}(T)$ effect, a new parameter Z instead of $T^{\star}$ should be introduced, which is defined as:

$$Z = \frac{kT}{(1+\alpha)^T} \ln\left(\frac{\dot{\gamma}_0}{\dot{\gamma}}\right) = \frac{kT}{(1+\alpha)^T} \ln\left(\frac{\dot{\varepsilon}_0}{\dot{\varepsilon}}\right), \tag{6.43}$$

Furthermore, if the parameter F defined as below is introduced:

$$F = \left(1+\frac{\tau}{\tau_0}\right)^{1-m} - (2)^{1-m} \tag{6.44}$$

then Eq. (6.42) can be rewritten as the following simple equation in a linear form.

$$F = \frac{(m-1)k}{\tau_0 V_0} Z \tag{6.45}$$

This means that the experimental data of flow stress measured at different strain rates and temperatures will fall on a straight line represented by Eq. (6.45) in the F–Z coordinates, often referred to as the **master curve**.

The experimental results of Campbell–Ferguson for mild steel (Fig. 6.30A) are redrawn in the F–Z coordinates, as shown in Fig. 6.31.

It can be seen from the figure that all experimental points fall around the straight line characterized by Eq. (6.45). It not only shows that the experimental data of Campbell–Ferguson are governed by the dislocation dynamics mechanism based on the hyperbolic barrier, but also indicates that the new rate–temperature equivalent parameter Z (Eq. 6.43), rather than the traditional parameter $T^{\star}$ (Eq. 5.7), can better describe the complex rate–temperature coupling essence.

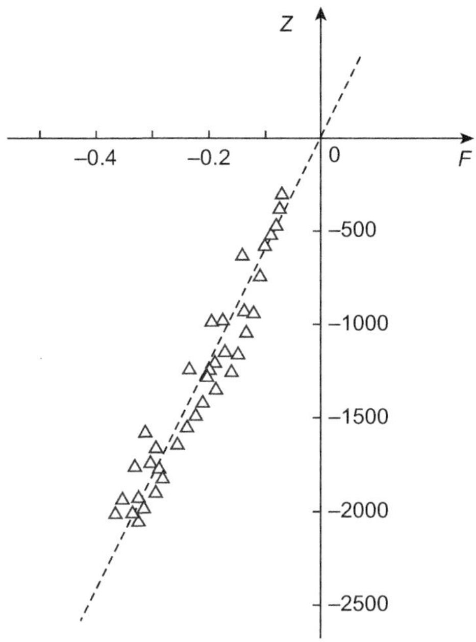

Figure 6.31 Comparisons between the experimental results of Campbell−Ferguson for mild steel with the regression curve of hyperbolic barrier. *From Wang, L.L., 1984. A Thermo-Viscoplastic Constitutive Equation Based on Hyperbolic Shape Thermo-Activated Barriers. Trans ASME. J Eng Mat Tech. 106 331−336., Fig. 5, p.335. Reprinted with permission of the publisher.*

6.4.5 Zerilli−Armstrong model

The Orowan equation indicates that the plastic strain rate $\dot{\gamma}^p$ depends on both the dislocation moving velocity v_d and the rate of dislocation density with respect to time $\dot{\rho}_m$. But up to now, we mainly consider the influence of the dislocation moving velocity v_d, and assume that the mobile dislocation density rate $\dot{\rho}_m$ changes slowly over time so that can be ignored. This essentially ignores the **strain-hardening effect** based on the dislocation multiplication mechanism.

Zerilli and Armstrong (1987, 1990) took the strain-hardening effect into account and also noticed that the strain-rate sensitivity of different lattice structures is different. For example, the body-centered cubic (BCC) metals exhibit much higher strain-rate sensitivity than the face-centered cubic (FCC) metals. Furthermore, the strain hardening has different effects on the BCC metals and FCC metals. For the FCC metals, such as oxygen-free high-conductivity copper (OFHC), the activation area A depends on the strain ε ($A \propto \varepsilon^{1/2}$), so that its strain rate−temperature sensitivity of flow stress increases strongly with strain. For the BCC metals, on the contrary, such as industrial pure iron (Armco iron), the activated area does not depend on the strain, thus its strain rate—temperature sensitivity of flow stress is independent of strain.

With regard to the thermal activation stress τ^* itself, Zerilli and Armstrong (1987) started from the following $U(\tau)$ relation (adopting symbols of the present book):

$$U(\overline{\tau}) = -U_0 \ln(\overline{\tau}) \tag{6.46a}$$

Substituting the above equation into the dimensionless plastic Arrhenius equation (Eq. 6.23), after the formula is expanded, there is:

$$\ln\left(\frac{\tau^*}{\tau_0}\right) = \frac{kT}{\tau_0 V^*}\ln\frac{\dot{\gamma}}{\dot{\gamma}_0} \tag{6.46b}$$

Note that it actually can be regarded as a special case of hyperbolic barrier when $m = 1$ (Eq. 6.32b). Or, it could be rewritten into the following form:

$$\tau^* = \tau_0 \exp\left(-\frac{k \ln\dot{\gamma}_0}{\tau_0 V^*} T + \frac{k}{\tau_0 V^*} T \ln\dot{\gamma} \right) \tag{6.46c}$$

For convenience, in the following, C_1, C_3, C_4, and so on are used, respectively, instead of the preexponential coefficient τ_0, the coefficient of T and the coefficient of $T \ln\dot{\gamma}$ both within the exponential term.

On this basis, Eq. (6.46a) is further modified to take into account of strain-hardening effect, and then the thermal activation stress σ^* in the form of normal stress, for the BCC metals and FCC metals, can be written, respectively, as:

$$\sigma^* = C_1 \exp(-C_3 T + C_4 T \ln\dot{\varepsilon}) \quad \text{(for BCC)} \tag{6.47a}$$

$$\sigma^* = C_2 \varepsilon^{1/2} \exp(-C_3 T + C_4 T \ln\dot{\varepsilon}) \quad \text{(for FCC)} \tag{6.47b}$$

Zerilli and Armstrong also considered the effect of grain size d on flow stress, which is expressed as $\sigma = \sigma_0 + kd^{-1/2}$ according to the famous Hall—Petch formula, where σ_0 and k are material constants. Finally, take account of the long-range athermal stress σ_G, thus:

$$\sigma = \sigma_G + C_1 \exp(-C_3 T + C_4 T \ln\dot{\varepsilon}) + C_5 \varepsilon^n + kd^{-1/2} \quad \text{(for BCC)} \tag{6.48a}$$

$$\sigma = \sigma_G + C_2 \varepsilon^{1/2} \exp(-C_3 T + C_4 T \ln\dot{\varepsilon}) + kd^{-1/2} \quad \text{(for FCC)} \tag{6.48b}$$

which are called the **Zerilli–Armstrong** equations, or the **ZA** equation for short.

Zerilli and Armstrong verified the validity of the proposed constitutive equation by using the Taylor bar impact experiment. The details of Taylor bar experiments will be discussed in the next chapter (see Section 7.2.1). Originally, it was mainly used for the inverse determination of the yield stress and constitutive relation of materials. In recent years, because the Taylor bar appears a serious inhomogeneous plastic distribution after an impact, so that through only one impact test of Taylor bar, it provides rich information of constitutive response in a wide strain range and also in a wide strain-rate range (across the two to three orders of magnitude), thus it becomes also a convenient, sensitive, and useful method and is used as a validation test of different material constitutive models. This impact experimental method has aroused people's extensive interest again.

Zerilli and Armstrong (1987) carried out the Taylor bar impact test at 190 m/s impact velocity on the oxygen-free high-conductivity copper (OFHC), an FCC metal, with an original bar radius R_0. After the experiment, the measured radial strain, $\log(R/R_0)$, varies with the distance from the impact surface, as shown in the dot-and-dash line in Fig. 6.32. In the same figure, the theoretical prediction curves by Zerilli–Armstrong

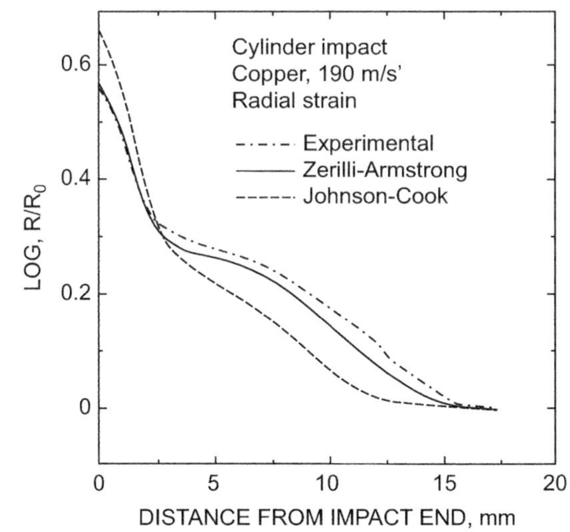

Figure 6.32 Comparisons between the measured profile shape and the theory predictions by the Taylor impact test for OFHC. *From Zerilli, F.J, Armstrong, R.W., 1987. Dislocation-mechanics-based constitutive relations for material dynamics calculations. J. Appl. Phys. 61 (5), 1816–1825., Fig. 3, p.1822. Reprinted with permission of the publisher.*

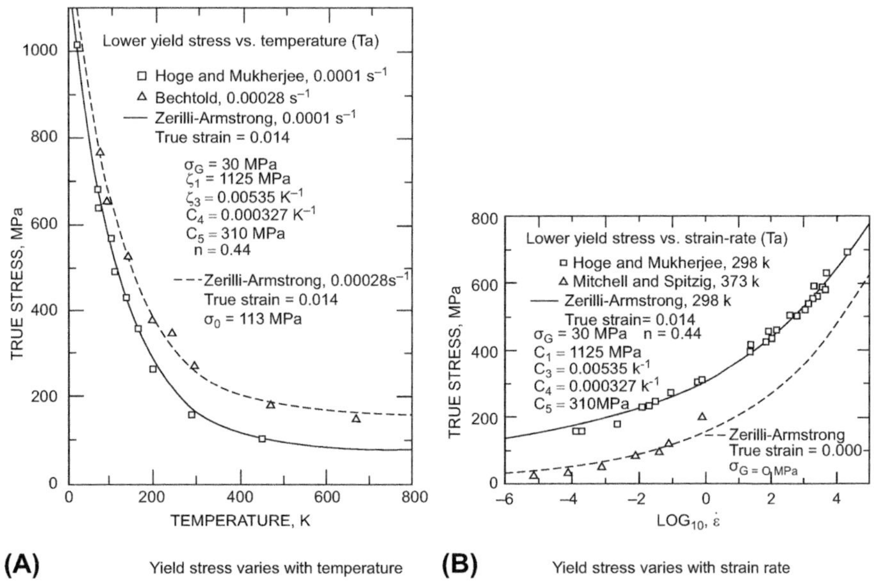

Figure 6.33 Comparisons of the Zerilli–Armstrong equation with the experimental data for tantalum. (A) Yield stress varies with temperature; (B) Yield stress varies with strain rate. *From Zerilli, F.J, Armstrong, R.W., 1987. Dislocation-mechanics-based constitutive relations for material dynamics calculations. J. Appl. Phys. 61 (5), 1816–1825., Fig. 7 and Fig. 9, p.1586. Reprinted with permission of the publisher.*

equation presented in solid line and by Johnson–Cook equation (Eq. 5.14 in Chapter 5) presented in dotted line are given too. Clearly, the Zerilli–Armstrong equation is closer to the experimental results.

Fig. 6.33 gives how the lower yield stress varies with temperature and strain rate for tantalum (a BCC metal), where the predictions of Zerilli–Armstrong equation are compared with the experimental results by other researchers (Zerilli and Armstrong, 1990; Meyers, 1994). As can be seen from the figure, whether the yield stress changes with the temperature (Fig. 6.33A) or the yield stress changes with the strain rate (Fig. 6.33B), the Zerilli–Armstrong equation fits well with the experimental data in a large experimental range.

The main advantages of the Zerilli–Armstrong equation are that strain-hardening effects and grain size effects are taken into account for different lattice structures. Regarding the thermal activation barrier shape, its $U(\tau)$ relation is described by Eq. (6.46a) and expressed as a linear relation in the $\ln\tau$-$\ln\dot{\gamma}$ coordinates (Eq. 6.46b), which actually can be regarded as a special case of hyperbolic barrier when $m = 1$ (Eq. 6.32b), namely there exists a linear relation between $\ln\overline{\tau}$ and $\ln\overline{\dot{\gamma}}$.

6.4.6 Mechanical threshold stress model

In what has been discussed above all sorts of thermal activation models, the barrier peak stress τ_0, as shown in Fig. 6.23, generally has been assumed as a constant, which in physics refers to the mechanical threshold stress required to overcome the dislocation potential barrier without thermal activation help, namely under 0°K temperature. The premise of this assumption is that there is no change of microstructural state. However, in fact, with the evolution of microstructure, τ_0 changes too.

On the other side, a large number of experimental data on the variation of flow stress with strain rate (see Chapter 5) are generally expressed as stress—strain curves under different strain rates and on this basis expressed as $\ln \tau - \ln\dot{\gamma}$ or $\tau - \ln\dot{\gamma}$ curves under a given strain. The premise of all those expressions is that the distortional law of materials can be generally expressed as:

$$\tau = \tau(\gamma, \dot{\gamma}, T), \tag{5.10b}$$

Note that the strain history effect and the strain-rate history effect are both ignored in this form of distortional law (see Section 5.1.3 in Chapter 5). In fact, for the irreversible viscoplastic deformation, if take into account of the strain history effect which is a path correlation process, according to the theory of internal variables, the constitutive distortional law of materials should be expressed as:

$$\tau = \tau(\gamma, \dot{\gamma}, T, \xi_i), \tag{6.49}$$

where ξ_i ($i = 1,2,3 \ldots n$) are internal variables which are generally functions of $\gamma, \dot{\gamma}, T$, and $\xi_i = \xi_i(\gamma, \dot{\gamma}, T)$. In such case, a given strain value does not correspond to a given microstructural state of materials.

It is based on the above situation that Follansbee and Kocks pointed out that the strain is not an effective state parameter and then suggested to use the mechanical threshold stress τ_0 as an internal variable reflecting the change of microscopic state (Follansbee, 1986; Follansbee and Kocks, 1988; Meyers, 1994). Regarding the basic thermal activation model, they adopted the Kocks—Argon—Ashby nonlinear model $\overline{U}(\overline{\tau})$ (Eq. 6.28a), which is expressed as (in terms of normal stress):

$$U = U_0\left[1 - \left(\frac{\sigma}{\sigma_0}\right)^p\right]^q \tag{6.49a}$$

and suggested that $P = 1/2$ and $q = 3/2$. Considering that the activation energy when stress is zero, U_0, has the dimension of the product of stress and activation volume, introduce dimensionless normalized activation energy $g_0 = U_0/(G(T)b^3)$; the above equation can be rewritten as:

$$U = G(T)b^3 g_0 \left[1 - \left(\frac{\sigma}{\sigma_0}\right)^p\right]^q \tag{6.49b}$$

where $G(T)$ is the elastic shear modulus (generally a weak function of temperature T) and b is the magnitude of Burger's vector. Substituting the above equation into the Arrhenius equation for plastic strain rate (Eq. 6.17b), after reorganizing and expressing in the form of normal stress, it can be written as the following dimensionless form.

$$\left(\frac{\sigma}{G(T)}\right)^p = \left(\frac{\sigma_0}{G(T)}\right)^p \left[1 - \left(\frac{kT}{G(T)b^3 g_0} \ln\frac{\dot{\varepsilon}_0}{\dot{\varepsilon}}\right)^{1/q}\right] \tag{6.50a}$$

This is called the mechanical threshold stress (MTS) model. From the view of irreversible thermodynamics, the mechanical threshold stress σ_0 (or τ_0) here is an internal variable reflecting the microstructure evolution of materials, equivalent to taking $\xi_1 = \sigma_0, \xi_2 = \xi_3 = \ldots = \xi_n = 0$ in Eq. (6.49). The mechanical threshold stress σ_0 varies with the microstructure evolution induced by strain history and strain-rate history process and is generally a function of strain and strain rate $\sigma_0 = \sigma_0(\varepsilon, \dot{\varepsilon})$.

The key and difficulty of the next step is how to determine the evolution relation of mechanical threshold stress, $\sigma_0 = \sigma_0(\varepsilon, \dot{\varepsilon})$, by experiments.

According to the definition of mechanical threshold stress, the σ_0 can be determined by measuring the flow stress at temperature 0 K. But, it is difficult to realize in practice, the only way is to measure the flow stresses at a series of low temperature and then to obtain the σ_0 at 0 K temperature by the extrapolation approach.

By using the split Hopkinson pressure bar (SHPB) technique, Follansbee et al. designed a series of "dynamic preloading—unloading—reloading" tests for OFHC, where the measurements involved different strain histories and strain-rate histories in order to better reflect the microstructure evolution (Follansbee, 1986; Follansbee and Kocks, 1988; Meyers, 1994). All specimens were dynamic preloaded to a given strain under high strain rate (10^4 s^{-1}) and room temperature. After unloading, they were reloaded to different final strain under different strain rates and different low temperatures, respectively.

If the mechanical response of specimen satisfies the MTS equation (Eq. 6.50a, where $P = 1/2$ and $q = 3/2$), then there is:

$$\left(\frac{\sigma}{G(T)}\right)^{1/2} = \left(\frac{\sigma_0}{G(T)}\right)^{1/2}\left[1 - \left(\frac{kT}{G(T)b^3 g_0}\ln\frac{\dot{\varepsilon}_0}{\dot{\varepsilon}}\right)^{2/3}\right] \tag{6.50b}$$

It indicates that for a given strain rate $\dot{\varepsilon}$, the curve of the dimensionless flow stress $\widetilde{\sigma}\left(= (\sigma/G(T))^{1/2}\right)$ against the dimensionless temperature $\widetilde{T}\left(= \left(kT/G(T)b^3\right)^{2/3}\right)$can be plotted. If there exists a linear relation, the intercept of this straight line with the ordinate is the dimensionless mechanical threshold stress $\widetilde{\sigma}_0\left(= (\sigma_0/G(T))^{1/2}\right)$.

The typical experimental results are shown in Fig. 6.34 (Follansbee, 1986; Meyers, 1994). Fig. 6.34A shows the experimental results of the dimensionless flow stress $\widetilde{\sigma}$ against the dimensionless temperature $\widetilde{T}$ when the specimen is reloaded to different final strains at a constant strain rate (1.4×10^{-4} s^{-1}). As expected, the $\widetilde{\sigma}-\widetilde{T}$ relations for different terminal strains all satisfy linear relation, and the intercepts of the ordinate with each straight line increase with the final strain, which means that the mechanical threshold stress σ_0 is an increasing function of strain.

Fig. 6.34B shows the experimental results of the dimensionless flow stress $\widetilde{\sigma}$ against the dimensionless temperature $\widetilde{T}$ when the specimen is reloaded to

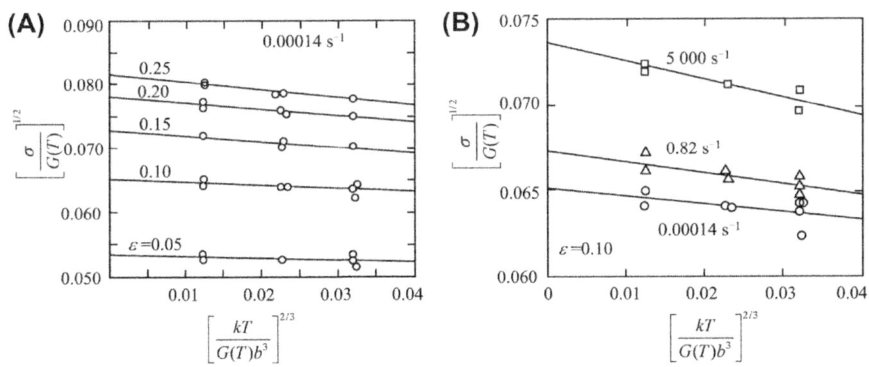

Figure 6.34 Experimental results of the dimensionless flow stress $\widetilde{\sigma}$ against the dimensionless temperature $\widetilde{T}$ for OFHC. *From Follansbee, P.S., 1985. High strain rate deformation in FCC metals and alloys. No. LA-UR-85-3026; CONF-850770-8. Los Alamos National Lab, NM (USA). doi: https://www.osti.gov/servlets/purl/5276223, Fig. 11 and 13.*

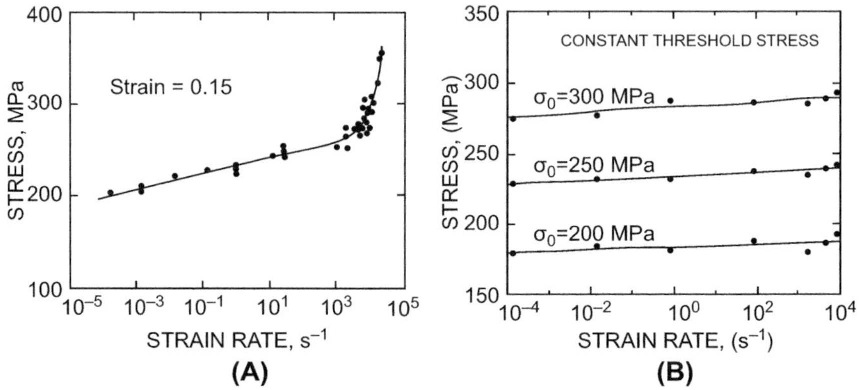

Figure 6.35 Experimental results of $\sigma - \ln\dot{\varepsilon}$ for OFHC under the condition of (A) constant strain and (B) constant mechanical threshold stress. *From Follansbee, P.S., Kocks, U.F., 1988. A constitutive description of the deformation of copper based on the use of the mechanical threshold stress as an internal state variable. Acta Metall. 36 (1), 81–93., Fig. 1 and Fig. 2, p.82. Reprinted with permission of the publisher.*

the same final strain 0.01 at different strain rates. The obtained $\widetilde{\sigma}-\widetilde{T}$ relations are linear also, and the intercepts of the ordinate with each straight line increase with the strain rate, which means that the mechanical threshold stress σ_0 is an increasing function of strain rate. Thus, the evolution equation of mechanical threshold stress, $\sigma_0 = \sigma_0(\varepsilon, \dot{\varepsilon})$, can be finally determined.

From the experimental data for OFHC given by Follansbee et al (Follansbee, 1986; Follansbee and Kocks, 1988; Meyers, 1994), another important result should be noticed. If a graph of σ against $\ln\dot{\varepsilon}$is plotted for a given strain $\varepsilon = 0.15$ from the data, as shown in Fig. 6.35A (the same figure as shown in Fig. 5.8 in Chapter 5), it is roughly divided into two regions bounded by $\dot{\varepsilon} = 10^3\ \mathrm{s}^{-1}$. Traditionally, it is accepted that the region of $\dot{\varepsilon} \leq 10^3\ \mathrm{s}^{-1}$ is correspondent to the linear $\sigma - \ln\dot{\varepsilon}$ relation dominated by dislocation thermal activation mechanism, while the region of $\dot{\varepsilon} \geq 10^3\ \mathrm{s}^{-1}$ is correspondent to the $\sigma - \dot{\varepsilon}$ viscous relation dominated by dislocation drag mechanism (see Section 5.8 in Chapter 5). In accordance with the MTS model, however, for the same experimental data, if a graph of σ against $\ln\dot{\varepsilon}$ is plotted for a given constant mechanical threshold stress σ_0, as shown in Fig. 6.35B, the slope of $\sigma - \ln\dot{\varepsilon}$ curve, which characterizes the strain-rate sensitivity, under the constant threshold stress remains constant value until $\dot{\varepsilon} = 10^4\ \mathrm{s}^{-1}$, which means that there is no any sign of turning to dislocation drag mechanism. It indicates that the dramatic increase in strain-rate sensitivity near $\dot{\varepsilon} = 10^3\ \mathrm{s}^{-1}$ in Fig. 6.35A is not a change of dislocation mechanism. However, as mechanical stress threshold σ_0 increases

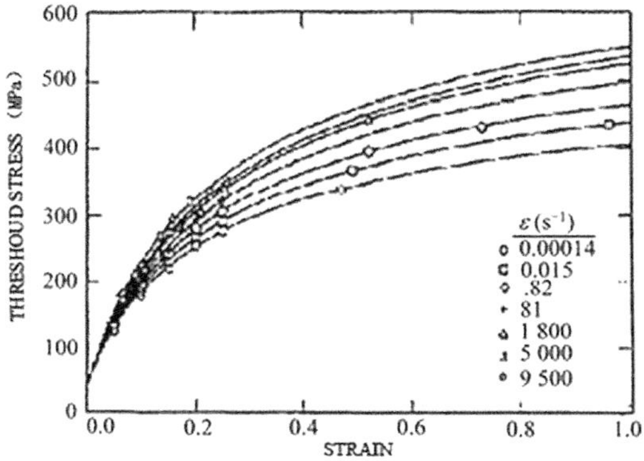

Figure 6.36 The $\sigma_0 = \sigma_0(\varepsilon, \dot{\varepsilon})$ relation obtained after fitting with experimental data for OFHC. *From Follansbee, P.S., Kocks, U.F., 1988. A constitutive description of the deformation of copper based on the use of the mechanical threshold stress as an internal state variable. Acta Metall. 36 (1), 81–93., Fig. 7, p.87. Reprinted with permission of the publisher.*

with the material microstructure evolution, the rate sensitivity improves. The main reason of this misunderstanding is that for irreversible viscoplastic flow, the strain is not a valid state parameter, while the mechanical stress threshold σ_0 is an appropriate internal variable reflecting microscopic state evolution.

Considering that the mechanical stress threshold σ_0 itself consists of two parts, namely the athermal long-distance stress σ_{0G} and the rate-dependent short-distance stress $\sigma_0^{\star}$, $\sigma_0 = \sigma_{0G} + \sigma_0^*$; and moreover considering the strain-hardening effect, Follansbee and Kocks (1988) further modified Eq. (6.50) as:

$$\sigma = \sigma_{0G} + (\sigma_0 - \sigma_{0G})\left[1 - \left(\frac{kT}{G(T)b^3 g_0}\ln\frac{\dot{\varepsilon}_0}{\dot{\varepsilon}}\right)^{1/q}\right]^{1/p} \tag{6.51}$$

They also suggested that the P and q values in the formula should be set as $P = 2/3$ and $q = 1$ in order to match the experimental results better. The evolution relationship of mechanical threshold stress $\sigma_0 = \sigma_0(\varepsilon, \dot{\varepsilon})$ obtained after fitting with experimental data is shown in Fig. 6.36.

Fig. 6.37 shows the experimental results of "dynamic preloading–unloading–reloading" tests for OFHC in room temperature, namely the specimen is first dynamic preloaded to a given strain 0.15 under high strain rate ($10^4\ s^{-1}$), after unloading, then is reloaded to the final strain 0.25 under low strain rate ($10^{-3}\ s^{-1}$). The figure also shows the numerical predictions of the MTS model, the Johnson–Cook model (Eq. 5.14 in Chapter 5), and the Zerilli–Armstrong model (Eq. 6.48). The comparison results show that the

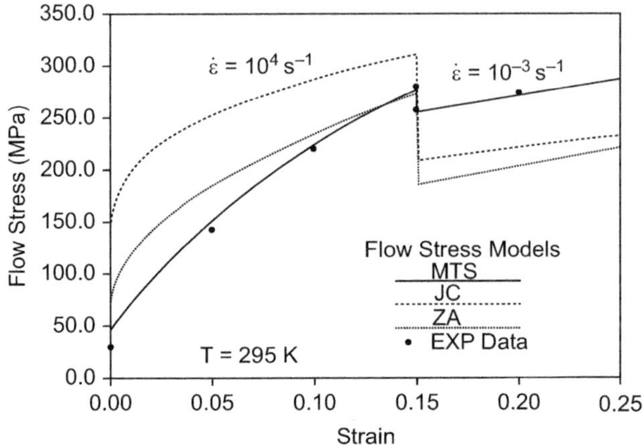

Figure 6.37 Comparisons between the preloading–reloading experimental data for OFHC with the predictions by the MTS model, the JC model, and the ZA model. *From Maudlin, P.J., Davidson, R.F., Henninger R.J, 1990. Implementation and assessment of the mechanical-threshold-stress model using the EPIC2 and PINON computer codes, No. LA-11895-MS, Los Alamos National Lab, NM (USA).*

Johnson–Cook (JC) model overestimates the preloading flow stress by 25% –30%, while underestimates the reloading flow stress by 25%–30%; the Zerilli–Armstrong model overestimates the preloading flow stress by 5% –10%, while underestimates the reloading flow stress by 30%; only the MTS model correctly predicts the preloading flow stress and the reloading flow stress. Obviously, this is because the MTS model has taken into account of the strain/strain-rate historical effect via the internal variable $\sigma_0 = \sigma_0(\varepsilon, \dot{\varepsilon})$, while the other two models lack this function.

As mentioned earlier, the Taylor bar experiment, through only one test, can obtain very rich constitutive information (in a large range of strain and a wide range of strain rate in the two to three orders of magnitude) from the inhomogeneous plastic strain distribution of the impact bar, so it often becomes an effective method to evaluate comprehensively the constitutive model.

Fig. 6.38 shows the experimental results of the Taylor bar (0.76 cm in diameter and 2.54 cm in length) for OFHC, i.e., the shape of the test bar after the experiment in the cylindrical coordinate z–r. The impact velocity of the three graphs is (a) 130 m/s, (b) 146 m/s, and (c) 190 m/s, respectively. In the same figure, the numerical predictions of the MTS model (in dotted line) and the Johnson–Cook model (in dashed lines) calculated by EPIC2

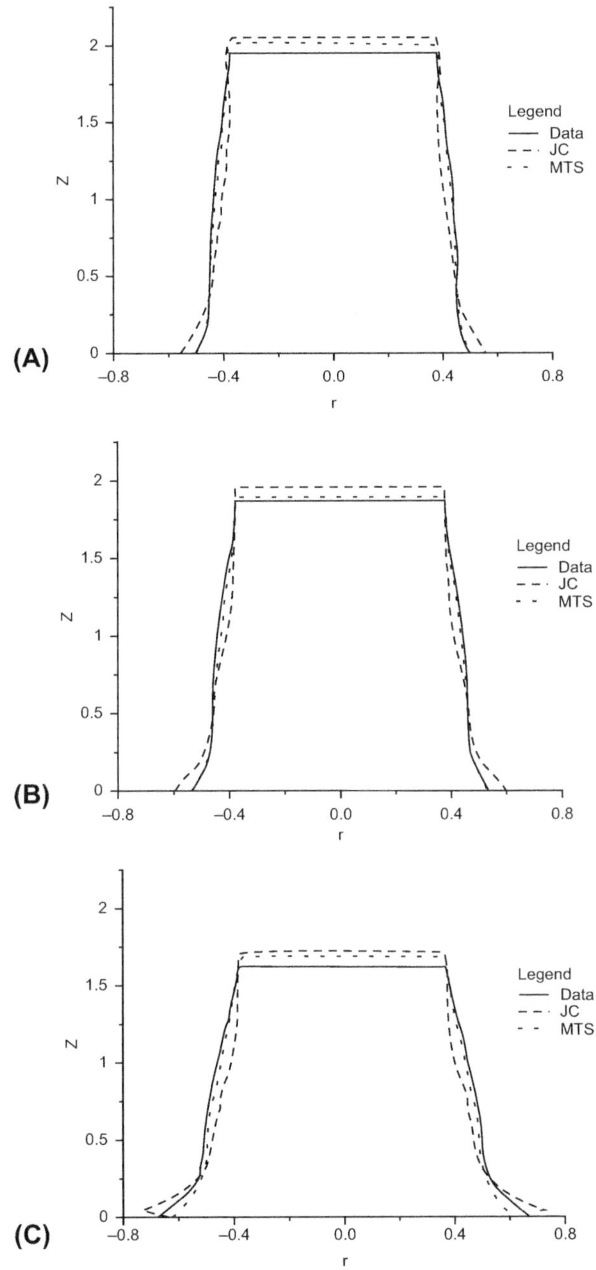

Figure 6.38 Comparisons between the Taylor bar experimental results for OFHC with the predictions of MTS model and JC model. *From Maudlin, P.J., Davidson, R.F., Henninger R.J, 1990. Implementation and assessment of the mechanical-threshold-stress model using the EPIC2 and PINON computer codes, No. LA-11895-MS, Los Alamos National Lab, NM (USA).*

code are given too, respectively, to compare with the experimental result (in solid line) (Maudlin et al., 1990).

As can be seen from the comparison in the figure, the MTS model is much closer to the experimental results than the Johnson–Cook model, in terms of the final length, the size of the impact head ($z = 0$), and the size of the bulging part (around $z = 1$) of the test bar.

From the comparison results shown in Figs. 6.37 and 6.38, the MTS model has obvious advantages. However, it is more complex and difficult to determine the material parameters of the model, including the low temperature experiments that will affect the wide application of the model in engineering.

Research on the thermoviscoplastic distortional law based on the dislocation dynamics at present is still a hot topic concerned by the scientists of both mechanics and materials science, and the related new models and literature reviews are still in continue to emerge (Xu-hong et al., 2007). This section focuses on the basic principles and representative models in this aspect, hoping to help readers follow and participate in new developments on this basis.

CHAPTER SEVEN

Dynamic experimental study on the distortional law of materials

There are two core issues in dynamic experiments of materials to be emphasized here: (1) what is the key difficulty in the experimental study of dynamic distortional law of materials under explosion/impact loading and (2) what is the main difference compared with the quasistatic experimental study on material properties.

Before answering these questions, we must first consider the often-overlooked fact that experimental studies on mechanical behavior of materials are usually realized through the corresponding experimental studies on mechanical response of structures. At present, it is difficult to measure at the same Lagrange material point and at the same time the variation with time of the stress, strain, strain rate, temperature, and other mechanical and thermal quantities that constitute the constitutive relationship of the material, so people have to perform experiments on the specific structure (specimen) made of the material to infer the constitutive characteristics of the material. For example, in the traditional quasistatic uniaxial experiment, the specimen as a special structure is designed to satisfy the requirement that the stress and strain are uniformly distributed along its gauge length in a load-unload process. Thus, the measured stress at any point A within the gauge length can be correlated to the strain measured at any other point B within the same gauge length. Because the stress at point B is equal to the stress at point A, and the strain at point A is equal to the strain at point B, so that the stress—strain relationship of the material can be obtained independent of the measurement point. It has been implicitly assumed here that the stress wave propagation process is allowed to be neglected in this experimental process; otherwise the requirement of uniform distribution of stress/strain within the gauge length cannot be satisfied at all times. It can be clearly seen that the mechanical responses of materials and the mechanical responses of structures are often coupled together, and the so-called **material experiments** are virtually all implemented in the form of **structural experiments**.

Therefore, the experimental research on the dynamic constitutive behavior of materials under explosion/impact loading becomes very difficult

Dynamics of Materials
ISBN: 978-0-12-817321-3
https://doi.org/10.1016/B978-0-12-817321-3.00007-3

and complex. Explosion/impact loading is characterized by rapid loading and unloading in milliseconds, microseconds, or even much shorter durations. If the time scale of the rapid loading—unloading variation of explosion/impact loading is characterized by T_L, and the characteristic time of the dynamic response of the structure is $T_W = L_s/C_w$, where C_w represents the characteristic velocity of stress wave and Ls the characteristic length of the structure, a dimensionless time $\overline{T} = T_L/T_W$ can be introduced to describe the stress wave effects. If $\overline{T} = T_L/T_W >> 1$, then in the time scale where the external load has not changed significantly, the stress wave has been propagated back and forth for many times in the structural characteristic scale, thus reaching the state of static equilibrium. At this time, it is no longer necessary to analyze the stress wave propagation process in the specific structure under investigation. This is exactly what the quasi-static experiment corresponds to, and thus allows for ignoring the stress wave effect. On the contrary, if $\overline{T} = T_L/T_W < 1$ or its order of magnitude is about 10^0, then the stress wave propagation must be taken into account, because the number of back and forth times of the stress waves in the characteristic scale of the structure is not enough to achieve the static equilibrium state. Thus, the stress wave effect in specimens and the related experimental devices should be taken into account under explosion/impact loading.

Lindholm (1971) has classified mechanical tests of materials into four classes according to strain rate, as shown in Fig. 7.1.

Quasistatic experiments using conventional material testing machines generally have strain rates in the range of 10^{-5} to 10^{-1} s^{-1}. Creep experiments have even less strain rates. Those with strain rate between 10^{-1} and 10^1 s^{-1} are called dynamic or medium strain rate experiments. Those with higher strain rate are known as impact or high strain rate experiments. Those with strain rate over 10^5 s^{-1} are sometimes called ultrahigh strain rate experiments. The inertia effect (stress wave effect) should be considered except in quasistatic experiments and creep testing.

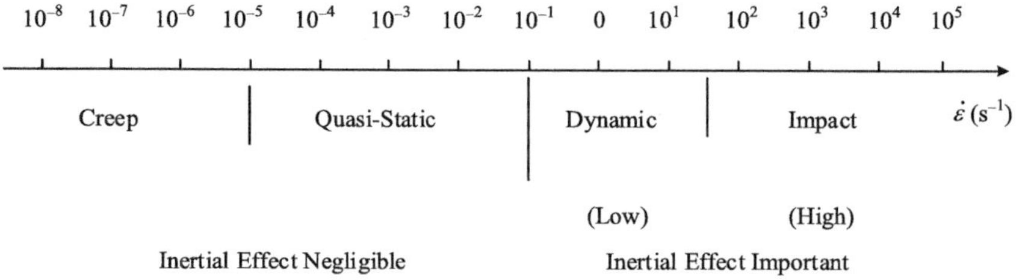

Figure 7.1 Mechanical experiments of materials classified according to strain rates.

As such, two basic dynamic effects must be considered when studying the dynamic distortion law of materials under impact loading, namely the material strain rate effect and the structure inertia effect (various types of stress wave propagation effects) (Wang Lili, 2007). The former is the object of discussion here, i.e., various types of strain rate—dependent (rate-type) constitutive relations of materials. The latter would couple to the former, which is an effect that we have to consider and has to be decoupled from the former when studying the former.

The key difficulty is that these two effects are often interconnected, interdependent, and intercoupled, which make the problem much more complex. In fact, on the one hand, in the study of stress wave propagation, the dynamic constitutive relation of materials is an indispensable part of the basic governing equations. On the one hand, in the study of stress wave propagation, the material dynamic constitutive equation is an indispensable part of the basic control equations. In other words, wave propagation presupposes that the dynamic constitutive relation of material is known. On the other hand, the stress wave propagation in the specimen and the test device must be taken into account in the study of the dynamic constitutive relationship under high strain rate. In other words, wave propagation in structures cannot be analyzed without knowing the corresponding constitutive relation of materials. On the other hand, in the experimental study of dynamic constitutive relation of materials at high strain rates, the stress wave propagation and interaction in either the specimens or the experimental device should not be neglected. Therefore, in the study of these two kinds of dynamic effects, people came across the vicious circle of "dog bites tail" or "egg or chicken comes first."

How to resolve this problem? In terms of the study of dynamic constitutive relation of materials, there are two kinds of methods at present. The first is the wave propagation inverse analysis (WPIA). In this method, the specimen is designed as a simple structure on which stress wave propagation analysis can be easily performed. Under a known explosion/impact loading, the dynamic constitutive relation of materials can then be deduced by the wave propagation information or the residual effects on the specimen (such as residual deformation distribution). For examples, the residual deformation of a straight rod impact test, the wave propagation in a long rod test, and the wave propagation in a plate impact by gas gun test, proposed, respectively, by Taylor (1948), Whiffin (1948), and Lenski (Ленский) (1951). In principle, all these methods are used to determine the dynamic constitutive relation of materials inversely from wave propagation information. Mathematically, it belongs to solving the so-called "second kind of inverse problem."

The second method tries to decouple the structural stress wave effect and material strain rate effect in experiments. The most typical and widely used is the split Hopkinson pressure bar (SHPB) experiment, which is made of long incident and transmitted bars with a short specimen sandwiched between them. For a short specimen that satisfies the uniform stress/strain distribution, SHPB experiment is equivalent to a "quasistatic" experiment at a high strain rate, so that the effect of stress waves can be neglected. For the long compression bar considering the stress wave propagation, this is equivalent to determine the dynamic constitutive response of the adjacent short specimen material inversely from the wave propagation information.

Regardless of the first or second method used to study the dynamic constitutive behavior of materials, the analysis of stress wave propagation plays a key role, and it is the most significant distinction from quasistatic experiments. Accordingly, high-speed loading devices and dynamic diagnostic techniques with high-frequency response must be used in the dynamic experiments to capture the wave propagation information, which is another distinct difficulty from quasistatic experiments.

This chapter elaborates on two aspects. The first part discusses the widely used SHPB technique at present (although the residual deformation method of the straight-bar impact test dates back to World War II) and the second part discusses various methods of WPIA. This chapter focuses on the discussion of one-dimensional (1D) stress experiments, including combined stress impact tests, and so on. As for the 1D strain experiments under plate impact loading, it has been discussed in detail in the Chapter 4 of Part I.

7.1 The Split Hopkinson Pressure Bar technique

The origin of the 1D SHPB technique can be traced back to the milestone works of John Hopkinson and Bertram Hopkinson in the field of explosion/impact dynamics. One hundred years ago, the wire impact tensile experiment (see Fig. 5.1) proposed by Hopkinson J (1872) revealed two basic effects in impact dynamics: inertial (stress wave) and strain rate effects. His son Hopkinson B (1914) designed a Hopkinson pressure bar device (Fig. 7.2) to divide the long ballistic pendulum measuring the impulse into a long bar and a short bar, which can be used to measure the actual waveform of impact/explosion load. This was an innovation as there was no test instrument such as an oscilloscope at that time.

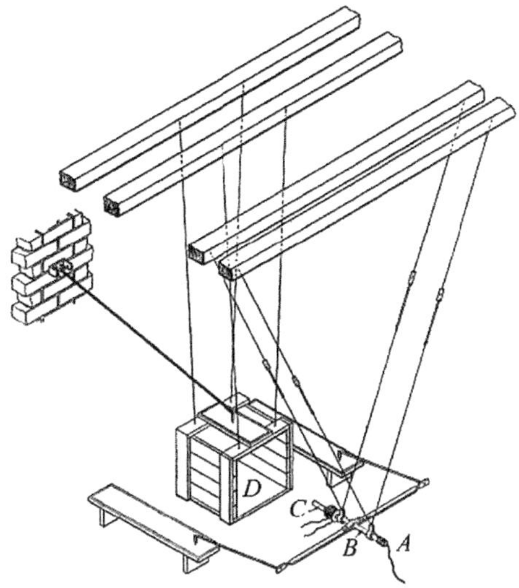

Figure 7.2 The Hopkinson pressure bar experimental device designed by Hopkinson B (1914).

By the late 1940s, the SHPB technique was developed and can be used to study the high strain rate behavior of materials. The main pioneers that must be mentioned are Taylor (1946), Volterra (1948), Davies (1948), and Kolsky (1949). The development of SHPB is a greater innovation because it presents a direct measurement for the dynamic stress—strain curve of materials. The SHPB experimental device has become the most basic means to study the dynamic properties of materials for its novel design principle, ingenious measurement, and convenient operation.

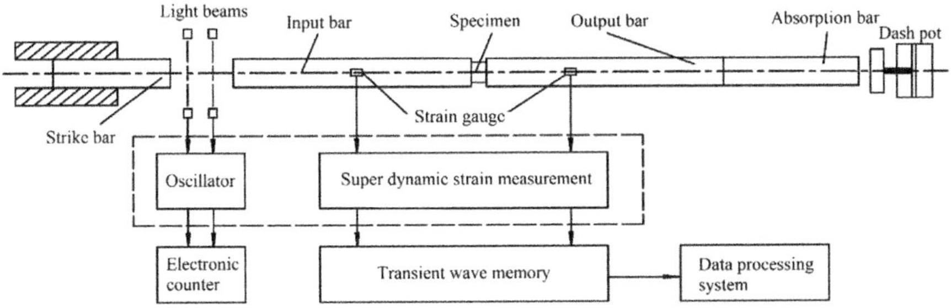

Figure 7.3 Schematic of a split Hopkinson pressure bar device.

7.1.1 The basic principle of SHPB

A schematic of a typical SHPB experimental device is shown in Fig. 7.3, where the strike bar (bullet), input bar (incident bar), and output bar (transmission bar) made of high-strength metal are all in the elastic state, and generally have the same diameter and material, namely the elastic modulus E, wave speed C_0, and wave impedance $\rho_0 C_0$ are the same, and the bars are thin enough maintaining to propagate only 1D stress elastic waves. During the test, a short specimen is sandwiched between the input bar and the output bar. When the strike bar driven by the air-gun impacts the input bar at velocity $v^\star$, an incident pulse $\sigma_I(t)$ is initiated and propagates in the input bar. The short specimen is deformed in high strain rate under the action of this incident pulse. Meanwhile, a reflection pulse $\sigma_R(t)$ is generated and propagates back to the input bar and transmits to the output bar a transmission pulse $\sigma_T(t)$. The signals of these pulses are measured and recorded by a system consisting of strain gauges, a super dynamic strain amplifier, a transient wave recorder, etc., while the impact velocity of the strike bar is measured by a speed measuring system consisting of a pair of parallel light beams and photoelectric cells, an amplification circuit, an electronic time counter, etc. The absorption bar works to capture the transmission pulse. When the transmission pulse is reflected from the free end of the absorption bar, it flies away with the entire momentum of the transmission pulse, and its energy vanishes eventually by hitting the damper. Thus, the output bar is not moving anymore, and it is also possible to prevent reflected waves that may cause secondary interference to the specimen from the other end of the output bar when there is no more absorption bar.

It should be emphasized that the SHPB technique is based on two basic assumptions: (1) the assumption of 1D stress wave in the bar and (2) the assumption that the stress/strain is uniformly distributed along the length of the short specimen (dynamic equilibrium).

Under the first assumption, once the stress $\sigma(X_1, t)$ and particle velocity v (X_1, t) at the interface X_1 (Fig. 7.4) between the specimen and the input bar, as well as $\sigma(X_2, t)$ and v (X_2, t) at the interface X_2 (Fig. 7.4) between the

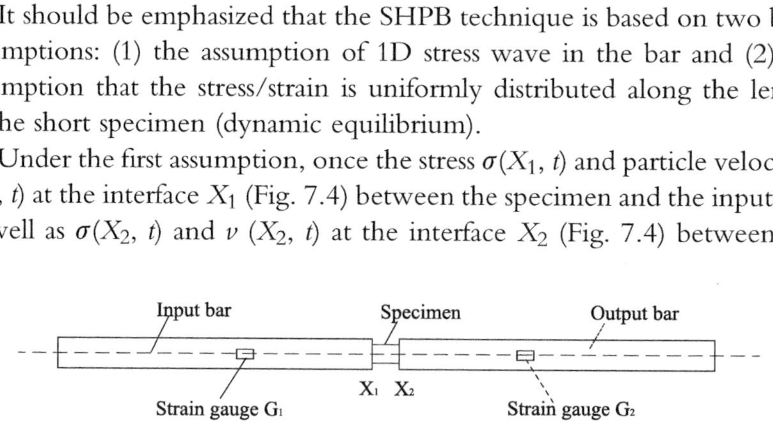

Figure 7.4 The position of input bar, specimen, and output bar in split Hopkinson pressure bar.

specimen and output bar are measured, the average stress $\sigma_s(t)$, strain rate $\dot{\varepsilon}_s(t)$, and strain $\varepsilon_s(t)$ of the specimen can be determined according to the following equations, respectively

$$\sigma_s(t) = \frac{A}{2A_s}[\sigma(X_1,t) + \sigma(X_2,t)] = \frac{A}{2A_s}[\sigma_I(X_1,t) + \sigma_R(X_1,t) + \sigma_T(X_2,t)] \tag{7.1a}$$

$$\dot{\varepsilon}_s(t) = \frac{v(X_2,t) - v(X_1,t)}{l_s} = \frac{v_T(X_2,t) - v_I(X_1,t) - v_R(X_1,t)}{l_s} \tag{7.1b}$$

$$\varepsilon_s(t) = \int_0^t \dot{\varepsilon}_s(t)dt = \frac{1}{l_s}\int_0^t [v_T(X_2,t) - v_I(X_1,t) - v_R(X_1,t)]\,dt \tag{7.1c}$$

where A is the section area of pressure bars, A_s is the section area of the specimen, and l_s is the length of the specimen. In the case of the elastic pressure bar, there are four linear relationships among the strain, the stress, and the particle velocity according to the 1D elastic wave analysis (Wang, 2007), and they can be written as follows

$$\left.\begin{aligned} \sigma_1 &= \sigma(X_1,t) = \sigma_I(X_1,t) + \sigma_R(X_1,t) = E[\varepsilon_I(X_1,t) + \varepsilon_R(X_1,t)],\\ \sigma_2 &= \sigma(X_2,t) = \sigma_T(X_2,t) = E\varepsilon_T(X_2,t),\\ v_1 &= v(X_1,t) = v_I(X_1,t) + v_R(X_1,t) = C_0[\varepsilon_I(X_1,t) + \varepsilon_R(X_1,t)],\\ v_2 &= v(X_2,t) = v_T(X_2,t) = C_0\varepsilon_T(X_2,t); \end{aligned}\right\} \tag{7.2}$$

As such, the problem is focused on how to measure the incident strain wave $\varepsilon_I(X_1,\ t)$ and the reflection strain wave $\varepsilon_R(X_1,\ t)$ at the interface X_1 and the transmitted strain wave $\varepsilon_T(X_2,\ t)$ at the interface X_2. Obviously, as long as the pressure bar remains in the elastic state and the waveforms at different positions in the pressure bar are the same and without distortion, the incident strain wave $\varepsilon_I(X_1,\ t)$ and the reflection strain wave $\varepsilon_R(X_1,\ t)$ at interface X_1 can be replaced by the incident strain wave $\varepsilon_I(X_{G1},\ t)$ and the reflection strain wave $\varepsilon_R(X_{G1},\ t)$ measured by the strain gauge G_1 mounted at X_{G1} on the input bar, and the transmitted strain wave $\varepsilon_T(X_2,\ t)$ at the interface X_2 can be replaced by the transmitted strain wave $\varepsilon_T(X_{G2},\ t)$ measured by the gauge G_2 mounted at X_{G2} on the output bar.

It should be pointed out that when a reflected wave generated at the interface X_1 and a transmitted wave generated at the interface X_2 propagate in the input and output bars, respectively, the stress waves also propagate back and forth in the specimen between the interfaces X_1 and X_2. It can be imagined that if the specimen is short enough and the wave speed is fast enough, the stress/strain can be soon distributed uniformly along the length of the specimen and the dynamic balance will be achieved, so that the stress wave effect in the specimen can be ignored. This is the second basic assumption on which the SHPB technique is based. According to this assumption of uniform stress/strain distribution, $\sigma_{X_1} = \sigma_{X_2}$, and according to the 1D stress wave theory, we can further obtain

$$\sigma_I + \sigma_R = \sigma_T, \quad \varepsilon_I + \varepsilon_R = \varepsilon_T \tag{7.3}$$

Then, Eq. (7.1) can be expressed as

$$\sigma_s(t) = \frac{EA}{A_s}\varepsilon_T(X_{G2}, t) = \frac{EA}{A_s}[\varepsilon_I(X_{G1}, t) + \varepsilon_R(X_{G1}, t)] \tag{7.4a}$$

$$\dot{\varepsilon}_s(t) = -\frac{2C_o}{l_s}\varepsilon_R(X_{G1}, t) = \frac{2C_o}{l_s}[\varepsilon_I(X_{G1}, t) - \varepsilon_T(X_{G2}, t)] \tag{7.4b}$$

$$\varepsilon_s(t) = -\frac{2C_o}{l_s}\int_0^t \varepsilon_R(X_{G1}, t)\, dt = \frac{2C_o}{l_s}\int_0^t [\varepsilon_I(X_{G1}, t) - \varepsilon_T(X_{G2}, t)]\, dt \tag{7.4c}$$

Hence, the dynamic stress—strain curve $\sigma_s - \varepsilon_s$ of the specimen under high strain rates can be obtained by eliminating the time parameter t.

The above equation also shows that when the assumption of "uniform stress/strain distribution" is satisfied, any two between the incident strain wave $\varepsilon_I(X_1, t)$, the reflected strain wave $\varepsilon_R(X_1, t)$, and the transmitted strain wave $\varepsilon_T(X_2, t)$ are enough to determine the dynamic relation between stress $\sigma_s(t)$ and strain $\varepsilon_s(t)$ of the specimen from Eq. (7.4).

The method to obtain the dynamic stress—strain of the specimen directly from the measured incident strain wave $\varepsilon_I(X_1, t)$, reflection strain wave $\varepsilon_R(X_1, t)$, and transmitted strain wave $\varepsilon_T(X_2, t)$ according to Eq. (7.1) is known as "three-waves method," whereas the method using any two measured waves according to Eq. (7.4) is known as "two-waves method." There are three types of two-waves methods: (a) the "RT two-waves method," based on the reflected strain wave $\varepsilon_R(X_1, t)$ and the transmitted strain wave $\varepsilon_T(X_2, t)$, (b) the "IT two-waves method," based on the incident strain wave $\varepsilon_I(X_1, t)$ and the transmitted strain wave $\varepsilon_T(X_2, t)$, and (c) the "IR two-waves method," based on the incident strain wave $\varepsilon_I(X_1, t)$ and the reflected strain wave $\varepsilon_R(X_1, t)$. In principle, the results of two-wave methods and three-wave methods are consistent as long as the stress/strain is uniformly distributed along the length of specimen (Assumption 2). However, the uniformity assumption is not always valid. Results from the two-wave and three-wave methods can be used to cross-check for the validity of this assumption.

From the above discussion and analysis, the ingenious aspect of the SHPB technique lies in that it decouples the stress wave effect and the strain rate effect. On the one hand, for the input and output bars, which play the dual roles of both impact loading and dynamic measurement, because they are always in the elastic state, they are allowed to ignore the strain rate effect and only consider the stress wave propagation in the pressure bars. Thus, the theory of 1D stress waves can be applied to analyze these experiments as long as the bar diameter is small enough and the friction effect between the 2 bars and the specimen can be neglected. On the other hand, for the specimen sandwiched between the input bar and the output bar, as long as the length is short enough and the wave velocity is fast enough, so that the duration for a stress wave propagating between the two ends of the specimen is much shorter compared with the total loading duration, then the specimen can be approximately regarded as in a state of uniform deformation. Thus, only the strain rate effect should be considered and the stress wave effect in the specimen can be ignored. In this way, the stress wave effect and the strain rate effect in the pressure bars and the specimen are decoupled, respectively, and the rate-dependent mechanical response of specimen material can be determined by analyzing the information of stress wave propagation in elastic bars. For the specimen, this is equivalent to a "quasistatic" experiment at high strain rates; for the pressure bar, this is equivalent to solving an inverse problem, i.e., deducing the constitutive relation of the adjacent short specimen inversely from the wave propagation information.

Taking the SHPB test of metal material as an example, the typical incident strain wave $\varepsilon_I(X_{G1}, t)$, reflection strain wave $\varepsilon_R(X_{G1}, t)$, and transmitted strain wave $\varepsilon_T(X_{G2}, t)$ measured by the input and output bars, respectively, are shown in Fig. 7.5A. In the case of an ideal 1D stress wave, the incident strain wave $\varepsilon_I(X_{G1}, t)$ is a trapezoidal wave with a steep front. Its amplitude ($=v^\star/2C_0$) can be controlled by the impact velocity $v^\star$ of the strike bar, and its duration ($=2L_0/C_0$) can be controlled by adjusting the length L_0 of the strike bar. From Eq. (7.4b), the reflected strain wave $\varepsilon_R(X_{G1}, t)$ describes exactly the history of the strain rate of the specimen $\dot{\varepsilon}_s(t)$, and the enclosed area of strain rate curve (time integral) is the history of the strain of the specimen $\sigma_S(t)$. According to Eq. (7.4a), the transmitted

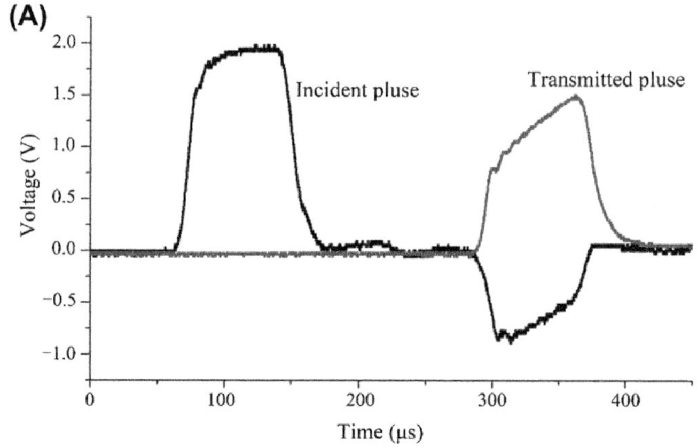

Typical wave signals of metallic materials measured by SHPB.

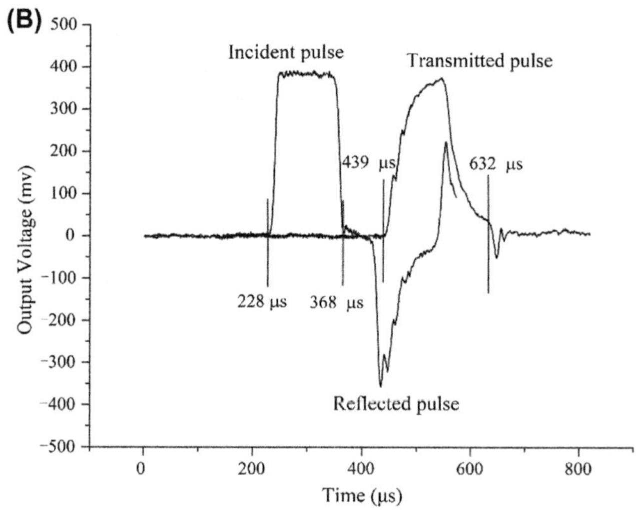

Typical wave signals of polymers measured by SHPB.

Figure 7.5 Typical measured wave signals in split Hopkinson pressure bar (SHPB) experiments with a bar diameter of 14.5 mm(A) Typical wave signals of metallic materials measured by SHPB. (B) Typical wave signals of polymers measured by SHPB.

strain wave $\varepsilon_T(X_{G2}, t)$ describes exactly the stress history curve of the specimen.

For a given incident strain wave $\varepsilon_I(X_{G1}, t)$, the corresponding reflection strain wave $\varepsilon_R(X_{G1}, t)$ and transmitted strain wave $\varepsilon_T(X_{G2}, t)$ vary with different measured material, which reflects different dynamic mechanical behaviors of different specimen materials. For instance, the typical waveform measured for polymer materials is shown in Fig. 7.5B, one major difference compared with that of metallic materials (Fig. 7.5A) is that the duration of the transmitted waves is significantly greater than the duration of the incident wave, indicating that the polymer has a significant constitutive viscous effect.

As long as the assumptions of "1D stress waves in bars" and "uniform distribution of stress/strain along the specimen length" can be satisfied, it is also possible to obtain the dynamic stress—strain relation of specimen without the input bar. In fact, if strike bar (bullet) is used to directly impact the specimen, and if the particle velocity $v_I(X_1, t)$ at the incident end of the specimen can be directly measured by a dynamic particle velocity meter, then together with the $\varepsilon_T(X_{G2}, t)$ measured on the output bar, the transmitted stress $\sigma(X_2, t)$ and particle velocity $v(X_2, t)$ at the transmitted end of the specimen can be derived according to Eq. (7.2). With reference to Eqs. (7.1) and (7.4), the average stress $\sigma_s(t)$, strain rate $\dot{\varepsilon}_s(t)$, and strain $\varepsilon_s(t)$ of the specimen can also be determined. By this method, we can investigate the dynamic behavior of materials under higher strain rates (in the order of $10^4 \sim 10^5\ \mathrm{s}^{-1}$).

7.1.2 Split Hopkinson bar experiments under different stress states

Based on the traditional SHPB technique, people have further developed a variety of split Hopkinson bar (SHB) experimental techniques to study the dynamic stress—strain behavior of materials under different stress conditions. These SHB techniques are mainly divided into tensile SHB technique, torsional (shear) SHB technique, and combined stress SHB technique. The main difference between them is that the specimens are subjected to impact loads in different stress states. For this purpose, two key problems need to be addressed in the experimental technique: the generation and application of impact loads in different stress states and the connection of the specimens to the Hopkinson bars. However, the basic principles of these techniques are similar to that of the traditional SHPB technique.

1. Tensile SHB technique

The key challenge of the tensile SHB technique is to generate a tensile pulse for the specimen, as the high-speed collision usually produces a compressive stress wave. The schematic diagram of the earliest tensile SHB experimental device designed by Harding et al. (1960) is shown in Fig. 7.6, where a hollow short circular tube is used as the striker tube, and it impacts a long circular tube to generate a compressive stress wave, which is converted into

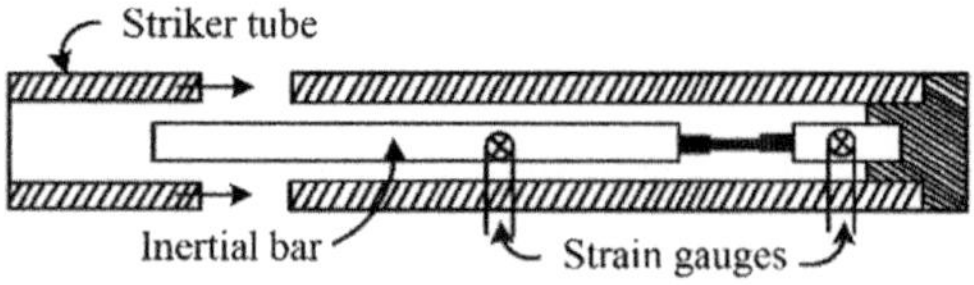

Figure 7.6 Schematic diagram of the tensile split Hopkinson bar experimental device designed by Harding et al.

tensile wave through the yoke seat at the free end of the tube. Thus, an impact tensile load can be applied to the specimen that is placed in the hollow tube and connected to the two Hopkinson bars in the figure.

Nicholas (1981) proposed a much simpler method to apply dynamic tensile loadings for the specimen, and its principle is shown in Fig. 7.7A. The specimen is connected with the input and output bars by threads, respectively, and is surrounded by a collar. Here, the collar needs to be specially designed so that the compression wave generated by the striker can be directly transmitted to the output bar through the collar and converted from the free end of the output bar to a tensile wave for the impact loading. However, the specimen will in fact inevitably be subjected to partial compression impact loadings before being subjected to a tensile impact loading.

Ogawa (1984) integrated the advantages of the above two methods, using a short hollow tube as the striker to impact the end flange of the input bar and directly generating a tensile wave to apply impact loading to the specimen. The schematic is shown in Fig. 7.7B.

The dynamic tensile loads generated by the aforementioned type of tensile SHBs are all through "collision." There exists another method to generate a dynamic tensile wave, which is by suddenly releasing a preloaded

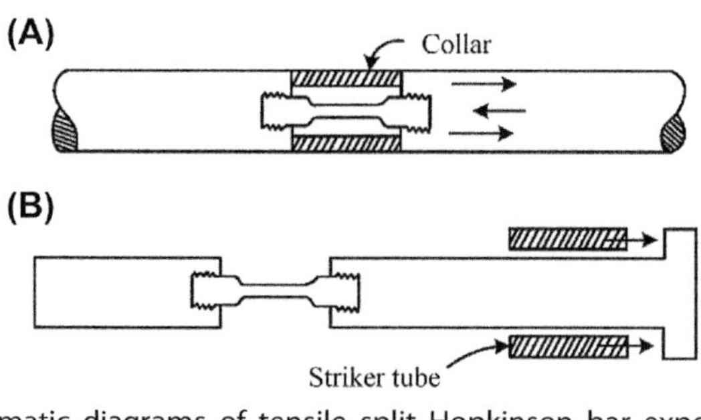

Figure 7.7 Schematic diagrams of tensile split Hopkinson bar experimental devices designed by (A) Nicholas (1981) and (B) Ogawa (1984).

tensile load. For example, a locking device is set at a certain section of the input bar to statically load the upstream part of the bar. When the locking device is suddenly released, a dynamic tensile wave is generated and propagates in the downstream part of the bar. The shape of the tensile wave (and consequently the strain rate of the specimen) is obviously affected by the speed at which the locking device is released. Usually an explosive bolt is used to control the release of the locking device. Based on this principle, Albertini et al. (1985) established a large tensile SHB device (as shown in Fig. 7.23 below).

2. Torsional/Shear SHB technique

Baker and Yew (1966) first designed a torsional SHB experimental device, as shown in Fig. 7.8, where a torsional wave is generated by the suddenly release of the preloaded torque, then the specimen is subsequently subjected to the torsional impact loading and its shear stress/strain curves can be obtained.

Duffy et al. (1971) used the explosion loading method to generate torsional waves, and its principle is shown in Fig. 7.9. This method can produce torsional impact loading with higher strain rate, but the technical requirements are very high. In this method, it is necessary to detonate simultaneously two explosives that can generate the same explosion pulse at the

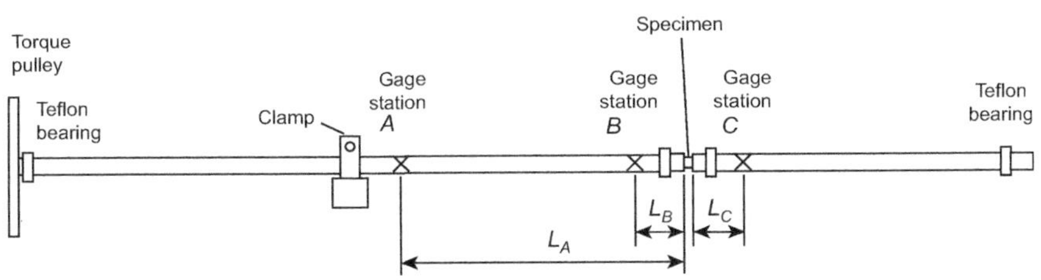

Figure 7.8 Schematic diagrams of torsional split Hopkinson bar experimental device designed by Baker and Yew (1966).

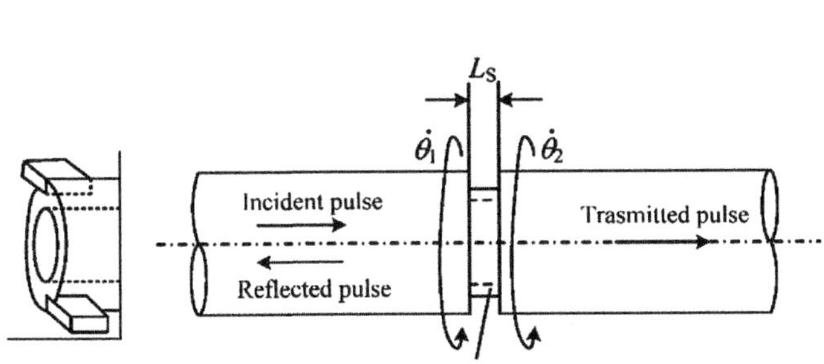

Figure 7.9 Schematic diagrams of torsional split Hopkinson bar experimental device designed by Duffy et al. (1971).

same time on the two wings at the input bar end so that only pure torsional waves are generated. If the two explosion pulses detonated on the two wings are not the same, or are detonated not at the same time, there will be bending waves generated in the input bar.

In addition to the torsional SHB experiment, various "cap-type" specimens were used to study the dynamic shear stress—strain curve of materials under high strain rates on the conventional SHPB device. A representative schematic is shown in Fig. 7.10. However, it should be noticed that in such case the basic assumption of "uniform distribution of stress/strain along the length of specimen" is no longer satisfied, because the structure factors such as stress concentration and so on have been introduced. The experimental results reflect the coupling response of the material with the specimen structure.

Compared with the traditional SHPB technique, a common advantage of the tensile SHB technique and the torsional SHB technique is that there is no friction between the specimen and the Hopkinson bars, thus there is no influence of the three-dimensional stress state caused by friction force on the basic assumption of "one-dimensional stress wave in the bar." However, the threaded connection between the specimen and pressure bars in the tensile or torsional SHB will also cause some interference and error in the propagation of 1D stress waves, which needs to be carefully examined.

3. SHB technique under combined stress states

To study the high strain rate behavior of materials under combined stress states, various kinds of combined stress SHB experimental devices have been developed and they are particularly important for the following two types of research. One is to study whether the high strain rate behavior of materials under general three-dimensional stress state can be characterized by introducing the "effective stress σ_{eff}," the "effective strain ε_{eff}" and the "effective strain rate $\dot{\varepsilon}_{eff}$," similar to that of the von Mises yield condition under general three-dimensional stress state in the classical plastic mechanics (see Eqn, 5.8 in Chapter 5). The other is to study the influence of the multiaxial stress state

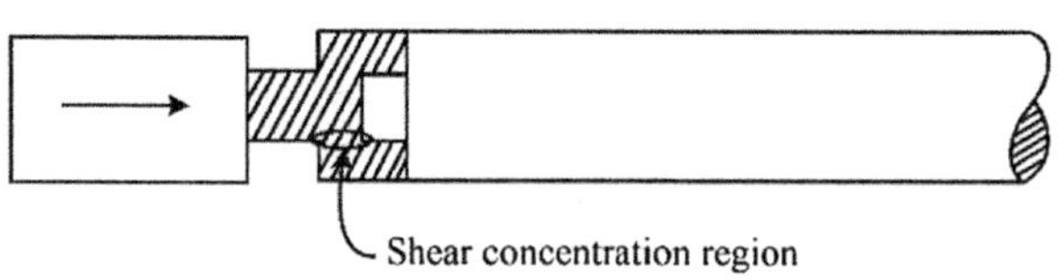

Figure 7.10 Split Hopkinson pressure bar experimental schematic of cap-type specimens designed by Meyer and Manwaring (1986).

on the dynamic failure of materials, especially for brittle materials (see the Part 3 of this book).

In principle, such devices can be divided into the following types:

(1) Combination of torsional SHB and compressive/tensile SHB

Based on the aforementioned torsional SHB and compressed/tensile SHB, the combination can apply a combined torsional-compressive/tensile impact load to the specimen. It should be noted that the wave velocities of the axial compressive/extensile wave and the torsional wave are different (Wang, 2007). A typical schematic of a SHB for simultaneous torsion and compression is shown in Fig. 7.11, where the torsional wave is generated by a sudden release of statically preloaded torque, and the compression wave is simultaneously generated by the impact of the striker. These two waves propagate in the Hopkinson bar on either sides of the specimen, respectively, and in the direction of the relative, within 10 μs arrive at the specimen at the same time.

(2) Combination of axial impact and confining pressure

Based on the compressive SHB (SHPB), a test under triaxial stress state can be realized by applying additional confining pressure to the specimen. In such triaxial stress experiments, a sleeve is usually added on the specimen to produce a confining pressure. According to the principle of confining pressure formation, it can be divided into two categories, active and passive confining techniques.

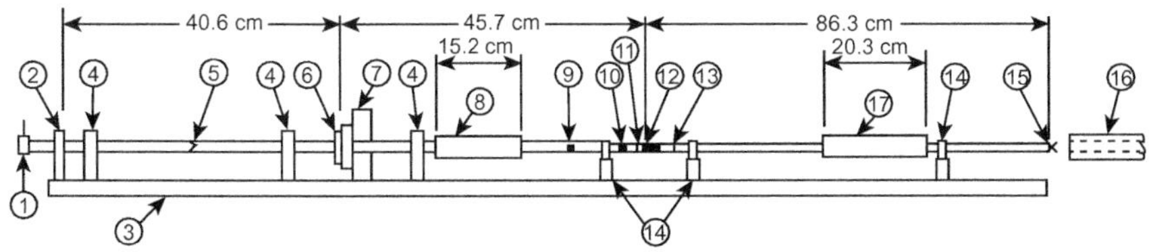

Figure 7.11 A biaxial split Hopkinson bar for simultaneous torsion and compression designed by Lewis and Goldsmith. 1-Torque handle; 2-clamping block; 3-machine bed plate; 4-bearing block; 5-static torque gauge; 6-holding flange; 7-holding frame; 8-torsional bending suppressor; 9-bending strain gauge; 10-longitudinal strain gauge; 11-test specimen; 12-specimen strain gauge; 13-torsion strain gauge; 14-alignment post; 15-crossed wire trigger; 16-air gun; 17-longitudinal bending suppressor. *From Lewis, J.L., Goldsmith, W., 1973. A biaxial split Hopkinson bar for simultaneous torsion and compression. Rev. Sci. Instrum. 44 (7), 811–813, Fig. 1, p.811. Reprinted with permission of the publisher.*

The confining pressure formation of the active confining technique is independent of the specimen deformation and is actively applied onto the specimen. The confining pressure of the passive confining technique is passively caused by the resistance of the sleeve to the radial expansion (Poisson effect) of the specimen.

A typical schematic of an SHB device with the active confining technique is shown in Fig. 7.12. As shown in the figure, the specimen and the end of the pressure bars are together covered with a latex membrane and placed in a medium with hydrostatic pressure within the outer sleeve. The latex membrane can cut off the direct contact between the specimen and the pressure medium and can also prevent any movement of the pressure bar under active confining pressure so as to keep the specimen in close contact with the pressure bar. The O-ring is sealed between the sleeve and the pressure bar, allowing for both tight sealing and unobstructed stress wave propagation in the bar. When water is used for the pressure medium, the confining pressure can reach 10 MPa, and when an oil medium is used, the confining pressure can reach 50 MPa. Shi et al. (2009) used a silicone oil medium, and the active confining pressure can reach 50 MPa.

The confining pressure in the abovementioned active confining technique is a preset hydrostatic pressure. Thus, the triaxial stress state in this device is essentially a combination of uniaxial impact pressure and dual-axial static pressure and is not a dynamic triaxial stress state in a strict sense. Moreover, due to the limitations of the sleeve strength and sealing technique, it is difficult to achieve a higher confining pressure.

The confining pressure in the passive confining technique is caused by the resistance of the sleeve to the radial expansion (Poisson effect) of the specimen on an axial impact compression. Therefore, the passive confining pressure is controlled and affected by many factors, such as the tightness of

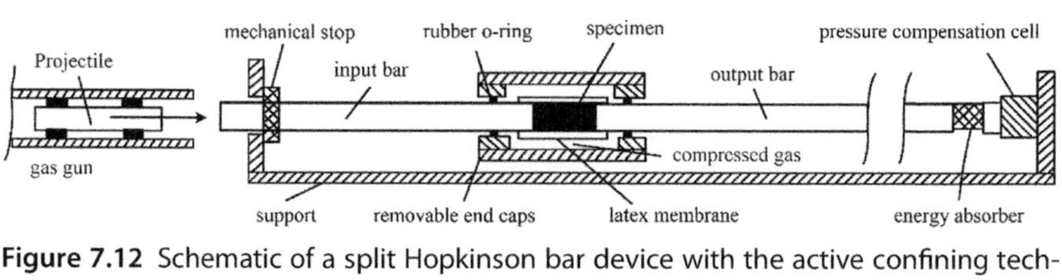

Figure 7.12 Schematic of a split Hopkinson bar device with the active confining technique designed by Gary and Bailly. *From Gary, G., Bailly, P., 1998. Behavior of quasi-brittle material at high strain rate. Experiment and modeling. Eur. J. Mech. A-Solid. 17 (3), 403–420, Fig. 4, p.407. Reprinted with permission of the publisher.*

the fit between the specimen and the sleeve, the surface smoothness of the sleeve, the size of the gap, the material, and the thickness of the sleeve.

The early passive confining technique proposed by Gong and Malvern (1990) used a snug fit between the specimen and the elastic sleeve, i.e., it was assumed that there was no gap between the specimen and the sleeve (Fig. 7.13). Therefore, by measuring circumferential strain on the outer wall of the elastic sleeve, the dynamic confining pressure of the specimen can be estimated from the mechanics of elasticity. It was found that the confining pressure applied on the concrete specimen by using aluminum sleeve was about 40–50 MPa.

However, as pointed out by Gong and Malvern, the passive confining test with the seamless "snug fit" requires high-precision processing of the specimen and the sleeve. Except the granular material such as sediment, the out-of-roundness and the roughness of both the outer surface of solid material specimen (such as concretes) and the inner surface of the metal sleeve, and consequently the resulting frictional effect between them, greatly hinder the instantaneous generation of the radial confining pressure and the uniform distribution of the confining pressure along the circumference and length of the specimen. These will affect the repeatability and reliability of the passive confining test.

To solve this contradiction, researchers used a "moving fit" between the specimen and the sleeve, and the gap is filled with a coupling medium that can transmit the confining pressure (Fig. 7.14).

For example, Shi and Wang (2000) proposed an improved SHPB device with passive confining technique and reserved a gap with the tolerance of ±0.05 mm between the outer diameter of the specimen and the inner diameter of the sleeve. The No. 2 petroleum grease type antirust grease (SY1575-80) was filled evenly into the gap and acted as a coupling agent for the confining pressure transfer between the inner surface of the sleeve and the outer surface of the specimen. Because the oil membrane is relatively

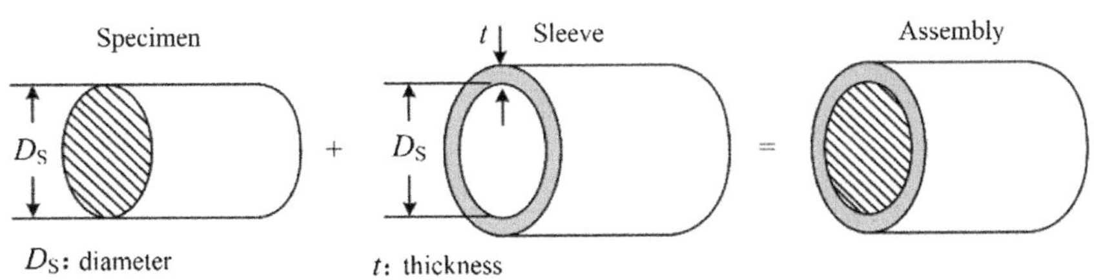

Figure 7.13 One of the passive confining techniques: snug fit (no gap) between the specimen and the sleeve.

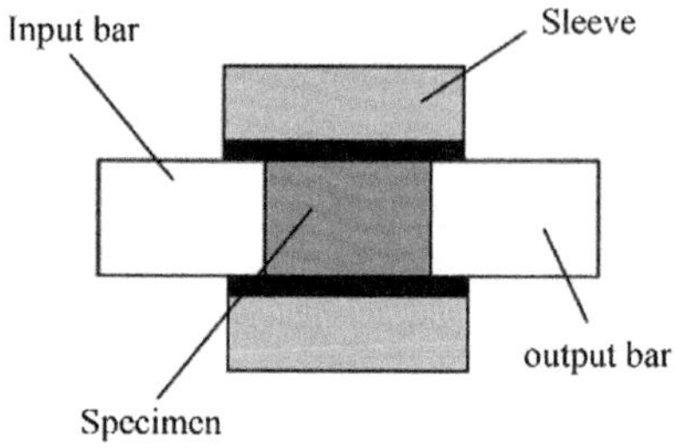

Figure 7.14 One of the passive confining techniques: moving fit between the specimen and the sleeve.

difficult to be compressed, the stress wave propagates back and forth several times in the oil membrane within a very short time (in the order of μs), hence the radial passive confining pressure can be instantaneously and uniformly generated and transmitted. Following this method, not only the machining accuracy requirements of the specimen and the sleeve can be reduced but also the friction between the out surface of specimen and the inner wall of the sleeve can be reduced due to the lubricating property of the oil membrane.

Forquin et al. (2008) used epoxy resin to fill a gap of about 0.2 mm between the concrete specimen and the metal sleeve, and the dynamic confining pressure can reach 600–900 MPa.

The magnitude of the confining pressure mentioned above consequently depends on the radial expansion (due to Poisson effect) of the specimen and the degree of restraint of the sleeve, and such confining pressure is completely passively generated. Rome et al. (2004) combined the active and the passive confining techniques. During the axial impact compression of the specimen, an impact-pressurized loading is simultaneously applied to the medium between the specimen and the sleeve, and thus they achieved a higher confining pressure. The detailed experimental device is shown in Fig. 7.15, where the incident bar consists of three parts: the large incident bar (B), the small incident bar (C), and the incident tube (F). The outer diameter of the incident tube (F) is the same as the outer diameter of the large incident bar (B), and the inner diameter of the incident tube (F) is equal to the diameter of the small incident bar (C). The diameter of the small transmission bar (E) and the inner/outer diameters of the transmission tube (I) are the same as those of the small incident bar (C) and the incident tube (F), respectively, and the specimen (D) is sandwiched between the small incident bar (C) and the small transmission bar (E). A Teflon slotted sleeve (G) is filled between the specimen (D) and the sleeve (H) as a coupling

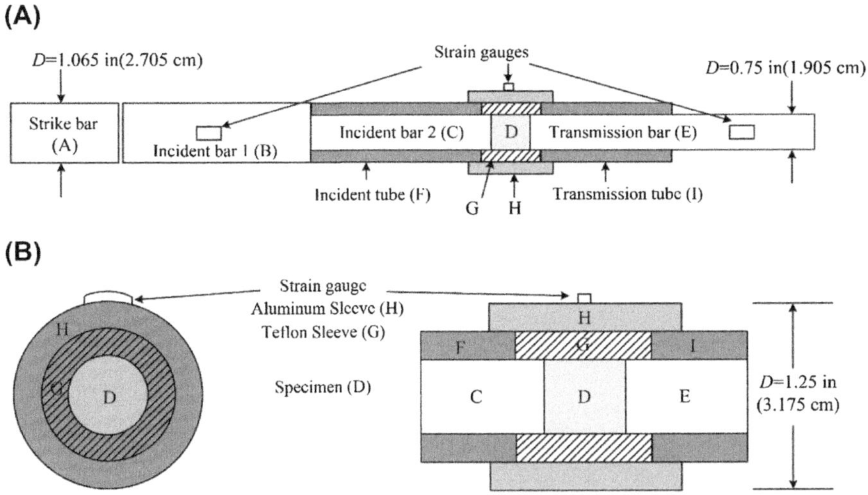

Figure 7.15 Dynamic experimental technique combining (A) passive confining technique and (B) active confining technique. *From Rome, J., Isaacs, J., Nemat-Nasser, S., 2004. Hopkinson Techniques for Dynamic Triaxial Compression Tests. Recent Advances in Experimental Mechanics, Springer, Netherlands, 3–12, Fig. 5, p.7. Reprinted with permission of the publisher.*

medium and is sandwiched between the incident tube (F) and the transmission tube (I) at the same time. During the test, when the strike bar (A) impacts the large incident bar (B), the axial impact compressive loading of the specimen (D) is performed by the small incident bar (C), and the coupling medium Teflon is impact-compressed through the incident tube (F) at the same time. Because the coupling medium Teflon is, respectively, restrained by the expanding specimen and the metal sleeve in the radial direction, the higher radial confining pressure can be generated. By changing the material and thickness of the sleeve (H), different confining pressures can be obtained.

(3) 2D/3D orthogonal SHPB

An impact loading with a combined stress state on the specimen can be achieved by combining two or three sets of orthogonal SHPB devices. The typical schematics of a two-dimensional orthogonal SHPB device and a three-dimensional orthogonal SHPB device are shown in Figs. 7.16 and 7.17.

These devices are mainly used for a combination of two-axis static load and one-axis dynamic load. To carry out a true triaxial dynamic loading experiment, the key technical issue is to ensure the orthogonal propagating pressure/tensile stress waves are able to reach the specimen at the same time.

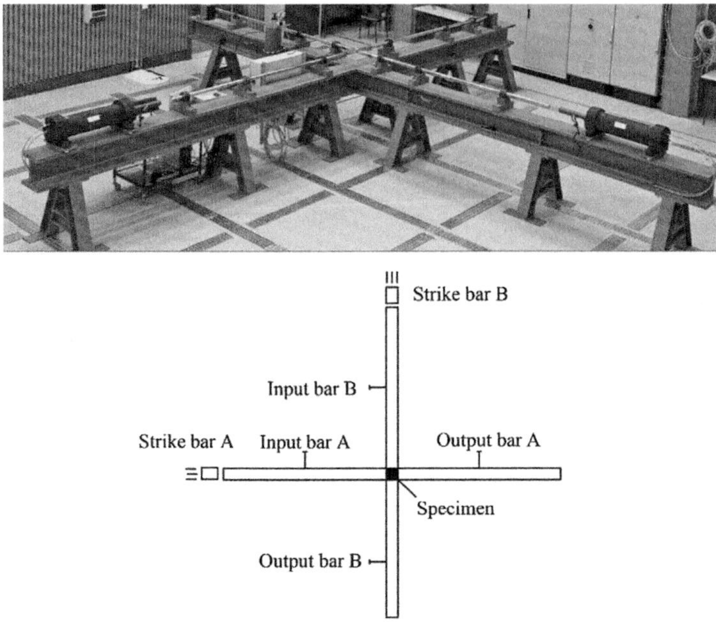

Figure 7.16 Two-dimensional orthogonal split Hopkinson pressure bar designed by Hummeltenbery and Curbach *From Hummeltenberg, A., Curbach, M., 2012. Design and construction of a biaxial Split-Hopkinson-Bar, Beton- Stahlbetonbau 107 (6), 394–400, Fig. 1, p.395 and Fig. 6, p.399. Reprinted with permission of the publisher.*

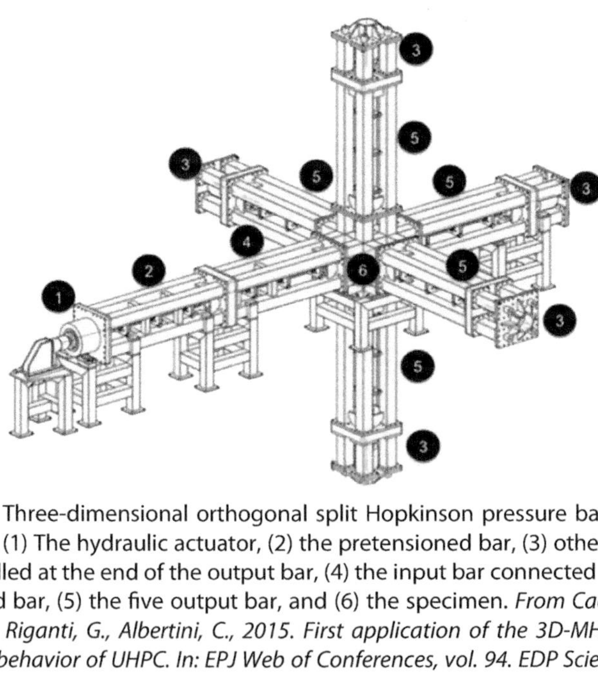

Figure 7.17 Three-dimensional orthogonal split Hopkinson pressure bar designed by Cadoni et al. (1) The hydraulic actuator, (2) the pretensioned bar, (3) other hydraulic actuators installed at the end of the output bar, (4) the input bar connected directly to the pretensioned bar, (5) the five output bar, and (6) the specimen. *From Cadoni, E., Dotta, M., Forni, D., Riganti, G., Albertini, C., 2015. First application of the 3D-MHB on dynamic compressive behavior of UHPC. In: EPJ Web of Conferences, vol. 94. EDP Sciences, p. 01031, Fig. 1, 01031-p.2. Reprinted with permission of the author.*

The basic principle of the SHB impact test under different stress states is identical to that of the traditional 1D SHPB technique. They are both restricted by the two basic assumptions, which are "one-dimensional stress wave in bars" and "uniform stress/strain distribution along the length of specimen." The assumptions also ensure the validity and reliability of the SHPB results. In the following, we will further discuss the related issues of these two basic assumptions based on the stress wave principle.

7.1.3 Discussion on the assumption of "one-dimensional stress state in bars"

It should be pointed out that the 1D stress wave theory used to derive Eqs. (7.1)−(7.3) is established under the condition of ignoring the transversal (lateral) movement (the effect of transversal (lateral) inertia) of the particle in the bar, usually called the primary (elementary) theory or the engineering theory.

In fact, under a uniaxial stress $\sigma_X(X, t)$, the axial strain of the bar is

$$\varepsilon_X = \frac{\partial u_X}{\partial X} = \frac{\sigma_X(X,t)}{E}$$

In addition to this strain component, there must be transversal (lateral) strain due to Poisson effect

$$\varepsilon_Y = \frac{\partial u_Y}{\partial Y} = -\nu\varepsilon_X(X,t), \varepsilon_Z = \frac{\partial u_Z}{\partial Z} = -\nu\varepsilon_X(X,t)$$

where u_X, u_Y, and u_Z are the displacement components along X, Y, and Z axes, respectively, ν is the Poisson ratio. Correspondingly, there also exist transversal (lateral) particle velocity v_Y, v_Z and transversal (lateral) particle accelerations a_Y, a_Z, etc. So strictly speaking, the motion state in the pressure bar is in fact not a simple 1D state, but a three-dimensional problem.

The Pochhammer-Chree solution considering the transversal (lateral) inertia (Kolsky, 1953) shows that the elastic longitudinal wave no longer propagates in the bar at a constant speed C_0 as mentioned in the primary (elementary) theory, but as harmonic waves with different frequencies f (or wavelength $\lambda = C/f$) propagating at different wave velocity (phase velocity) C. In a dimensionless form, the dimensionless wave velocity C/C_0 is a function of r/λ and Poisson ratio ν

$$\frac{C}{C_0} = f\left(\frac{r}{\lambda}, \nu\right) \tag{7.5}$$

where r is the radius of the bar. When $\nu = 0.29$, the result of Eq. (7.5) shows that the high-frequency wave (short wave) velocity approaches the Rayleigh wave velocity C_R, and the low-frequency wave (long wave) approaches C_0, as shown in Fig. 7.18 (Rayleigh, 1885).

In the range $r/\lambda \le 0.7$, Rayleigh (1887) gives the following approximate solution

$$\frac{C}{C_0} \approx 1 - \nu^2\pi^2\left(\frac{a}{\lambda}\right)^2 \tag{7.6}$$

From the above two equations, it can be concluded that high-frequency (short wavelength) waves propagate slowly, and low-frequency (long wavelength) waves propagate fast. Only when the ratio of the radius of the bar to the wavelength $r/\lambda << 1$, would there be $C = C_0$, and as a result the assumption of "one-dimensional stress wave in the bar" can be satisfied. Taking a steel bar as an example, its elastic wave velocity in the bar C_0 is in the order of 5×10^3 m/s. If the high-frequency component of the stress wave in the bar is 100 KHz, namely the corresponding wavelength of the stress wave is 50 mm, then the reasonable radius of the pressure bar should advisably be $r \le 5$ mm.

It should be noted that apart from the transversal (lateral) inertia, the friction between the specimen and the pressure bar can also directly introduce the three-dimensional stress effect, destroying the assumption of "one-dimensional stress wave in the bar." To reduce the friction, on the one hand, the lubricant with the lowest possible friction coefficient (such

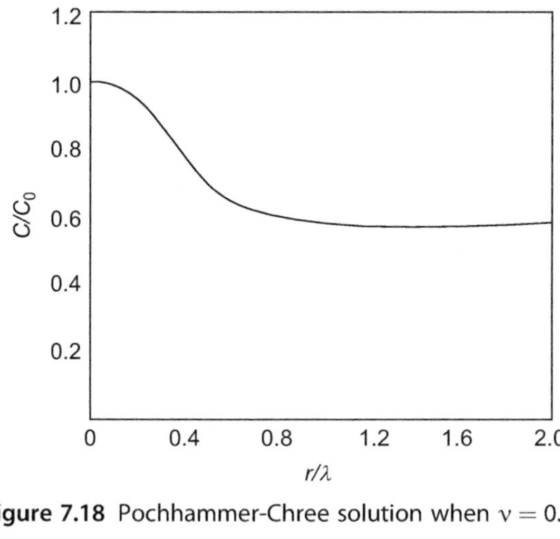

Figure 7.18 Pochhammer-Chree solution when ν = 0.29.

as molybdenum disulfide) can be used. On the other hand, the ratio of length to diameter L/D of the specimen must be reasonably designed (for example, the ratios of metal specimens, according to Davies and Hunter's analysis (Davies and Hunter, 1963), are often taken as $L/D = 0.5$), making the specimen long enough to disregard the friction at the end of the specimen and short enough to disregard the stress wave effect.

In China, the widely used SHPB device is one with pressure bars of diameter 14.5 mm, where the assumption of "one-dimensional stress wave in the bar" can be satisfied for the majority of homogeneous metal materials, although small high-frequency oscillations are often superimposed on the trapezoidal wave platform of the measured incident wave, as shown in Fig. 7.5, and this is exactly what weak transversal (lateral) inertia looks like.

However, for heterogeneous materials, SHPB technique has encountered new challenges. Heterogeneous composites can be regarded as the materials consisting of "matrices" and generalized "inclusions," such as short fibers, grains, and aggregates. For the measured specimen to reflect the macroscopic average mechanical response of the heterogeneous material, the diameter of the specimen should be one order of magnitude larger than the size of the heterogeneous "inclusions" in the material, so a large-diameter SHPB device should be used for heterogeneous composites. At this time, any elastic wave in the large-diameter pressure bar can be regarded as the superimposition of harmonic components with different frequencies, which will now propagate at their own phase velocities (Eq. 7.6). Therefore, the waveform no longer maintains its original shape in the propagation process, i.e., the so-called wave dispersion phenomenon occurs. Such dispersion caused by the geometric factors of the structure is called geometric dispersion, which is distinguished from the nonlinear constitutive dispersion and the constitutive viscosity dispersion. The latter two are caused by the nonlinearity and the viscosity, respectively, in the constitutive relation of materials.

The phenomena of wave dispersion in the large-diameter SHPB are mainly manifested in the following aspects:

(1) The high-frequency oscillation superimposed on the main wave.

Here, we consider a two-dimensional (axisymmetric) numerical analysis of elastic wave propagation in steel pressure bars, of which the details can be found in the works of Wang (2007), Liu and Hu (2000), and Wang and Wang (2004). The material parameters of the steel are Young's modulus $E = 200$ GPa, density $\rho_0 = 7.8 \times 10^3$ kg/m^3, Poisson ratio $\nu = 0.3$. A trapezoidal pulse is applied at the end of the steel bar $X = 0$, and the loading parameters are wave amplitude $\sigma_0 = 800$ MPa, total loading duration 120 μs including the rise time and fall time of 10 μs, respectively. For the four steel bars with different diameters $D = 5$, 14.5, 37, and 74 mm, the stress wave profiles obtained from LS-DYNA numerical simulation at different propagation distance ($L = 0$, 10, 20, 30, 40, and 50 mm) are summarized in Fig. 7.19.

It can be seen that with the increase of the bar diameter D and the propagation distance, the high-frequency oscillation superimposed on the main waveform is significantly enhanced. This will have an adverse effect on the accuracy of data processing and experimental results.

(2) The nonuniform stress distribution on the cross section of the bar.

The numerical results by LS-DYNA also show that the effect of transversal (lateral) inertia will cause nonuniform two-dimensional axial stress distribution on the cross section of the bar. Consider a steel cylindrical bar with diameter $D = 2R = 37$ mm, the axial stress wave profiles at three radii ($r = 0$, $0.5R$, and R) on a cross section 0.5D away from impact end are

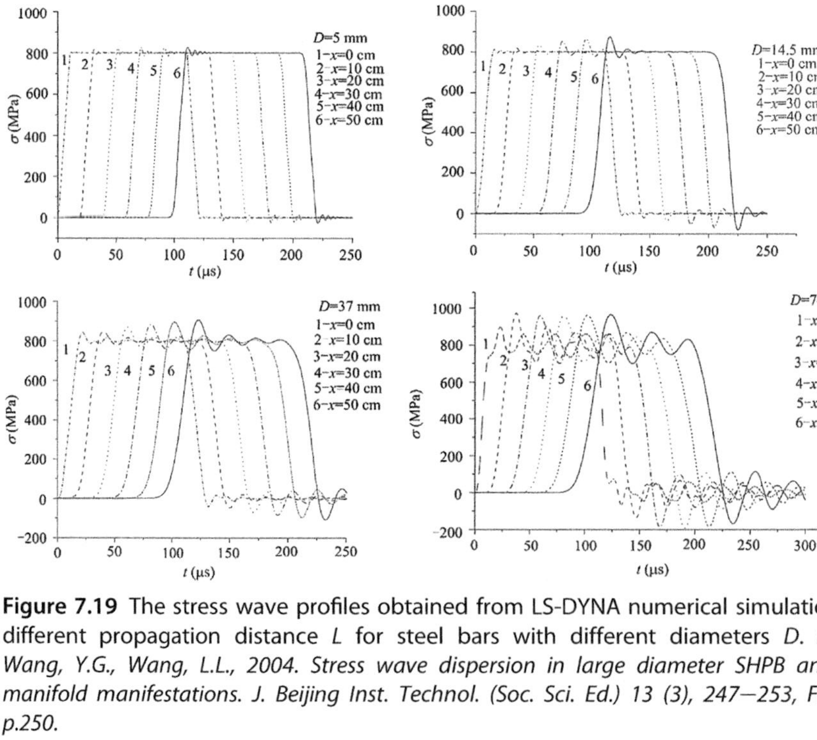

Figure 7.19 The stress wave profiles obtained from LS-DYNA numerical simulation at different propagation distance *L* for steel bars with different diameters *D*. *From Wang, Y.G., Wang, L.L., 2004. Stress wave dispersion in large diameter SHPB and its manifold manifestations. J. Beijing Inst. Technol. (Soc. Sci. Ed.) 13 (3), 247–253, Fig. 5, p.250.*

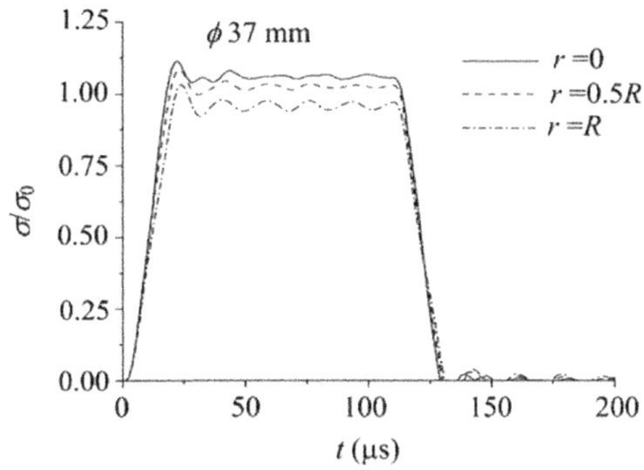

Figure 7.20 The stress wave profiles at different radii $r = 0$, $0.5R$, and R, respectively, on the cross section 0.5D away from the impact end. *From Wang, Y.G., Wang, L.L., 2004. Stress wave dispersion in large diameter SHPB and its manifold manifestations. J. Beijing Inst. Technol. (Soc. Sci. Ed.) 13 (3), 247–253, Fig. 2, p.249.*

shown in Fig. 7.20. The axial stress gradually decreases from the center to the outer surface along the radius, and the stress at the center of the bar is the largest (the material is in an approximately 1D strain state), followed by $0.5R$, and the outer surface R is the smallest (the material is in an approximately 1D stress state). Because the primary (elementary) theory of 1D stress wave in the bar is based on the assumption that the stress on the plane cross section is uniformly distributed, the transversal (lateral) inertia effect destroys this assumption, which will affect the accuracy of the experimental results.

(3) Attenuations of stress peaks during propagation

Another important manifestation of the geometric dispersion of the stress wave in the bar is that the peak of the stress pulse decreases with the propagation distance. Because a trapezoidal stress pulse propagating in a large bar results in significant wave oscillations, which is not conducive to the analysis of stress wave attenuation, we consider a triangular pulse applied at the bar end $X = 0$ with a peak magnitude $\sigma_0 = 800$ MPa, and its rise time and fall times are 150 μs, respectively. For the three cases of pressure bars with diameters of 37 mm, 74 , and 100 mm, how does the stress peak value attenuates with the propagation distance X are plotted in Fig. 7.21. It can be seen that the larger the bar diameter, the more severe the stress peak attenuation, and thus the strain signal measured by the strain gauge G (see Fig. 7.4) can no longer represent the strain signal at the interface between the pressure bar and the specimen.

(4) The increase of wave front rise time

As can be seen from Fig. 7.19 that the wave front edge of the stress pulse actually gradually slows down with the increase of the propagation distance

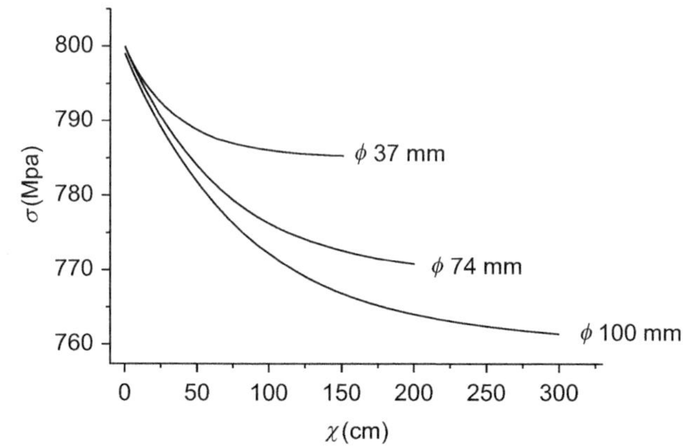

Figure 7.21 Attenuations of stress peaks of a triangular stress pulse propagating in bars with diameter of 37 mm, 74 , and 100 mm, respectively. *From Wang, Y.G., Wang, L.L., 2004. Stress wave dispersion in large diameter SHPB and its manifold manifestations. J. Beijing Inst. Technol. (Soc. Sci. Ed.) 13 (3), 247–253, Fig. 7, p.251.*

due to the transversal (lateral) inertia effect. In other words, the rise time t_s of the stress pulse front, which is referring to the duration from the starting point to the peak of the stress pulse, increases with the propagation distance; and the larger the bar diameter, the more significant the change in rise time.

For different bar diameters, the variations of the rise time t_s with the propagation distance X are plotted in Fig. 7.22. The results show that the

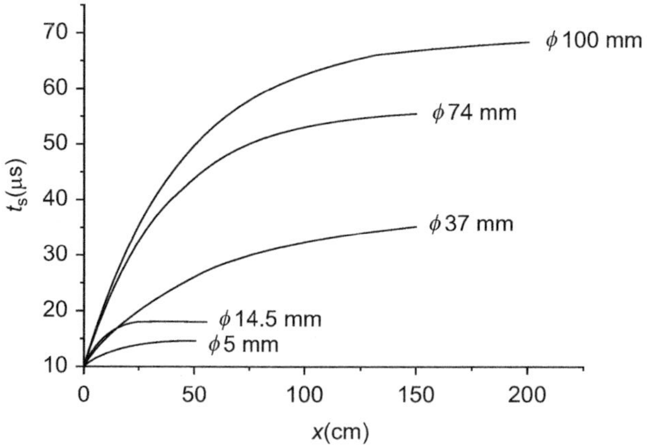

Figure 7.22 The variations of stress wave front rising time (t_s) with the propagation distance (X) in different bars. *From Wang, Y.G., Wang, L.L., 2004. Stress wave dispersion in large diameter SHPB and its manifold manifestations. J. Beijing Inst. Technol. (Soc. Sci. Ed.) 13 (3), 247–253, Fig. 6, p.251.*

increase of the rise time t_s during propagation is more significant with the increase of the pressure bar diameter.

The impact load is characterized by short duration, especially short rise time. In the traditional SHPB experiment, when the incident stress pulse has a short rise time (for example, $t_s < 20$ μs), the strain rate of the specimen can generally reach the order of 10^3 s^{-1}. However, when a large-diameter pressure bar (e.g., $D = 74$ mm) is used, the rise time of the incident stress pulse is greatly increased ($t_s \sim 50$ μs), consequently the strain rate of the specimen can only reach the order of 10^2 s^{-1}. This is the main reason why it is difficult to reach the strain rate of 10^3 s^{-1} when heterogeneous materials such as concretes are tested with large-diameter SHPB.

It is important to avoid or correct the transversal (lateral) inertia effect when using the large-diameter SHPB device, and the following are some of the methods currently used:

(1) SHPB bundle device

The pressure bar of the SHPB bundle device consists of several small-sized pressure bars instead of one large-sized pressure bar, thus its transversal (lateral) inertia effect is negligible for each small-sized pressure bar. Fig. 7.23 shows the 200 × 200 mm horizontal square SHPB bundle device at the ISPRA Joint Research Centre in Italy, and Fig. 7.24 shows the 150 × 150 mm vertical SHPB bundle device at Ningbo University in China (Dong et al., 2011).

(2) The pulse shaper technique

As shown in Fig. 7.25, the pulse shaper technique is a technique to change or adjust the shape of incident waveform in the input bar by adding a material with a lower wave impedance and better plasticity (such as copper) at the impact end of the input bar. Similar effects can also be achieved by the strike bars with variable cross section or gradient materials.

Duffy et al. (1971) first applied this technique in the torsional SHB experiment to study the mechanical behavior of the aluminum, and the technique was subsequently used by Christensen et al. (1972) to investigate rock materials with SHPB experiments.

The comparison of the measured incident/reflected waves before and after using the pulse shaper technique (Chen and Song, 2011) is depicted in Fig. 7.26. It can be seen that the pulse shaper technique can effectively reduce or even eliminate wave oscillations caused by the transversal (lateral) inertia. It is because that the pulse shapers can essentially "filter out" high-frequency components in the incident/reflected waves.

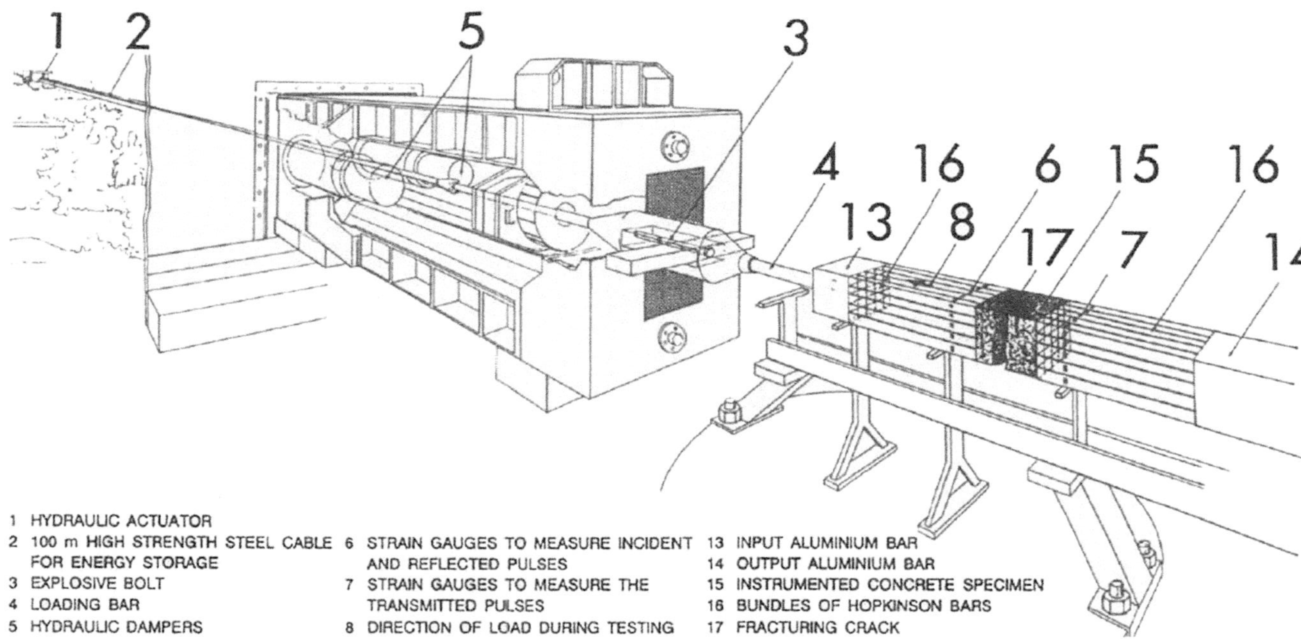

Figure 7.23 200 × 200 mm horizontal square split Hopkinson pressure bar bundle device at the ISPRA Joint Research Centre. *From Albertini, C., Montagnani, M., 1994. Study of the true tensile stress-strain diagram of plain concrete with real size aggregate; need for and design of a large Hopkinson bar bundle. J. Phys. IV Colloq. 04 (C8), pp.C8-113—C8-118. doi: 10.1051/jp4:1994817>. <jpa-00253372>, Fig. 6, p.C8-117. Reprinted with permission of the publisher.*

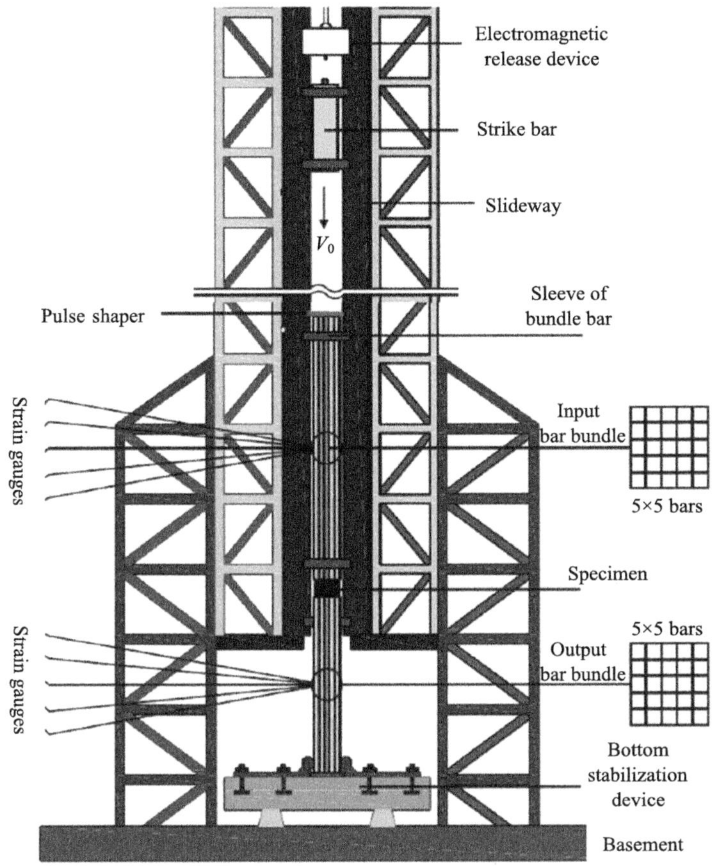

Figure 7.24 150 × 150 mm vertical split Hopkinson pressure bar bundle device at Ningbo University in China. *From Dong X.L., Zhang S.L., Yuan H.L., Wang L., 2011. Hopkinson bar bundle technique and experimental studies on dynamic behavior of concretes. Proceedings of 10th National Symposium on Impact Dynamics, July 26–31, 2011, Taiyuan, China (in Chinese).*

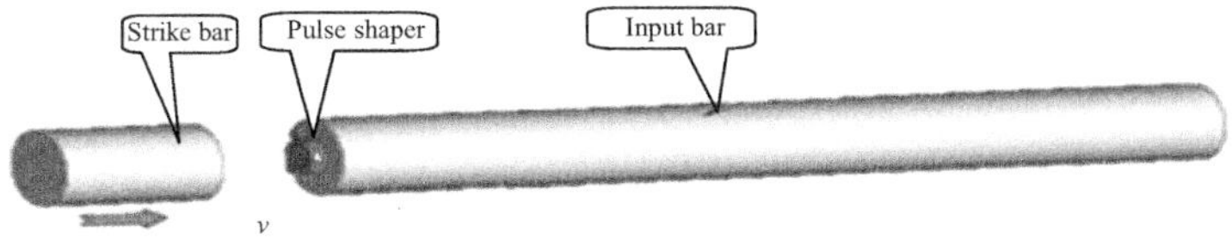

Figure 7.25 Schematic of the pulse shaper technique.

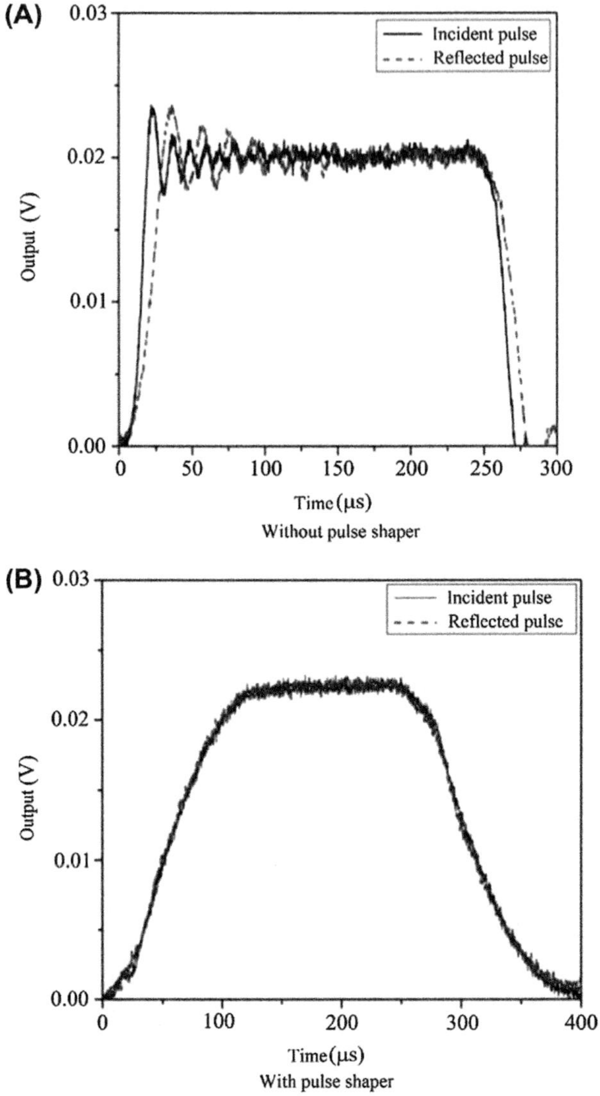

Figure 7.26 The comparison of the measured incident/reflected waves (A) before and (B) after using the pulse shaper technique. *From Chen, W.N., Song, B., 2011. Split Hopkinson (Kolsky) Bar — Design, Testing and Applications, Springer, Fig. 2.4, page 41 and Fig. 2.7, page 44. Reprinted with permission of the publisher.*

The spectral distributions of the incident wave before and after using the pulse shaper technique (Chen and Song, 2011) are shown in Fig. 7.27, and the results indicate that the pulse shaper can "filter out" high-frequency components above 40 KHz. From the comparison of Fig. 7.26A,B, it can

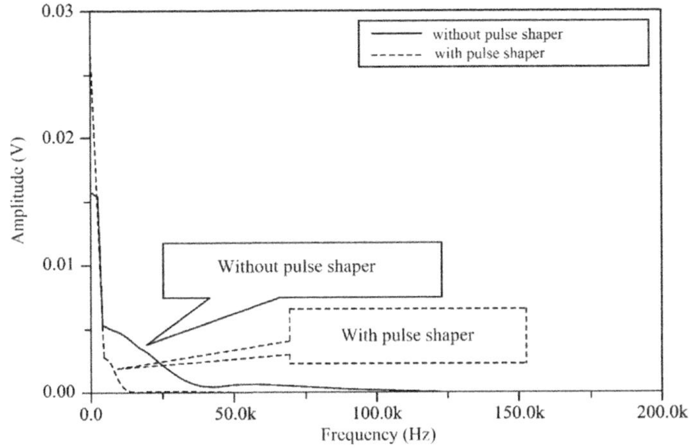

Figure 7.27 The spectral distributions of the incident wave before and after using the pulse shaper technique. *From Chen, W.N., Song, B., 2011. Split Hopkinson (Kolsky) Bar – Design, Testing and Applications, Springer, Fig. 2.6, page 44. Reprinted with permission of the publisher.*

be seen that the rise time t_s of the incident wave increases from about 20 μs to about 120 μs, and this inevitably leads to the decrease of strain rate of the specimen. In other words, eliminating the wave oscillation by the pulse shaper is at the expense of reducing or sacrificing the strain rate of specimen in experiments. To eliminate wave oscillations caused by transversal (lateral) inertia, a viscous filter can also be used to achieve the same effect.

(3) Inverse-analytic correction of geometric dispersion

The basic principle of the SHPB experiment is to determine the dynamic stress–strain curve of material by the incident strain wave $\varepsilon_I(X_1, t)$, the reflection strain wave $\varepsilon_R(X_1, t)$, and the transmission strain wave $\varepsilon_T(X_2, t)$, where X_1 and X_2 are the interfaces between the specimen and the input bar, and that between the specimen and the output bar, respectively. However, the strain waves at X_1 and X_2 are difficult to be measured directly, and as such it can be indirectly done by measuring the strain waves at X_{G1} in the input bar and at X_{G2} in the output bar based on the assumption of "one-dimensional stress wave in the bar."

However, when a large-diameter pressure bar is used, the dispersion effect is not negligible, and the waveforms at different positions in the bar are distorted. To deduce the correct wave information of $\varepsilon_I(X_1, t)$,

$\varepsilon_R(X_1, t)$, and $\varepsilon_T(X_2, t)$, transversal (lateral) inertial corrections need to be performed on the measured waves $\varepsilon_I(X_{G1}, t)$, $\varepsilon_R(X_{G1}, t)$, at X_{G1} on the input bar and $\varepsilon_T(X_{G2}, t)$ at X_{G2} on the output bar. Where, the deduction from the $\varepsilon_I(X_{G1}, t)$ to the $\varepsilon_I(X_1, t)$ located in the downstream of wave propagation belongs to "positive analysis," which can be easily realized by general numerical calculation with commercial software, while the deduction from the $\varepsilon_R(X_{G1}, t)$ and the $\varepsilon_T(X_{G2}, t)$, respectively, to the $\varepsilon_R(X_1, t)$ and the $\varepsilon_T(X_2, t)$ in the upstream of wave propagation belong to "inverse analysis," which is challenging.

Under impact load conditions, there are many established methods for performing inverse analysis from indirect measurements, referring to the review by Inoue et al. (2001). For the inverse analysis of elastic waves in a straight bar involved in SHPB, based on the Pochhammer-Chree solution (Miklowitz, 1978), Fourier analysis is a convenient and practical method (see, for instance, Wang and Xia (1998)). Because any elastic wave in a bar can always be regarded as a superposition of harmonic components with different frequencies, the basic steps of Fourier analysis method to modify waves are as follows. Firstly, the original elastic wave measured in time domain is transformed into harmonic components with different frequencies, amplitudes, and phases in frequency domain by Fourier transform. Secondly, according to the Pochhammer-Chree solution taking into account the transversal (lateral) inertia, the harmonic components with different frequencies propagate at different phase velocities, so as in the frequency domain, by various harmonic components, the elastic waves taking into account the dispersion can be reconstructed at a specified position (such as the interface between the specimen and the pressure bar). Finally, these signals in frequency domain are converted into the required signals in time domain by using the inverse Fourier transform, to obtain the $\varepsilon_I(X_1, t)$, $\varepsilon_R(X_1, t)$, and $\varepsilon_T(X_2, t)$ at the interface X_1 and X_2 that we need. Fig. 7.28 shows the dynamic stress—strain curve for cement mortar obtained by using an SHPB device of diameter 37 mm, where the solid line represents the result directly obtained by Eq. (7.4) without dispersion correction, and the dotted line represents the modified result by the Fourier analysis method. The maximum relative error of those two results is about 20%, and the curve before correction gives a false strain softening phenomenon ($d\sigma/d\varepsilon < 0$). It can be seen that it is very important to conduct dispersion correction of the waveform to obtain true and reliable dynamic stress—strain relation when using large-diameter SHPB, which must not be ignored.

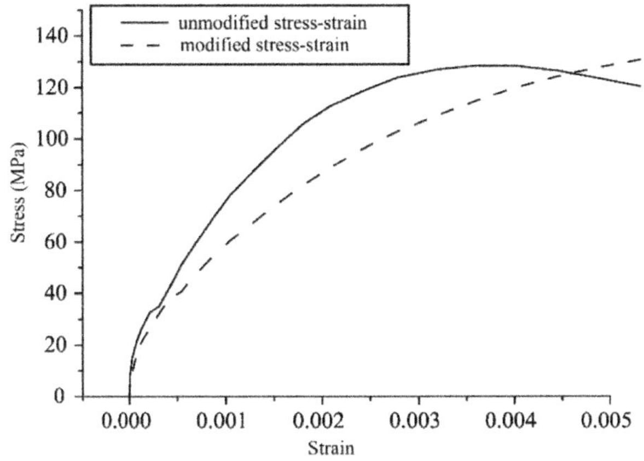

Figure 7.28 Dynamic stress–strain curves of the cement mortar with and without dispersion correction. *From Zhu Jue, 2006, Fig. 2.9.*

7.1.4 Discussion on the assumption of uniform distribution of stress/strain along the specimen length

In this section, we will discuss the factors influencing the assumption of uniform distribution of stress/strain along the specimen length (henceforth the "uniformity assumption" for short). Recently, some in-depth-related research (Yang and Shim, 2005; Zhu et al., 2006, 2009) has been carried out based on the stress wave theory.

First, we consider a simple case where the input bar, specimen, and output bar all have the same cross-sectional area and are in an elastic state (ref. Fig. 7.4). Obviously, once the specimen is loaded by an incident pulse, it would be perfect that if the "uniformity" can be realized when the specimen is still in the condition of small elastic deformation.

According to the theory of stress wave propagation (Wang, 2007), when an elastic wave is transmitted from the bar 1 (with the wave impedance of $(\rho_0 C_0)_1$) to the bar 2 (with the wave impedance of $(\rho_0 C_0)_2$), the relation between the reflected disturbance (stress $\Delta\sigma_R$ and particle velocity Δv_R) with the incident disturbance (stress $\Delta\sigma_I$ and particle velocity Δv_I), as well as the relation between the transmitted disturbance (stress $\Delta\sigma_T$ and particle velocity Δv_T) with the incident disturbance (stress $\Delta\sigma_I$ and particle velocity Δv_I), should satisfy the following equations:

$$\left.\begin{aligned} \Delta\sigma_{\mathrm{R}} &= F\Delta\sigma_I \\ \Delta v_{\mathrm{R}} &= -F\Delta v_I \end{aligned}\right\} \tag{7.7a}$$

$$\left.\begin{aligned} \Delta\sigma_T &= T\Delta\sigma_I \\ \Delta v_T &= nT\Delta v_I \end{aligned}\right\} \tag{7.7b}$$

where

$$\left.\begin{aligned} n &= (\rho_0 C_0)_1/(\rho_0 C_0)_2 \\ F &= \frac{1-n}{1+n} \\ T &= \frac{2}{1+n} \end{aligned}\right\} \tag{7.7c}$$

n is the ratio of wave impedance, F and T are the reflection coefficient and the transmission coefficient, respectively, which are completely determined by the ratio of wave impedance n. When the stress wave propagates from the pressure bar into the specimen, the n in the above formula should be taken as $n_{B-S}= (\rho C)_B/(\rho C)_S$, where $(\rho C)_B$ and $(\rho C)_S$ represent the elastic wave impedance of the pressure bar and the specimen. When the stress wave is transmitted from the specimen into the pressure bar, $n = n_{S-B} = (\rho C)_S/(\rho C)_B$, which is the reciprocal of n_{B-S}. This indicates that the behavior of wave reflection and transmission in the input bar-specimen-output bar system are mainly relying on the wave impedance ratio of the pressure bar to the specimen.

According to Eq. (7.7) and correspondingly on the physical plane ($X-t$ plane) and the velocity plane ($\sigma-v$ plane), it is not difficult to determine the reflection process and the transmission process of elastic waves in the input bar-specimen-output bar system, and the state of stress σ and particle velocity v in each step, as shown in Fig. 7.29A,B. In the figures, the strike bar impinges normally with a velocity v_0 onto the input bar, and a strong discontinuous elastic wave (rectangle wave front) having an amplitude of σ_A $(=-(\rho C)_B \cdot v_0/2)$ is generated.

Therefore, when an incident wave σ_A propagates with an elastic wave velocity C_B in the input bar and reaches the interface X_1, the first elastic wave reflection and transmission occur. The transmitted wave propagates in the specimen at the elastic wave velocity C_S and results in a strong discontinuity stress disturbance

$$\Delta\sigma_1 = \sigma_1 - 0 = T_{B-S}\sigma_A$$

where the transmission coefficient T_{B-S} is calculated according to Eq. (7.7c)

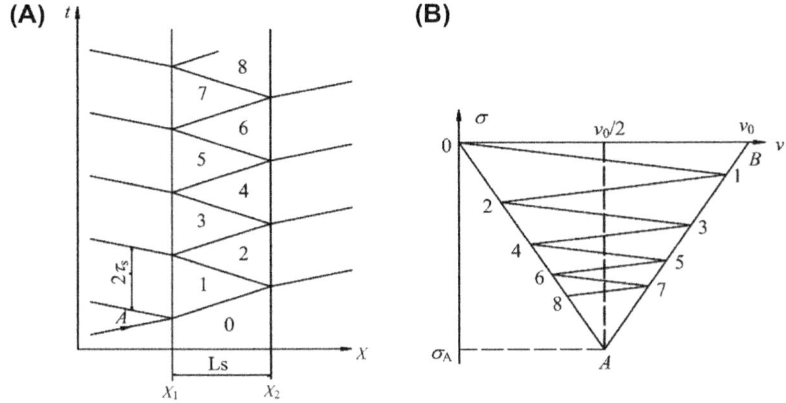

Figure 7.29 (A) Reflection and (B) transmission of elastic waves in the system of input bar-specimen-output bar (*rectangle* wave front).

$$T_{B-S} = \frac{2}{1 + n_{B-S}}, \quad n_{B-S} = \frac{(\rho C)_B}{(\rho C)_S}.$$

Here, the subscript "$B - S$" means that the stress waves propagate from the bar B to the specimen S. After time $\tau_S = L_S/C_S$ (where L_S is the length of the specimen), the secondary transmission and reflection occur at the interface X_2. According to Eq. (7.7), a strong discontinuity stress disturbance caused by the reflected wave propagating back to the specimen is

$$\Delta\sigma_2 = \sigma_2 - \sigma_1 = F_{S-B}\Delta\sigma_1$$

where the reflection coefficient F_{B-S} is calculated according to Eq. (7.7)

$$F_{S-B} = \frac{1 - n_{S-B}}{1 + n_{S-B}}$$

and the subscript "S—B" means that the stress waves propagate from the specimen S to the bar B. Noticing n_{B-S} and n_{S-B} are reciprocal to each other, and if n_{S-B} is rewritten as β, it is not difficult to prove that T_{B-S} and F_{S-B} have the following relationship and can be expressed as

$$\left.\begin{aligned} T_{B-S} &= \frac{2}{1 + n_{B-S}} = \frac{2\beta}{1+\beta} \\ F_{B-S} &= \frac{1-\beta}{1+\beta} \\ 1 - F_{B-S} &= 1 - \frac{1-\beta}{1+\beta} = \frac{2\beta}{1+\beta} = T_{B-S} \end{aligned}\right\} \tag{7.8}$$

When the reflected wave continues propagation and returns to the interface X_1, the third transmission and reflection occur, which causes a new strong discontinuity stress disturbance in the specimen, and it can be expressed as

$$\Delta\sigma_3 = \sigma_3 - \sigma_2 = F_{S-B}\Delta\sigma_2 = F_{S-B}^2\Delta\sigma_1$$

By this analogy, when the *k*th transmission and reflection occur, a new induced strong discontinuous stress disturbance is

$$\Delta\sigma_k = \sigma_k - \sigma_{k-1} = F_{S-B}\Delta\sigma_{k-1} = F_{S-B}^{k-1}\Delta\sigma_1 \tag{7.9}$$

After the *k*th transmission and reflection, the final stress state σ_k of the *k*-zone (ref. Fig. 7.29) is

$$\sigma_k = \sum_{i=1}^{k}\Delta\sigma_i = \left(1 + F_{S-B} + F_{S-B}^2 + F_{S-B}^3 + \ldots + F_{S-B}^{k-1}\right)\Delta\sigma_1 \tag{7.10a}$$

By using the binomial expansion as follows

$$1 - x^k = (1-x)\left(1 + x + x^2 + x^3 + \ldots + x^{k-1}\right)$$

and taking into account the relation between T_{B-S} and F_{S-B} given by Eq. (7.8) and the expression of β, Eq. (7.10a) can finally be written as

$$\begin{aligned}\sigma_k &= \frac{1 - F_{S-B}^k}{1 - F_{S-B}}\Delta\sigma_1 = \frac{1 - F_{S-B}^k}{1 - F_{S-B}}T_{B-S}\sigma_A = \left(1 - F_{S-B}^k\right)\sigma_A \\ &= \left[1 - \left(\frac{1-\beta}{1+\beta}\right)^k\right]\sigma_A\end{aligned} \tag{7.10b}$$

This indicates that after the stress wave propagating in the specimen *k* times of back and forth (transmissions-and-reflection), the stress σ_k depends on not only the number of times, *k*, but also the wave impedance ratio between the specimen and the pressure bar, namely β. Notice that the number *k* is actually the same thing as the dimensionless time $\bar{t}(= t/\tau_S = tC_S/L_S)$.

For a given β, when *k* is an even value (refer to Fig. 7.29), Eq. (7.10) gives the transmission stress at interface X_2, varying with number of times of transmission-reflection *k* or the dimensionless time $\bar{t}(= tC_S/L_S)$.

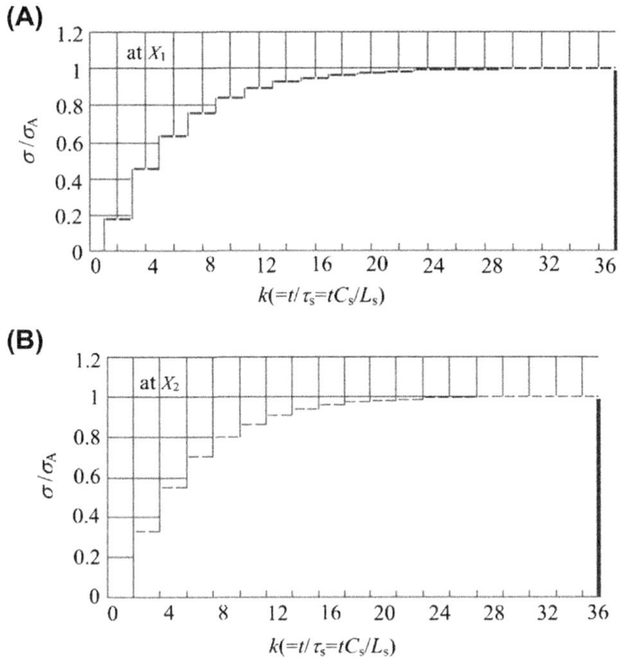

Figure 7.30 Variations of the dimensionless stress σ/σ_A with k (or $\bar{t}$) when $\beta = 1/10$: (A) at the interface X_1, (B) at the interface X_2.

When k takes an odd value, Eq. (7.10) gives the reflection stress at interface X_1, varying with number of times of transmission-reflection k or the dimensionless time $\bar{t}$.

As an example, when $\beta = 1/10$, the variations of the dimensionless stress σ/σ_A with k (or $\bar{t}$) at interfaces X_1 and X_2, given by Eq. (7.10), are shown in Figs. 7.30A,B, and they both gradually tend to 1. It means that the distribution of stress along the length of the specimen is a gradual uniform process, which is dependent on both β and k.

Note that Eq. (7.9) gives the stress difference between interface X_1 and interface X_2 (refer to Fig. 7.29), and a dimensionless stress difference (relative stress difference) at both ends of the specimen can then be defined as

$$\alpha_k = \frac{\Delta\sigma_k}{\sigma_k} \tag{7.11}$$

For a rectangular strong discontinuous incident wave, substituting Eqs. (7.9) and (7.10) into Eq. (7.11), we have

$$\alpha_k = \frac{\Delta\sigma_k}{\sigma_k} = \frac{F_{S-B}^{k-1}}{\dfrac{1-F_{S-B}^{k}}{1-F_{S-B}}} = \frac{\left(\dfrac{1-\beta}{1+\beta}\right)^{k-1}\left(1-\dfrac{1-\beta}{1+\beta}\right)}{1-\left(\dfrac{1-\beta}{1+\beta}\right)^{k}}$$

$$= \frac{2\beta(1-\beta)^{k-1}}{(1+\beta)^k-(1-\beta)^k} \tag{7.12}$$

This formula analytically describes the variation law of the relative stress difference α_k at both ends of the specimen with the changes of the wave impedance ratio β and the number of transmission-reflection times k.

For different values of β ($\beta = 1/2$, 1/4, 1/6, 1/10, 1/25, 1/100), the relation between α_k and k calculated by Eq. (7.12) are shown in Fig. 7.31. It can be found that the α_k increases with decreasing β for a given k. In other words, with the decrease of the impedance ratio β, stress waves in the specimen have to undergo more round-trip reflection process to meet the requirements of the uniformity assumption.

As suggested by Zhou et al. (1992), Ravichandran and Subhash (1994), when $\alpha_k \leq 5\%$, the stress/strain distribution along the specimen length can be approximately considered to satisfy the requirements of uniformity assumption. Consequently, as shown in Fig. 7.31, the corresponding

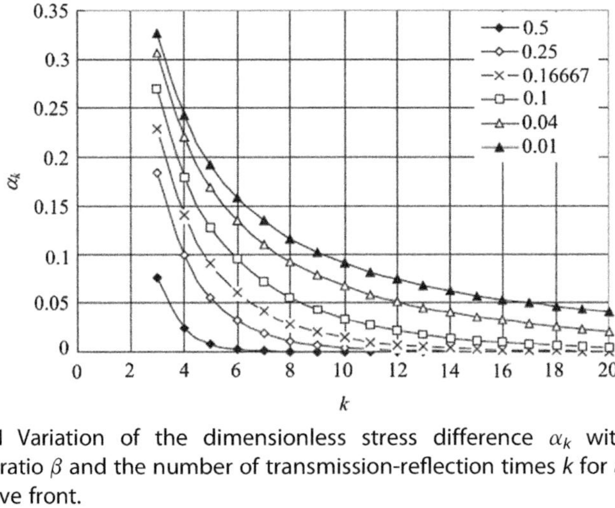

Figure 7.31 Variation of the dimensionless stress difference α_k with the wave impedance ratio β and the number of transmission-reflection times k for a rectangular incident wave front.

minimum number of back and forth reflective time k_{min} is equal to 4 when $\beta = 1/2$. The corresponding minimum number of back and forth reflective time k_{min} is increased to 18 when $\beta = 1/100$. Thus, it is a common misconception that reflecting 2–3 times back and forth in the specimen is able to satisfy the uniformity assumption.

The above results are based on the rectangular strongly discontinuous incident wave. However, the incident waves encountered in most SHPB experiments are trapezoidal with a certain rise time. In this regard, analysis and discussion can also be carried out using methods similar to those described above. Assuming the rise time of the trapezoidal waveform is just the time the elastic wave propagates around the specimen, i.e., $2\tau_S = (2L_S/C_S)$, Yang and Shim (2005) indicated an analytical solution of α_k after the elastic wave propagates back and forth once in the specimen ($k > 2$)

$$\alpha_k = \frac{2\beta^2(1-\beta)^{k-2}}{(1+\beta)^k - (1-\beta)^{k-2}} \tag{7.13}$$

For different values of β ($\beta = 1/2$, 1/4, 1/6, 1/10, 1/25, 1/100), the results of α_k with varying k calculated by Eq. (7.13) are shown in Fig. 7.32. It can be seen that, contrary to the case of the rectangular wave (Fig. 7.31), the current α_k-k curves move down with the decrease of β, which means the α_k decreases with decreasing β under a given k. In the range of β discussed in this example, the results demonstrate that the minimum number of reflection times needed for satisfying the uniformity assumption k_{min} is only 3–4 times.

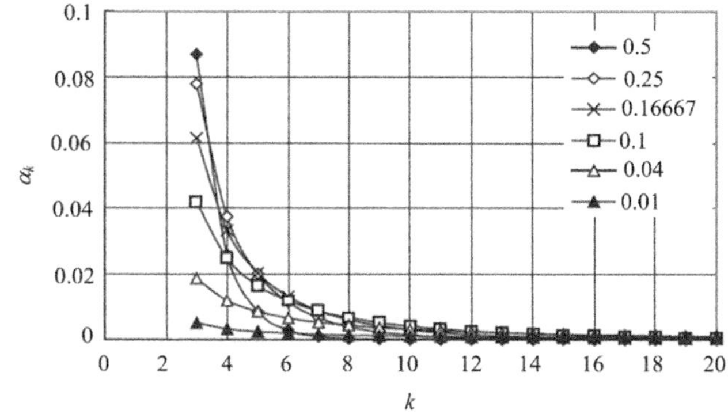

Figure 7.32 Variation of the dimensionless stress difference α_k with the wave impedance ratio β and the number of transmission-reflection times k for a trapezoidal incident wave front.

If the incident wave has a wave front with a longer rise time and a linear increase with time, which is called linear ramp wave $\sigma_I(t)$, and it can be expressed as

$$\sigma_{\mathrm{I}}(t) = \frac{\sigma^* t}{\tau_S} = \frac{\sigma^* C_S t}{L_S}$$

where $\sigma^\star$ is the incident wave amplitude at $t = \tau_S = L_S/C_S$, Yang and Shim (2005) gave the following analytical results (for $k \geq 3$)

$$\alpha_k = \frac{2\beta^2\left[1-\left(-\dfrac{1-\beta}{1+\beta}\right)^k\right]}{2k\beta - 1 + \left(\dfrac{1-\beta}{1+\beta}\right)^k} \tag{7.14}$$

For different values of β ($\beta = 1/2, 1/4, 1/6, 1/10, 1/25, 1/100$), the results of α_k with varying k calculated by Eq. (7.14) are shown in Fig. 7.33. It can be seen that the α_k-k curves of the linear ramp wave also move downward with the decrease of β, but curves oscillate markedly with the decrease of β. In addition, in the range of β discussed in this example, the minimum number of reflection times needs to be higher to satisfy the uniformity assumption than that of the trapezoidal incident wave. This means that the use of linear ramp incident wave with longer rise time is actually not conducive to satisfying the uniformity assumption.

In the case of the same β (=0.5, 0.25, 0.01), the comparison of α_k-k curves when the incident wave is the rectangular wave (curve A), the trapezoidal wave (curve B), and the linear ramp wave (curve C) is

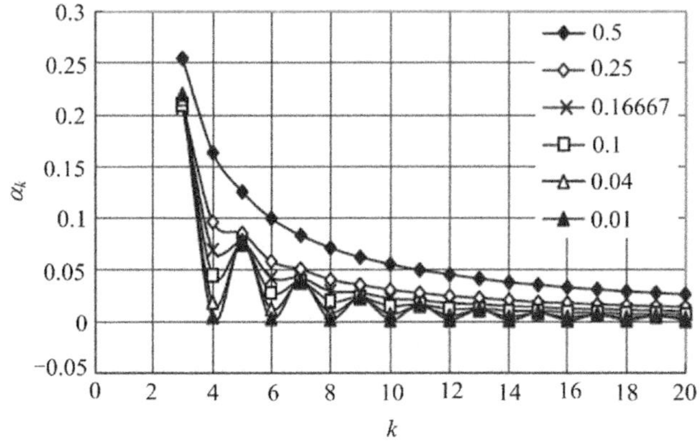

Figure 7.33 Variation of the dimensionless stress difference α_k with the wave impedance ratio β and the number of transmission-reflection times k for a linear ramp incident wave front.

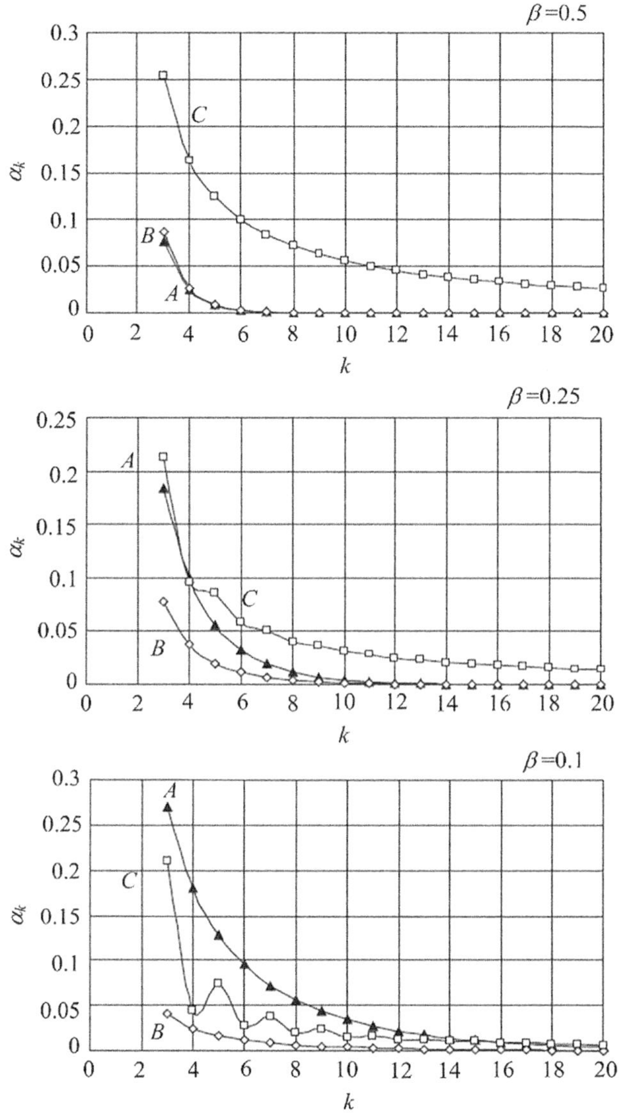

Figure 7.34 Comparison of α_k-k curves between the rectangular wave (curve A), the trapezoidal wave (curve B), and the linear ramp wave (curve C) under the same β.

depicted in Fig. 7.34. It can be seen that, when $\beta = 0.5$, there is no significant difference between the rectangular wave and the trapezoidal wave, and both are more superior to the linear ramp wave with respect to satisfying the uniformity assumption. However, this is contrary to the general thinking that "the use of pulse shaper to adjust the incident wave into a linear ramp wave will favor the stress/strain uniformity." With the decrease of

β, the α_k-k curves of the trapezoidal and linear ramp waves move downward, while that of the rectangular wave rises. Thus, for $\beta = 0.1$, the rectangular wave is the least favorable choice, while the trapezoidal wave is usually best for satisfying the uniformity assumption.

From the above analysis and discussion on the propagation process of stress waves in the input bar-specimen-output bar system, it is known that both the wave impedance ratio β and the waveform of the incident wave (especially the rise time) will significantly affect the minimum number of reflection times $k_{\min}$ required for satisfying the uniformity assumption. This is something that we should pay attention to when designing SHPB experiments for different materials.

In addition to the abovementioned uniformity analysis of the specimen in the elastic deformation stage, the uniformity process of viscoelastic materials such as polymers was further analyzed (Zhu et al., 2006). Studies have shown that compared with the stress uniformity analysis of elastic specimens, the stress uniformity process of viscoelastic specimens depends on not only the relative rise time τ_s/t_L of the incident wave and the instantaneous impedance ratio $R_i(=1/\beta)$ but also the high-frequency relaxation time of the material θ_2. In terms of the influence of the relative rise time τ_s/t_L, it is consistent with the conclusion of the stress uniformity analysis of the elastic specimen, that it is much easier for the specimen to achieve stress uniformity when $\tau_s/t_L = 2$. In terms of the influence of the high-frequency relaxation time θ_2, generally, the smaller the θ_2, the more difficult it is to achieve stress uniformity (except in the case of short rise time $\tau_s/t_L = 1$). In addition, unlike the elastic specimen, the strain uniformity and the stress uniformity of the viscoelastic specimen are different. In the case of short rise time, the strain is more easily uniformized than the stress, but as the rise time increases, it becomes more difficult for the strain to achieve uniformity than the stress.

It should also be pointed out that for brittle materials that undergo brittle failure under small elastic strain without plastic deformation, the speed of achieving uniformity of specimens in SHPB experiment is of special significance. Because "uniformity" is a time process in which stress waves propagate back and forth with time, if before "uniformity" is realized the specimen has already been fractured, the uniformity assumption is destroyed, and the validity and reliability of the SHPB experiment cannot be guaranteed. For concrete-like rate-dependent brittle materials, people may image that the longer the rise time τ_s/t_L, the better the stress uniformity of the specimen. However, studies by Zhu et al. (2009) show that similar to other viscoelastic and elastic specimens, it is not as the situation that people may image but show that it is in fact easier to achieve uniformity when $\tau_s/t_L = 2$, and the strain rate of the specimen will decrease with the increase of τ_s/t_L. For a concrete-like material with a dynamic fracture strain of 0.5%, even if the specimen is fractured after achieving uniformity by reducing the amplitude of the incident wave, the strain in the nonuniform stage has been as high as 0.2%–0.25%. Therefore, the measured stress–strain curve still does not fully satisfy the uniformity assumption, though the dynamic fracture stress value measured after achieving uniformity is effective.

7.1.5 SHPB experiment on soft materials with low wave impedance

The pressure bar of traditional SHPB device is generally made of high-strength steel to ensure that the pressure bar is always in the elastic state during the test. The density and the elastic wave velocity of the steel are $\rho_0 = 7.8 \times 10^3$ kg/m^3 and $C_0 = 5.19$ km/s, so the elastic wave impedance ($\rho_0 C_0$) is as high as 40 MPa/m/s. However, when the traditional SHPB technique is used to study soft materials with low wave impedance, such as solid propellants, explosives, foams, and organisms, because the wave impedance of these soft materials is only about 0.1–1 MPa/m/s, the transmitted signal measured by the output bar becomes too weak, and its amplitude is only a few tenths or less of the incident wave amplitude, which can be compared with the external interference signal, so that it is difficult to ensure the measurement accuracy.

There are two methods to solve this problem. The first method is to improve the sensitivity of the strain gauge. For example, a semiconductor strain gauge with high sensitivity is used, where the strain gauge sensitivity coefficient is about 50 times of that of the resistance strain gauge (Song and Hu, 2006), or a high-sensitivity thin film type quartz is used to directly measure the stresses at both ends of the specimen. Another method is to use a pressure bar with lower wave impedance, such as titanium alloy and magnesium alloy pressure bars (Gray and Blumenthal, 2000), which have low wave impedance for their low density compared with the high-strength steel. To use lower wave impedance bars, the SHPB technique of elastic steel bars has been extended to that of polymer bars (Wang et al., 1992; Wang et al., 1994; Zhao and Gary, 1995; Zhao et al., 1997). For example, a pressure bar made of polymethyl methacrylate (PMMA) has a wave impedance of about 1 MPa/m/s. However, in this case, the constitutive viscous dispersion effect of the viscoelastic wave propagating in the polymer bar should be taken into account (Wang, 2007).

Once viscoelastic bars are used, due to the strain rate dependence of the viscoelastic waves (manifested as dispersion and attenuation characteristics), the linear propositional relations between the stress, strain, and particle velocity, as well as the nondistortion character, no longer exist. Then the problem cannot be directly processed by the method as mentioned above for the linear elastic bar. Thus, the key points of the current problem are attributed (refer to Fig. 7.4) (1) to determine $\sigma_I(X_1, t)$ and $v_I\,(X_1, t)$ at interface X_1 from the measured strain signal $\varepsilon_I(X_{G1}, t)$ at G1, (2) to determine $\sigma_R(X_1, t)$ and $v_R(X_1, t)$ at interface X_1 from the measured strain signal $\varepsilon_R(X_{G1}, t)$ at G1, and (3) to determine $\sigma_T(X_2, t)$ and $v_T(X_2, t)$ at interface X_2 from the measured strain signal $\varepsilon_T(X_{G2}, t)$ at G2, and subsequently determine the dynamic stress $\sigma_s(t)$ and strain $\varepsilon_s(t)$ of the specimen according to Eqs. (7.1) or (7.4). Among them, the first issue is essentially solving a direct problem of viscoelastic wave propagation, while the latter two are attributed to solving the second kind of inverse problem of viscoelastic wave propagation.

Here, we take the SHPB bar made of PMMA as an example. Because the deformation of the pressure bar is usually small when a soft material is tested, the problem can be directly simplified as a linear viscoelastic wave propagation analysis. Studies (Wang et al., 1992; Wang et al., 1994; Wang et al., 1995) have shown that the constitutive relation of PMMA pressure bars at high strain rates can be well described by the three-element linear solid model (ref. Eq. 5.27b) consisting of a spring element and a Maxwell unit in parallel, namely we have

$$\frac{\partial \varepsilon}{\partial t} - \frac{1}{E_a + E_M}\frac{\partial \sigma}{\partial t} + \frac{E_a \varepsilon}{(E_a + E_M)\theta_M} - \frac{\sigma}{(E_a + E_M)\theta_M} = 0 \tag{7.15}$$

where E_a is the elastic constant of the parallel spring element, E_M and θ_M are the high-frequency elastic constant and high-frequency relaxation time of the parallel Maxwell unit, respectively. Eq. (7.15) together with the following motion equation and continuity equation constitute the governing equations for this problem

$$\begin{cases} \rho_0 \dfrac{\partial v}{\partial t} - \dfrac{\partial \sigma}{\partial X} = 0 \\ \dfrac{\partial \varepsilon}{\partial t} - \dfrac{\partial v}{\partial X} = 0 \end{cases}$$

By using the characteristics method, multiplying each of the above three equations by undetermined coefficients N, M, and L, respectively, and followed by their summation, we have

$$(L+N)\frac{\partial \varepsilon}{\partial t}+\left(M\rho_0\frac{\partial}{\partial t}-L\frac{\partial}{\partial X}\right)v-\left(\frac{N}{E_a+E_M}\frac{\partial}{\partial t}+M\frac{\partial}{\partial X}\right)\sigma$$
$$+\frac{N}{(E_a+E_M)\theta_M}(E_a\varepsilon-\sigma)$$
$$=0$$

To satisfy the requirement that the above equation consists of only the directional differential along characteristics $D(X, t)$, the undetermined coefficients N, M, and L must satisfy the following relations

$$\left.\frac{dX}{dt}\right|_c=\frac{0}{L+N}=-\frac{L}{M\rho_0}=\frac{M(E_a+E_M)}{N}$$

Evidently, there are two sets of solutions of N, M, and L. One is determined by the following equations

$$\left.\begin{aligned}&L+N=0\\&\rho_0(E_a+E_M)M^2=-LN\end{aligned}\right\} \tag{7.16}$$

We can then obtain the following two families of characteristic lines

$$\frac{dX}{dt}=\pm\sqrt{\frac{E_a+E_M}{\rho_0}}=C_v \tag{7.17a}$$

and the corresponding two sets of compatibility conditions along the characteristics

$$dv=\pm\frac{1}{\rho_0 C_v}d\sigma\pm\frac{\sigma-E_a\varepsilon}{\rho_0 C_v\theta_M}dt=\pm\frac{1}{\rho_0 C_v}d\sigma+\left[\frac{\sigma-E_a\varepsilon}{(E_a+E_M)\theta_M}\right]dX \tag{7.17b}$$

Here, the plus and minus signs correspond to the rightward-propagating and leftward-propagating waves, respectively.

Another set of solution of N, M, and L is determined by the following equation

$$\left.\begin{aligned}&L=M=0\\&N\neq 0\end{aligned}\right\} \tag{7.18}$$

Therefore, the third family of characteristics and the corresponding compatibility condition along it are, respectively,

$$dX = 0 \tag{7.19a}$$

$$d\varepsilon - \frac{d\sigma}{E_a + E_M} - \frac{\sigma - E_a\varepsilon}{(E_a + E_M)\theta_M} dt = 0 \tag{7.19b}$$

Eq. (7.19a) is physically consistent with the particle motion trajectory, whereas Eq. (7.19b) is a special form of the viscoelastic constitutive equation along the particle motion trajectory.

Thus, when using the characteristics method to solve the problem, there are three characteristic lines at any point on the $X-t$ plane (Fig. 7.35). By using the characteristic relations in the difference form, the three unknown variables σ, v, and ε can be determined according to the known initial and boundary conditions.

Assuming that the polymer pressure bar is initially at rest and in a stress-free state

$$\sigma(X,0) = \varepsilon(X,0) = v(X,0) = 0$$

and the end of the bar ($X = 0$) suddenly suffered a constant stress $\sigma^\star$, then a strongly discontinuous wave propagates along the line OA at the wave velocity D (Fig. 7.35)

$$D = \sqrt{\frac{1}{\rho}\frac{[\sigma]}{[\varepsilon]}} = \sqrt{\frac{E_a + E_M}{\rho_0}} \tag{7.20}$$

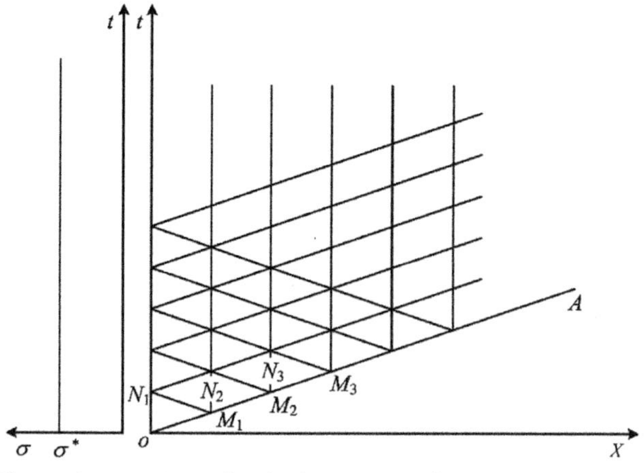

Figure 7.35 Three characteristics for the linear viscoelastic waves propagating in bars.

By using the kinematic and dynamic compatibility conditions across a strongly discontinuous interface,

$$[v] = -D[\varepsilon] \tag{7.21a}$$

$$[\sigma] = -\rho_0 D[v] \tag{7.21b}$$

where the symbol [] represents the difference of the mechanical quantity across the strongly discontinuous wave front, and considering that OA is a characteristic line and the corresponding characteristic compatibility condition (Eq. 7.17b) should be satisfied at the same time, it is not difficult to show that this strongly discontinuous wave attenuates according to the following exponential law

$$\sigma = \sigma^* \exp\left[-\frac{\rho_0 C_V}{2\eta_M \left(1 + \frac{E_a}{E_M}\right)^2} X \right] = \sigma^* \exp(-\alpha_a X) \tag{7.22}$$

Similar results can be found for the particle velocity v and strain ε along OA. Once the solution along the line OA is obtained, the solution in the AOt region in Fig. 7.35 can be further obtained. The problem is attributed to the boundary-value problem of characteristic lines for viscoelastic waves. This includes two basic types of operations: (a) determining the solutions at the boundary points and (b) determining the solutions at interior points.

Taking the arbitrary boundary point N_1 shown in Fig. 7.35 as an example, the particle velocity $v(N_1)$ and the strain $\varepsilon(N_1)$ can be determined by the two characteristic compatibility conditions along the characteristic lines M_1N_1 and ON_1, i.e., Eqs. (7.17b) and (7.19b). Writing in the finite difference form, we have

$$\left.\begin{aligned} v(N_1) - v(M_1) &= -\frac{1}{\rho_0 C_V}[\sigma(N_1) - \sigma(M_1)] \\ &\quad + \frac{E_a\varepsilon(M_1) - \sigma(M_1)}{\rho_0 C_V \theta_M}[t(N_1) - t(M_1)] \\ \varepsilon(N_1) - \varepsilon(0) &- \frac{1}{E_a + E_M}[\sigma(N_1) - \sigma(0)] \\ &\quad + \frac{E_a\varepsilon(0) - \sigma(0)}{(E_a + E_M)\theta_M}[t(N_1) - t(0)] = 0 \end{aligned}\right\} \tag{7.23}$$

While at an arbitrary interior point N_2 shown in Fig. 7.35, the particle velocity $v(N_2)$, stress $\sigma(N_2)$, and strain $\varepsilon(N_2)$ can be solved by three characteristic compatibility conditions along the characteristic lines N_1N_2, M_1N_2, and M_2N_2, and the finite difference forms of these conditions can be written as

$$
\left.\begin{aligned}
v(N_2) - v(N_1) &= \frac{1}{\rho_0 C_V}[\sigma(N_2) - \sigma(N_1)] \\
&\quad - \frac{E_a\varepsilon(N_1) - \sigma(N_1)}{\rho_0 C_V \theta_M}[t(N_2) - t(N_1)] \\
v(N_2) - v(M_2) &= -\frac{1}{\rho_0 C_V}[\sigma(N_2) - \sigma(M_2)] \\
&\quad + \frac{E_a\varepsilon(M_2) - \sigma(M_2)}{\rho_0 C_V \theta_M}[t(N_2) - t(M_2)] \\
\varepsilon(N_2) - \varepsilon(M_1) &= \frac{1}{E_a + E_M}[\sigma(N_2) - \sigma(M_1)] \\
&\quad - \frac{E_a\varepsilon(M_1) - \sigma(M_1)}{(E_a + E_M)\theta_M}[t(N_2) - t(M_1)]
\end{aligned}\right\} \tag{7.24}
$$

The above discussion provides the characteristic line numerical method to solve the propagation of viscoelastic waves from the given initial conditions and boundary conditions, that is, to solve the so-called direct problem. A similar method can also be used to solve the first kind of inverse problem, that is, to find the boundary condition from the known results of wave propagation and given initial conditions. For example, suppose the stresses, strains, and particle velocities at points M_2 and N_3 in Fig. 7.35 are known, and the stress, strain, and particle velocity at point M_1 have been given by the initial condition, then the stress, strain, and particle velocity at N_2 can be solved by the following three characteristic compatibility conditions along the characteristic lines N_3N_2, M_2N_2, and M_1N_2, respectively

$$\left.\begin{aligned} v(N_2) - v(N_3) &= \frac{1}{\rho_0 C_V}[\sigma(N_2) - \sigma(N_3)] \\ &\quad - \frac{E_a\varepsilon(N_3) - \sigma(N_3)}{\rho_0 C_V \theta_M}[t(N_3) - t(N_2)] \\ v(N_2) - v(M_2) &= -\frac{1}{\rho_0 C_V}[\sigma(N_2) - \sigma(M_2)] \\ &\quad + \frac{E_a\varepsilon(M_2) - \sigma(M_2)}{\rho_0 C_V \theta_M}[t(N_2) - t(M_2)] \\ \varepsilon(N_2) - \varepsilon(M_1) &= \frac{1}{E_a + E_M}[\sigma(N_2) - \sigma(M_1)] \\ &\quad - \frac{E_a\varepsilon(M_1) - \sigma(M_1)}{(E_a + E_M)\theta_M}[t(N_2) - t(M_1)] \end{aligned}\right\} \tag{7.25}$$

In the case of the split Hopkinson viscoelastic bar, assuming that the specimen is short enough such that the stress distribution is uniform along the length of the specimen, i.e., $\sigma(X_2, t) = \sigma(X_1, t)$, the dynamic stress–strain relationship of the specimen can then be determined by any two measurements of the incident, reflected, and transmitted waves. The whole problem can be summarized in the following four steps:

1. To determine the unknown incident stress $\sigma_I(X_1, t)$ and the particle velocity $v_I(X_1, t)$ at the incident interface X_1 from the incident strain wave signal $\varepsilon_I(X_{G1}, t)$ measured by the strain gauge G1 at X_{G1}.

This step is attributed to solving a direct problem, i.e., to determine the viscoelastic wave propagation (in the positive X direction) from the initial condition and the given strain boundary condition. Then the unknown stress and the particle velocity at the boundary point (X_i, t_j), as shown in Fig. 7.36A, can be determined by Eq. (7.23), or explicitly indicated as

$$\begin{aligned} \sigma(X_i, t_j) &= \sigma(X_i, t_{j-2}) + (E_a + E_M)[\varepsilon(X_i, t_j) \\ &\quad - \varepsilon(X_i, t_{j-2})] + \left[E_a\varepsilon(X_i, t_{j-2}) - \sigma\left(X_i, t_{j-2}\right)\right]\frac{2\Delta t}{\theta_M} \\ v(X_i, t_j) &= v(X_{i+1}, t_{j-1}) + \frac{[\sigma(X_{i+1}, t_{j-1}) - \sigma(X_i, t_j)]}{\rho_0 C_V} \\ &\quad + \frac{[E_a\varepsilon(X_{i+1}, t_{j-1}) - \sigma(X_{i+1}, t_{j-1})]\Delta t}{\rho_0 C_V \theta_M} \end{aligned} \tag{7.26}$$

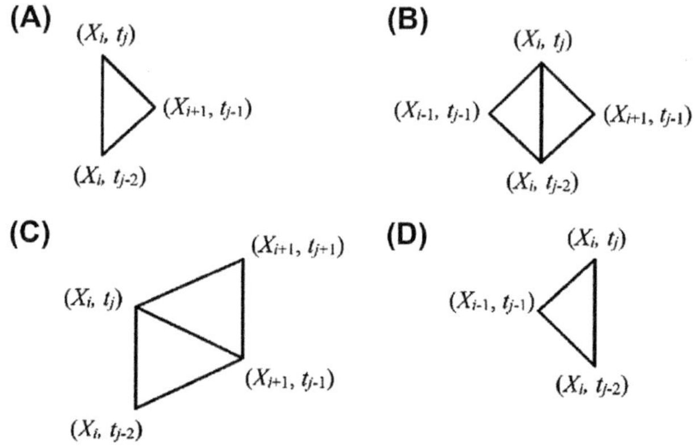

Figure 7.36 The characteristics solutions for different cases in polymer split Hopkinson pressure bar. (A) Boundary point for rightward waves, (B) interior point in a direct problem, (C) interior point in an inverse problem, (D) boundary point for leftward waves.

where $\Delta t = t_j - t_{j-1}$ is the time step in the numerical calculation. On the other hand, as shown in Fig. 7.36B, the unknown stress and particle velocity at an interior point (X_i, t_j) can be determined by Eq. (7.24), and their explicit form can be written as

$$
\left.
\begin{aligned}
\sigma(X_i, t_j) &= \frac{1}{2}\{\sigma(X_{i+1}, t_{j-1}) + \sigma(X_{i-1}, t_{j-1}) + \rho_0 C_V[v(X_{i-1}, t_{j-1}) \\
&\qquad -v(X_{i-1}, t_{j-1})] \\
&\quad +[E_a\varepsilon(X_{i+1}, t_{j-1}) - \sigma(X_{i+1}, t_{j-1}) + E_a\varepsilon(X_{i-1}, t_{j-1}) - \sigma(X_{i-1}, t_{j-1})]\frac{\Delta t}{\theta_M}\Big\} \\
v(X_i, t_j) &= \frac{1}{2}\Bigg\{\frac{\sigma(X_{i+1}, t_{j-1}) - \sigma(X_{i-1}, t_{j-1})}{\rho_0 C_V} + v(X_{i+1}, t_{j-1}) + v(X_{i-1}, t_{j-1}) \\
&\quad + \left(\frac{[E_a\varepsilon(X_{i+1}, t_{j-1}) - \sigma(X_{i+1}, t_{j-1})] - [E_a\varepsilon(X_{i-1}, t_{j-1}) - \sigma(X_{i-1}, t_{j-1})]}{\rho_0 C_V}\right)\frac{\Delta t}{\theta_M}\Bigg\} \\
\varepsilon(X_i, t_j) &= \varepsilon(X_i, t_{j-2}) + \frac{\sigma(X_i, t_j) - \sigma(X_i, t_{j-2})}{E_a + E_M} - \frac{E_a\varepsilon(X_i, t_{j-2}) - \sigma(X_i, t_{j-2})}{E_a + E_M}\frac{2\Delta t}{\theta_M}
\end{aligned}
\right\}
\tag{7.27}
$$

2. To determine the unknown transmitted stress $\sigma_T(X_{G2}, t)$ and the particle velocity $v_T(X_{G2}, t)$ from the transmitted strain wave signal $\varepsilon_T(X_{G2}, t)$ measured by the strain gauge G2 at X_{G2}.

This step is similar to the first step, and Eqs. (7.26) and (7.27) are still suitable for solving the propagation of the viscoelastic wave. Note that this step is a necessary precursor for the next step.

3. To determine the unknown transmitted stress $\sigma_T(X_2, t)$, strain $\varepsilon_T(X_2, t)$, and particle velocity $v_T(X_2, t)$ at the interface X_2 from the transmitted wave signals $\varepsilon_T(X_{G2}, t)$, $\sigma_T(X_{G2}, t)$, and $v_T(X_{G2}, t)$ measured by the strain gauge G2 at X_{G2}.

This step is attributed to solving an inverse problem, i.e., to determine the boundary condition from the initial condition and the known results of the viscoelastic wave propagation at a given point. Thus, the stress, strain, and particle velocity at an interior point (X_i, t_j) shown in Fig. 7.36C can be determined by Eq. (7.25), or explicitly indicated as

$$\left.\begin{aligned}
\sigma(X_i,t_j) &= \frac{1}{2}\{\sigma(X_{i+1},t_{j-1}) + \sigma(X_{i+1},t_{j+1}) \\
&\quad +\rho_0 C_V[v(X_{i+1},t_{j-1}) - v(X_{i+1},t_{j+1})] \\
&\quad +[E_a\varepsilon(X_{i+1},t_{j-1}) - \sigma(X_{i+1},t_{j-1}) + E_a\varepsilon(X_{i+1},t_{j+1}) - \sigma(X_{i+1},t_{j+1})]\frac{\Delta t}{\theta_M}\right\} \\
v(X_i,t_j) &= \frac{1}{2}\left\{\frac{\sigma(X_{i+1},t_{j-1}) - \sigma(X_{i+1},t_{j+1})}{\rho_0 C_V} + v(X_{i+1},t_{j-1}) + v(X_{i+1},t_{j+1})\right. \\
&\quad \left.+\left(\frac{[E_a\varepsilon(X_{i+1},t_{j-1}) - \sigma(X_{i+1},t_{j-1})] - [E_a\varepsilon(X_{i+1},t_{j+1}) - \sigma(X_{i+1},t_{j+1})]}{\rho_0 C_V}\right)\frac{\Delta t}{\theta_M}\right\} \\
\varepsilon(X_i,t_j) &= \varepsilon(X_i,t_{j-2}) + \frac{\sigma(X_i,t_j) - \sigma(X_i,t_{j-2})}{E_a+E_M} - \frac{E_a\varepsilon(X_i,t_{j-2}) - \sigma(X_i,t_{j-2})}{E_a+E_M}\frac{2\Delta t}{\theta_M}
\end{aligned}\right\} \tag{7.28}$$

Thus, $\sigma_I(X_1, t)$, $\varepsilon_I(X_1, t)$, and $v_I(X_1, t)$ at the incident interface and $\sigma_T(X_2, t)$, $\varepsilon_T(X_2, t)$, and $v_T(X_2, t)$ at the transmitted interface of the specimen have all been determined. According to the stress uniformity assumption $(\sigma_I + \sigma_R = \sigma_T)$, the reflected stress $\sigma_R(X_1, t)$ can be immediately determined by

$$\sigma_R(X_1,t) = \sigma_T(X_2,t) - \sigma_I(X_1,t) \tag{7.29}$$

The particle velocity and strain of the reflected wave at the incident interface X_1 are determined by the next step.

4. To determine the reflected particle velocity $v_R(X_1, t)$ and the reflected strain $\varepsilon_R(X_1, t)$ at the incident interface X_1 of the specimen from the reflected stress wave $\sigma_R(X_1, t)$.

This step can be attributed to solving a direct problem of the viscoelastic wave for negative propagation under a given stress boundary condition. Thus, the strain and particle velocity at the boundary point (X_i, t_j) shown in Fig. 7.36D can be determined by Eq. (7.23), but it should be noted that as the wave propagation direction is reversed, the symbols in the equation also need to be correspondingly changed, and they are

$$\left.\begin{aligned}
\varepsilon(X_i, t_j) &= \varepsilon(X_i, t_{j-2}) + \frac{\sigma(X_i, t_j) - \sigma(X_i, t_{j-2})}{E_a + E_M} \\
&\quad - \frac{E_a\varepsilon(X_i, t_{j-2}) - \sigma(X_i, t_{j-2})}{E_a + E_M}\frac{2\Delta t}{\theta_M} \\
v(X_i, t_j) &= v(X_{i-1}, t_{j-1}) + \frac{[\sigma(X_{i-1}, t_{j-1}) - \sigma(X_i, t_j)]}{\rho_0 C_V} \\
&\quad + \frac{[E_a\varepsilon(X_{i-1}, t_{j-1}) - \sigma(X_{i-1}, t_{j-1})]\Delta t}{\rho_0 C_V \theta_M}
\end{aligned}\right\} \tag{7.30}$$

However, for the interior point shown in Fig. 7.36B, it can still be solved by Eq. (7.24) as these equations are derived from the compatible conditions along the rightward and the leftward characteristic lines, and thus are independent of the wave propagation direction.

In this way, we have obtained the incident, reflected, and transmitted stresses and particle velocities at two interfaces of the specimen. According to Eq. (7.1), the dynamic stress, strain rate, and strain of the specimen can then be determined, and finally the dynamic stress—strain relationship of the specimen at high strain rate is determined. For researchers who are not familiar with the stress wave characteristic line, the dispersion effect correction process mentioned above seems rather complicated. In fact, for a given device, a special software can be programmed in practice, which can be easily and repeatedly applied.

Dong et al. (2008) applied this method to the polymer mini-SHTB device to study the dynamic tensile properties of nitrocellulose (the scale of length, width, and height are 4.62 mm, 1.82 , and 0.38 mm). Typical results are shown in Fig. 7.37. The incident waves measured at three different positions of the input bar in sequence and the three corresponding transmitted waves measured on the transmission bar are depicted in Fig. 7.37A. It can be seen that the viscoelastic waves in the polymer bar have evident dispersion and attenuation characteristics. Fig. 7.37B shows the comparison between the corrected dynamic stress—strain curve by the viscoelastic wave and the uncorrected curve, and a significant difference can be found, indicating that the relevant correction must be made when using the viscoelastic pressure/tensile bar. The quasistatic stress—strain curves are also depicted for comparison, and it indicates that the nitrocellulose is highly sensitive to the strain rate.

The above discussion is applicable to the viscoelastic thin rods with negligible transversal (lateral) inertia, so it is only necessary to correct the constitutive viscous dispersion effect of viscoelastic waves. For large-diameter viscoelastic bars, it is also necessary to consider the correction of the geometric dispersion effect as discussed in Section 7.1.3. This research can be referred to the works of Zhao and Gary (1995) and Liu et al. (2002).

It should be pointed out that in addition to low wave impedance, soft materials often have the characteristics of low wave velocity, which directly affects the "uniformity" process. This problem will be further discussed in the next section, namely "Section 7.2 Wave Propagation Inverse Analysis (WPIA) Experimental Technique."

7.2 Wave propagation inverse analysis experimental technique

Fundamentally speaking, the governing equations for studying stress wave propagation in continuum consist of three conservative equations (mass, momentum, and energy conservation equations) and the material constitutive equation. The three conservation equations reflect the universal commonality, and the material constitutive equation reflects the special characteristics of subdisciplines of mechanics. Therefore, the propagation characteristics of stress waves in different media/materials intrinsically depend on and reflect these constitutive relationship of materials. Because stress waves propagate with the constitutive properties of these materials, it is possible, in turn, to deduce the constitutive relationship of materials from a series of stress wave propagation information, which is called "wave propagation inverse analysis," hereinafter called WPIA for short. In mathematical mechanics, to determine the initial and boundary conditions by the stress wave propagation information and the known material

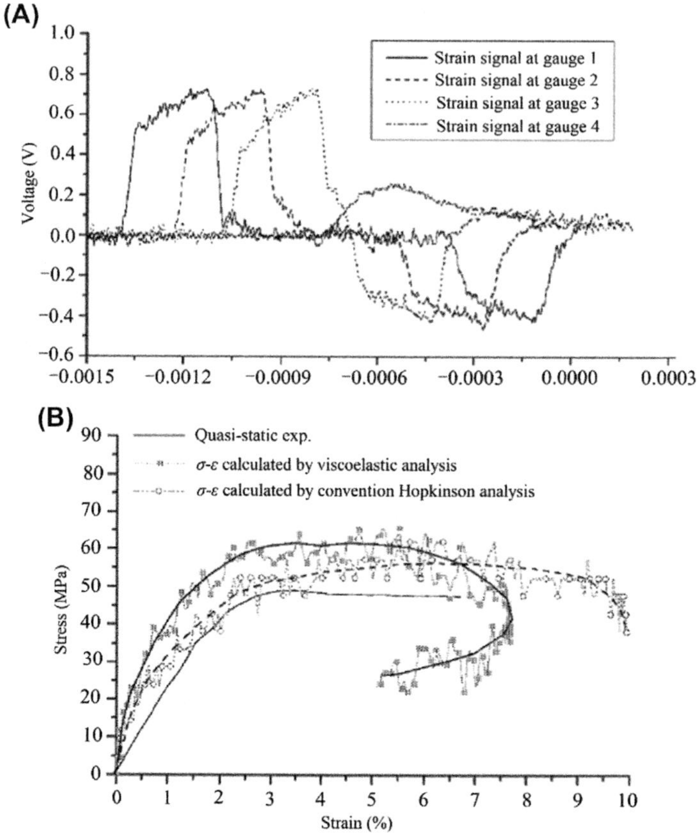

Figure 7.37 The experimental results of nitrocellulose measured by the polymer mini-SHTB device, where solid lines represent quasi-static curve, dotted lines with circles represent dynamic curve analyzed according to elastic bars, and solid lines with boxes represent dynamic curve analyzed according to viscoelastic bars. *From Dong, X.L., Leung, M.Y., Yu, T.X., 2008. Characteristics method for viscoelastic analysis in a Hopkinson tensile bar, Int. J. Mod. Phys. B. 22 (9–11), 1062–1067, Figs. 5 and 6, page 1066–67. Reprinted with permission of the publisher.*

constitutive relationship is considered as solving the "first kind of inverse problem," and to determine the material constitutive relationship by the stress wave propagation information and initial and boundary conditions is considered as solving the "second kind of inverse problem."

In this sense, the problem discussed in the Chapter 4 of the first Part "Dynamic experimental study of the high-pressure equation of state for solids" would be considered as the "second kind of inverse problem," that is, using the shock wave information to determine the equation of state for materials under high pressure, which is primarily limited to 1D strain conditions (plate impact test). In this section, we mainly discuss the inverse analysis of the wave propagation in the bar under 1D stress state.

7.2.1 Taylor bar

Before the adoption of SHPB techniques, Taylor (1948) had developed a simple method to inversely determine the dynamic yield strength of ductile metal materials by measuring the residual deformation of the cylindrical bar after impinging normally onto a rigid target in accordance with Eq. (7.31).

$$\sigma_{yd} = \frac{\rho V^2 (L - x)}{2(L - L_1)\ln(L/x)} \tag{7.31}$$

where L, L_1, and x are the original length of the cylindrical bar, the final length after impact deformation, and the length of the undeformed segment (see Fig. 7.38) in terms of Euler coordinates. Later, researchers also explored the inverse method to determine the constitutive relationship of materials based on the residual deformation distribution.

According to the theory of elastoplastic wave propagation (Wang, 2007), the above Taylor impact problem can be attributed to the 1D impact of a finite bar on a rigid target. At the beginning of the impact, the elastic precursor wave first propagates from the impact interface toward the free end of the bar at an elastic wave velocity $C_0 = (E/\rho_0)^{1/2}$, followed by a series of plastic waves propagating at a slower plastic wave velocity $C_p = \left(\frac{1}{\rho_0}\frac{d\sigma}{d\varepsilon}\right)^{1/2} < C_0$. When the elastic precursor wave reaches the free end, it is reflected at the free end as an unloading wave propagating in the opposite direction and interacts with the forward-propagating plastic waves. Therefore, this is a complex issue that the elastic unloading wave reflected at the free end continuously unloads the plastic loading waves formed at the impact end. It is not difficult to prove that the plastic waves can never reach the free end. In other words, there must be an interface between the plastic zone and the elastic zone in the bar, namely the interface between the residual deformation zone and the undeformed zone after impact. This is the theoretical mechanism of the Taylor impact problem (for more details, see Foundations of Stress Wave, Chapter 4, Section 4.9 (Wang, 2007)).

Evidently, in the 1D elastoplastic wave propagation problem, real-time axial plastic strain distribution directly represents the constitutive relationship of materials. If the constitutive relationship of materials is to be indirectly determined by the radial strain distribution after impact, it would involve the unknown dynamic Poisson's ratio and the unloading constitutive relation of materials. Even if it is only intended to inversely determine from Eq. (7.31) the dynamic yield stress of the material, it is actually difficult to accurately determine the interface position between the plastic zone and the elastic zone in the bar after impact. Especially under a high-speed impact, the impact end of the bar forms a highly localized nonuniform large deformation zone (mushroom head). Therefore, although researchers have made many improvements to the Taylor bar impact test, with the rise of SHPB technique and the following Lagrangian inverse analysis, Taylor bar impact test has gradually lost its appeal.

However, Taylor bar impact experiments have reemerged in recent years, because the nonuniform plastic distribution zone in the Taylor bar test can provide constitutive response information with a large range of strain distribution under the strain rate of the cross order of magnitude. It is very convenient, sensitive, and useful to use it as a verification test for different constitutive models of materials (Field et al., 2004).

7.2.2 The classic Lagrangian inverse analysis

The Lagrangian inverse analysis method (hereinafter called as Lagrangian method for short) was first proposed and developed by Fowles (1970), Cowperthwaite and Williams (1971), and Seaman (1974) in the early 1970s.

The basic idea is that the dynamic stress—strain curve of materials is directly determined by a series of wave propagation signals (such as stress, strain, or particle velocity) measured at different Lagrangian positions of the specimen based on the conservative equations, and the rate-dependent

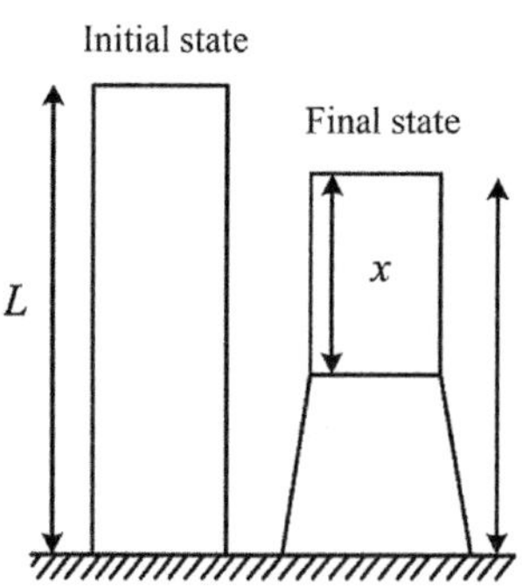

Figure 7.38 The initial and final states of Taylor bar.

constitutive relationship of the material is investigated further. Among many inverse analysis methods, the greatest advantage of this method is that it does not require any constitutive assumptions in advance.

For the 1D stress (or 1D strain) wave, the relationship between the partial derivative of the stress σ and the partial derivative of the particle velocity v is directly connected by the momentum conservative equation (Eq. 7.32), whereas the relationship between the partial derivative of the strain ε and the partial derivative of the particle velocity v is established by the mass conservative equation or the continuity equation (Eq. 7.33)

$$\rho_0 \frac{\partial v}{\partial t} = \frac{\partial \sigma}{\partial X} \tag{7.32}$$

$$\frac{\partial v}{\partial X} = \frac{\partial \varepsilon}{\partial t} \tag{7.33}$$

It can be seen that the relationship between the dynamic stress $\sigma(X, t)$ and the strain $\varepsilon(X, t)$ can be built with the aid of velocity field $v(X, t)$. However, the variables connected by the conservative equations are not the σ, v, and ε themselves but their first-order derivatives. Thus, to obtain the relation between σ and ε themselves, differential and integral operations are inevitable and initial and boundary conditions are required for determining the integral constants.

As such, the solution of the problem will have different degrees of difficulty according to whether a series of measured wave profiles are of stress waves, particle velocity waves, or strain waves.

When a series of stress wave profiles $\sigma(X_i, t)$ are measured at different Lagrangian coordinates $X_i (i = 1, 2, \ldots n)$ using stress (pressure) gauges, the problem is easily solved. The first-order partial derivatives $\partial\sigma/\partial t$ and $\partial\sigma/\partial X$ can be obtained by the numerical differential calculation, and $\partial v/\partial t$ can then be determined with the aid of the momentum conservative equation (Eq. 7.32). Because an initial condition in experiments can often be given as $v = 0$ at $t = 0$, it is not difficult to find $v(X_i, t)$ by integrating $\partial v/\partial t$. Subsequently, the first-order partial derivative $\partial v/\partial X$ can be determined by numerical differential calculation, and $\partial\varepsilon/\partial t$ can be further obtained by the mass conservative equation (Eq. 7.33). Similarly, with the aid of an initial condition $\varepsilon = 0$ at $t = 0$, $\varepsilon(X_i, t)$ can be found by integrating $\partial\varepsilon/\partial t$. Thus, the relationship between $\sigma(X_i, t)$ and $\varepsilon(X_i, t)$ can be finally established.

However, when a series of particle velocity profiles $v(X_i, t)$ measured at different Lagrangian coordinates $X_i(i = 1, 2, \ldots n)$ are used for the Lagrangian analysis, the problem is not so simple. In this case, the first-order partial derivatives $\partial v/\partial t$ and $\partial v/\partial X$ can still be determined by numerical differential calculation, and $\partial \varepsilon/\partial t$ can be obtained by the mass conservation equation (Eq. 7.33), and the strain $\varepsilon(X_i, t)$ is then obtained by the operation of integration and the zero initial condition. However, by the momentum conservation equation (Eq. 7.32), we can only obtain the first-order partial derivative of the stress $\partial \sigma/\partial X$, from which the stress $\sigma(X_i, t)$ to be obtained must require a stress boundary condition (for example, the $\sigma(X_j, t)$ at X_j). This means that a combination of boundary stress $\sigma(X_j, t)$ and particle velocity $v(X_j, t)$ should be measured simultaneously at a certain Lagrangian coordinate X_j, which is challenging.

Similarly, when a series of strain waves $v(X_i, t)$ are measured for the Lagrangian analysis, the problem is more complicated. For determining both the particle velocity by the mass conservation equation and the stress by the momentum conservation equation, there must be corresponding strain and stress boundary conditions (such as $\varepsilon(X_j, t)$ at X_j and $\sigma(X_j, t)$ at X_j) to obtain the integral constant accordingly.

As pointed out by Cowperthwaite and Williams (1971), the inability to simultaneously measure the stress and the particle velocity at more than one Lagrangian position in one test is the key problem.

In the past, even if the combined gauge for simultaneously measuring the particle velocity and the stress has not been solved, people have tried various approximation methods dealing with $v(X_i, t)$, such as the "curve or surface fitting" method, by which the fitting function is expanded into Taylor series, and the parameters are then determined from the measured data. In this method, some assumptions are made about the order of the Taylor series and so on. In essence, some implicit assumptions are also made about the stress boundary condition. Apparently, such assumptions inevitably introduce errors because these attempts merely evade, but do not address the essential problem of the necessary measurements for boundary conditions.

7.2.3 Modified Lagrangian inverse analysis

Generally, the measurement of a series of particle velocity waves $v(X_i, t)$ or strain waves $\varepsilon(X_i, t)$ is more convenient than that of a series of stress wave profiles $\sigma(X_i, t)$ in dynamic tests. When the particle velocity waves $v(X_i, t)$ or strain waves $\varepsilon(X_i, t)$ is measured by n velocity meters or n strain gauges at different Lagrangian coordinates X_i(i = 1, 2, … n), the key is how to determine the boundary condition $\sigma(X_j, t)$ necessary for integrating $\partial\sigma/\partial X$. In this regard, two improved Lagrangian inverse analysis methods have been developed in recent years (Wang et al., 2014).

The first method is to combine the Hopkinson pressure bar technique with the Lagrangian inverse analysis, where the Hopkinson pressure bar technique provides a combined meter that can simultaneously measure $v(X_0, t)$ and $\sigma(X_0, t)$ at the interface X_0 between the pressure bar and the specimen.

The second method is to switch the partial derivative $\partial\sigma/\partial X$ containing variable X in the conservation equation to the partial derivative $\partial\sigma/\partial t$ containing variable t by the total differentiation, and the stress boundary condition required for integration is then transformed into the initial stress condition, while it is usually known in most experiments that the stress has zero initial condition, i.e., $\sigma = 0$ at $t = 0$, such that the problem can be solved easily.

These two modified Lagrangian inverse analysis methods are discussed in detail below.

1. Combination of the Lagrangian method and the Hopkinson pressure bar technique

The schematic of the combination of the Lagrangian method and the Hopkinson pressure bar technique is shown in Fig. 7.39, and the interface between the Hopkinson pressure bar and the specimen is represented by $X = X_0$. As the stress $\sigma(X_0, t)$ and the particle velocity $v(X_0, t)$ can be determined by Eqs. (7.34) and (7.35), where the incident strain wave ε_I and the reflected strain wave ε_R are measured by the strain gauge at $X = X_G$ on the input bar (see Eq. 7.2), the problem of simultaneously measuring the stress and particle velocity at the same Lagrangian position is solved.

$$\sigma(X_0, t) = E[\varepsilon_I(X_G, t) + \varepsilon_R(X_G, t)] \tag{7.34}$$

$$v(X_0, t) = C_0[\varepsilon_I(X_G, t) - \varepsilon_R(X_G, t)] \tag{7.35}$$

In other words, the Hopkinson pressure bar now plays a dual role: it transmits the impact load to the specimen and acts as a "particle velocity stress" combining gauge at the interface between the pressure bar and the specimen (hereinafter the composite meter is simply referred to as "1sv").

(1) 1sv + nv inverse analysis

Hence, if n profiles of particle velocity wave, $v(X_i, t)$, where i = 1,2, …,n, are measured by n velocity gauges mounted at n different Lagrangian positions X_i ($i = 1, 2, \ldots, n$), then by utilizing the "particle velocity—stress" combined gauge at the bar—specimen interface, the difficulty in the lack of stress boundary condition mentioned above can be overcome. Such a new method is called "1sv + nv" inverse analysis method for short (Wang et al., 2011).

In the specific calculus operation, the path-line method, which was first proposed by Grady (1973), is more convenient. As shown in Fig. 7.40, a set of curves of the physical quantity $\phi(X_i, t)$ measured at different Lagrangian points X_i can be simultaneously plotted in the three-dimensional space of ϕ, X, and t, and they can be partitioned by loading, unloading, and characteristic inflection points on the curve. In each area, nodes of each measured curve are selected at equal time interval, and the corresponding nodes on each gauge line are connected by a smooth curve, which is the path line. If there are N nodes on each gauge line, we can connect N path lines (dashed lines in the figure), and these path lines can link the entire mechanical field information. The following is an example of the combination of the Hopkinson pressure bar technique with the Lagrangian analysis based on path-line method.

Once the stress wave $\sigma(X_0, t)$ and the particle velocity wave $v(X_0, t)$ are determined by Eqs. (7.34) and (7.35) with the Hopkinson pressure bar technique, their first-order partial derivatives with regard to time t, $(\partial\sigma/\partial t)_{X0}$, and $(\partial v/\partial t)_{X0}$ can be obtained. Furthermore, $(\partial\sigma/\partial X)_{X0}$ is determined by

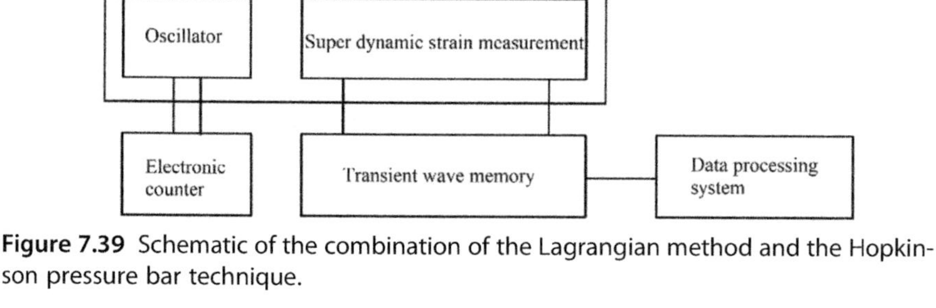

Figure 7.39 Schematic of the combination of the Lagrangian method and the Hopkinson pressure bar technique.

the momentum conservative equation (Eq. 7.32). By using the total differential along the path line, we have the following relationship

$$\left.\frac{d\sigma}{dX}\right|_p = \left.\frac{\partial\sigma}{\partial X}\right|_t + \left.\frac{\partial\sigma}{\partial t}\right|_X \left.\frac{dt}{dX}\right|_p = \left.\frac{\partial\sigma}{\partial X}\right|_t + \left.\frac{\partial\sigma}{\partial t}\right|_X \left.\frac{1}{X'}\right|_p \tag{7.36a}$$

where the subscript p denotes the total derivative along the path line, and $X' = \left.\frac{dX}{dt}\right|_p$ refers to the slope of the path line. Thus, according to the above equation, the stress history curve $\sigma(X_1, t)$ at the point X_1, adjacent to the point X_0 along the path line, can be obtained by the following difference equation

$$\sigma_{i,j} = \sigma_{i-1,j} + \left[-\rho_0 \frac{\partial v_{i-1,j}}{\partial t} + \frac{1}{2}\frac{\partial \sigma_{i-1,j}}{\partial t}\left(\frac{\mathrm{d}t_{i-1,j}}{\mathrm{d}X} + \frac{\mathrm{d}t_{i,j}}{\mathrm{d}X}\right)\right](X_i - X_{i-1}) \tag{7.36b}$$

where the slope of the path line is the average of that at $X = X_0$ and X_1. By analogy, once the stress $\sigma(X_{i-1}, t)$ and the particle velocity $v(X_{i-1}, t)$ at the position X_{i-1} are obtained, the stress $\sigma(X_i, t)$ at the next position X_i can be obtained. This equation can be used not only for loading but also for unloading stage.

On the other hand, when the particle velocity field $v(X_i, t)$ has been detected by the particle velocity meter at n Lagrangian positions $(X = X_i)$, their first-order partial derivative $(\partial v/\partial t)_{Xi}$ as well as the total derivatives along the path lines $(\mathrm{d}v/\mathrm{d}X)_p$ can be directly obtained, and $(\partial v/\partial X)_t$ can then be determined according to the definition of total differential. Therefore, the first-order partial derivative of strain with regard to time, $(\partial\varepsilon/\partial t)_{Xi}$, is obtained by the continuity equation (Eq. 7.33). With the zero initial condition ($\varepsilon = 0$

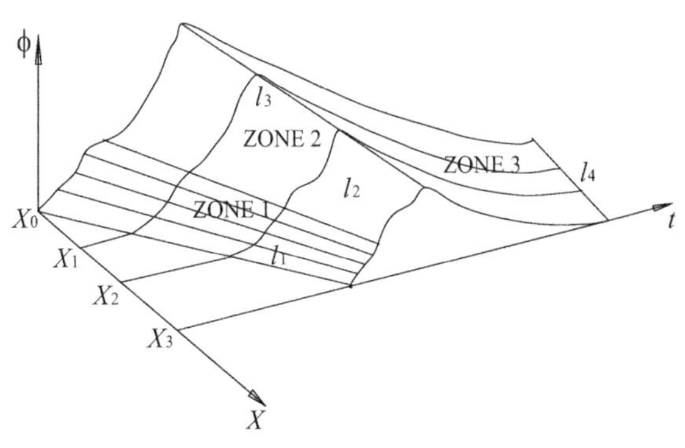

Figure 7.40 Schematics for path-line method.

at $t = 0$), it is not difficult to obtain the strain field $\varepsilon(X_i, t)$ by integrating $(\partial\varepsilon/\partial t)_{Xi}$ with regard to time t.

Through these two steps, the stress field $\sigma(X_i, t)$ and the strain field $\varepsilon(X_i, t)$ of the specimen are obtained, respectively, and a set of dynamic stress—strain curves can then be finally obtained by eliminating the time t.

(2) 1sv + nε inverse analysis

Similarly, if the stress and the particle velocity at the boundary X_0 are obtained by utilizing the Hopkinson pressure bar, and a set of strain waves $\varepsilon(X_i, t)$ are measured by n strain gauges mounted at n different Lagrangian positions, it is also possible to inversely determine the stress field and the particle velocity field at n different Lagrangian positions on the specimen.

When the stress wave $\sigma(X_0, t)$ and the particle velocity wave $v(X_0, t)$ at the interface $X = X_0$ are determined by the Hopkinson pressure bar technique, the partial derivatives $\partial\sigma(X_0, t)/\partial t$ and $\partial v(X_0, t)/\partial t$ can be obtained, and $\partial\sigma(X_0, t)/\partial X$ can be subsequently obtained according to the momentum conservative equation. Thus, according to Eq. (7.36a), the stress $\sigma(X_1, t)$ at the next Lagrangian position along the path line can be obtained. By analogy, once the stress $\sigma(X_{i-1},t)$ and the particle velocity $v(X_{i-1}, t)$ at the X_{i-1} position are obtained, $\sigma(X_i, t)$ can be obtained. Now the problem lies in how to determine $v(X_{i-1},t)$.

Because the strain $\varepsilon(X_i, t)$ at X_i has been measured, the corresponding partial derivative $\partial\varepsilon(X_i, t)/\partial t$ can be obtained by differential operation with regard to time t and the total derivative $d\varepsilon(X_i, t)/dt|_{xi}$ is determined along the path line. Then the partial derivative $\partial\varepsilon(X_i, t)/\partial X$ can be determined according to the definition of total derivative along the path line, and we have

$$\left.\frac{\partial\varepsilon}{\partial X}\right|_{X_i} = \frac{\left.\frac{d\varepsilon}{dt}\right|_{X_i} - \left.\frac{\partial\varepsilon}{\partial t}\right|_{X_i}}{\left.\frac{dX}{dt}\right|_{X_i}} \tag{7.37}$$

Similar to Eq. (7.36b), the strain at X_0, $\varepsilon(X_0, t)$ can be inferred from $\varepsilon(X_1,t)$, $\partial\varepsilon(X_1, t)/\partial t$, and $\partial\varepsilon(X_1, t)/\partial X$ by the following equation

$$\varepsilon(X_0, t) = \varepsilon(X_1, t) + \left[\frac{\partial\varepsilon(X_1, t)}{\partial X} + \frac{1}{2}\frac{\partial\varepsilon(X_0, t)}{\partial t}\left(\frac{dt_{1,j}}{dX} + \frac{dt_{0,j}}{dX}\right)\right](X_0 - X_1) \tag{7.38}$$

After obtaining the strain curve $\varepsilon(X_0, t)$ at the boundary point, we can obtain $\partial\varepsilon(X_0, t)/\partial t$ by differential operation with regard to time t and further determine $\partial v(X_0, t)/\partial X$ according to the continuity equation. So far, the stress, strain, and particle velocity at the boundary point X_0 and their first-order partial derivatives are all obtained.

Similar to Eq. (7.36b), the velocity history curve $v(X_1, t)$ at X_1 can be obtained from $v(X_0, t)$, $\partial v(X_0, t)/\partial t$, and $\partial v(X_0, t)/\partial X$ by the following equation

$$\varepsilon(X_1, t) = \varepsilon(X_0, t) + \left[\frac{\partial\varepsilon(X_0, t)}{\partial X} + \frac{1}{2}\frac{\partial\varepsilon(X_0, t)}{\partial t}\left(\frac{\mathrm{d}t_{1,j}}{\mathrm{d}X} + \frac{\mathrm{d}t_{0,j}}{\mathrm{d}X}\right)\right](X_1 - X_0) \tag{7.39}$$

By analogy, the particle velocity $v(X_i, t)$ at $X = X_i$ can be determined by the following difference equation

$$v_{i,j} = v_{i-1,j} + \left[\frac{\partial v_{i-1,j}}{\partial X} + \frac{1}{2}\frac{\partial v_{i-1,j}}{\partial t}\left(\frac{dt_{i-1,j}}{dX} + \frac{dt_{i,j}}{dX}\right)\right](X_i - X_{i-1}) \tag{7.40}$$

Thus, by eliminating time t from $\sigma(X_i, t)$ and $\varepsilon(X_i, t)$, a set of dynamic stress—strain curves are obtained without any assumptions about constitutive relations and boundary conditions, and because all derivatives are along the path line, no experimental data can be lost.

An example of studying the dynamic mechanical response of polyamide (Nylon) by using the 1sv + nv method is depicted in Figs. 7.41—7.43. Fig. 7.41A shows a set of particle velocity wave profiles measured by the NdFeB high-sensitivity particle velocity meters, and Fig. 7.41B gives the stress profile and the particle velocity profile at the boundary location X_0 measured by the Hopkinson pressure bar technique. By using the 1sv + nv method, the strain profiles and the stress profiles are obtained and shown in Figs. 7.42A,B, respectively. After eliminating the time t, a set of dynamic stress—strain curves of Nylon under the strain rate of 10^2 s^{-1} is obtained, as shown in Fig. 7.43.

2. The Lagrangian method based on zero initial condition (nv + T_0)

When a series of particle velocity waves $v(X_i, t)$ are measured at different Lagrangian positions $X_i (i = 1, 2, \ldots n)$ by n particle velocity meters, the first-order partial derivative of stress with regard to the Lagrangian coordinate X, $\partial\sigma/\partial X$, can be obtained by the momentum conservative equation (Eq. 7.32). However, there must be a stress boundary condition when $\sigma(X_i, t)$ is integrated with X, which requires the "particle velocity—stress" combined gauge. This promotes the combination of the Lagrangian method

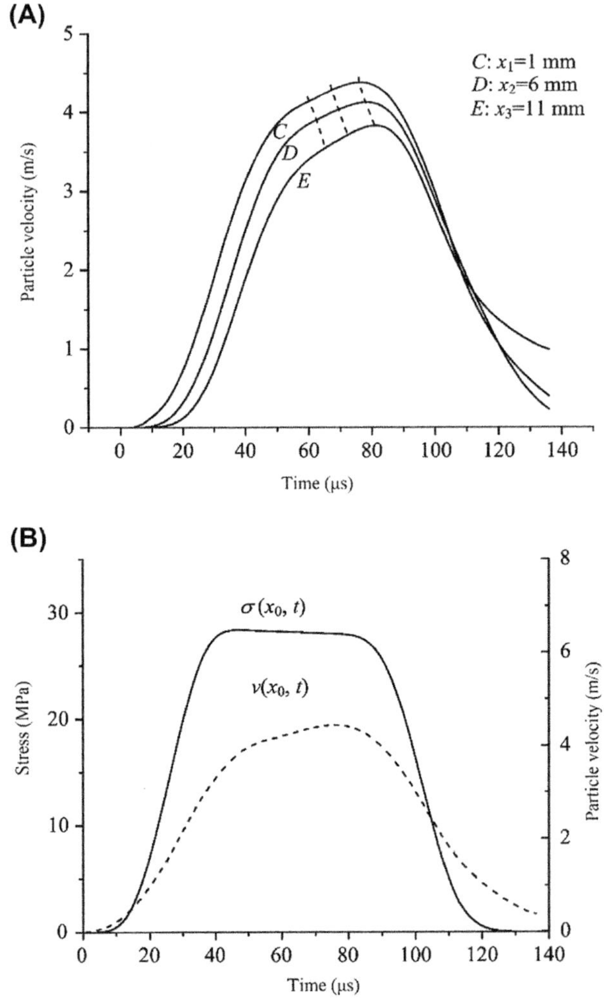

Figure 7.41 (A) Particle velocity wave profiles measured by the electromagnetic method; (B) the stress profile and the particle velocity profile at the boundary location X_0 measured by the Hopkinson pressure bar. *From Lai, H.W., Wang, L.L., 2011. Studies on dynamic behavior of Nylon through modified Lagrangian analysis based on particle velocity profiles measurements. J. Exp. Mech. 26 (2), 221–226 (in Chinese), Fig. 6, page 224.*

and the Hopkinson pressure bar technique as mentioned above. Nonetheless, there is no difficulty to determine the strain wave $\varepsilon(X_i, t)$ from the particle velocity $v(X_i, t)$ by the mass conservative equation (Eq. 7.33) and the zero initial condition ($\varepsilon = 0$ at $t = 0$).

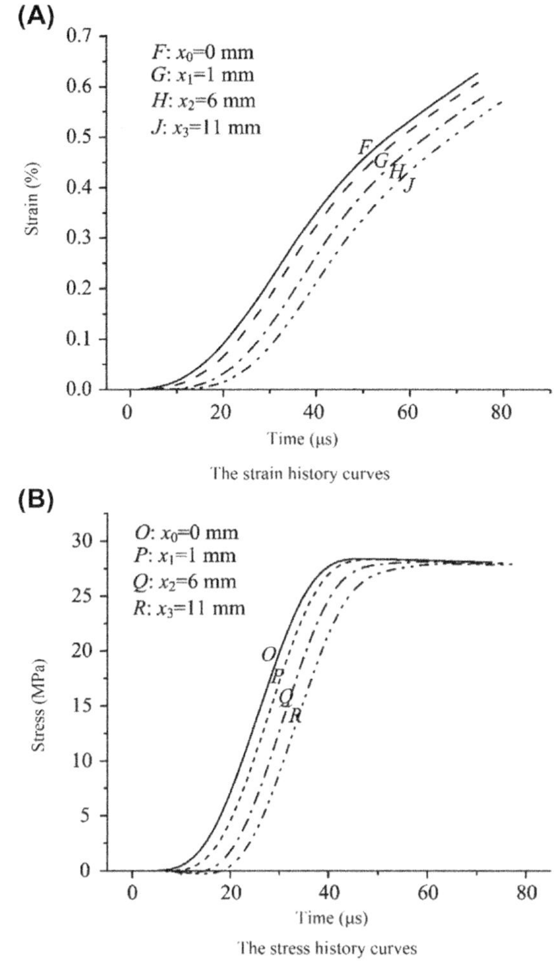

Figure 7.42 (A) The strain profiles and (B) the stress profiles are obtained by using the 1sv + nv method. *From Lai, H.W., Wang, L.L., 2011. Studies on dynamic behavior of Nylon through modified Lagrangian analysis based on particle velocity profiles measurements. J. Exp. Mech. 26 (2), 221–226 (in Chinese), Fig. 7, page 225.*

In fact, there also exist zero initial condition for the stress $\sigma(X_i, t)$, and if $\partial\sigma/\partial X$ in the momentum conservative equation (Eq. 7.32) can be transformed into $\partial\sigma/\partial t$, the problem can also be readily solved. Thus, the "Lagrangian method based on the zero initial condition" (hereinafter called as"nv + T_0" method for short) (Ding et al., 2012; Wang et al., 2013) has been developed.

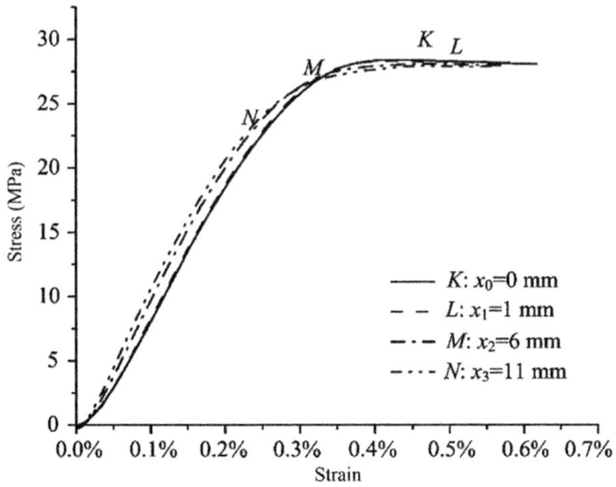

Figure 7.43 A set of dynamic stress–strain curves of Nylon obtained by the 1sv + nv method (strain rate: $10^2\ s^{-1}$). *FromLai, H.W., Wang, L.L., 2011. Studies on dynamic behavior of Nylon through modified Lagrangian analysis based on particle velocity profiles measurements. J. Exp. Mech. 26 (2), 221–226 (in Chinese), Fig. 8, page 225.*

By using the total differential relationship of stress along the path line (Eq. 7.36a), there is

$$\left.\frac{\partial\sigma}{\partial X}\right|_t=\left.\frac{d\sigma}{dX}\right|_p-\left.\frac{\partial\sigma}{\partial t}\right|_X\left.\frac{dt}{dX}\right|_p=\left.\frac{d\sigma}{dX}\right|_p-\left.\frac{\partial\sigma}{\partial t}\right|_X\left.\frac{1}{X'}\right|_p$$

The momentum conservative equation (Eq. 7.32) can then be rewritten as

$$\left.\frac{\partial\sigma}{\partial t}\right|_X=\left(\left.\frac{d\sigma}{dX}\right|_p-\rho_0\frac{\partial v}{\partial t}\right)X'\Big|_p \tag{7.41}$$

For the first path line p_0 which connects the points $(X_i, 0)$ at initial time $t=0$, we have

$$\sigma|_{t=0}=\varepsilon|_{t=0}=v|_{t=0}=\left.\frac{d\sigma}{dX}\right|_{p_0}=0$$

Therefore, for the first path line p_0, we can determine $\partial\sigma/\partial t$. By numerical integral operations, the stress σ on the second path line p_1 and its total derivative along the path line $(d\sigma/dX)_{p1}$ can be determined. By analogy, the stresses on all the path lines can be determined step by step as follows

$$\sigma_{i,j+1}=\sigma_{i,j}+\left(\left.\frac{d\sigma_{i,j}}{dX}\right|_p-\rho_0\frac{\partial v_{i,j}}{\partial t}\right)\left.\frac{dX_{i,j}}{dt}\right|_p(t_{j+1}-t_j) \tag{7.42}$$

As for the determination of the strain field, in addition to the method described in the "1sv + nv" Lagrangian inverse analysis method, a method similar to Eq. (7.42) can also be applied. That is, the strain is determined stepwise with the increasing time t according to the following equation

$$\varepsilon_{i,j+1} = \varepsilon_{i,j} + \left(\frac{dv_{i,j}}{dX} - \frac{\partial v_{i,j}}{\partial t}\frac{dt_{i,j}}{dX}\right)(t_{j+1} - t_j) \tag{7.43}$$

Through these two steps, the dynamic strain field $\varepsilon(X_i, t)$ and the dynamic stress field $\sigma(X_i, t)$ are determined. By eliminating the time t, a set of dynamic stress—strain curves are finally obtained.

An example of studying the dynamic mechanical response of aluminum foam by using the nv + T_0 method is depicted in Figs. 7.44—7.46. During the test, an aluminum foam specimen is launched by the gas gun and impinges onto a Hopkinson pressure bar, as shown in Fig. 7.44. The particle velocity profiles $v(X_i, t)$ at three different Lagrangian coordinates on the specimen are measured by a high-speed camera (FASTCAM-APX RS 250K) combined with digital image correlation technique, as shown in Fig. 7.45A. Thus, according to the nv + T_0 Lagrangian method, the dynamic strain field $\varepsilon(X_i, t)$ and the dynamic stress field $\sigma(X_i, t)$ can be obtained from the measured particle velocity waves, and are shown in Fig. 7.45B,C, respectively. By eliminating the time t, the dynamic stress—strain curve (strain rate is about $10^3\ s^{-1}$) is obtained and shown in Fig. 7.45D. The quasistatic stress—strain curve (strain rate is $10^{-3}\ s^{-1}$) is also given to illustrate the strain rate sensitivity of the specimen.

In fact, the nv + T_0 Lagrangian inverse analysis method does not require the measurement of the stress boundary condition. In Fig. 7.43, the Hopkinson pressure bar placed behind the specimen mainly plays the role of the target, but at the same time it can also serve as a measurer for the boundary

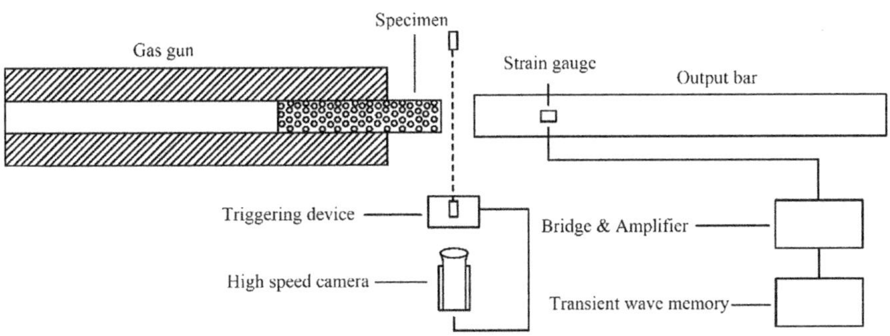

Figure 7.44 Schematic of experimental device for the "nv + T_0" Lagrangian method.

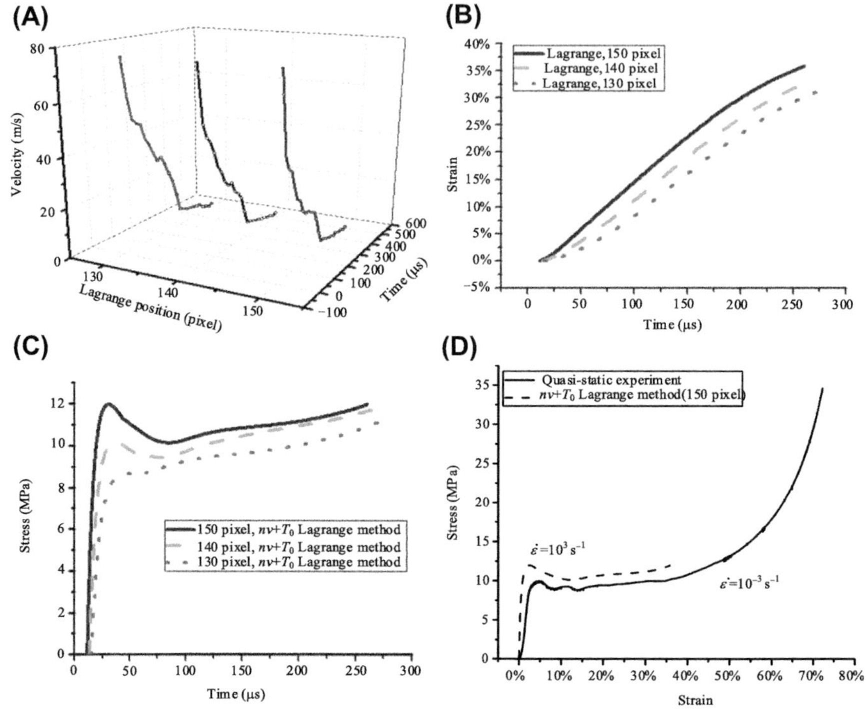

Figure 7.45 The dynamic response of aluminum foam based on the "nv + T_0" Lagrangian method. (A) The particle velocity profiles at different Lagrangian positions; (B) the dynamic strain profiles obtained by the "nv + T_0" method; (C) the dynamic stress profiles obtained by the "nv + T_0" method; (D) the comparison of the dynamic stress—strain curve with the quasistatic one. *From Wang L.-L., Ren H.-Q., Yu J.-L., Zhou F.-H., Wu X.-Y., Tang Z.-P., Hu S.-S., Yang L.-M., Dong X.-L., 2013a. Development and application of the theory of nonlinear stress wave propagation, Chin. J. Solid Mech. 34 (3), 217—240 (in Chinese); Wang, L.L., Ding, Y.Y., Yang, L.M., 2013b. Experimental investigation on dynamic constitutive behavior of aluminum foams by new inverse methods from wave propagation measurements. Int. J. Impact Eng. 62, 48—59, Figs. 3, 5, 6 and 8, page 51—53. Reprinted with permission of the publisher.*

stress at the impact end. The measured boundary stress wave is not indispensable for the $nv + T_0$ method but can be used to check the reliability of the results obtained. The comparison of the boundary stresses obtained by FE method (ABAQUS), where the material model is derived from the dynamic stress—strain curve in Fig 7.45D, and the Hopkinson pressure bar is shown in Fig. 7.46. The boundary stresses obtained from the two methods are in good agreement, verifying the effectiveness of the $nv + T_0$ Lagrangian inverse analysis method.

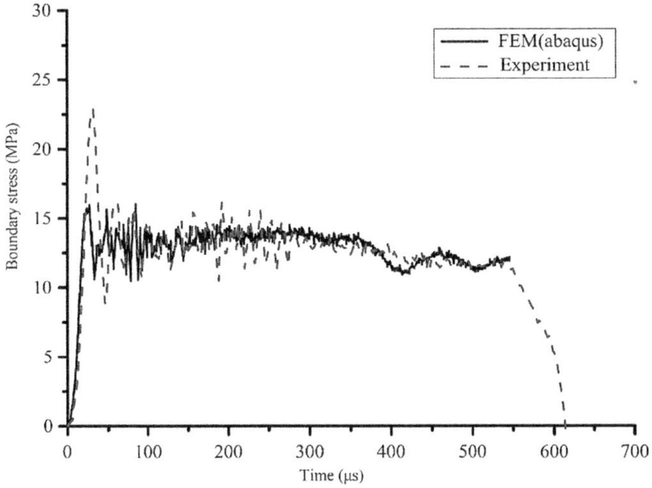

Figure 7.46 The comparison of the boundary stresses obtained by the strain gauge in the experiment and the finite element method based on the dynamic stress–strain curve in Fig. 7.45D. *From Wang L.-L., Ren H.-Q., Yu J.-L., Zhou F.-H., Wu X.-Y., Tang Z.-P., Hu S.-S., Yang L.-M., Dong X.-L., 2013a. Development and application of the theory of nonlinear stress wave propagation, Chin. J. Solid Mech. 34 (3), 217–240 (in Chinese); Wang, L.L., Ding, Y.Y., Yang, L.M., 2013b. Experimental investigation on dynamic constitutive behavior of aluminum foams by new inverse methods from wave propagation measurements. Int. J. Impact Eng. 62, 48–59, Fig. 12, page 56. Reprinted with permission of the publisher.*

In summary, the SHPB technique and WPIA technique each has its own advantages and limitations. The key is to master the stress wave propagation principles of each technique, optimizing their pros and circumventing their cons. A suitable method, or a combination of them, can be selected according to the characteristics of the materials to be measured. The techniques can complement each other, leading to the development of new experimental methods.

PART THREE

Dynamic failure of materials

In general, the dynamic response of materials under explosion/impact loading covers the entire rate–temperature dependent flow/deformation process till final dynamic failure, as shown in Fig. 5.15. The former is characterized by dynamic constitutive relations and is therefore sometimes referred to as constitutive responses of materials, and the latter is related to the dynamic failure criteria of materials.

In the previous two parts, we have discussed the dynamic constitutive relation of materials, in which Part 1 describes the volumetric deformation law and Part 2 is about the distortional deformation law. In this part, the dynamic failure of materials will be discussed.

Compared with the dynamic constitutive relation of materials, more complicated case will be encountered in the discussion of dynamic failure of materials. Especially the following three points should be noticed in advance.

Firstly, dynamic response of materials is often coupled with dynamic response of structure, but the two are different from each other (Wang Lili, 2003). In the previous two parts discussing the material constitutive response, we can often distinguish the material response from the structural response. After separating the structural response, we can "extract" the constitutive response completely belonging to the intrinsic nature of material itself. However, the two responses are usually inseparable in the study of dynamic failure. For example, spalling, which will be discussed in the following chapter (Chapter 8), occurs when the loading part of the incident compression pulse reflects as the unloading wave on the free surface and then interacts with the unloading part of the incident compression pulse to form a tensile stress sufficient to meet the failure criterion. The spalling process is inseparable from the reflection and interaction of unloading stress wave (at this point, it is shown as the dynamic response of structure occurring in the structure) and but on the other side it must meet a failure criterion of materials (at this point, it is shown as the dynamic response of materials occurring at the spalling point). Therefore, it is difficult to say that spalling is simply the dynamic response of materials, and of course it is not simply the dynamic response of structure.

As mentioned in Chapter 7, when studying the dynamic constitutive relationship of materials, the specimen itself, as the research carrier, is a structure with specific design. Because it is difficult to measure at the same time at the same material point all the mechanical quantities required to establish the dynamic constitutive relation of materials, the dynamic response of materials is deduced from the structural dynamic response of the specific structure (specimen) used in the material experiment. For example, in the SHPB test of dynamic behavior of materials, the overall dynamic response of the specimen measured generally includes the coupling of both response of structure and response of material. However, as long as the second basic assumption of the SHPB experiment is satisfied, that is, the stress/strain of the short specimen is uniformly distributed along its length, the effect of stress wave propagation within the specimen (the reflection of the structure dynamic response) can be neglected, therefore we can "extract" the alone material dynamic response from the overall dynamic response of the tested specimen. However, once the dynamic failure occurs in the process of test, as the failure is often localized, and it is impossible to simultaneously destroy everywhere along the length of the specimen, this must make it difficult to continue to meet the second basic assumptions of the SHPB experiment. In such case, the overall dynamic response of specimen (structure) is no longer the alone material dynamic response.

In fact, even in the quasistatic uniaxial tensile test, if localized necking occurs in the specimen before failure, it will subsequently cause stress concentration and the transformation of stress state from uniaxial tension to triaxial one, because then the data measured have included effect of structure factors and no longer represent the alone material response under uniaxial stress.

It can be seen that in the study of dynamic failure of materials, once the occurrence and development of failure involve the stress wave propagation and interaction in the specimen structure, once the failure occurs in the specimen structure in the form of localization, it is no longer a simple problem of material dynamic response, but the coupled structure dynamic response must be taken into account, which greatly complicates the problem. The dynamics of cracks and the dynamics of fragmentation which will be introduced in Chapter 9 are all such kind of complicated problems of dynamics.

Secondly, in the previous Part 1 and Part 2 concerning the dynamic constitutive relation of materials, we have generally assumed that the volumetric deformation law does not include irreversible viscoplastic flow/deformation, and conversely, the distortional deformation law of materials is not affected by the hydrostatic stress (spherical stress). It means that the volumetric deformation law and the distortional deformation law are decoupled (Wang Lili, 2003). However, in the study of dynamic failure of materials, a large number of experimental observations show that the spherical stress (hydrostatic pressure or hydrostatic tension) generally plays an important role in the process of materials failure on the basis of distortion deformation. This can be illustrated vividly by the bell-like fracture surface in the principal stress space suggested by Liu Shuyi (1954), as shown in Fig. III-1.

In this figure, the Mises yielding surface (and the surface of subsequent yield) is represented as a cylindrical surface with axis of line (111) which is isoclinic to the three principal stress axes (see the cylinder surface (1) in the Fig. III.1), indicating that the yield is independent on the hydrostatic

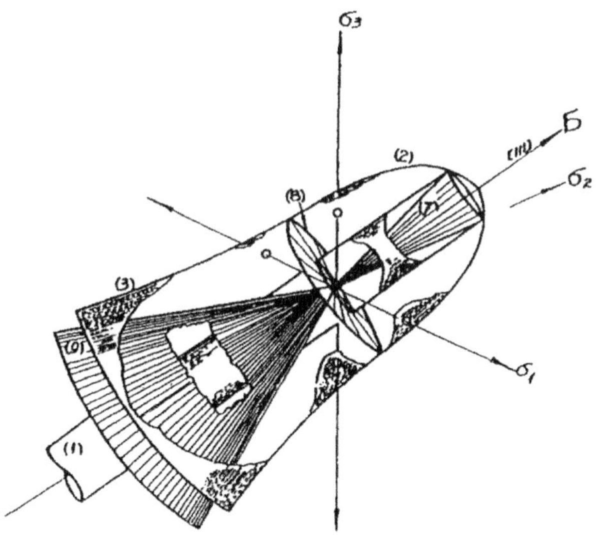

(1) Yield cylinder (Hencky-Mises), (2) and (3) Liu's bell-like rupture surface,

(7) Brittle fracture cone, (8) plane of pure shear, (9) Liu's non-fracturing cone

Fig. III.1 Mises yielding cylinder and Liu's bell-like fracture surface in principal stress space (Liu Shuyi, 1954).

stress, and reflecting the assumption of the decoupling of the volumetric deformation law and the distortional deformation law. The stress states within the Mises yielding cylindrical surface are all in the elastic state. In the hydrostatic tension quadrant of the stress space, the increase of hydrostatic spherical tension would lead to the deformation process directly transited from elastic deformation to fracture, the stress trajectory will fall on the surface of brittle fracture cone within the Mises cylinder (the cone (7) in the figure), corresponding to the brittle fracture without plastic deformation. On the contrary, in the stress space outside the brittle fracture cone, the deformation process will first undergo plastic deformation (arriving and then passing through the Mises yielding cylinder) and then transition to failure, and the stress trajectory will fall on the Liu's bell-like rupture surface (the elliptical bell-like rupture surface (2) and (3) in the Fig. III.1). This corresponds to ductile rupture undergoing plastic deformation. As the spherical stress varies from hydrostatic tension to hydrostatic pressure (bounded by the pure shear π plane (8) in the figure), the bottom of Liu's bell-like rupture surface (the single leaf hyperbolic-type or bell-like rupture surface (3) in the figure) enlarges with the increase of hydrostatic pressure, which means that the plasticity of materials increases with the increase of hydrostatic pressure and the failure of materials becomes less and less likely to occur. In the hydrostatic pressure quadrant of the stress space, the line (111) never intersects with the Liu's bell-like rupture surface, which means that in the triaxis constant pressure under hydrostatic pressure, the material has no shear strain and in principle could withstand an extreme hydrostatic pressure without failure. There is an asymptotic cone (the conical surface (9) in the figure) within the hyperbolic-type or Liu's bell-like rupture surface (the bell-like rupture surface (3) in the figure), which called Liu's nonfracturing cone. In principle, as long as the hydrostatic pressure component is large enough and its stress trajectory falls in the stress space below the Liu's nonfracturing cone, failure of the material will not occur. As for the characteristics of the Liu's brittle fracture cone surface and the Liu's bell-like rupture surface as well as the Liu's nonfracturing cone, such as its shape and size, respectively, depend on different materials, and like that of Mises yield cylinder must be determined by experiments.

In the following discussion concerning the spalling under the state of one-dimensional strain and the unstable propagation of crack under plane strain condition, it will be seen that the triaxial hydrostatic tension component plays a role in promoting dynamic failure.

Thirdly, the dynamic failure of materials, regardless of the specific mechanism, is essentially a time/rate-dependent process of occurrence and development, which could be, for example, characterized by the intrinsic failure characteristic time T_F of materials. In the case of quasistatic loading, T_F is negligible compared with the time scale T_L used to characterize the variation of external load with times. Thus, the quasistatic failure can often be regarded as time/rate independent and will occur as soon as the failure criterion is met. However, for a short pulse and high strain rate loading such as explosion/impact loading, T_F is not negligible compared with the time scale T_L characterized by the short duration of the dynamic loading and must be considered in accordance with the time/rate-dependent process.

It is further recognized that, from a mesoscopic perspective, the dynamic failure of materials is a time-dependent process in which the different forms of meso-damage (such as meso-cracks, meso-voids, meso-shear bands, etc.) evolve at a finite rate. Therefore, from the perspective of mechanism, the study of dynamic failure of materials cannot be separated from discussions of the dynamic evolution laws of meso-damage (Wang Li-Li and Dong Xin-Long, 2004). Moreover, the initiation and growth of meso-damage itself will induce secondary stress waves, which can have a feedback effect on the evolution of meso-damage and thus affect the process of dynamic failure.

The complexity of the problem lies in that, on the one hand, meso-damage develops with the flow/deformation process of the material, and the evolution of meso-damage depends on the material constitutive mechanical variables such as stress, strain, strain rate, etc. On the other hand, damage evolution will in turn affect the mechanical behavior of materials, including the apparent constitutive relationship. Thus, the study of the rate-dependent constitutive relation taking account of dynamic damage evolution under high strain rate and the study of dynamic failure criterion are often mixed, interrelated, and intercoupled with each other, which has become one of important research topics, common concerned by the scientists of mechanics and material science (Wang Lili et al., 2010).

Taking these characteristics into consideration, the following topics will be discussed in three chapters:

Chapter 8, firstly taking spalling as an example, discusses the unloading failure caused by unloading waves. Under quasistatic load, the failure usually occurs in the loading stage, which belongs to the so-called loading failure. However, the dynamic process characterized by stress wave propagation includes both loading wave propagation and unloading wave propagation.

Therefore, the unloading failure associated with interaction of unloading wave will occur in certain conditions, which is one of the special forms of dynamic failure.

Chapter 9 focuses on the dynamics of cracks, discussing the unstable extension and dynamic propagation of an existing crack under dynamic loading. Furthermore, the problem of fragmentation will be discussed based on dynamic evolution of multisource cracks.

Chapter 10 firstly focuses on adiabatic shearing, one of the most typical meso-damages of dynamic failure of materials, including the interaction between shear bands and cracks. Furthermore, the dynamic evolution law of meso-damage in general forms and the rate-dependent macroscopic constitutive model taking account of dynamic evolution of continuous damage as well as other related problems are discussed.

CHAPTER EIGHT

Spalling and other unloading failure

8.1 Spalling

8.1.1 Phenomenon of spallation

Spalling (or scabbing) is a typical phenomenon of dynamic failure, which was first observed by B. Hopkinson (1914) in the steel plates subjected to the contact explosion (Hopkinson, 1910, 1914); it is therefore sometimes called **Hopkinson fracture**. In the paper, Hopkinson had indicated clearly that the main features of spalling are: (1) spall fracture does not occur at the contact explosion surface (front surface) between the explosive and the steel plate, but near the free surface (back surface). (2) spalling is actually caused by a short-duration tensile loading, although the steel plate is originally subjected to an explosion pressure pulse, which is attributed to the unloading rarefaction wave formed by the reflection of pressure pulse on the free surface of steel plate. (3) the spalling piece or the so-called scab flies away at high speed, indicating that the steel plate has previously been subjected to explosive compressive waves loading and thus stored a considerable amount of energy, (4) the original ductile steel plate displays a brittle characteristic during spalling, showing a crystalline cleavage fractographic feature. These phenomena, which can still be distinguished between the spalling and the failure under quasistatic loading, are the characteristics of **dynamic unloading failure**. This is also hard not to deeply admire B. Hopkinson, who had such penetrating insights even in more than 100 years ago. It was not until 1949 that the spalling phenomenon was subsequently studied in some depth by many other researchers.

Fig. 8.1 shows a typical spalling phenomenon of metallic targets (Rinehart, 1951, 1952), where Fig. 8.1A shows the spalling on the free surface of a 24S-T4 aluminum alloy plate, and Fig. 8.1B shows multilayer spalling on the longitudinal section of a mild steel plate.

As an example of spalling phenomenon in nonmetallic materials, Fig. 8.2A and B (Dong et al., 2009) show the spalling fracture on the free surface of thick concrete slabs under contact explosion, and Fig. 8.2C is a

Dynamics of Materials
ISBN: 978-0-12-817321-3
https://doi.org/10.1016/B978-0-12-817321-3.00008-5

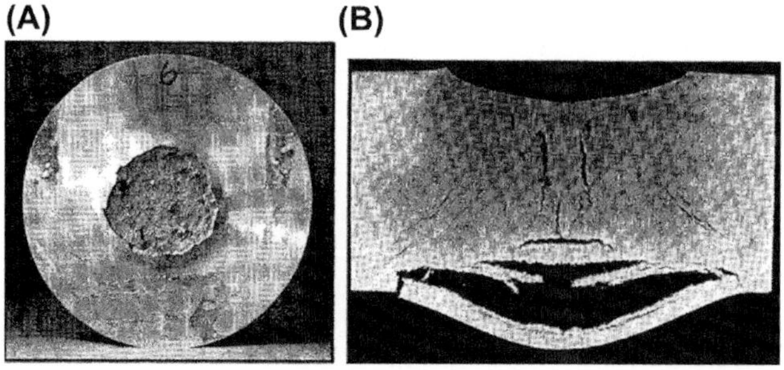

Figure 8.1 Typical spalling photos of metallic materials. *From Rinehart, J.S., 1951. Some quantitative data bearing on the scabbing of metals under explosive attack. J. Appl. Phys. 22 (5), 555–560., Fig. 5, p.557; 1952, Fig. 1, p.1229. Reprinted with permission of the publisher.*

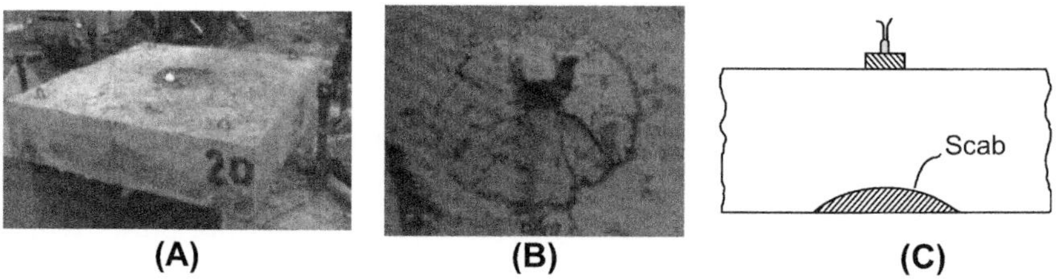

Figure 8.2 (A) Test of thick concrete slabs under contact explosion, (B) spalling on free surface of specimen, (C) schematic of experiment setup (Dong et al., 2009).

schematic of the experimental setup. Fig. 8.3 (Dong et al., 2004, 2009) shows that the one-dimensional stress spallation of a long cement bar specimens occurs when one end of the specimen is loaded by the input bar of a Hopkinson device; the incident pressure wave is reflected as an unloading wave at the other end (free end) and then superposes with the incident unloading wave leading to different forms of spalling, including e.g., the single-layer spalling (Fig. 8.3A) and the double-layer spalling (Fig. 8.3B).

Spalling is related to the propagation and interaction of unloading rarefaction waves. Therefore, the strict definition of spallation should be: when two unloading rarefaction waves interact, a tensile region will be generated within the structure, where the spalling fracture will occur once the material dynamic tensile failure criterion is met.

One of the typical experimental configurations used to study spallation is the plate impact experiment under one-dimensional strain condition (see Chapter 4 and Fig. 8.4) by impacting of a flyer plate against a target

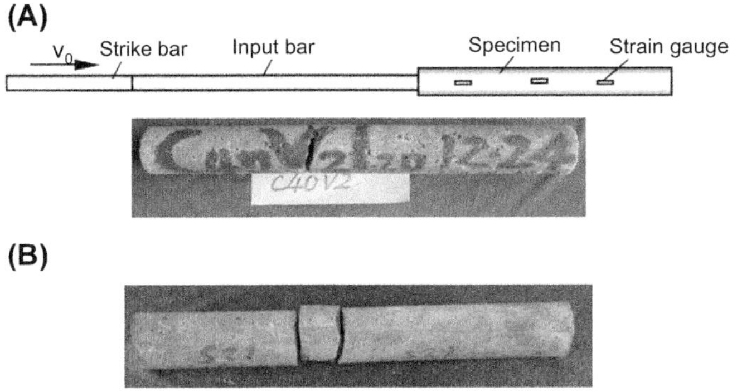

Figure 8.3 Spalling test of long cement rod with Hopkinson technique shows (A) single-layer spalling and (B) double-layer spalling (Dong et al., 2004, 2009).

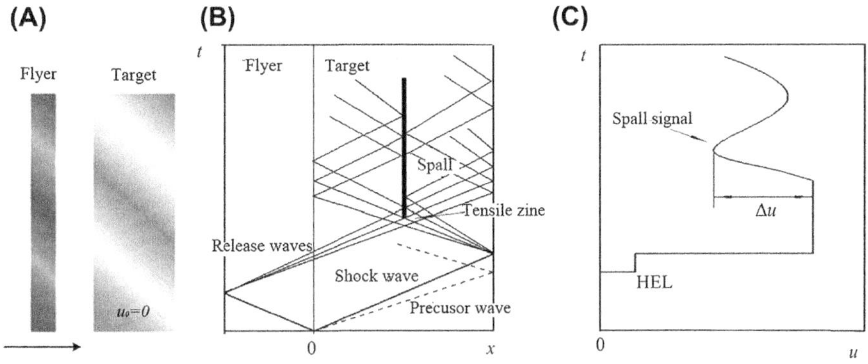

Figure 8.4 Schematic and wave dynamics of plate impact experiment.

specimen. When the flyer launched by a gas gun and strikes the stationary target at a high velocity (see Fig. 8.4A), two compression waves are generated at the interface and propagate in the flyer leftward and in the target rightward, respectively, as shown in Fig. 8.4B. It is worth noting, however, that those compressive waves contain, strictly speaking, the elastic precursor waves propagating with the fastest wave velocity (as shown by the dotted line in Fig. 8.4B), while their amplitudes are much weaker than that of the plastic shock waves behind them, thus are generally neglected in the spalling analysis. Once the two compressive waves propagate to the free surfaces of the flyer and the target, respectively, they are reflected into an unloading rarefaction wave propagating in a fan-like pattern in $x-t$ plot. In case these two waves meet head-on and superpose at a certain position inside the target (sample), a tensile stress zone with a certain high amplitude and high strain rate will be generated. For typical plate impact tests, the strain rate can reach about $10^4-10^6\ s^{-1}$. Once the dynamic tensile failure criterion is satisfied, spalling will occur at this location, as shown in Fig. 8.4B. The curve of particle velocity (u) versus time (t) at the free surface of the specimen is usually measured experimentally, as shown in Fig. 8.4C. The characteristic of the $u-t$ curve could be generally used to distinguish whether spallation occurs in the target and the drop of the free surface velocity from the peak value Δu is used to study its spalling strength of the tested material accordingly (see details below).

Therefore, it is not difficult to understand that the spalling scab will fly away with the momentum trapped therein, which happen to the contact explosion of the metal plate in Fig. 8.1 and the thick concrete plate in Fig. 8.2, also in the long cement rod subjected to impact load at one end in Fig. 8.3, where spalling or scabbing occurs.

It is not hard to imagine that once the first spalling occurs, a new free surface will also be formed at the same time. The subsequent incident pressure wave will be reflected on the new free surface, which may possibly cause the second spallation. By analogy, multiple spalling can be formed under certain conditions, resulting in a series of multiple scabs, as shown in Fig. 8.1B and Fig. 8.3B.

Rinehart (1952) found when the incident pulse is a "triangular" pulse, i.e., it has a steep shock wave front and a wave tail that decays immediately, and the pulse amplitude is several times higher than the critical spall strength σ_c of material, the multilayered spalling could be formed, as shown in Fig. 8.5.

In Fig. 8.5B, the case (a) shows the situation of an incident "triangular" pulse with an amplitude σ_0 before being reflected from the free surface; the case (b) shows the situation of "triangular" pulse when the first spalling piece formed after the incident pulse reflected from the free surface, where the amplitude of the incident pulse has been reduced to σ_1 ($=\sigma_0-\sigma_c$). In this way, the amplitude of the incident pulse decreases by σ_c in turn as each spalling layer formed. If the initial amplitude (σ_0) of the "triangular" pulse is n times greater than the critical spalling strength (σ_c) of materials, i.e., $\sigma_0 > n\sigma_c$, n spalling pieces will be formed. Of course, it has been assumed here that the occurrence of spalling depends on the critical spalling strength (σ_c) of materials (detailed below).

It should be emphasized that it does not necessarily immediately form a tensile wave when the pressure wave is reflected on the free surface, and whether a tensile stress occurs in the specimen depends on the result of

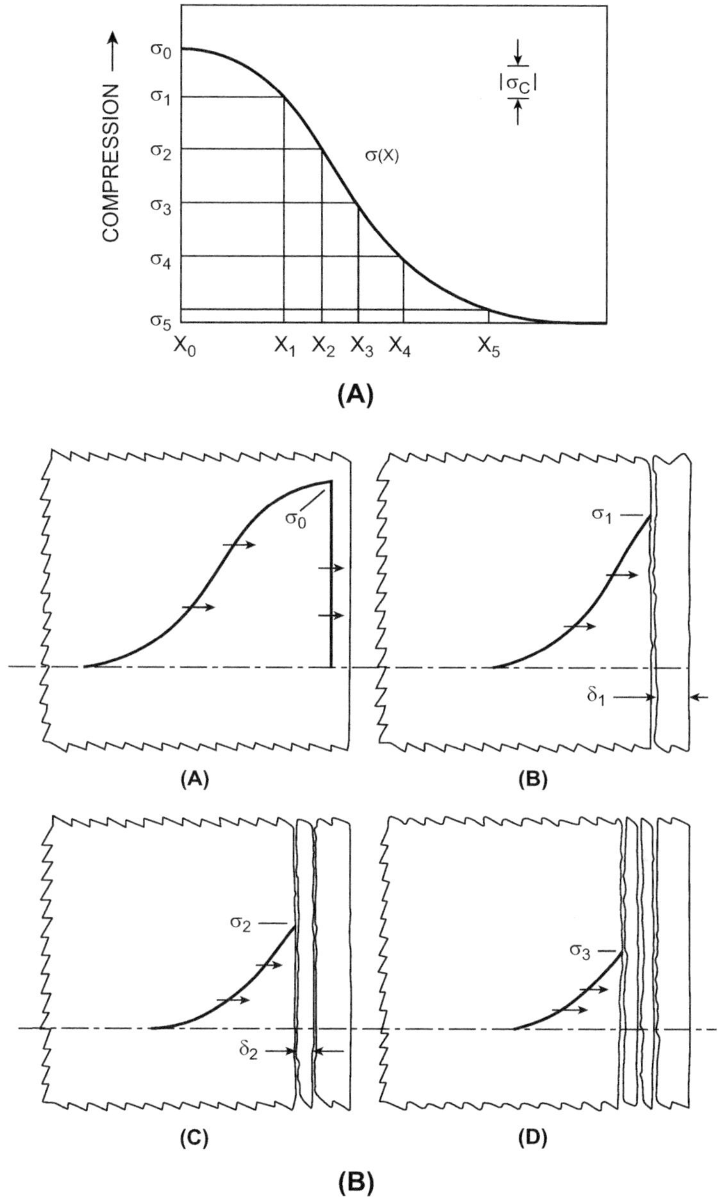

Figure 8.5 Development of reflection of a "triangular" pulse on the free surface to form multilayers spallation. *From Rinehart, J.S., 1952. Scabbing of metals under explosive attack: multiple scabbing. J. Appl. Phys. 23 (11), 1229–1233., Fig. 2 and Fig. 3, p.1130. Reprinted with permission of the publisher.*

interaction between the reflected unloading wave and the rarefaction wave tail of the original incident pulse continuing to incident. An incident one-dimensional pressure pulse is normally composed of a compressive loading part at the head of the pulse and a subsequent unloading tail part. Most engineering materials can undergo relatively strong compressive stress waves without fracture, but cannot undergo tensile stress of the same intensity. The occurrence of spalling is attributed to the interaction between the part of unloading rarefaction wave of the incident compressive pulse and the unloading wave formed by the incident pulse reflected on the free surface, or in short, the spallation is caused by the interaction between the incident unloading wave and the reflected unloading wave.

Similar to the spallation near the back surface previously discussed, where the tensile stress is generated due to the incident unloading wave tail interacting with a reflection unloading wave under the one-dimensional incident longitudinal wave propagating normally to the free surface, when a spherical (or cylindrical) compressive stress wave propagates toward a corner enclosed by two free surfaces in a certain angle, the unloading rarefaction waves reflected by the two free surfaces will also form a tensile stress and may lead to fracture, which is called **corner spalling** (Rinehart, 1975), as shown in Fig. 8.6. Furthermore, if the unloading rarefaction waves reflected by the spherical or cylindrical pressure stress waves on the two free surfaces meet at the center of the body, it may lead to a so-called **central spalling** (Rinehart, 1975), as shown in Fig. 8.7. Either the back spalling or the corner fracture and the central fracture are all related to the reflective unloading phenomenon of stress waves, which can be collectively referred to as **reflective unloading failure**.

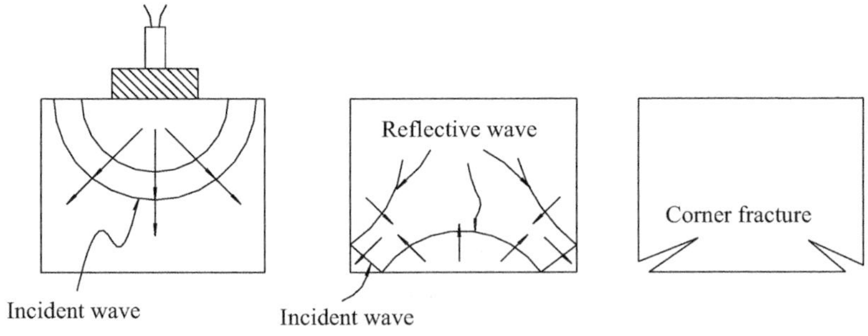

Figure 8.6 Illustration of corner spalling caused by unloading rarefaction waves reflected by two free surfaces.

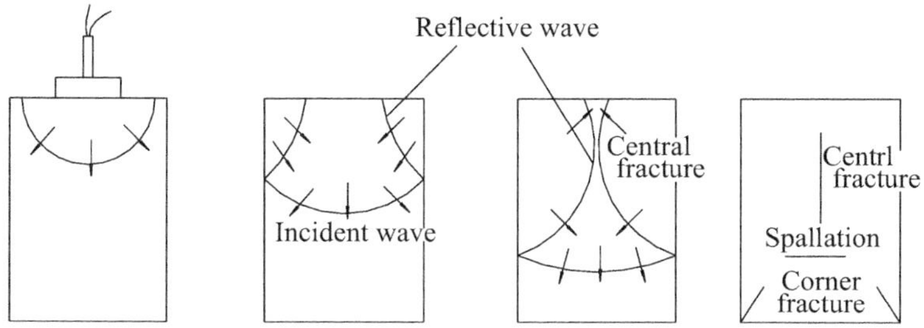

Figure 8.7 Illustration of central fracture caused by two unloading rarefaction waves reflected by free surface.

It should be noted that when a pressure pulse propagates obliquely to a free surface, the interaction between the reflected unloading wave and the unloading wave at the tail of the incident pressure pulse will also form tensile stress as discussed above for normally incident plane wave. However, the direction of the maximum principal tensile stress is related to the angle of oblique incidence. Fig. 8.8 shows a schematic diagram of the formation of spalling in isotropic materials due to the reflection of obliquely incident pressure longitudinal pulses (P waves) on free surfaces (Rinehart, 1975). The spalling will develop with the reflection process of incident pulse, and the orientation of the microcrack is related to the incident angle.

In addition, phase transformation spalling may occur in materials that undergo a shock-induced phase transformation. The pioneering researches on phase transformation spalling were carried out by Ivanov and Novikov (1961) and Erkman (1961) for iron and steel with shock-induced phase transition. As stated in the Section 4.6 "shock phase transformation" of Chapter 4, iron undergoes a phase transformation from α-phase (body-centered

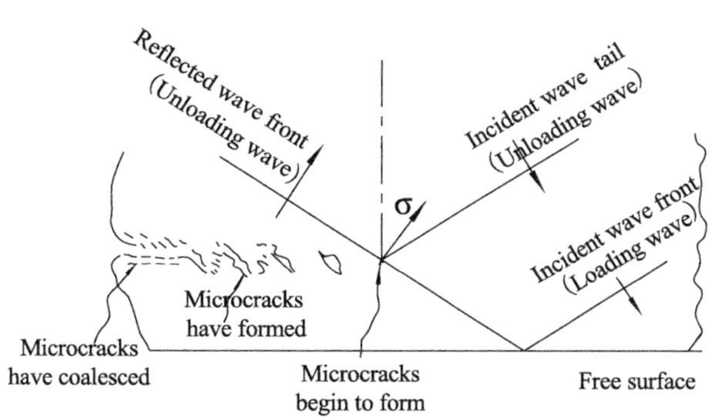

Figure 8.8 Schematic of spalling by an obliquely incident pulse (p wave) reflecting on the free surface.

cubic crystal) to ε-phase (closely packed hexagonal crystal) at 13 GPa pressure (corresponding to the point B in Fig. 8.9). If the pressure of the explosion/shock loading is higher than 13 GPa (but lower than the pressure at the point D in Fig. 8.9), a double shock wave structure will be formed, namely, the first shock wave propagating with a faster wave velocity and the subsequent second shock wave propagating with a slower wave velocity. The wave velocity of the former is determined by the slope of the shock adiabat (Rayleigh chord) AB, and the wave velocity of the latter is determined by the slope of the shock adiabat (Rayleigh chord) BC in Fig. 8.9. If the pressure of the explosion/impact loading is higher than the pressure at the point D in Fig. 8.9 (called overdrive pressure and where the point D is on the extension line of the straight line AB), the shock will change to single waveform again.

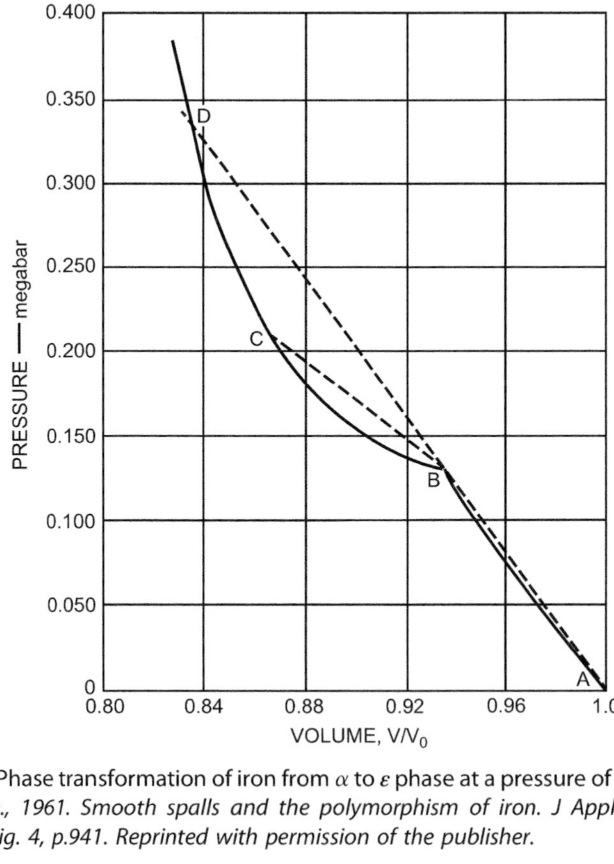

Figure 8.9 Phase transformation of iron from α to ε phase at a pressure of 13 GPA. *From Erkman, J.O., 1961. Smooth spalls and the polymorphism of iron. J Appl Phys. 32 (5), 939–944., Fig. 4, p.941. Reprinted with permission of the publisher.*

According to the fundamental theory of shock wave stated in Chapter 4, the propagation characteristics of unloading rarefaction wave, which corresponding to the unloading part of pressure pulse, are first determined by the compression isentropic CE shown in Fig. 8.10. The velocity of unloading rarefaction wave decreases with the decrease of pressure. After that a shock wave will be formed due to a reversible phase transition $\varepsilon \rightarrow \alpha$ at the pressure of 9.8 GPA, i.e., at the point E in Fig. 8.10, which is the so-called unloading shock waves. This shock wave velocity is determined by the slope of Rayleigh chord EA.

Ivanov and Novikov (1961) and Erkman (1961) pointed out that when an unloading shock wave, which is formed by an incident shock wave with pressure larger than 13 GPA reflected on the free surface, interacts with the unloading shock wave at the tail of the incident pressure pulse wave, a sudden rise in tensile stress pulse will occur in a very narrow region. Therefore, the spall fracture region is very localized and smooth, resulting in smooth fracture morphology called smooth spall.

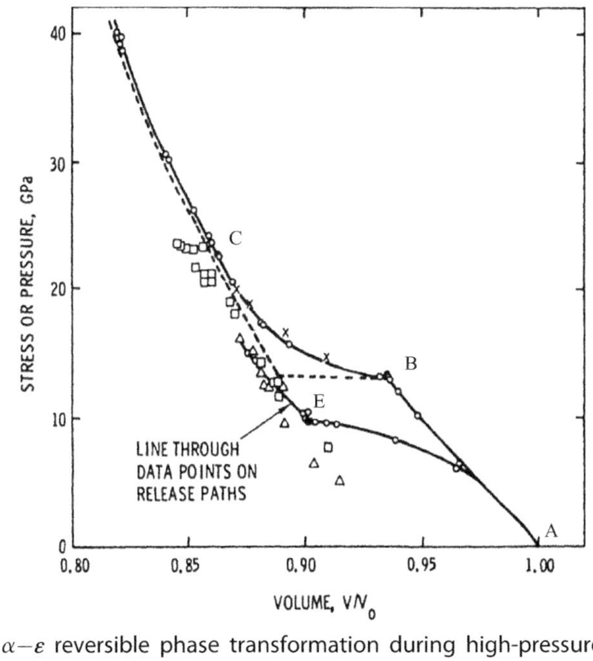

Figure 8.10 $\alpha-\varepsilon$ reversible phase transformation during high-pressure loading and unloading for iron. *From Duvall, G.E., Graham, R.A., 1977. Phase transitions under shock-wave loading. Rev. Mod. Phys. 49 523., Fig. 18, p. 542. Reprinted with permission of the publisher.*

Based on characteristic line method, the distance—time plot of wave propagation calculated numerically by Erkman (1961) is shown in Fig. 8.11. As can be seen from the plot, the incident wave system contains two loading shock waves and a rarefaction (unloading) shock wave. The first shock has a higher velocity than that of the second shock; they are subsequently followed by a rarefaction (unloading) shock with a slower velocity. When the first shock wave propagates to the free surface, it is reflected leftward firstly as a series of rarefaction unloading waves in fan shape (in x—c_0t plot) with different velocities, and then the reflection rarefaction (unloading) shock wave is formed. The location where it meets the incident

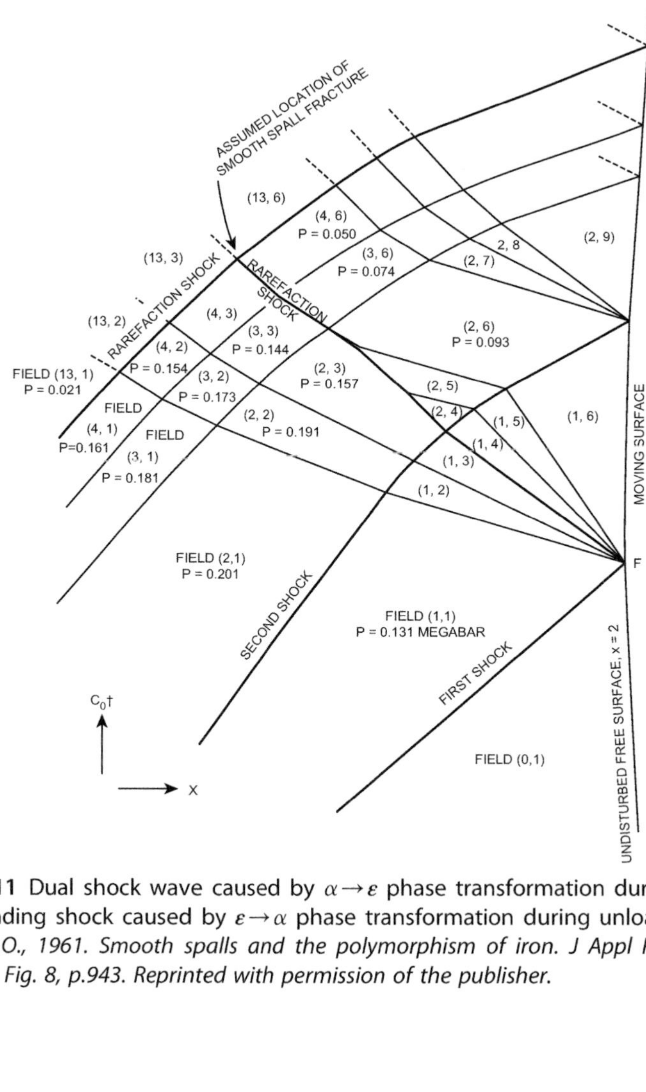

Figure 8.11 Dual shock wave caused by $\alpha \rightarrow \varepsilon$ phase transformation during loading and unloading shock caused by $\varepsilon \rightarrow \alpha$ phase transformation during unloading. *From Erkman, J.O., 1961. Smooth spalls and the polymorphism of iron. J Appl Phys. 32 (5), 939–944., Fig. 8, p.943. Reprinted with permission of the publisher.*

rarefaction shock wave traveling rightward is where the smooth spall fracture occurs.

On the other side, in case of the loading pulse pressure being below the phase transformation pressure (<13 GPa), there is no phase transformation in irons and steels; the tensile stress increases more gradually, allowing the initiation and propagation of spall over a wider region, and thus results in a rough fracture morphology, called **rough spall**.

As a comparison, Fig. 8.12 and Fig. 8.13 show the micrographs of the etched cross-section of the smooth spall and the rough spall in 4340 steel under a shock of 15 GPa (with phase transformation) and 10 GPa (without phase transformation), respectively, (Zurek et al., 1992).

In addition to Armco irons and steels, in which the unloading rarefaction shock wave due to the $\alpha \rightarrow \varepsilon$ phase transformation under the loading shock higher than 13 GPa will result in phase transformation spall, Dremin et al. (1992) found that in 18–8 austenitic stainless steels, the martensitic transformation occurs in the tensile stress region, which will induce stress relaxation effect and thus cause an abnormal increase of the spalling strength.

It should be pointed out that although spalling occurs locally in a body within a very short time, it is related to a process of damages evolution. A large number of experimental observations and numerical simulations provide information which indicates that the spall process is related to the

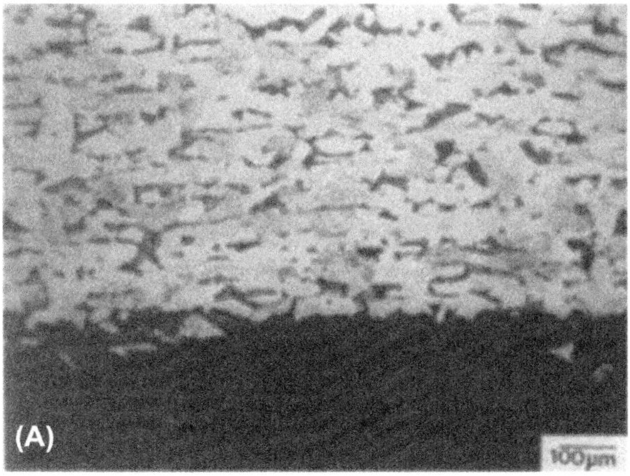

Figure 8.12 Smooth spall in AISI 1010 steel shown here as an etched section under a pressure of 125 GPA (with phase transformation). *From Meyers, M.A., Aimone, C.T, 1983. Dynamic fracture (spalling) of metals. Prog. Mater. Sci. 28 1., Fig. 48, p.69. Reprinted with permission of the publisher.*

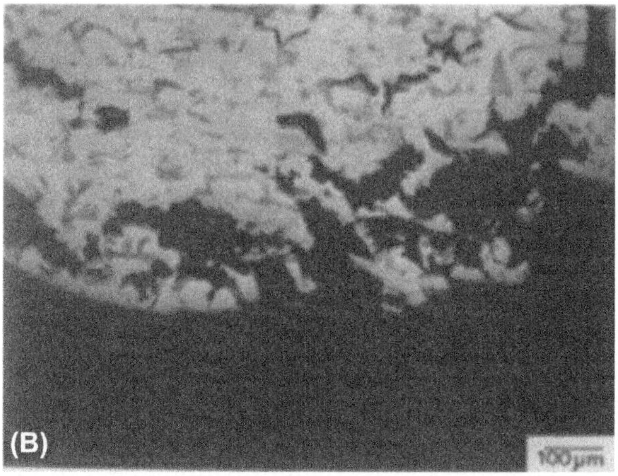

Figure 8.13 Rough spall in AISI 1010 steel shown here as an etched section under a pressure <13 GPA (without phase transformation). *From Meyers, M.A., Aimone, C.T, 1983. Dynamic fracture (spalling) of metals. Prog. Mater. Sci. 28 1., Fig. 48, p.69. Reprinted with permission of the publisher.*

nucleation, evolution, and coalescence of microdamages. For brittle spalling, microcracks are the main form of microdamage, while for ductile spalling, microvoids are the main form of microdamage. As an example of the ductile spalling, the microdamage evolution of the metal target impacted by a flyer at high velocity is given in Fig. 8.14 (Molinari et al., 2014). Experimental observations show how microvoids are nucleated along the spall plane and undergo a rapid development with increasing impact velocity v, and finally spalling occurs at $v = 306$ m/s. It can be seen that the spalling is related to the mechanisms of nucleation, growth, and coalescence of the microdamage under the interaction of the related reflective rarefaction waves.

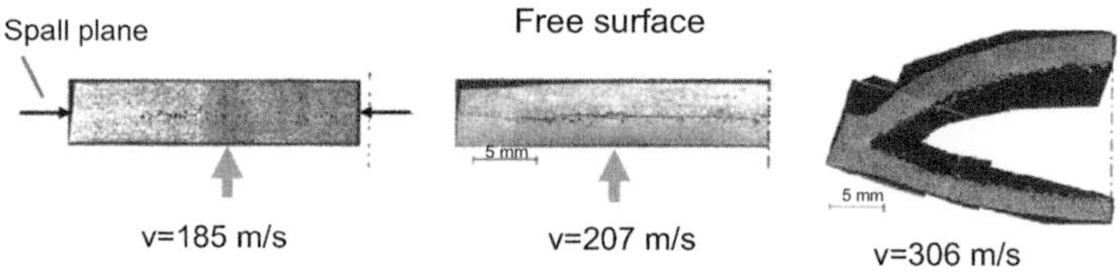

Figure 8.14 Development of microvoids along the spall plane shown damages increasing with the impact velocity *v*. *From Molinari, A., Mercie, S., Jacques, N., 2014. Dynamic failure of ductile materials, 23rd International Congress of Theoretical and Applied Mechanics. Procedia IUTAM. 10 201–220. Fig. 5, p.209. Reprinted with permission of the publisher.*

In order to quantitatively assess the damage degree caused by the evolution of spalling, Oscarson and Graff (1968) of Sandia National Laboratory suggested that a small piece could be cut out from the spalling damaged zone as a static tensile specimen; the damage degree could be judged according to the measured residual tensile strength, as shown in Fig. 8.15. The results show that the higher the impact pressure is, the greater the tensile stress generated in the unloading wave interaction zone is and the smaller the residual strength is, and it means that the greater the damage degree is. In Fig. 8.15, point A is the starting point of the drop in residual strength, and its corresponding stress σ_A is called the initial threshold stress of spalling. Point B is the end point of the decrease in the residual strength, after which the residual strength will no longer decrease even if the shock load increases. The corresponding σ_B is called the intermediate threshold stress of spalling. The macroscopic cracking is observed at point C, and the corresponding stress σ_C is called the threshold stress of completed spall. Of those three spalling threshold stresses, σ_A is the most significant; therefore, in the early days, σ_A is usually used as a main judgment for the occurrence of spalling.

More discussions on dynamic damage evolution would be set forth in the Chapter 10 of this book.

8.1.2 Spalling criterion

In the previous section, the general qualitative description of spalling phenomena was introduced, while the core question arises as to the quantitative spalling models, i.e., under what kind of critical condition does spall fracture occur, that is, the study of spalling criterion.

Rinehart (1951) first proposed maximum tensile stress criterion to describe dynamic spalling quantitatively. His research work is of milestone significance, although the criterion has not been taken into consideration that spalling is a process related to time or strain rate. Whiteman (1962)

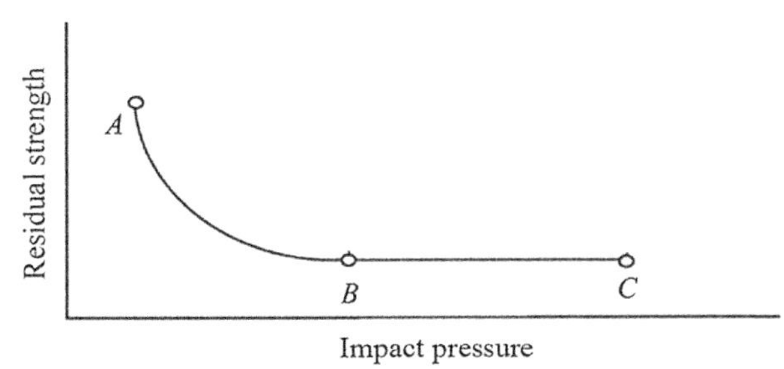

Figure 8.15 Residual strength as a function of impact pressure.

and Skidmore (1965) suggested a tensile stress-rate criterion where the time/strain-rate—dependent correlation of spalling is taken into account. Breed et al. (1967) developed the tensile stress-gradient criterion based on Whiteman's stress-rate criterion. Thurstan and Mudd (1968) further improved the Breed's stress-gradient criteria. Tuler and Butcher (1968) realized that the dynamic failure of materials should be related to the accumulation and development of damage in materials and put forward the damage accumulation criterion. Since then, a series of studies on the damage evolution of dynamic failure have arisen, which has become the main current research area of spallation.

The following is a brief introduction to the above representative macroscopic (engineering) spalling criteria, respectively.

8.1.2.1 Maximum normal tensile stress criterion

The maximum normal tensile stress criterion for spall fracture was proposed by Rinehart (1951). It is assumed that spall occurs instantaneously when the normal tensile stress σ reaches or exceeds a critical value σ_c, namely

$$\sigma \geq \sigma_c \tag{8.1}$$

where σ_c is a material constant characterizing the resistance to dynamic fracture, called dynamic fracture strength. It seems as a formal extension of the maximum normal stress criterion under quasistatic loading condition and to the dynamic case. It is assumed that the fracture happens instantaneously once the criterion is satisfied, which is a type of time/rate-independent fracture model. However, the σ_c here is determined by the dynamic spalling test and is usually higher than the quasistatic strength limit σ_b. In this sense, it implies that the time/rate-dependent correlation of fracture is taken into account.

Rinehart conducted spall tests by plate impacts on several materials; the measured dynamic fracture strength σ_c for five metals is listed in Table 8.1. It is worth noting that the different values of σ_c were measured for the same material but with different thickness of specimens, respectively. Rinehart pointed out that such difference of σ_c can be explained in terms of the dependence of spall strength on different stress states. The deformation and failure characteristics of materials should be related to the stress triaxiality. Differing from the strength under the uniaxial tension state, the stress triaxiality of plate impact test is related to the specimen thickness. The results mean that the stress state has a noticeable influence on spall strength σ_c. This is consistent with the discussion in Fig. III.1 at the beginning of part 3.

Table 8.1 The measured dynamic fracture strength σ_c for five metals.

Material	Thickness of plates		σ_c	
	in	mm	kip	MPa
24S-T4 aluminum alloy	1.5	38.1	140	965
Copper	1.5	38.1	430	2965
	2	50.8	400	2758
Brass	1.5	38.1	310	2137
1020 Steel	2	50.8	230	1586
	2.5	63.5	130	896
4130 Steel	1.5	38.1	440	3034

When σ_c reaches the cohesive stress (theoretical ultimate tensile strength) of crystals, the material breaks immediately without the process of damage accumulation (McQueen and Marsh, 1962; Zhu Zhao-xiang et al., 1981), which is called cohesive stress criterion or ultimate spall. This means that the tensile shock generated by the interaction of unloading waves is large enough to directly abrupt the bond of crystals. Therefore, such spall fracture can only occur under the action of extremely high-stress pulses.

8.1.2.2 Tensile stress-rate criterion and tensile stress-gradient criterion

The maximum normal tensile stress criterion is a model of instantaneous fracture, i.e., spalling is time/rate independent. Whiteman (1962) proposed a tensile stress-rate criterion, which is an extension of Rinehart's criterion, to take into account of the time/rate effect on spall fracture, i.e., the spalling strength is proportional to the square root of the stress rate $\frac{\partial \sigma}{\partial t}$.

$$\sigma = \sigma_0 + A\left(\frac{\partial \sigma}{\partial t}\right)^{\frac{1}{2}} \tag{8.2}$$

where σ_0 is a tensile stress threshold, which is usually taken as the quasistatic tensile ultimate strength σ_b of materials and A is the material constant.

Breed et al. (1967) proposed a tensile stress-gradient criterion based on Whiteman's stress-rate criterion.

$$\sigma = \sigma_0 + B\left(\frac{\partial \sigma}{\partial X}\right)^{1/2} \tag{8.3}$$

where B is the material constant. Noting $\Delta X = U\Delta t$, we have $\frac{\partial \sigma}{\partial X} = \frac{\partial \sigma}{U \partial t}$, where U is the unloading wave velocity. So Eqs. (8.3) and (8.2) are actually equivalent. However, Breed et al. believed that the form of Eq. (8.3) is more convenient for computer calculation and processing.

Breed et al. (1967) conducted spall tests on 1100-s aluminum and copper; the experimental results, as shown in Fig. 8.16, show that the predicted curves by Eq. (8.3) are in good agreement with the experimental data.

Thurstan and Mudd (1968) summarized a number of previous experimental data of spalling and found that the power exponent of stress gradient

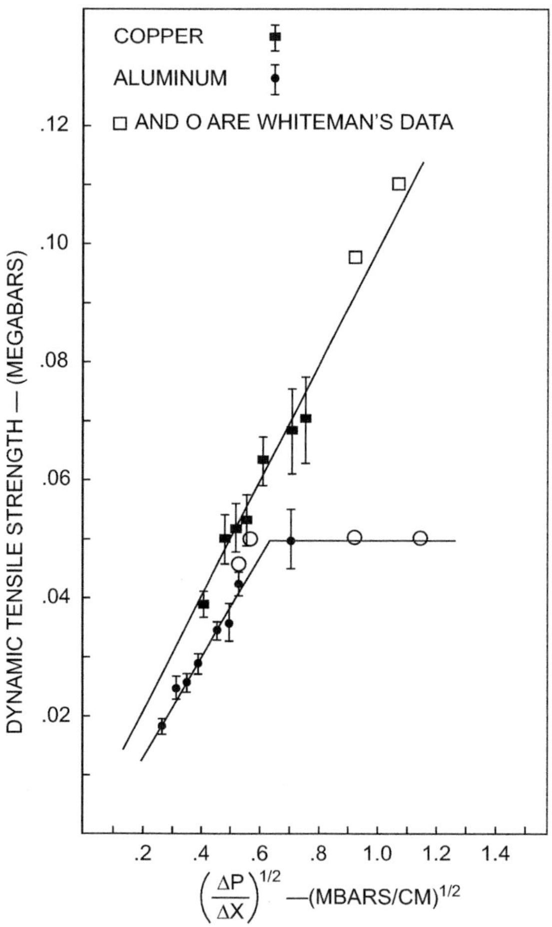

Figure 8.16 Comparison of test results and tensile stress-gradient model (Eq. (8.3)) for aluminum and copper. *From Breed, R.B., Mader, C.L., Venable, D., 1967. Technique for the Determination of Dynamic Tensile Strength Characteristics. J. Appl. Phys. 38 (8), 3271–3275., Fig. 4, p.3273. Reprinted with permission of the publisher.*

in formula (8.3) is not necessarily limited to 1/2; it can be expressed more generally as follows.

$$\sigma = \sigma_0 + B\left(\frac{d\sigma}{\partial X}\right)^n \tag{8.4}$$

where n is the material constant. This means that there is a linear relationship between $\ln(\sigma - \sigma_0)$ and $\ln\left(\frac{\partial\sigma}{\partial X}\right)$, and its slope n could be taken as different values for different materials. Eq. (8.4) is a general form of stress-gradient criterion.

8.1.2.3 Damage accumulation criterion

As shown in Fig. 8.14, the spall fracture actually does not occur instantaneously but rather develops at a finite speed, showing a significant fracture hysteresis especially at high strain rates. The occurrence of fracture is not only related to the applied stress amplitude, but also to the duration of the applied stress. Therefore, an instantaneous fracture criterion (e.g., Eq. 8.1) should be replaced by the time-dependent dynamic fracture criterion.

Tuler and Butcher (1968) proposed a damage accumulation criterion in integral form, in which the continuous measurement of dynamic damage accumulation and the lower bound condition of macroscopic spall fracture are given. As a function of time t, the stress $\sigma(t)$ should satisfy the following equation.

$$K(t) = \int_0^t (\sigma(t) - \sigma_0)^\alpha dt = K_c \tag{8.5}$$

where α, K_c, and σ_0 are the material constants, σ_0 denotes the lower boundary stress (threshold stress) required for fracture. If $\sigma(t) < \sigma_0$, no fracture will occur even if the action lasts longer. When $\alpha = 1$, Eq. (8.5) is reduced to the so-called momentum condition for spalling, which means that the fracture occurs when the stress impulse reaches a certain value. When $\alpha = 2$, Eq. (8.5) is equivalent to the so-called energy criterion of spalling.

If the evolution rate of the phenomenological macroscopic continuous damage D is introduced (regardless of its definition and microscopic mechanism) with respect to time t, $\dot{D}\left(\equiv\frac{\partial D}{\partial t}\right)$ and it is assumed to be proportional to $(\alpha(t) - \sigma_0)^\alpha$.

$$\dot{D}(t) \propto (\sigma(t) - \sigma_0)^\alpha \tag{8.6a}$$

then Eq. (8.5) is equivalent to the following critical damage rule:

$$D(t) = \int_0^t \dot{D}dt = \int_0^t (\sigma(t) - \sigma_0)^\alpha dt = D_c \tag{8.6b}$$

It can be seen that the K_c in Eq. (8.5) corresponds to the critical damage D_c.

Li Yong-chi et al. (2003) concretized the macroscopic continuous damage D as the ratio of the void damage volume V_d to the total volume V, i.e., it is defined as:

$$D(t) = \frac{V_d(t)}{V(t)} = \frac{V_d(t)}{V_s(t) + V_d(t)} \tag{8.7a}$$

where $V_s(= V - V_d)$ is the volume of solid substance, then the evolution rate of D with respect to time t, $\dot{D}\left(\equiv \frac{\partial D}{\partial t}\right)$ is:

$$\dot{D} = \frac{\dot{V}_d V_s - \dot{V}_s V_d}{(V_s + V_d)^2} \approx \frac{\dot{V}_d V_s}{(V_s + V_d)^2} \tag{8.7b}$$

The last approximate equal sign of the above equation has neglected small quantities of higher order term $\dot{V}_s V_d$. $\dot{V}_d$ could be further classified as the sum of nucleation rate $\dot{V}_n$ and growth rate $\dot{V}_g$.

$$\dot{V}_d = \dot{V}_n + \dot{V}_g \tag{8.7c}$$

And similar to Eq. (8.6a), it is assumed that the nucleation rate $\dot{V}_n$ and growth rate $\dot{V}_g$, respectively, obey the following rules:

$$\dot{V}_n = N\left(\frac{\sigma}{\sigma_0} - 1\right)^\alpha V_s \tag{8.7d}$$

$$\dot{V}_g = G\left(\frac{\sigma}{\sigma_0} - 1\right)^\alpha V_d \tag{8.7e}$$

where N is the nucleation characteristic frequency and G is the growth characteristic frequency. Substituting Eqs. (8.7c) and (8.7e) into Eq. (8.7b) and further considering Eq. (8.7a), the damage evolution equation can be obtained as follows:

$$\dot{D} = \left[N(1 - D)^2 + GD(1 - D)\right]\left(\frac{\sigma}{\sigma_0} - 1\right)^\alpha \tag{8.8a}$$

Similar to Eq. (8.6b), the corresponding critical condition for spalling is:

$$D = D_c \tag{8.8b}$$

Li Yong-chi et al. (Cao and Li, 2006) further proved that by taking different specific histories of tensile stress, Eq. (8.8) can be reduced, respectively, to those spalling criteria discussed above, such as the stress-rate criterion, stress-gradient criterion, and Tuler—Butcher integral damage accumulation criterion. In other words, either the stress-rate criterion, stress-gradient criterion, or the Tuler—Butcher integral damage accumulation criterion can be regarded as a special simplified case of Eq. (8.8) derived from the damage evolution, all of them have implicitly the meaning of damage evolution.

The evolution of damage will be discussed in more detail in Chapter 10, where adiabatic shear bands, one of the important forms of dynamic mesoscopic damage, will be first discussed, and then the more general rules of damage evolution will be discussed in succession.

8.1.3 Experimental measurement of spalling strength

The experimental study about spalling involves two aspects: loading method and measurement diagnostics technique. Loading methods are mainly the flyer plate impact driven by light-gas gun and the chemical explosion loading (see Section 4.3 in Chapter 4). Recently, a variety of innovative loading techniques such as laser loading, laser-driven flyer impact, and particle beam loading have been introduced. In plate impact tests (see Fig. 8.4), in which the flyer plate impacts a stationary target (sample), axial compression plane waves under one-dimensional strain condition (three-dimensional stress state) are produced to load the sample. According to impact speeds experimentally required, the flyer can be launched by light-gas guns (one stage and two-stage gas guns), explosive plane lenses, lasers, etc. The plate impact test can be classified into symmetrical collision and asymmetrical collision according to the different materials of the target (sample) plate and the flyer plate. The spallation experiments performed with the flyer of the same material as the target is called symmetric impact, while the other cases are asymmetric impact.

The diagnostics techniques used to measure spalling mainly include: (1) velocity interferometer system for any reflector (VISAR) is used to measure the particle velocity histories $u_{fs}(t)$ on the free surface of the sample, by which we could distinguish whether the spalling occurs or not, and calculate the spalling strength, the thickness of the scabbing, the damage evolution, and other relevant information; (2) the soft-recovered samples are dissected to analyze the mark, microstructure, and macro- and micro-spall appearance on the cross-sections by means of optical microscopy, scanning electron microscopy, transmission electron microscopy, etc.; (3) pressure sensors used to measure the stress history $\sigma_{int}(t)$ at the interface of sample window, when a low-wave impedance window is placed behind the sample; and (4) analyzing by means of high-speed camera and X-ray imaging.

So far, the diagnostic methods widely used in spalling tests are to combine VISAR to measure the particle velocity $u_{fs}(t)$ on the free surface of the sample with the soft recovery of sample to analyze the spall appearance microscopically.

Fig. 8.17 illustrates how spalling occurs after a triangular compression shock reflects from the free surface of the flat sample. Figs. 8.17A and B show, respectively, the time—distance (*t-x*) diagram and the pressure—velocity (*P—u*) diagram, illustrating the interaction process of waves when a triangular pulse reflected from a free surface of a plate sample.

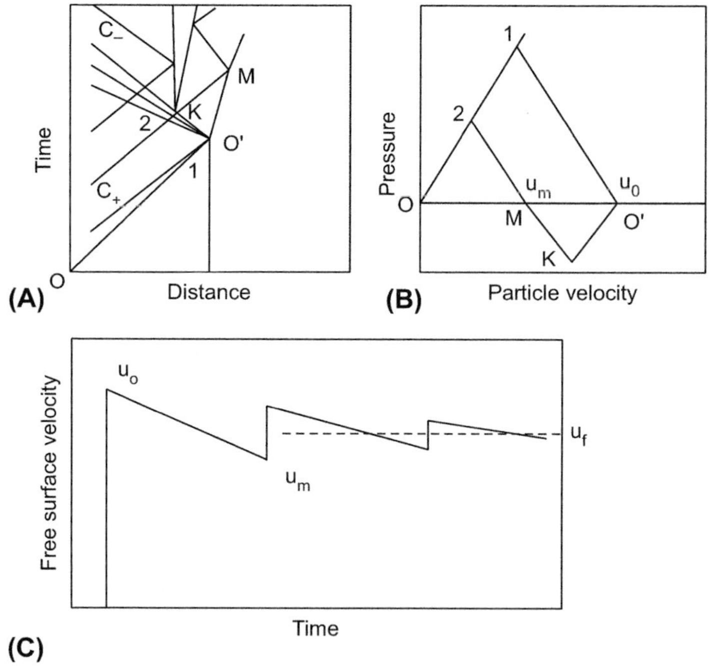

Figure 8.17 Schematic diagram of triangular shock wave pulse reflected on free surface to form spalling surface.

In the t—x diagram, the shock front trajectory is described by oo'. A series of rightward C^+ characteristic lines represent the subsequent triangular unloading waves. When the compression shock front reaches the free surface o' of the target, the particle velocity on the surface undergoes a jump from 0 to $u_0 = 2u_s$ (corresponding to the state point 1 on the P—u diagram), where u_s is the particle velocity behind the shock front. The reflection of the shock wave on the free surface forms a series of centered rarefaction waves described by fan shape leftward C^- characteristic lines on the t—x diagram. The corresponding free-surface velocity $u_{fs}(t)$ decreases as shown in Fig. 8.17C.

The maximum tensile stress appears at the point k on the t—x diagram, where the reflected rarefaction wave o'k intersects with the incident rarefaction wave 2k. Correspondingly, the value of tensile stress in the spalling plane is determined by the state point k on the P—u diagram of Fig. 8.17B, which is the intersecting point of the trajectories o'k and 2k.

Once spalling occurs, a new free surface will be created and allows the tensile stress to decrease rapidly to zero. As a result, reflection of the remaining incident compressive waves on the new free surface (cf. Fig. 8.5) will generate a spalling pulse, displaying as a velocity pullback in free-surface velocity profile: $\Delta u_{fs} = u_0 - u_m$, where u_m is the particle velocity of the free surface just before the spalling pulse (corresponding to the point M in the P—u diagram) and u_0 is the peak free-surface velocity. The stress wave trapped between the spall plane and the free surface reflects back and forth between them, causing $u_{fs}(t)$ attenuated oscillation, as shown in Fig. 8.17C. The thickness of the spalling piece $h_s(=C_l\tau/2)$ can be calculated from the oscillation period τ, where C_l is the axial longitudinal wave velocity. In the case of that the maximum tensile stress is not enough to cause spalling, the particle velocity of free surface will drop directly from u_0 to zero.

The peak velocity u_0 of the free surface and the particle velocity u_m just before spalling can be determined immediately from the measured velocity—time history of the free surface. According to the linear approximate analysis of shock waves (refer to Chapter 4.2), the maximum tensile stress just before spalling is (Novikov et al., 1966):

$$\sigma^* = \frac{1}{2}\rho_0 c_0 \Delta u_{fs} \tag{8.9}$$

Therefore the spalling strength $\sigma^\star$ can be determined, where ρ_0 is the material density and C_0 is the speed of sound.

It should be noted that the nonlinear compression characteristics of the material should be taken into consideration in the calculation of the spalling strength, which can be obtained from the isentropic extrapolation on the P—u diagram. The difference caused by such correction does not exceed 10% in practice if the impact speed is not very high.

As an example, Fig. 8.18 shows the free-surface velocity profiles $u_{fs}(t)$ for titanium alloy BT6 (Ti—6Al—4V) obtained by plate impact experiments (Kanel and Petrova, 1981). As can be seen that when the impact load intensity is relatively small (curve 1: impact speed 450 ± 20 m/s), the free-surface velocity profiles $u_{fs}(t)$ record the elastic and plastic compression pulse waves and the subsequent complete unloading wave shape. The material underwent elastic and plastic deformation, but no spalling occurred. As the impact speed increases, the tensile stress induced from the reflection of the compression shock on the free surface increases. When the peak tensile stress reaches the threshold value of spalling, the nucleation, growth, and coalescence of damage take place. Then the tensile stress in the damage accumulation region relaxes and decreases with the development of spalling. As a result, a compressive disturbance called "spalling pulse" appears on the free-surface particle velocity profile. Thereafter, the spalling pulse reflected back and forth between the free surface and spalling plane causes the oscillation on the free-surface velocity profile. As discussed above, the thickness of the spalling piece can be estimated by the period of velocity oscillation, and the spalling strength of the material can be calculated by

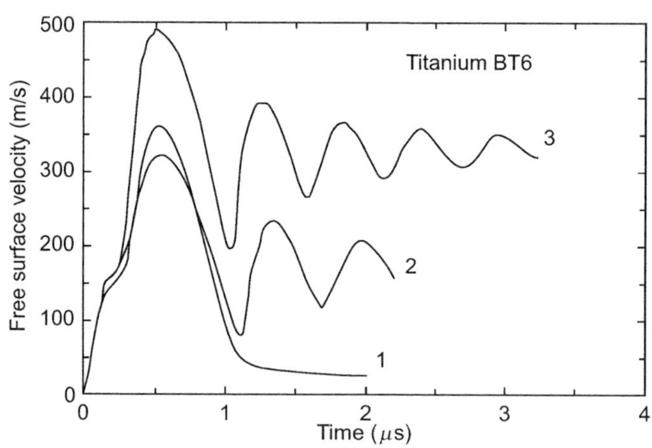

Figure 8.18 Free-surface particle velocity profile for Ti—6Al—4V alloy under impact velocity 445 m/s (profile 1), 600 m/s (2), and 700 m/s (3) (Kanel and Petrova, 1981).

velocity pullback Δu_{fs}. The experimental results show that the increase of the shock amplitude does not affect the amplitude of Δu_{fs}, as shown in Fig. 8.18.

Eq. (8.9) is based on a simplified shock wave analysis of the hydrodynamic model, neglecting the influence of material distortion. For most solids, the free-surface velocity profile exhibits elastic—plastic characteristics. In the case of one-dimensional strain state, from the free-surface particle velocity curve, it can be seen the characteristics of different wave velocities based on a hydro-perfect elastoplastic model, such as the longitudinal wave velocity C_l and the bulk wave velocity C_b ($C_b < C_l$), and so on, as shown in Fig. 8.18. Under these circumstances, analysis for the incident and reflection of the shock pulses should be carried out on the basis of the hydro-perfect elastoplastic model, as shown in Fig. 8.19 (see Section 5.4 of Chapter 5), which is more complicated than that in Fig. 8.17 (Wang Li-li, 2005). Its complexity mainly lies in the following two aspects:

(1) **Firstly, the plastic compression shock waves do not reach the free surface for reflection.** Because the elastic wave propagates faster than that of the plastic wave, precursor elastic shock will reach the free surface and is reflected firstly. The reflected leftward elastic unloading wave will interact with the subsequent rightward incident plastic compression wave, attenuating the intensity of plastic shock and producing a secondary rightward precursor incident elastic shock, which will be reflected again on the free surface. By analogy, the intensity of incident plastic wave will be weakened again and again until its pressure amplitude is less than the yield stress, and then it becomes an

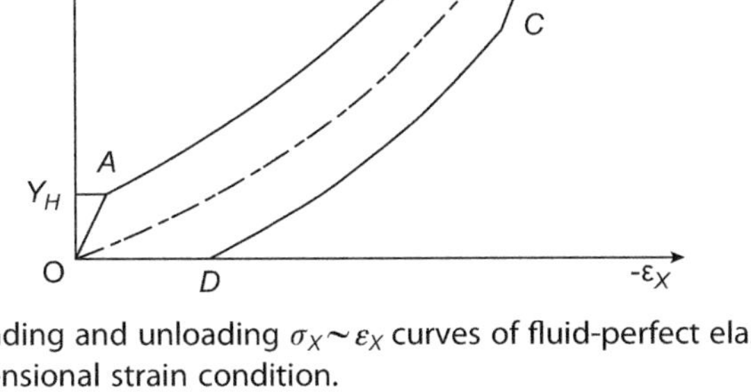

Figure 8.19 Loading and unloading $\sigma_X \sim \varepsilon_X$ curves of fluid-perfect elastoplastic model under one-dimensional strain condition.

elastic wave. In other words, the front of plastic compression wave can never reach the free surface. All spalling analysis based on the plastic compression shock reflected on the free surface is only applicable to hydrodynamic models that neglect elastic precursor waves.

(2) Secondly, the incident unloading wave will include the precursor elastic unloading wave and the subsequent plastic unloading wave. For a triangular pulse loading which has a shock front followed by an attenuated wave tail (similar to a contact explosion loading), as shown in Fig. 8.20, the elastic unloading wave will be reflected back and forth between the plastic shock wave and the plastic unloading wave (i.e., the reverse plastic loading wave) so that the intensity of incident plastic shock is reduced continuously because the elastic unloading waves propagate faster than that of plastic unloading wave and the plastic loading wave (Wang Li-li, 2005).

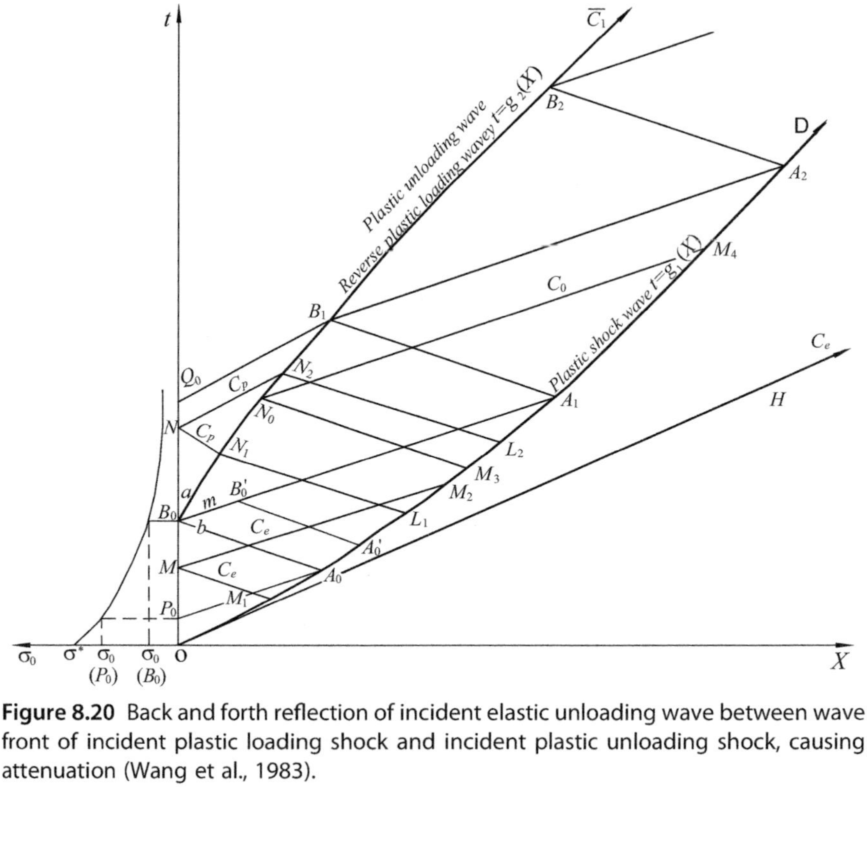

Figure 8.20 Back and forth reflection of incident elastic unloading wave between wave front of incident plastic loading shock and incident plastic unloading shock, causing attenuation (Wang et al., 1983).

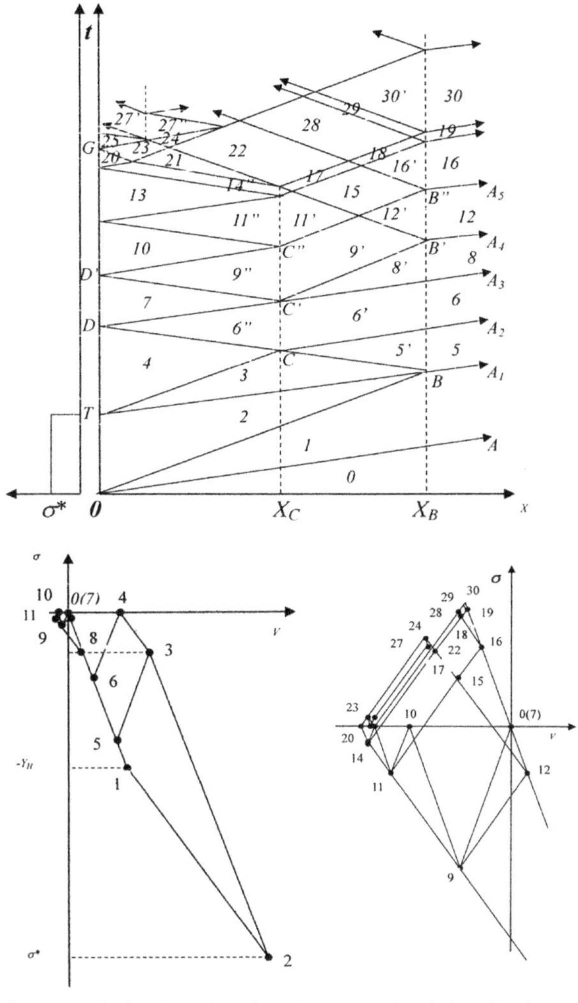

Figure 8.21 Influence of plastic unloading (reverse plastic loading) on propagation of elastic–plastic wave in one-dimensional strain state (Wang, 1982).

For a rectangular pulse loading, as shown in Fig. 8.21, before the incident compression pulse is reflected on the free surface, the incident elastic unloading wave described by the rightward characteristic line *TB* may first catch up the incident plastic compression loading wave described by the characteristic line *OB*. Then the internal reflection unloading wave *BC* will further interact with the incident plastic unloading wave *TC*, which makes the characteristics of wave system very complicated. Under certain conditions, there may even be another unloading failure occurred near the impact surface, it is the so-called "front spalling" (Wang Li-li, 1982).

For such complicated wave system, it is obviously difficult to use a simple equation such as Eq. (8.9), or a slightly modified formula, to estimate the spalling strength. Instead, it is necessary to use the computer numerical simulation according to each specific situation.

Readers interested in further study of spalling can refer to, for example, the monograph *Spall Fracture* (Antoun et al., 2003) jointly coauthored by famous scholars in Russia and the United States, which makes authoritative comments on the history and current research status of spallation.

8.2 Erosion

One of the important inspirations from the study of spalling is that unlike the structural failure under quasistatic load, which usually occurs in the loading stage, the so-called unloading failure caused by the interaction of unloading waves will occur in the dynamic process characterized by stress wave propagation.

In case of the structure subjected to dynamic loading such as explosion/impact, stress waves are often the main load forms that cause structural failure. The dynamic process characterized by stress wave propagation includes both loading wave propagation and unloading wave propagation, which involves the interaction between loading wave and loading wave, also between loading wave and unloading wave, or between unloading wave and unloading wave. Although impact pulse loading is usually to produce compressive wave first, but the impact pulse itself contains the unloading wave tail. Furthermore, the compressive wave will be reflected as unloading wave at the low-impedance interface (including the free surface) and the interaction of those unloading waves will lead to the appearance of tensile stress. Due to the failure strength of materials in tensile state is significantly less than that in compressive state, the dynamic fracture in structures loaded by stress waves is often exhibited as an unloading tensile failure. The common backside spalling is only one of the typical unloading failure modes, and the front side spalling shown in Fig. 8.21 is another manifestation of unloading failure.

The crater formed by throwing blasting, which exhibits a hopper-form complete destruction on the side near the free surface called blasting crater in the blasting engineering, is also a manifestation of unloading failure. The blasting charge is buried at a certain distance from the free surface (called resistance line in the blasting engineering). A spherical wave with a shock front and a subsequent unloading wave tail (with an attenuating pressure) is first generated during detonation of the charge, then reflects a spherical unloading wave on the free surface, which interacts with the unloading wave tail leading to a tensile stress zone. When the unloading failure criterion is satisfied, similar to the spalling mechanism, the material within the explosive crater zone will be destructed and fly away with momentum trapped in it.

Erosion to be discussed below is another type of unloading failure induced by surface unloading waves. All dynamic tensile failures caused by unloading waves have similar inherent laws, so these are collectively referred to as unloading failures (Wang Li-li, 2004).

8.2.1 Erosion—unloading failure induced by Rayleigh surface wave

When discussing spalling above, the attention is focused mainly on the back-side of the target. In fact, serious deformation and damage can also be observed on the frontage of loading side, for example, as can be seen in Fig. 8.1B and Fig. 8.1A. The formation of this kind of crater damage is related to a variety of factors, but cannot be separated from the influence of rarefaction waves from the outerside free surface, which is also related to unloading failure.

It is noteworthy that even if the intensity of pulse load is not sufficient to cause spalling near the free surface, the damage and failure formed on the loading surface are enough to threaten the overall safety of the structure during service. Such a typical example is erosion (Field and Hutchings, 1987). For example, the erosion of high-speed aircrafts by raindrops (the so-called rain erosion), by sands (the so-called sand erosion), or by hails (the so-called hail erosion) is sufficient to endanger the safety of aircrafts during service. The combined action of erosion and corrosion of the inner wall of the pipeline by impurity particles in two-phase flow at the bend of the pipe is also often one of the important causes of pipeline destroy.

Fig. 8.22 shows fractures produced in plate glass by the impact of a 3 mm diameter liquid jet at 650 m/s impact velocity (Magn. x 7-3.) (Bowden and Field, 1964). It is particular noteworthy that there are many short, circumferential wave-like ripples within the edge region of saucer-shaped crater.

Fig. 8.23 shows a sequence of high-speed photographs of a soda-lime grass specimen (12 $\times$50 $\times$5 mm) being impacted by 3 mm diameter steel sphere at a speed of 250 m/s (Field et al., 1989; Walley and Field, 2005).

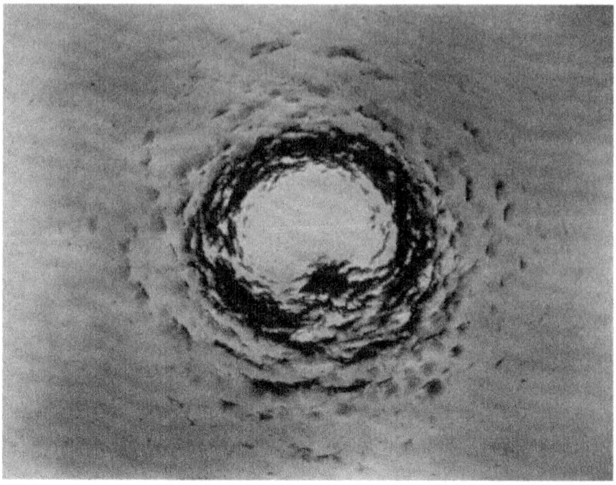

Figure 8.22 Dished fractures produced in plate glass by the impact of a 3 mm diameter liquid jet at 650 m/s impact velocity (Magn. x 7-3). *From Bowden, F.P., Field, J.E., 1964. The brittle fracture of solids by liquid impact, by solid impact and by shock. Proc. Roy. Soc. London, A282 331–352., Fig. 3, Plate 29, p.334. Reprinted with permission of the publisher.*

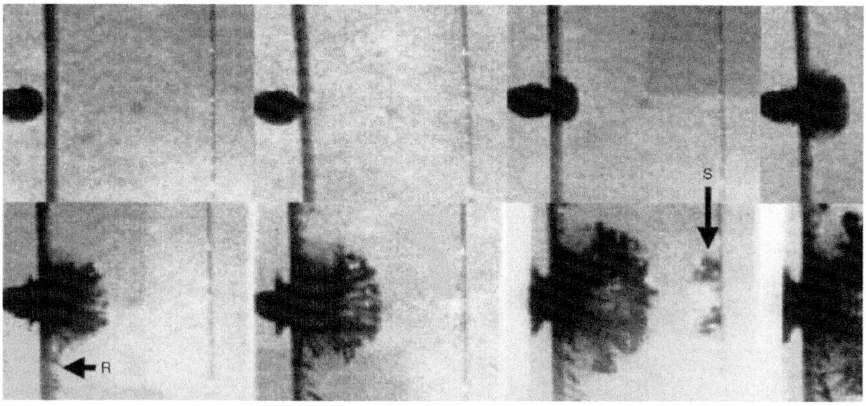

Figure 8.23 High-speed photographic sequence of a soda-lime grass specimen being impacted by a 3 mm diameter steel sphere at a speed of 250 m/s, and the interframe time is 0.95 μs. *From Walley, S.M., Field, J.E., 2005. The contribution of the Cavendish Laboratory to the understanding of solid particle erosion mechanisms. Wear 258 552–566. Fig. 7, p.556. Reprinted with permission of the publisher.*

There are totally eight frames from left to right and then top to bottom; the interframe time is 0.95 μs. Starting from the fifth frame (the first one from the left of the bottom row), the circumferential surface cracks marked R can be seen, which increase gradually with loading time. In the seventh frame (the second one from the right of the bottom row), the spalling damage marked S can be seen.

One feature of erosion is the nonplanar impact of spherical projectiles on planar targets. The collision of the projectile against the target will form two zones: the "contact zone" of the target with the spherical projectile and the "noncontact zone". For a constant velocity V_0 impact of spherical projectile onto the plate target, the contact area develops over time. In the meridian profile of the projectile, it appears as the propagation of the contact boundary (Bowden and Field, 1964), as schematically shown in Fig. 8.24.

The propagation velocity of the contact boundary is different on spherical projectile and planar target. For spherical projectiles with radius R, the direction of contact velocity $\overrightarrow{Q}_e$is tangent to the circumference R at the contact point S. The basic relationship between the $\overrightarrow{Q}_e$, its vertical component V_e, its horizontal component U_e and the impact velocity V is as follows.

$$V_e = Q_e \sin \beta = V \tag{8.10a}$$

$$U_e = Q_e \cos \beta = V/\tan \beta \tag{8.10b}$$

For a rigid target, the contact velocity is just the horizontal component U_e along the X-axis.

Obviously, when the propagation velocity of the contact boundary is faster than that of the stress wave, the "contact zone" is formed. At this moment, the effect of the stress wave is covered up and suppressed due to the superwave velocity propagation of the contact boundary. However, when the stress wave in the target propagates faster than the expanding contact boundary, the "noncontact zone" is formed. At this time, the stress wave plays a key role to form the damage zone consisting of many rings of the circumferential surface cracks. Typical high-speed photograph sequences

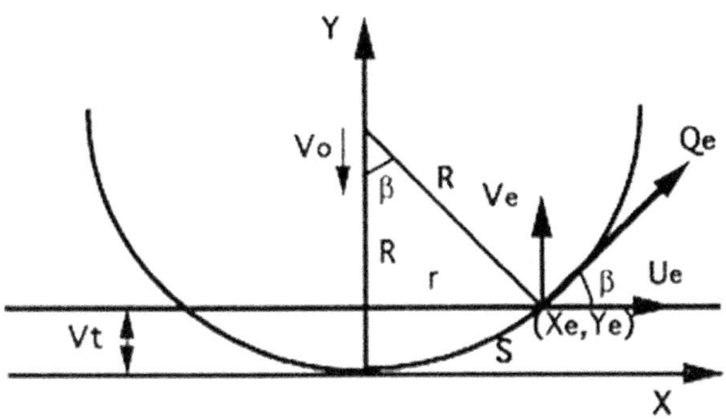

Figure 8.24 Propagation of the contact boundary when the target impacted by projectile at a constant velocity V_0.

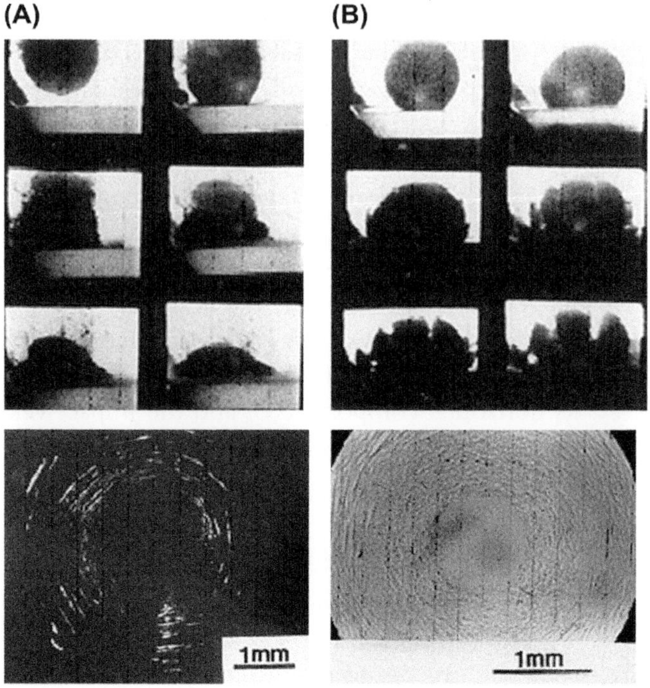

Figure 8.25 High-speed photograph (interframe time is 20 μs) and the damages characteristics for (A) ice ball impacted onto a PMMA plate at a velocity 143 m/s and (B) nylon ball impacted onto a PMMA plate at a velocity 392 m/s. *From Wang. Li.Lih., Field, J.E., Sun, Q., Liu, J., 1995. Surface damage of polymethylmethacrylate plates by ice and nylon ball impacts. J. Appl. Phys. 78 (3), 1643–1649, Fig. 1, p.1643; Fig. 2 to Fig. 4, p.1644. Reprinted with permission of the publisher.*

(with interframe time 20 μs) are shown in Fig. 8.25, where Fig. 8.25A is for diameter 14.3 mm ice sphere impacted onto a PMMA target at a speed of 392 m/s and Fig. 8.25B is for diameter 14.3 mm nylon sphere impacted onto a PMMA target at a speed of 392 m/s. Note that the both targets suffer surface damage, including the clear "contact zone" and the "noncontact zone" composed of circumferential surface cracks (Wang et al., 1994, 1995).

The condition of distinguishing the contact zone and the noncontact zone, i.e., the critical conditions for the occurrence of damage zone, is of great significance for the assessment of erosion severity. In fact, in case of grazing incidence of stress waves on the target surface, there are three possible reflected stress: the longitudinal wave (P-wave), the transverse wave (S-wave), and the Rayleigh surface wave, among which only the Rayleigh surface wave could induce surface tensile stress. The surface cracks caused by Rayleigh surface waves have also been observed experimentally, as shown in Fig. 8.23. Thus, the critical condition for the occurrence of damage zone is that the propagating velocity of stress wave in the target exceeds the contact velocity. According to Eq. (8.10b), this situation takes place when

$$C_R = U_e = V/\tan\beta \tag{8.11}$$

where C_R is the Rayleigh surface wave velocity and the impact velocity V in the above equation. Fig. 8.24 depends on different situations: in the case of a rigid target, the V is the initial velocity of the projectile V_0. However if both the projectile and the target are deformable materials, the influence of the wave impedance ratio $\frac{(\rho C)_t}{(\rho C)_p}$ and the radius ratio $\frac{R_t}{R_p}$ for the target and projectile should be taken into account, which can be generally expressed as:

$$V = \alpha V_0, \quad \alpha = \alpha\left(\frac{(\rho C)_t}{(\rho C)_p}, \frac{R_t}{R_p}\right) \tag{8.12}$$

where the subscripts t and p denote the target and projectile, respectively. There are two extremes for α (Wang Li-li, 2005):

$$\begin{cases} \text{if } (\rho C)_t \to \infty \text{ (rigid target)}, \alpha = \alpha_1 = 1 \\ \text{if } R_t = R_p = \infty \text{ (plane impact)}, \alpha = \alpha_2 = \dfrac{(\rho C)_t}{(\rho C)_t + (\rho C)_p} \end{cases} \tag{8.13}$$

For the case of an elastic spherical projectile impacting onto an elastic plate target, the average value α of the above two extremes can be taken as the first-order approximation.

$$\alpha = \frac{\alpha_1 + \alpha_2}{2} = \frac{1}{2}\left[1 + \frac{(\rho C)_t}{(\rho C)_t + (\rho C)_p}\right] \tag{8.14}$$

Substitute it into Eq. (8.12), then we have:

$$V = \frac{1}{2}\left[1 + \frac{1}{1+n}\right]V_0, \quad n = \frac{(\rho C)_p}{(\rho C)_t} \tag{8.15}$$

Moreover, from the geometric relationship shown in Fig. 8.24:

$$\tan\beta = \frac{r}{R - Vt}, \quad r = R\sin\beta \tag{8.16}$$

Noting that $\sin\beta = \tan\beta\Big/\left(1 + \tan^2\beta\right)^{1/2}$, and substituting it and Eq. (8.16) into Eq. (8.11), after calculation and taking into account Eq. (8.15), the critical condition for the occurrence of the damaged zone characterized by the critical radius r_c of the undamaged circle on the target is given by:

$$r_{cr} = R \sin \beta_{cr} = \frac{RV}{C_R}\left[1 + \left(\frac{V}{C_R}\right)^2\right]^{-1/2},$$

$$\frac{V}{C_R} = \frac{\alpha V_0}{C_R} = \frac{1}{2}\left[1 + \frac{1}{1+n}\right]\frac{V_0}{C_R}, \quad n = \frac{(\rho C)_p}{(\rho C)_t} \tag{8.17a}$$

In the case of $\frac{V}{C_R} \leq 0.6$, expand the above equation into Taylor series, we have approximately:

$$r_{cr} = \frac{RV}{C_R}\left[1 - \frac{1}{2}\left(\frac{V}{C_R}\right)^2\right] \quad \text{for} \quad \frac{V}{C_R} \leq 0.6 \tag{8.17b}$$

Although Eq. (8.17) derived above is for normal impact, it is not difficult to extend to the oblique impact. In fact, Eq. (8.17) can be generalized to the oblique impact by using $V_0\cos\theta$, instead of V_0 in Eq. (8.17), where θ is the impact angle with respect to the normal line of the target plane. The comparison between the theoretical predictions given by Eq. (8.17) and the experimental results reported by McNaughton et al. (1970) for ice balls obliquely impacting onto PMMA targets are shown in Fig. 8.26.

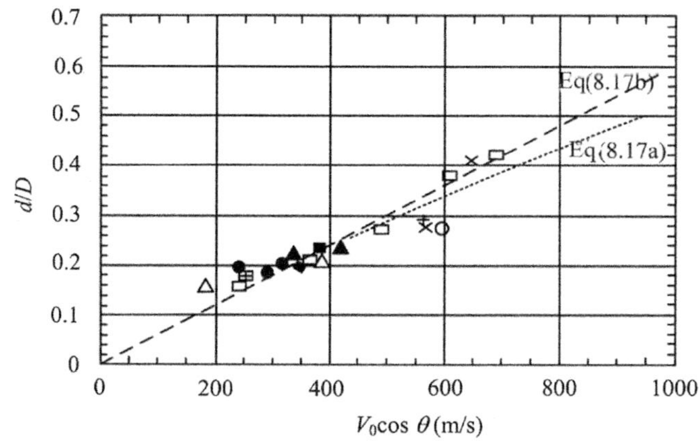

Figure 8.26 Comparisons of the undamaged diameters r_c between the predictions by Eq. (8.17) and the experimental data for PMMA targets impacted by ice balls at varied angles. (O) $D = 17.7$ mm,$\theta = 0$ degree; (□) $D = 19.0$ mm,$\theta = 0$ degree; (◆) $D = 25.4$ mm, $\theta = 0$ degree; (X)$D = 12.7$ mm, $\theta = 15$ degrees; (+) $D = 12.7$ mm,$\theta = 30$ degrees; (△)$D = 19.0$ mm,$\theta = 30$ degrees; () $D = 25.4$ mm, $\theta = 45$ degrees; (■) $D = 19.0$ mm, $\theta = 45$ degrees; (◆)$D = 25.4$ mm, $\theta = 45$ degrees; (▲) $D = 19.0$ mm, $\theta = 60$ degrees; (□) $D = 25.4$ mm, $\theta = 60$ degrees. *From Wang. Li.Lih., Field, J.E., Sun, Q., Liu, J., 1995. Surface damage of polymethylmethacrylate plates by ice and nylon ball impacts. J. Appl. Phys. 78 (3), 1643–1649, Fig. 9, p.1647. Reprinted with permission of the publisher.*

Considering that the impact angles involved in the experiments varied from 0 to 60 degrees, the used ice ball had the diameters from 12.7 to 25.4 mm, and the impact velocity range was from 240 up to 835 m/s, the theoretical predictions are in good agreement with the experimental results, and it is sufficient to illustrate the validity of Eq. (8.17).

Introduce two dimensionless parameters, the "relatively effective impact velocity" $\overline{V}_{eff}$ defined as:

$$\overline{V}_{eff} = \frac{V_{eff}}{C_R} = \frac{\alpha V_0 \cos\theta}{C_R} \tag{8.18b}$$

and the "dimensionless undamaged diameter" $\overline{d}$ defined as:

$$\overline{d} = \frac{d}{D} = \frac{r}{R} \tag{8.18b}$$

where $d_{cr} = 2r_{cr}$ is the critical diameter of undamaged zone and $D = 2R$ is the diameter of spherical projectile, so Eq. (8.17) for oblique impact can be rewritten in dimensionless form as:

$$\overline{d} = \overline{V}_{eff}\left[1 + \overline{V}_{eff}^2\right]^{-1/2} \tag{8.19a}$$

$$\overline{d} \approx \overline{V}_{eff}\left[1 - \frac{1}{2}\overline{V}_{eff}^2\right] \quad \text{for } \overline{V}_{eff} \leq 0.6 \tag{8.19b}$$

The dimensionless equation discussed above could theoretically be applied to erosion analysis on plate targets of different materials impacted by the spherical projectiles of different materials. Fig. 8.27 shows the comparison of the theoretical predictions given by Eq. (8.19) to those experimental results from the different researchers including the targets impacted by ice balls, raindrops, nylon balls, etc., where the solid line represents Eq. (8.19a) and the dashed line represents Eq. (8.19b) (Rickerby, 1977; Field et al., 1979; Wang et al., 1995). The satisfactory agreement between the many experimental results and the dimensionless theoretical predictions provides a sufficient proof for the above theoretical analysis and establishes the common theoretical analysis basis between these similar phenomena of erosion.

For a given material, the erosion damage of materials is usually increased with increasing impact velocity and the erosion frequency. The residual strength of the specimens after the erosion test is generally measured to assess quantitatively the severity of erosion damage, such as the fracture stress σ_f measured by a hydraulic burst test (Matthewson and Field, 1980).

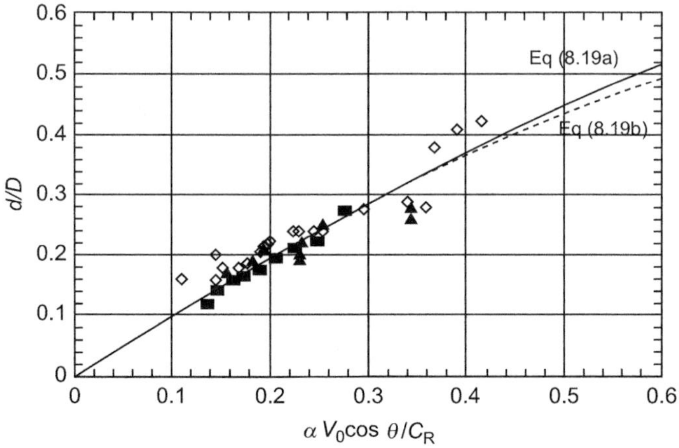

Figure 8.27 The unified relation between the dimensionless undamaged diameter $\overline{d}$ and the dimensionless relative effective impact velocity $\overline{V}_{eff}$, and the comparison between experimental results and theoretical predictions. *From Wang. Li.Lih., Field, J.E., Sun, Q., Liu, J., 1995. Surface damage of polymethylmethacrylate plates by ice and nylon ball impacts. J. Appl. Phys. 78 (3), 1643–1649, Fig. 10, p.1648. Reprinted with permission of the publisher.*

The typical erosion damage patterns on sodium glass targets impacted by water droplets with the equivalent diameter of 2 mm at various impact velocities are shown in Fig. 8.28, where Figs. (A), (B) and (C) are corresponding to impact velocities of 300 m/s, 450 m/s and 700 m/s, respectively (Zwaag and Field, 1983).

For sodium glass targets impacted by water droplets with equivalent diameter of 4 mm at various velocities, how the residual fracture stresses σ_f vary with the impact velocity are shown in Fig. 8.29 (Zwaag and Field, 1983). The experiments show that the residual fracture stress σ_f, remains unchanged when the impact velocity of water droplets is less than about 150 m/s, then it decreases with the increasing impact velocity. This indicates that there is a critical impact velocity, and the residual fracture stress decreases rapidly in the range of impact velocity from 150 to 300 m/s, which can be defined as a "transition region". Under a higher impact velocity, the residual fracture stress is only about 20% of the original strength of materials, although it had only experienced single erosion impacted by water droplets.

In fact, erosion of most structures will never stop at a single erosion, such as aircraft would be encountered rain erosion, hail erosion, or sand erosion, where the impact is often repeated for many times and more harmful to the structures. Fig. 8.30 shows the erosion damage pattern on the sodium glass

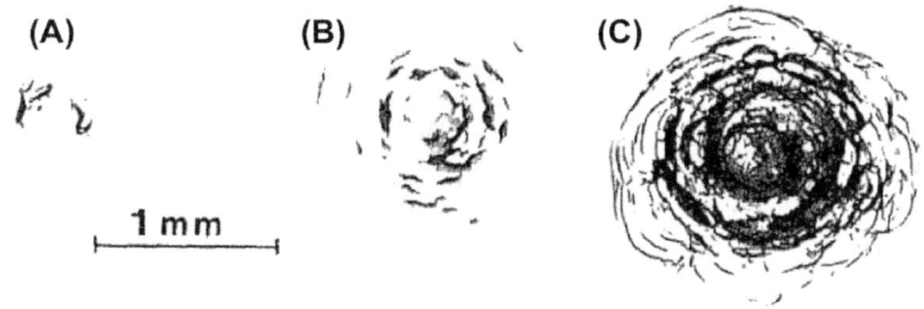

Figure 8.28 Erosion damage on sodium glass targets impacted by water droplets of 2 mm equivalent diameter; (A), (B), and (C) correspond to impact velocities 300 m/s, 450 m/s, and 700 m/s, respectively. *From Zwaag, S.Van.Der., Field, J.E., 1983. Rain erosion damage in brittle materials. Engineering Fracture Mechanics 17 (4), 367–379., Fig. 1, p.369. Reprinted with permission of the publisher.*

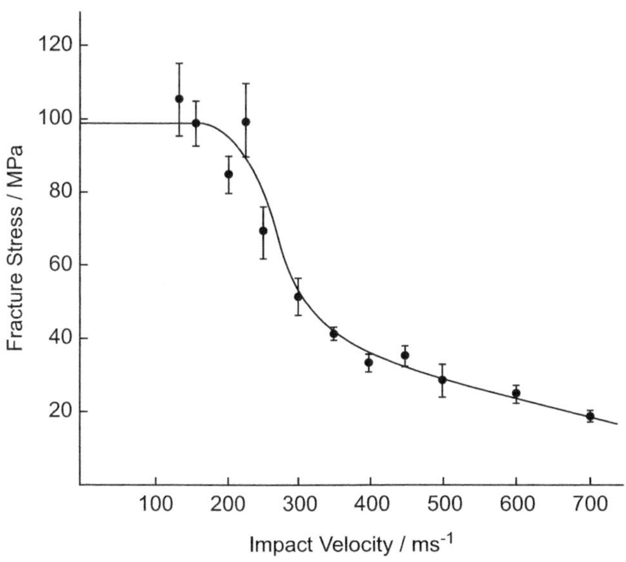

Figure 8.29 Residual fracture stress σ_f of sodium glass targets varied with impact velocity of water droplets with 4 mm equivalent diameter. *From Zwaag, S.Van.Der., Field, J.E., 1983. Rain erosion damage in brittle materials. Engineering Fracture Mechanics 17 (4), 367–379., Fig. 2, p.368. Reprinted with permission of the publisher.*

target impacted by water droplets of 4 mm equivalent diameter; it shows that erosion damages increase with increasing impact times at speed 250 m/s (Zwaag and Field, 1983).

As can be seen from Fig. 8.30, the erosion damage becomes more and more serious with the increase of impact times. Correspondingly, the

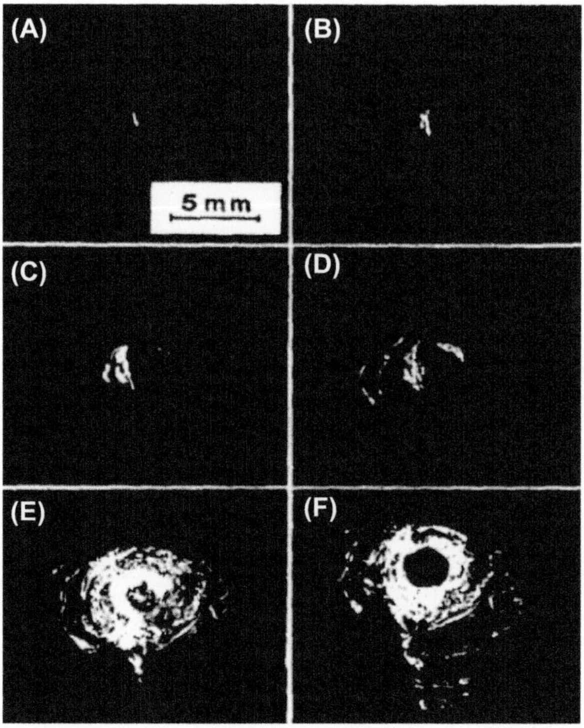

Figure 8.30 Erosion damage on sodium glass target impacted by water droplets with equivalent diameter of 4 mm at a speed of 250 m/s under different times, in which (A) 2 times, (B) 3 times, (C) 5 times, (D) 8 times, (E) 15 times, and (F) 23 times. *From Zwaag, S.Van.Der., Field, J.E., 1983. Rain erosion damage in brittle materials. Engineering Fracture Mechanics 17 (4), 367–379., Fig. 7, p.369. Reprinted with permission of the publisher.*

residual fracture stress decreases obviously with the increase of impact times, and the "transition region" becomes narrow when the impact times increase, as shown in Fig. 8.31 (Zwaag and Field, 1983).

Once the residual fracture stress σ_f is measured and if the fracture toughness K_{Ic} of the material is known, the damages degree of erosion can be converted to an equivalent crack size according to the crack/fracture mechanics theory (Zwaag and Field, 1983).

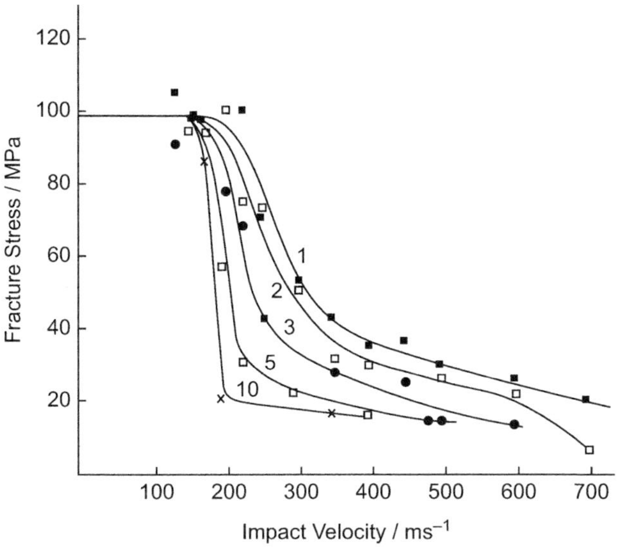

Figure 8.31 Residual fracture stress σ_f changes with impact velocity under different impact times. *From Zwaag, S.Van.Der., Field, J.E., 1983. Rain erosion damage in brittle materials. Engineering Fracture Mechanics 17 (4), 367–379., Fig. 6, p.372. Reprinted with permission of the publisher.*

For example, if the equivalent crack size of damage is considered as the crack with a length $2c$, the equivalent crack size c can be calculated by the following formula.

$$c = \frac{1}{\pi}\left(\frac{\sigma_f}{K_{Ic}}\right)^2 \tag{8.20}$$

This can be used to evaluate the safety of the structure subjected to erosion.

The involved knowledge of crack mechanics (also called fracture mechanics) will be discussed in the next chapter.

CHAPTER NINE

Crack dynamics and fragmentation

At the beginning of this part, it was mentioned that from the mesoscopic perspective, the dynamic failure of materials is a time-dependent process in which different types of mesodamage (such as mesocracks, mesovoids, meso-shear bands, etc.) evolve at a definite rate. Therefore, in a broad sense, all failure of materials is a dynamic time process by nature, a rate-dependent damage evolution process. The study of material failure is inseparable from the study of the dynamic evolution law of mesoscopic damage until the mesoscopic damages coalesce to form macroscopic crack and ultimately leads to material failure. From the perspective of macromechanics, we generally face two types of problems: if the mesodamages have not been coalesced to form the macrocrack, our research object is the "crack-free body" without the macrocrack, even if it contains mesodamage; On the contrary, our research object is the "body with macrocracks" (here and hereinafter referred to as "cracked body" or "crack body"). Classical continuum mechanics focuses on mainly the study of crack-free bodies, in where the displacement at any particle in the body is a continuous single-valued function of time and space. However, once the macrocrack appears, the displacement is no longer limited to the continuous single value function, because the crack can be expressed as the strong discontinuity of the displacement in mathematics, and the singularity appears, which greatly complicates the problem.

An interesting and important result from the study of crack bodies is that the strength of the crack body depends on the characteristics of mechanical field in a small neighborhood at the crack tip. Therefore, many mechanics researchers focused on the study of the mechanics in relation of macroscopic cracks, forming a new branch of mechanics—mechanics of solids containing cracks, some scholars call it *fracture mechanics* (e.g., Brock, 1982; Huang Ke-zhi and Yu Shou-wen, 1985), some scholars call it *crack mechanics* (e.g., Slepyan, 1981). Since the term "fracture" can also be applicable to describe the break of crack-free bodies (as shown in Fig. 5.14), thus, in a broad sense, the fracture mechanics is not limited to describe crack bodies but has a wider meaning. So this book is inclined to use the nomenclature

Dynamics of Materials
ISBN: 978-0-12-817321-3
https://doi.org/10.1016/B978-0-12-817321-3.00009-7

of "crack mechanics". When it is necessary to follow the "fracture mechanics" that has been widely adopted in many papers, it is understood in its narrow sense (i.e., for crack body).

The deformable body containing crack is a special kind of structures. When studying the fracture phenomena of cracked body from the perspective of crack mechanics, two core problems must be concerned: the first is focused on quantitative description of the mechanics field (i.e., time—space distribution of mechanical variable such as stress, strain, displacement, etc.) in a small neighborhood at the crack tip under various loading conditions, which is the structural response aspect of the problem; the second is focused on determining the material's ability to resist failure induced by the crack unstable extension under various loading conditions (fracture toughness), which is the material response aspect of the problem.

Under explosive/impact dynamic loading, the issue of crack dynamics is further complicated. On the one hand, in terms of the dynamic response of structure, the stress wave effect (inertial effect) should be taken into account, which includes the influence of stress waves on the crack-tip dynamic mechanical field in the case of a stationary crack under rapid loading, and also the influence of kinetic energy and inertia of a moving crack on the crack-tip dynamic mechanical field. On the other hand, in terms of the material dynamic response, the influence of loading rate on material fracture toughness should be taken into account. The complexity of the problem also lies in the fact that the process of dynamic initiation and extension of crack is accompanied by the emission of unloading waves and their interaction, which is of particular concern in the study of fragmentation (a kind of dynamic failure from multicrack sources).

Thus, this chapter focuses on two major categories of problems:

(1) The crack dynamics (or dynamic fracture mechanics) of a cracked body dominated by macroscopic main cracks, including structural dynamic response, material dynamic response, and dynamic failure criterion of stationary cracks and moving cracks under various dynamic loading.

(2) The so-called fragmentation in which multiple cracks appear and extend simultaneously and break into multiple fragments: the crack-free bodies with mesoscopic damage, through the nucleation, growth, and coalescence of the multisource mesoscopic damage, will eventually form multiple macroscopic cracks and lead to the fragmentation.

The process of fragmentation of a crack-free body is actually a process in which crack-free body is transformed into a multicracked body by means of mesodamage evolution. The early stage of the process is the evolution process of mesodamage (to be discussed separately in Chapter 10), while the final stage of the process is the dynamic failure of multicrack bodies.

For the convenience of discussion, let us review some basic knowledge of crack statics under quasistatic loading, but it is not derived in detail. Readers interested in relevant details can refer to relevant works (e.g., Brock, 1982; Knott, 1973; Lawn and Wilshaw, 1975; Huang Ke-zhi and Yu Shou-wen, 1985; Slepyan, 1981).

9.1 Crack dynamics

9.1.1 Basic knowledge of crack statics

Many conceptual frameworks of crack dynamics are developed on the basis of crack statics, so let us briefly review some basic concepts of crack statics.

There are three basic modes of crack extension in crack mechanics: opening mode (Mode-I), in-plane shear mode (Mode-II), and out-of-plane shear or tear mode (Mode-III), as shown in Fig. 9.1. The crack is described by a displacement discontinuity, namely the displacement across a crack surface is discontinuous. If u, v, and w are the three components of the displacement vector in x, y, and z coordinate axes, respectively, then the discontinuities of the displacement components v, u, and w across crack correspond to Mode-I, Mode-II, and Mode-III, respectively. Among those, Mode-I is often the most dangerous basic fracture mode.

As a precursor to crack mechanics research, Inglis (1913) studied an infinite elastic plate with an elliptical hole subjected to uniform tensile stress σ_L, as shown in Fig. 9.2. The half-length of long axis and short axis of the elliptical hole are b and c, respectively. According to the classical solution of elasticity,

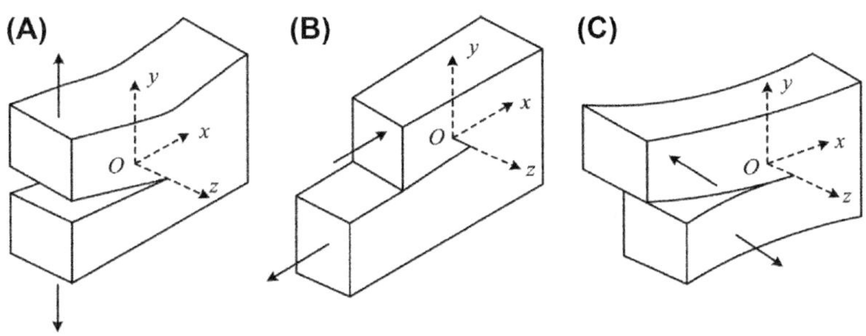

Figure 9.1 Modes of crack extension: (A) Mode-I, (B) Mode-II, (C) Mode-III.

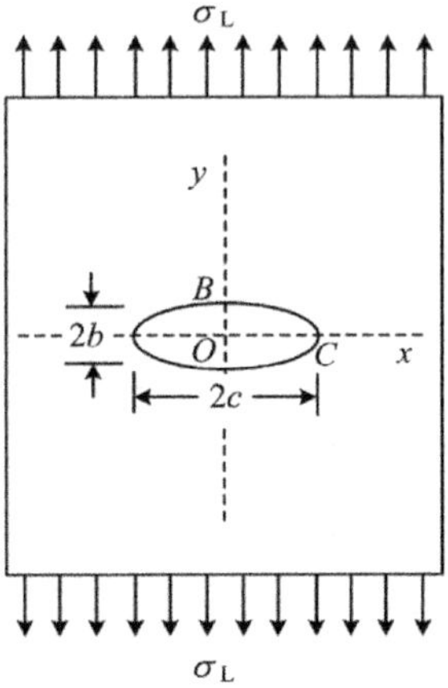

Figure 9.2 Schematic of infinite elastic plate with elliptical hole subjected to uniform tensile stress σ_L.

stress concentration occurs at point C of the long axis of the elliptical hole, and the maximum value $\sigma_{yy}(c,0)$ is:

$$\sigma_{yy}(c,0) = \sigma_L\left(1 + 2\frac{c}{b}\right) = \sigma_L\left[1 + 2\left(\frac{c}{\rho}\right)^{1/2}\right] \tag{9.1a}$$

It is much larger than the applied uniform tensile stress σ_L. In the formula, the second equal sign has taken into account the elliptical curvature radius ρ at point C, which can be expressed as:

$$\rho = \frac{b^2}{c} \tag{9.1b}$$

For the long and narrow elliptical hole, $c/b >> 1$, the 1 in Eq. (9.1a) can be ignored, and then it is approximately:

$$\frac{\sigma_{\max}}{\sigma_L} = \frac{\sigma_{yy}(c,0)}{\sigma_L} \approx 2\frac{c}{b} = 2\left(\frac{c}{\rho}\right)^{1/2} \tag{9.2}$$

The ratio of $\sigma_{\max}/\sigma_L$ is called stress concentration factor (SCF in short), SCF $= \sigma_{\max}/\sigma_L$. When $\rho \to 0$ (elliptical hole tends to a sharp crack), SCF $\to \infty$ or $\sigma_{yy}(c,0) \to \infty$, it means that the stress appears singularity when elliptical hole tends to crack.

If setting the elliptic hole $c = 3b$, the local stress distribution along the X-axis is shown in Fig. 9.3. The results show that the stress component $\sigma_{yy}(X - c,0)$ decreases from the peak with increase of the distance from point

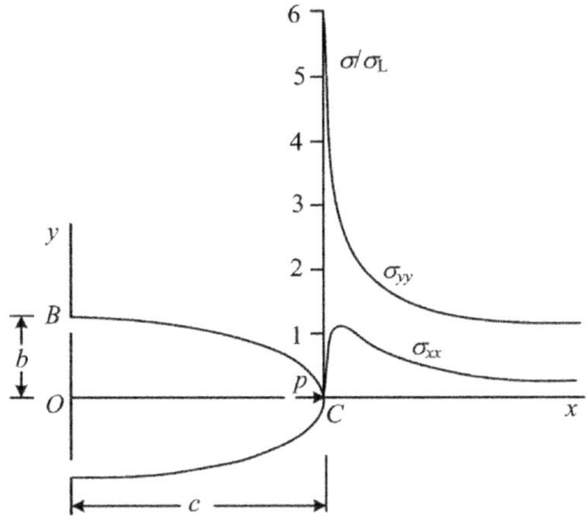

Figure 9.3 Local stress distribution along *X*-axis when the elliptic hole $c = 3b$.

C at the end of the long axis and will eventually approaches to σ_L when X is large. The stress component $\sigma_{xx}(X - c,0)$ quickly peaks near the end point C of the long axis of the elliptical hole and then declines similar to $\sigma_{yy}(X - c,0)$, while it approaches to zero when X is large. Fig. 9.3 gives two important revelations: (a) the interference of the elliptical hole with the external stress field σ_L is firstly manifested in the change from the original uniform stress field to the nonuniform, especially the stress concentration phenomenon appears, but this stress localization mainly occurs in a small area near the hole end; (b) the interference of the elliptical hole with the external stress field σ_L is also manifested in the transition from the one-dimensional stress state to the complex stress state, especially that the multidimensional tensile stress state would promote brittle failure of the materials, which is often overlooked. In fact, the effect of stress state on materials failure is of great significance (see Figure III.1 in the part 3).

However, Inglis analysis of the elliptical hole shows that, as shown in Eq. (9.1), the stress concentration depends only on the ratio of the major axis length to the minor axis length of the elliptical hole, c/b, and has nothing to do with the actual size of the hole. This is far from the fact that long cracks are more likely to fail under low stress than short cracks. Nevertheless, Inglis' works have provided useful inspiration to later researchers and promoted the development of crack mechanics.

From the development history of crack mechanics in the domain of continuum mechanics, researchers have explored the same crack extension problem from two different approaches or methods, but they have reached the same goal. One is energy approach and the other is mechanical field approach. These two approaches are often used in conjunction with each other or interchangeably in the study of other mechanical problems.

9.1.1.1 Griffith's energy approach—energy release rate criterion

The problem of unstable extension of crack was first analyzed by Griffith (1920, 1924) from the perspective of the energy balance of the cracked body.

By using Inglis' analysis of elliptical holes (Fig. 9.2), when the short half-axis b of the elliptical hole approaches 0, $b \to 0$, it turns into a problem of a linear crack with a length of $2c$ in an infinite elastic plate.

Under the external load of uniform tensile stress σ_L, if there is no crack in an infinite elastic plate, the strain energy per unit volume stored in the elastic plate, i.e., the strain energy density, is $\frac{\sigma\varepsilon}{2} = \frac{\sigma^2}{2E}$, where E is the Young's modulus. While once a linear crack with a $2c$ length is introduced, when the crack opens under the action of σ_L, on the one hand, the strain energy associated with the crack opening U_E is released, while on the other hand, the surface energy U_S is increased, which is required to create fresh new crack surfaces. Whether the crack extends or not depends on the balance between the gain and loss of energy of these two parts.

Using Inglis' solution of Eq. (9.1a), we are able to obtain strain energy U_E as follows:

$$U_E = \frac{\pi\sigma_L^2 c^2}{E'}$$
$$E' = E \text{ (plane stress)}, \quad E' = \frac{E}{(1-\nu^2)} \text{ (plane strain)} \tag{9.3a}$$

And if the surface energy per unit area is γ_S, considering that the crack has two free surfaces, the U_s can be expressed:

$$U_S = 4c\gamma_S \tag{9.3b}$$

Accordingly, the energy release rate of crack extension per unit area $\frac{dU_E}{dc}$ can be expressed as:

$$\frac{dU_E}{dc} = \frac{2\pi\sigma_L^2 c}{E'}$$
$$E' = E(\text{plane stress}), E' = \frac{E}{(1-\nu^2)}(\text{plane strain}) \tag{9.4a}$$

And the growth rate of surface energy required to extend crack per unit area, $\frac{dU_S}{dc}$, can be expressed as:

$$\frac{dU_S}{dc} = 4\gamma_S \tag{9.4b}$$

Obviously, the critical condition for crack extension is that the released strain energy by crack extension per unit area is equal to the surface energy required to create fresh new crack faces, i.e.,

$$\begin{aligned} &\frac{\pi\sigma_L^2 c}{E'} = 2\gamma_S \\ &E' = E(\text{plane stress}), E' = \frac{E}{(1-\nu^2)}(\text{plane strain}) \end{aligned} \tag{9.5a}$$

For the sake of convenience,

$G_I = \frac{\pi\sigma_L^2 c}{E'} = \frac{1}{2}\frac{dU_E}{dc}$ is often used instead of $\frac{dU_E}{dc}$ and instead of $\frac{dU_S}{dc}$, while its definition remains unchanged, $R_I = 2\gamma_S = \frac{1}{2}\frac{dU_S}{dc}$, thus the critical condition represented by Eq. (9.5a) can be abbreviated as:

$$\begin{aligned} &G_I = R_I, \text{ or } G_I = G_{Ic} \\ &G_I = \frac{\pi\sigma_L^2 c}{E'} = \frac{1}{2}\frac{dU_E}{dc} \\ &R_I = 2\gamma_S = \frac{1}{2}\frac{dU_S}{dc} \end{aligned} \tag{9.5b}$$

where the subscript "I" refers specifically to the mode-I crack extension. If the strain energy released is greater than the surface energy ($G_I > R_I$), the crack unstable extension will occur; If the released strain energy is less than the surface energy ($G_I < R_I$), the crack does not extend. The energy release rate G (unit N/m) acts as a driving power of crack extension, so it is also called crack extension force, which is a generalized energy force. R (also known as critical energy release rate G_c) acts as barrier to crack extension, so it is also called crack extension resistance, which is classified as a material property that can be measured by experiments.

Eq. (9.5a) can be rewritten as:

$$\begin{aligned} &\sigma_L c^{1/2} = \left(\frac{2E'\gamma_S}{\pi}\right)^{1/2}, \\ &E' = E\ (\text{plane stress}), \quad E' = \frac{E}{(1-\nu^2)}\ (\text{plane strain}) \end{aligned} \tag{9.5c}$$

Eq. (9.5) is the critical condition of crack unstable extension based on the concept of energy release rate G, which is called **energy release rate criterion** or **G criterion** for short.

The right-hand side of Eq. (9.5c) is the constant determined by material properties and the left-hand side corresponds to the function of applied stress and square root of crack characteristic size, which means that the critical condition of crack extension is not determined by stress alone, but by the combination $\sigma_L c^{1/2}$. The critical stress of long crack is smaller than that of short crack, which can lead to **low stress failure**. This is an important characteristic which distinguishes from the failure criterion of crack-free body.

Griffith's energy release rate criterion is only applicable to the brittle unstable extension of crack without plastic deformation. The Griffith energy criterion was extended by Irwin (1948) and Orowan (1949) independently to elastic—plastic materials such as metals with a small plastic zone at the crack tip. It was considered that the strain energy released during crack extension is converted not only to the surface energy of the crack surface, but also to the plastic strain energy at the crack-tip zone. Therefore, the surface energy γ_S in Eq. (9.5) is replaced by $(\gamma_S + \Gamma_p)$, where Γ_p is a plastic strain energy consumed per unit area of crack extension. For those plastic materials such as metals, Γ_p is usually larger than γ_S. According to the experimental results of low-carbon steel, Γ_p is more than three orders of magnitude larger than γ_S.

9.1.1.2 Irwin's force field approach—stress intensity factor criterion

Returning to the solution of the elliptic hole given in Eq. (9.1a), when the curvature radius of the long-axis end of the elliptic hole $\rho \rightarrow 0$, that is, when the elliptic hole tends to crack, the stress $\sigma_{yy}(c,0) \rightarrow \infty$ appears singularity. Hence, it brings difficulties to the analysis of the stress field (strain field, displacement field, etc.) in the crack neighborhood. Griffith derived the energy release rate criterion for crack unstable extension by energy approach, which in fact cleverly avoids the specific analysis of stress singularity.

In the theory of mathematical mechanics of elasticity, researchers have successfully solved the problem of stress singularity by using complex variable function. The mathematical solutions of stress field and displacement field near the $2c$ crack tip ($r/c << 1$) of an infinite elastic plate

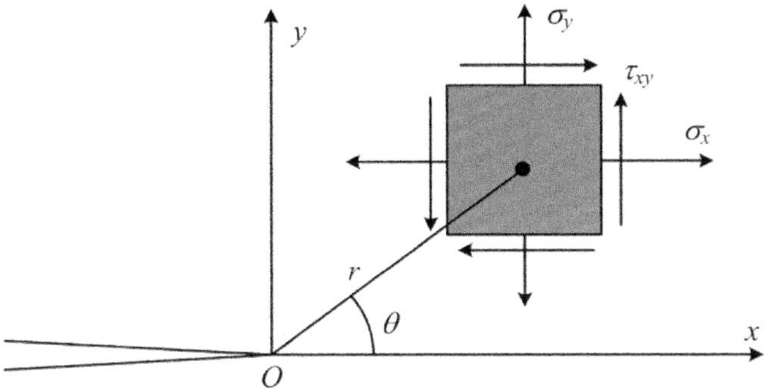

Figure 9.4 Schematic of crack-tip neighborhood ($r/c << 1$).

subjected to tensile stress σ at infinity (see Figs. 9.2 and 9.4) after neglecting the higher order small quantity were given by Irwin (1957):

$$\left.\begin{aligned} \sigma_{xx} &= \frac{K_{\mathrm{I}}}{\sqrt{2\pi r}}\cos\frac{\theta}{2}\left(1-\sin\frac{\theta}{2}\sin\frac{3\theta}{2}\right) \\ \sigma_{yy} &= \frac{K_{\mathrm{I}}}{\sqrt{2\pi r}}\cos\frac{\theta}{2}\left(1+\sin\frac{\theta}{2}\sin\frac{3\theta}{2}\right) \\ \sigma_{xy} &= \frac{K_{\mathrm{I}}}{\sqrt{2\pi r}}\cos\frac{\theta}{2}\sin\frac{\theta}{2}\cos\frac{3\theta}{2} \\ \sigma_{zz} &= \begin{cases} 0 & \text{(plane stress)} \\ \nu(\sigma_{xx}+\sigma_{yy}) & \text{(plane strain)} \end{cases} \end{aligned}\right\} \tag{9.6a}$$

$$\left.\begin{aligned} u_x &= \frac{K_{\mathrm{I}}}{E}(1+\nu)\sqrt{\frac{r}{2\pi}}\cos\frac{\theta}{2}\left(\kappa-1+2\sin^2\frac{\theta}{2}\right) \\ u_y &= \frac{K_{\mathrm{I}}}{E}(1+\nu)\sqrt{\frac{r}{2\pi}}\sin\frac{\theta}{2}\left(\kappa+1-2\cos^2\frac{\theta}{2}\right) \\ \kappa &= \begin{cases} \dfrac{3-\nu}{1+\nu} & \text{(plane stress)} \\ 3-4\nu & \text{(plane strain)} \end{cases} \end{aligned}\right\} \tag{9.6b}$$

where ν is Poisson's ratio and K_I is called stress intensity factor; its expression is as follow:

$$K_{\mathrm{I}} = \sigma\sqrt{\pi c} \tag{9.6c}$$

According to Eq. (9.6), one can find that the stress field in the crack tip has the following three important characteristics:

(1) The relations between the stress components near the crack-tip field and the radial distance r in the polar coordinates are all in the form of function of $r^{-1/2}$. When the radial distance $r \to 0$, all the stress components near the crack tip tend to infinity, indicating that the stress field at the crack-tip neighborhood has $r^{-1/2}$ order singularity.

(2) The stress field at the crack tip depends entirely on K_I. Since it is the only physical quantity characterizing the intensity of crack-tip stress field, it is called the **stress field intensity factor** or the **stress intensity factor**, sometimes referred to as the K factor for short. Although the crack-tip stress field exhibits $r^{-1/2}$ singularity, the K factor that controls the stress field intensity at the crack tip is a finite value. The dimension of K is [stress] × [length]$^{1/2}$ or [force] × [length]$^{-3/2}$. In SI unit system, its unit is $\text{MPa}\cdot\text{m}^{1/2}$ or $\text{MN}\cdot\text{m}^{-\frac{3}{2}}$. In metric system, its unit is kilogram/millimeter $^{-3/2}$.

(3) When $\theta \to 0$, we have $\sigma_{xx} = \sigma_{yy} = \frac{K_I}{\sqrt{2\pi r}}$. If it is in plane strain state, there is $\sigma_{zz} = \nu(\sigma_{xx} + \sigma_{yy})$. It can be seen that under the action of external one-dimensional tensile stress σ_{yy}, a three-dimensional tensile stress state with high intensity will appear in front of the crack tip (r is very small but not equal to 0), which will promote brittle failure.

Since the crack-tip stress field has $r^{-1/2}$ singularity, whether the crack extends unstably no longer depends on a certain maximum stress component like the quasistatic failure criterion, (in fact all stress components at the crack tip tend to be infinite), but depends on the K factor that controls the stress field intensity at the crack tip.

If the stress intensity factor of Mode-I crack K_I reaches a critical value K_{Ic}, the unstable extension of crack occurs, then the critical condition of crack unstable extension can be expressed as

$$K_I = K_{Ic} \tag{9.7}$$

which is called the **stress intensity factor criterion**. K_{Ic} characterizes the material property of resisting crack unstable extension, is called **fracture toughness**, whose unit is the same as that of K_I.

The measured K_{Ic} value of the material depends on the stress state in experiments. For example, the measured value under plane stress is higher than that under plane strain. It is known from Eq. (9.6a) that a three-dimension high-intensity tensile stress state at the crack tip will be formed in the state of plane strain state, which is conducive to brittle failure of materials. Experiments also confirmed that the critical stress intensity factor tends to the minimum when the specimen is close to the plane strain state. Thus, the K_{Ic} in Eq. (9.7) is generally defined as plane strain fracture toughness. According to American ASTM-E399 standard, in order to satisfy the plane strain condition in the experiment, the thickness B, width W, crack size a, and ligament width ($W - a$) for the tested specimen should satisfy the following relation:

$$B, W, (W-a) \geq 2.5\left(\frac{K_{IC}}{\sigma_Y^2}\right)^2 \tag{9.8}$$

where σ_y is the yield strength of the material. The values of fracture toughness K_{Ic} for several engineering materials are listed in Table 9.1 (e.g., c.f. Fracture toughness - Wikipedia).

The above is a brief discussion for pure Mode-I crack, similar discussions can be made for Mode-II and Mode-III cracks. The stress fields at crack-tip neighborhood for these three modes can be unified to be expressed as follows:

$$\sigma_{ij} = \frac{K_m}{\sqrt{2\pi r}} f_{ij}(\theta) \tag{9.9}$$

where K_m is the stress intensity factors for a given mode (subscript m = I, II, and III corresponding to Mode I, II, and III), and $f_{ij}(\theta)$ is a function of the polar angle θ in polar coordinate system, called the angular distribution function. Readers interested in details can refer to relevant monographs (e.g., Brock, 1982; Knott, 1973; Lawn and Wilshaw, 1975; Huang and Yu, 1985; Slepyan, 1981).

It should be noted that the K_I expression given in Eq. (9.6c) is derived for the case where an infinite plate with a center 2c crack subjected to the unidirectional uniform tensile stress σ at the infinite distance. In other cases, the expression of stress intensity factor K is different. More comprehensive

Table 9.1 Typical values of K_{Ic} of several engineering materials.

Material type	Material	K_{Ic}(MPa·m$^{1/2}$)
Metal	Aluminum alloy (7075)	24
	Steel alloy (4340)	50
	Titanium alloy	44–66
	Aluminum	14–28
Ceramic	Aluminum oxide	3–5
	Silicon carbide	3–5
	Soda-lime glass	0.7–0.8
	Concrete	0.2–1.4
Polymer	Polymethyl methacrylate	0.7–1.6
	Polystyrene	0.7–1.1

expressions of K_m for various shapes of crack bodies under different loading conditions can be referred to the related manual of stress intensity factor (e.g., China Institute of Aeronautical Research, 1981).

Above, the energy release rate G is derived from the energy approach and the stress intensity factor K is derived from the force field approach, respectively. The two actually describe the same crack problem and must have their inherent connection. In fact, comparing Eq. (9.5b) and Eq. (9.6b), the following expressions are obtained for the relationship between G and K:

$$\begin{aligned} G_{\mathrm{I}} &= \frac{K_{\mathrm{I}}^2}{E'} \\ E' &= E(\text{plane stress}), E' = \frac{E}{(1-\nu^2)}(\text{plane strain}) \end{aligned} \tag{9.10}$$

Confirming that both approaches lead to the same destination.

In many practical problems, the influence of plastic deformation near the crack tip should be taken into account, especially when the plastic zone is larger compared to the crack size or specimen dimension. On the basis of the above elastic analysis, considering the influence of plastic deformation near the crack tip, the elastic—plastic crack mechanics have been further developed, including the crack opening displacement (COD) criterion, the *J-integral* criterion, etc., which will not be described here. Readers who are interested in it can refer to relevant monographs (e.g., Brock, 1982; Knott, 1973; Lawn and Wilshaw, 1975; Huang and Yu, 1985; Slepyan, 1981).

9.1.2 Fundamental concepts of crack dynamics

In the crack statics, the critical conditions to crack unstable extension could be equivalently expressed as Eq. (9.5) and Eq. (9.7):

$$\left\{ \begin{aligned} G_{\mathrm{I}} &= \frac{\Pi\sigma^2 c}{E'} = G_{Ic} \\ K_I &= \sigma\sqrt{\Pi c} = K_{Ic} \end{aligned} \right\}$$

In the above two equations, G_I and K_I on the left side of the equal sign represent the structural response of cracked body near the crack tip under quasistatic loading condition, while G_{Ic} and K_{Ic} on the right side represent the material response of resistance to crack extension under quasistatic loading.

In the case of dynamic loading such as explosion/impact loading, the differences with crack statics under quasistatic loading are mainly manifested in: (1) G_I and K_I should be replaced by the dynamic structural response near the crack tip under dynamic loading, i.e., the dynamic stress intensity factor $K_I^d(t)$, which is a function of time and reflects the inertia effect; (2) G_{Ic} and K_{Ic} should be replaced by the dynamic responses of material resisting crack extension under dynamic loading, i.e., dynamic fracture toughness K_{Id}, which is a function of loading rate ($\dot{\sigma}$, $\dot{\varepsilon}$ or $\dot{K}$, etc.) and reflects the loading rate effect. How the dynamic fracture toughness K_{Id} depends on the loading rate is one of the main research fields of dynamic fracture resistance of cracked body in material dynamics. Those are further illustrated as follows.

9.1.2.1 On dynamic structure response of crack bodies

It is mainly concerned with the problems of dynamic structure response of crack mechanics where inertial effect cannot be ignored. Research to these questions can be divided into two categories: the first one deals with a stationary crack subjected to an external dynamic load varying very quickly with time. Introduce two characteristic times t_l and t_c, the former characterizes the short duration of the external dynamic loading, such as the loading duration or the rising time of a pulse, and the latter characterizes the time required for the stress wave to pass through the crack characteristic size (e.g., crack length) a at the wave velocity C, namely $t_a = a/C$. If $t_l < t_a$, then the stress wave effect on the mechanical field near crack tip must be taken into account. The second problem is that of a crack propagating (moving) with a constant or time-varying velocity, thus the kinetic energy and inertia effect due to crack propagation should be taken into account.

For these two kinds of problems, the inertia term obviously cannot be omitted from the motion equation. In the former case, the dynamic stress intensity factor near the crack tip under the action of stress waves is usually studied, that is, the precondition to study the initiation of dynamic crack extension. In the latter problem, attention is focused on the dynamic stress intensity factor changing with crack propagation, which is resulted from the variation of kinetic energy and inertia effect when the crack itself propagates rapidly, referred to as **propagating crack** or **moving/running crack** problems. Broadly speaking, it also includes crack branching and crack arrest of moving cracks. Crack branching is a type of dynamic instability problem, which involves bifurcation theory. These phenomena, as a special stage of crack movement, have recently been unified to be treated as part of propagating crack problems.

For the problem of crack propagation, crack branching, and crack arrest, since the crack surface itself as a part of boundary is in motion and the law of the crack movement is generally unknown in advance, it is a highly nonlinear problem even if the basic equations of the problem are linear. Such problem is the so-called "moving boundary" problems in mathematical physics. In the theory of mathematical physics, some progress has been made on moving boundary problem of the simplest parabolic equations (the so-called Stefan problem). The loading—unloading moving boundary problem for elastic—plastic waves of hyperbolic equations has also been studied. However, the moving boundary problem of second-order hyperbolic equations encountered in crack dynamics has not been analytically studied. Recently, numerical analysis has been often adopted to solve such problems, and the calculated results have reached a certain degree of agreement with the experimental results.

9.1.2.2 On the dynamic material response of crack bodies

It is mainly to study the dynamic fracture toughness, K_{Id} of materials, corresponding to the quasistatic fracture toughness, K_{Ic} in the crack statics. K_{Id} characterizes the material resistance to crack initiation and extension under dynamic loading, which is related to loading rate $\dot{\sigma}\left(=\frac{\partial\sigma}{\partial t}\right)$ or stress intensity factor rate $\dot{K}\left(=\frac{\partial K}{\partial t}\right)$, thus can be denoted as $K_{Id}(\dot{\sigma})$ or $K_{\mathrm{Id}}\left(\dot{K}\right)$. For a crack-free body, the loading rate is often defined by stress rate $\dot{\sigma}\left(=\frac{\partial\sigma}{\partial t}\right)$ or strain rate $\dot{\varepsilon}\left(=\frac{\partial\varepsilon}{\partial t}\right)$. For crack bodies, the intensity of stress field near crack tip is dominated by the stress intensity factor K, so the loading rate is defined by the rate of stress intensity factor $\dot{K}\left(=\frac{\partial K}{\partial t}\right)$ (its unit is $\mathrm{MPa{\bullet}m^{1/2}{\bullet}s^{-1}}$ or $\mathrm{MN{\bullet}m^{-3/2}{\bullet}s^{-1}}$).

In this way, the critical condition of dynamic initial extension of stationary crack under dynamic loading in crack dynamics, similar to that of quasistatics fracture, can be expressed as:

$$K_I^d(a,\sigma,t) = K_{Id}\left(\dot{K}\right) \tag{9.11}$$

Once the dynamic stress intensity factor K_I^d reaches the critical value of $K_{Id}\left(\dot{K}\right)$, the crack initiates to extend; $K_{Id}\left(\dot{K}\right)$ thus is called the **dynamic fracture toughness** or more accurately called the **dynamic crack initiation toughness** to distinguish with other kind of dynamic fracture toughness defined in other different situation.

In fact, for the continuous growth of propagating (moving/running) crack, there exists a **dynamic crack growth toughness** $K_{ID}(\dot{a})$ which is related to the crack propagating velocity $\dot{a}$. Thus, the **dynamic crack growth criterion** of moving cracks can be expressed as follows:

$$K_I^d(t, \sigma, a, \dot{a}) = K_{ID}(\dot{a}) \tag{9.12a}$$

And for the crack arrest of moving cracks, there similarly exists **dynamic crack arrest toughness**, K_{Ia}, the corresponding **dynamic crack arrest criteria** can be expressed as:

$$K_I^d(t, \sigma, a) < K_{Ia} \tag{9.12b}$$

It should be noted that $K_{Id}(\dot{K})$, $K_{ID}(\dot{a})$, and K_{Ia} are three distinct dynamic fracture toughness of materials.

However, the crack propagating velocity $\dot{a}$ itself is not one of the constitutive parameters of materials. Different crack propagating velocity $\dot{a}$ essentially corresponds to different stress intensity factor rates $\dot{K}$ under different conditions, while $\dot{K}$ is not necessarily dependent uniquely on $\dot{a}$. Therefore, strictly speaking, the $\dot{a}$ in Eq. (9.12) should be replaced by $\dot{K}$, which is a problem that needs to be further studied.

Each of these points will be further discussed below.

9.1.3 Dynamic stress intensity factor of stationary crack under stress wave loading

Assuming that a is the characteristic size of the crack and C is the stress wave velocity, then the t_a ($= a/C$) represents the characteristic time of the structural response of the cracked body under the stress wave loading, and if t_l is used to describe the characteristic time of external dynamic load (such as load duration or rise time), then in case of $t_l \geq t_a$, the interaction between stress wave and crack could be ignored; all equations for crack statics, such as Eq. (9.5) and Eq. (9.7), can be used, just replace the static load σ with the dynamic load $\sigma(t)$, which is the so-called quasistatic treatment. In such case, although the external load still changes with time, the time scale (t_l) to measure this change is much larger than the time scale (t_a) required by stress wave propagating in the characteristic size of cracked body; therefore each moment, the problem can be treated as quasistatic equilibrium state, which makes the problem much simpler. On the contrary, when $t_l \leq t_a$, the effect of stress wave on the mechanical field near the crack tip must be considered, which comes down to the effect of stress wave on dynamic stress intensity factor.

In fact, when $t_l \leq t_a$, the crack will interfere with the propagation of stress waves, and stress waves will diffract at the crack tips, thus affect directly the mechanical field at the crack tip. For example, consider an infinite elastic plate containing a crack with length $2a$ and a step uniform compressive stress σ_0 is suddenly applied on the upper and lower inner surfaces of the crack. By using Laplace transform and numerical inverse transformation, the change of dynamic stress intensity factor $K_I^d(t)$ at the crack tip with time t can be obtained, as shown as the solid curve marked by "normal impact" in Fig. 9.5 for steel ($\nu = 0.29$) (Sih et al., 1972; Chen and Sih, 1977). The figure is drawn in dimensionless form; the longitudinal coordinate is expressed as the dimensionless dynamic stress intensity factor $\overline{K}_I(t)$, i.e., the ratio of dynamic stress intensity factor $K_I^d(t)$ to static stress intensity factor $K_I^s(t)$; in the present case, $K_I^s(t) = \sigma_0\sqrt{a}$, so there is $\overline{K}_I(t) = K_I^d(t)/\sigma_0\sqrt{a}$, and the abscissa is expressed as the dimensionless time $\bar{t}\left(=\frac{t}{t_a}=\frac{C_S t}{a}\right)$, i.e., the ratio of time t to the characteristic time t_a $(= a/C_S)$ representing the time required for shear stress wave to propagate through the crack characteristic size a at the shear stress wave velocity $C_S = \left(\frac{G}{\rho_0}\right)^{1/2}$, where G is the shear elastic modulus.

As can be seen from the figure, when the step uniformly distributed pressure σ_0 is suddenly applied to the upper and lower surfaces of the crack, the dimensionless dynamic stress intensity factor $\overline{K}_I(t)$ increases rapidly and

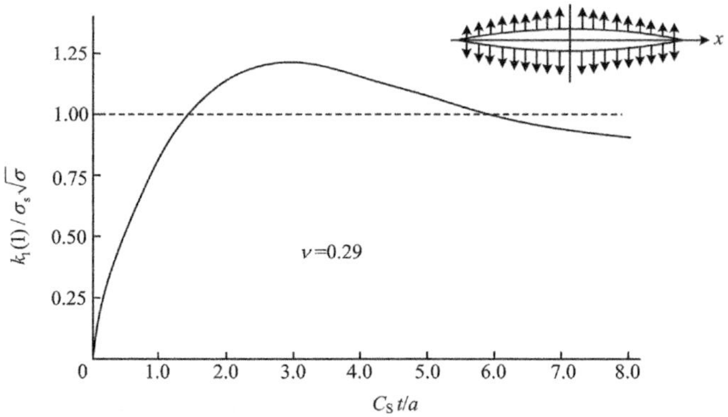

Figure 9.5 Plot of the dimensionless dynamic stress intensity factor versus the dimensionless time. *From Sih G.C., Embley G.T., Ravera R. S., 1972. Impact response of a finite crack in plane extension. Int. J. Solids Struct. 8 (7), 977–993, Fig. 5, p.991. Reprinted with permission of the publisher.*

at the dimensionless time $\bar{t} \approx 3.0$ reaches the maximum value $(\overline{K}_I)_{max}$, which exceeds the static stress intensity factor value by about 20% and then attenuated oscillates up and down at the static value. For steel, if the crack half-length is $a = 25.4$ mm, the time needed to reach the maximum value $(\overline{K}_I)_{max}$ is only about 24 μs.

Let us look at another example of a propagating tensile stress pulse acting on a crack and diffracting at the crack tip. Consider a crack with length $2a$ in an infinite elastic plate; a tensile stress pulse σ_0 is normally incident toward the crack at longitudinal wave velocity of C_L, i.e., the wave front is parallel to the crack surface, as shown in Fig. 9.6 (Meyers, 1994). Since the crack surface is displacement discontinuity in which tensile stress cannot be transmitted, the incident tensile stress wave is reflected on the lower crack surface on the one hand (Fig. 9.6b) and is diffracted at the crack tips on the other hand (Fig. 9.6c). The longitudinal wave with the wave velocity C_L and the shear wave with the wave velocity C_S radiate out with the left and right

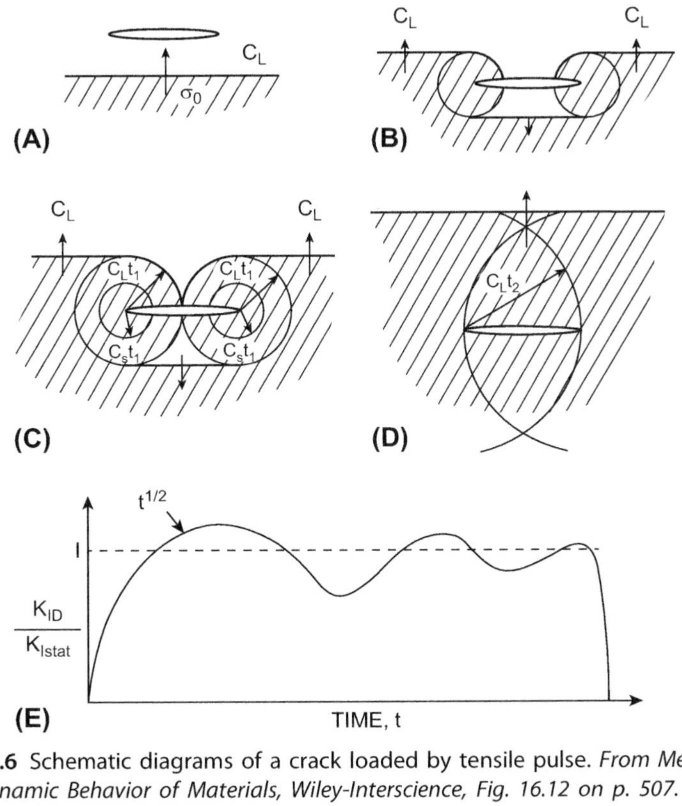

Figure 9.6 Schematic diagrams of a crack loaded by tensile pulse. *From Meyers M.A., 1994. Dynamic Behavior of Materials, Wiley-Interscience, Fig. 16.12 on p. 507. Reprinted with permission of the publisher.*

crack tips as the centers and are interacted each other (Fig. 9.6d). By Laplace transform and Wiener–Hopf method, the variation of the dynamic stress intensity factor $K_I^d(t)$ with time t can be obtained as (Achenbach, 1972):

$$K_I^d = 2\sigma_0 \frac{(1-\nu)^{1/2}}{1-\nu} \sqrt{\frac{C_L t}{\pi}} \tag{9.13}$$

Note that the stress wave loading region of the crack is characterized by $C_L t$ which is the distance swept by the longitudinal wave, and from the comparison between the dynamic Eq. (9.13) with the static Eq. (9.6c), $C_L t$ can be regarded as the effective length of the crack under the stress wave loading. In Eq. (9.13), σ_0, wave velocity C_L, and Poisson's ratio ν are all constants, thus $K_I^d(t)$ is proportional to $t^{1/2}$. The variation of the dimensionless dynamic stress intensity factor $\overline{K}_I(t) = K_I^d(t) \big/ K_I^s(t)$ with $t^{1/2}$ is illustrated in Fig. 9.6e. When the maximum value $\left(\overline{K}_I\right)_{\max}$ is reached, $\overline{K}_I(t)$ begins to oscillate up and down at the static stress intensity factor value due to the interaction of the secondary waves radiated from the two crack tips (Fig. 9.6d) and finally tends to the static stress intensity factor. It can be seen that after the stress wave acts on the crack for a long enough time, it can be treated as quasistatic. However, when the stress wave effect needs to be taken into account, in this case, the maximum value of the dynamic stress intensity factor $\left(\overline{K}_I\right)_{\max}$ is about 25% higher than the static value (Fig. 9.6e), which cannot be ignored.

The above discussion provided us with the main physical characteristics that distinguish dynamic stress intensity factor from static stress intensity factor. It is true that it is a highly complicated and difficult problem to solve the dynamic stress intensity factors for different cracks under the action of different stress waves by analytical method, except for a few simple cases. Fortunately, with the development of numerical calculation method as well as the development of dynamic experimental technique, it can be solved by numerical simulation and experimental measurement, or a combination of the two. The combination of mechanical analysis, numerical simulation, and experiment is the main approach to solve those problems in the future.

For a finite elastic plate (40 mm in length and 20 mm in width) containing a center crack of length $2a$ ($a = 2.4$ mm), loaded by a step uniform tensile load $\sigma(t)$, as shown in Fig. 9.7, it is really difficult to get an analytical solution at the crack tip. The relevant numerical solutions of stress dynamic intensity factor varying with time are obtained by the finite difference method shown as solid curve in Fig. 9.7 (Chen, 1975).

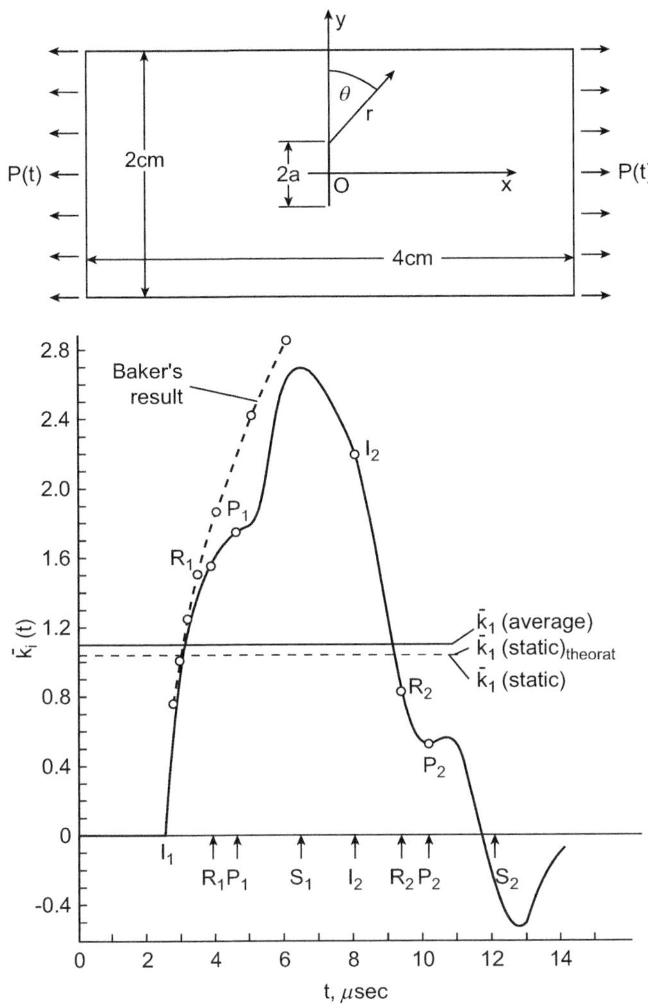

Figure 9.7 Numerical simulation of a finite plate with $2a$ center crack subjected to a uniform step tensile load $\sigma(t)$. *From Chen, Y.M., 1975. Numerical computation of dynamic stress intensity factors by a Lagrangian finite-difference method (the HEMP code). Eng. Fract. Mech. 7 (4), 653–660, Fig. 1, p.654; Fig. 5, p.656. Reprinted with permission of the publisher.*

In the figure, the finite difference solution can track the diffraction process of stress waves at the crack tip (similar to the diffraction process shown in Fig. 9.6A–D). The times that the stress wave arrives at this crack tip, including the longitudinal wave, the surface wave traveling from another crack tip, the pressure wave reflected from the nearest boundary, and the shear wave reflected from the nearest boundary, are indicated on the abscissa of Fig. 9.7 by the symbols *l*, *R*, *P*, and *S*, respectively, and are marked correspondingly with solid dots on the solid line curve, while the subscripts "1" and "2" of these symbols are used to indicate whether the wave is first arrival or second arrival, respectively.

Comparing Fig. 9.7 with Fig. 9.6, it can be seen that there are obvious differences between the finite plate solution and the infinite plate solution. Particularly, the reflected waves from the nearest free boundary (such as P_1 and P_2) of the cracked infinite plate cause an obvious change in the curve $\overline{K}_I(t)$, and the ratio of $K_I^d(t) \Big/ K_I^s(t)$ could reach 2.7. It is also worth noting that due to the influence of the wave reflected from the free boundary of the finite plate, the $\overline{K}_I(t)$ curve exhibits the phenomenon of tension—compression alternation (turning negative after the time S_2). Those phenomena are all important characteristics that distinguish $K_I^d(t)$ from $K_I^s(t)$.

For those who are interested in more detailed analysis of dynamic stress intensity factors under stress wave loading can refer to the related monographs (e.g., Fan Tian-you, 2006; Sih, 1977; Freund, 1990; Ravi-Chandar, 2004).

9.1.4 Loading rate dependence of crack initiation toughness

When the critical condition for the dynamic initial extension of a stationary crack subjected to dynamic loading (initial extension criterion), as shown in Eq. (9.11) below, is applied

$$K_I^d(a,\sigma,t) = K_{Id}\left(\dot{K}\right)$$

The left-hand side represents dynamic stress intensity factor $K_I^d(a,\sigma,t)$ (characterizing dynamic structural response of the cracked body), and the right-hand side is the rate-dependent dynamic fracture toughness $K_{Id}\left(\dot{K}\right)$ (characterizing the dynamic response of materials), both of which are indispensable, be short of one cannot. The determination of $K_I^d(a,\sigma,t)$ has been discussed previously, and how dynamic fracture toughness depends on the loading rate will be discussed in this subsection.

Considering that it is necessary to distinguish the dynamic fracture toughness between the stationary crack and the moving crack, the dynamic fracture toughness for initial extension of stationary crack is called **dynamic crack initiation toughness**, which will be discussed in this subsection; while since the dynamic fracture toughness of moving crack is also dependent on the crack propagating velocity, it is correspondingly called **dynamic crack growth toughness**, which will be discussed in the next section.

The dynamic initiation toughness of materials K_{Id} is generally determined by the critical dynamic stress intensity factor $K_I^d(t_f)$ at the crack initiation time (fracture time) $t = t_f$. Therefore, the critical condition for the dynamic initial extension of a stationary crack (initial extension criterion) can be more accurately expressed as:

$$K_I^d(t) = K_{Id}\left(\dot{K}_I^d(t)\right), \quad t = t_f \tag{9.14}$$

For crack-free bodies, the loading rate is often expressed as stress rate $\dot{\sigma}\left(=\frac{\partial \sigma}{\partial t}\right)$ or strain rate $\dot{\varepsilon}\left(=\frac{\partial \varepsilon}{\partial t}\right)$. For cracked bodies, the stress field at crack tip is dominated by the stress intensity factor K, so its loading rate should be expressed by the rate of stress intensity factor $\dot{K}\left(=\frac{\partial K}{\partial t}\right)$. Since $K_I \propto \sigma a^{1/2}$ (Eq. 9.6), for stationary cracks (a is constant) in linear elastic medium, there exists obviously the following relationship between $\dot{K}\left(=\frac{\partial K}{\partial t}\right)$, $\dot{\sigma}\left(=\frac{\partial \sigma}{\partial t}\right)$, and $\dot{\varepsilon}\left(=\frac{\partial \varepsilon}{\partial t}\right)$:

$$\begin{aligned}\dot{K}_{\rm I} &= \frac{\partial K_{\rm I}}{\partial t} = \frac{K_{\rm I}}{\sigma}\frac{\partial \sigma}{\partial t} = K_{\rm I}\frac{\partial \log \sigma}{\partial t} \\ &= \frac{K_{\rm I}}{\varepsilon}\frac{\partial \varepsilon}{\partial t} = K_{\rm I}\frac{\partial \log \varepsilon}{\partial t}\end{aligned} \tag{9.15a}$$

In some cases, assuming that $K_I(t)$ is approximately a linear function of t when $t \le t_f$, then there is

$$\dot{K}_I = \frac{K_I(t_f)}{t_f} \quad \text{or} \quad t_f = \frac{K_I(t_f)}{\dot{K}_I} \quad \text{or} \quad \frac{\dot{K}_I}{K_I(t_f)} = \frac{1}{t_f} \tag{9.15b}$$

That is, the loading rate can also be approximately expressed by the change of K_I^d with t_f (which is inversely proportional to $\dot{K}$). Different forms of loading rate are alternatively used by different researchers; readers can understand the inherent connection through the relationship given by Eq. (9.15).

The loading rates of some structures under operating conditions are listed in Table 9.2 (Fan Tian-you, 2006) for readers' reference.

In general, according to the order of magnitude of stress intensity factor rate $\dot{K}$, in crack dynamics, they can be divided into the following three cases (Nilsson, 1984):

(1) $0 < \dot{K} < 10^3\ \text{MPa}\cdot\text{m}^{1/2}\cdot\text{s}^{-1}$, low-speed loading, equivalent to quasistatic loading of universal material testing machine;

Table 9.2 Loading rates of some structures under operating conditions (Fan Tian-you, 2006).

Structures	$\dot{\varepsilon}$ s^{-1}	$\dot{\sigma}$ kgf·cm^{-2}·s^{-1}	$\dot{K}$ kgf·mm$^{3/2}$·s^{-1}
Construction buildings, bridges, cranes	$<10^{-3}$	—	10^3
Aircraft landing gears	—	10^2	10^4
Bulldozers and handling machinery	—	10^3	10^5
Ships collision	—	10^4	10^6
Facilities subject to explosion and gunfire	—	10^9	10^{11}
Artillery firing	Peak: 10^6—10^7 Average: 10^4—10^5		
Shaped charge jet	Peak: 10^6—10^7 Average: 10^4—10^5		
Self-forging fragment	Peak: 10^6—10^7 Average: 10^4—10^5		

(2) 10^3 MPa·m$^{1/2}$·s^{-1} $< \dot{K} <$ 10^5 MPa·m$^{1/2}$·s^{-1}, medium-speed loading, equivalent to the dynamic loading of Charpy test, and inertia effect in fact should be considered; and

(3) $\dot{K} > 10^5$ MPa·m$^{1/2}$·s^{-1}, high-speed loading, equivalent to dynamic loading of the split Hopkinson bar test and the plate impact test, where stress wave effect must be considered.

As for how K_{Id} depends on the loading rate, it mainly relies on experimental studies. However, a series of key problems need to be solved including: the experiment device with dynamic loading rate, the determination of the history of the rapid change of dynamic stress intensity factor $K_I(t)$ with time t, and the accurate determination of crack initiation time t_f, etc.; it is obviously a very challenging and complicated problem.

In the early studies using the Charpy impact test apparatus and drop-weight impact apparatus because the stress wave effect of the structure and the rate effect of the material could not be well distinguished, the measurement accuracy was not enough, and the procedures for the interpretation of drop-weight and instrumented impact tests are far from satisfactorily established, thus the results obtained by different researchers were often contradictory. It makes the analysis less reliable and is difficult to reach a consensus. Since 1980s, relying on the combination of advanced experimental techniques with advanced numerical simulation, many valuable research results have been achieved. Some representative results are given below.

The first directly experimental measurement of the loading rate dependence of dynamic initiation toughness was provided by Ravi-Chandar and Knauss (1984). The specimen configuration was a semiinfinite crack in an unbounded medium; a uniform pressure pulse was applied upon the upper and lower inner surfaces of the crack, which was generated by an electromagnetic loading device and its amplitude and duration are adjustable. A high-speed camera (200,000 frames/second) combined with optical caustics method was used to determine the dynamic stress intensity factor $K_I(t)$ and the crack initiation time t_f. A series of experiments were performed at different stress intensity factor rate $\dot{K}$ in the range of 10^4-10^5 $\text{MPa}\cdot\text{m}^{1/2}\cdot\text{s}^{-1}$. The variation of dynamic initiation toughness K_{Id} with loading rate is directly measured. The results of rate dependence on $K_{Id}(t)$ for brittle polyester (Homalite-100) are shown in Fig. 9.8.

Note that in Fig. 9.8, the dynamic initiation toughness K_{Id} is plotted against the initiation time, t_f, which is inversely proportional to $\dot{K}_I$ (Eq. 9.15b). Larger t_f corresponds to lower $\dot{K}_I$, while smaller t_f corresponds to higher $\dot{K}_I$. Therefore, the relationship of K_I^d and t_f actually reflects the loading rate dependence of K_{Id}. In the figure, the corresponding values of

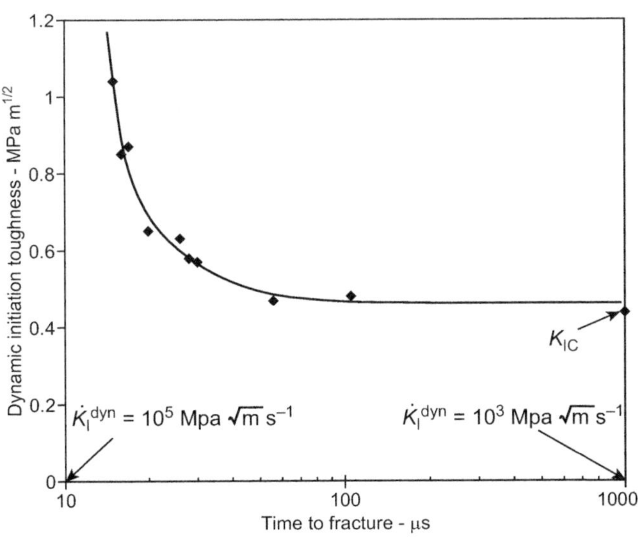

Figure 9.8 Dynamic initiation toughness K_{Id} for polyester (Homalite-100) variation with loading rate. *From Ravi-Chandar K., 2004. Dynamic Fracture, Elsevier Ltd., Oxford, Fig. 10.5, p.162. Reprinted with permission of the publisher.*

stress intensity factor rate have been given at both ends of the abscissa t_f, including the quasistatic stress intensity factor rate $\dot{K}_I = 10^3$ MPa·m$^{1/2}$·s^{-1} and the corresponding quasistatic plane strain fracture toughness of the material, K_{Ic}. Clearly, in the range of low loading rate, $\dot{K}_I^d \leq 10^4$ MPa·m$^{1/2}$·s^{-1}, the dynamic initiation toughness is almost equal to the quasistatic plane strain fracture toughness, while in the range of high loading rate, $\dot{K}_I^d > 10^4$ MPa·m$^{1/2}$·s^{-1}, K_{Id} increases rapidly with the loading rate. It is believed that such kind of rate-hardening effect of K_{Id} is related to the rate-hardening distortional process as well as the nucleation-growth-coalescence process of crack tip mesocracks controlled by the rate-dependent nonlinear viscoelastic behavior of the polymer material.

The relation between the K_{Id} and the t_f is actually a reflection of the relation between the dynamic initiation toughness K_{Id} and the stress intensity factor rate $\dot{K}_I$. If the same experimental results plotted in the K_{Id}–t_f coordinates, like those in Fig. 9.8, are replotted in the K_{Id}–$\dot{K}_I^d$ coordinates, or in the $K_{Id}/K_{Ic}^S - \dot{K}_I^d/K_{Ic}^S$ coordinates, the corresponding diagram will be shown schematically as the diagram in Fig. 9.9.

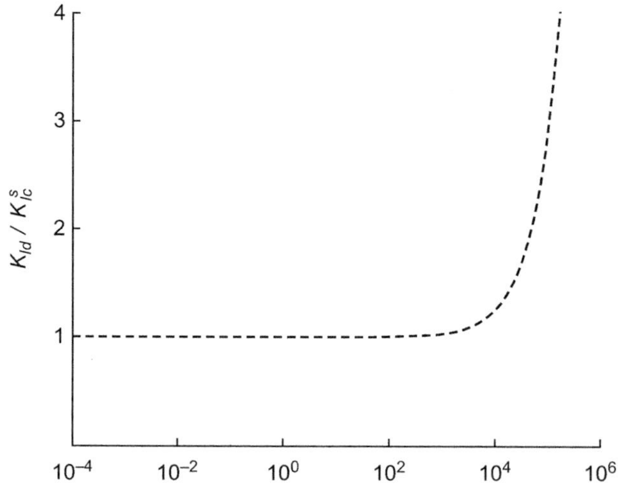

Figure 9.9 Schematics of the dependence of dimensionless dynamic initiation toughness K_{Id}/K_{Ic}^S on the loading rate $\dot{K}_I^d/K_{Ic}^S$, schematically corresponding to Fig. 9.8.

In fact, researches by Rosakis et al. (Rosakis et al., 1999; Bhat et al., 2012) do show that when the loading rate is expressed by the ratio of $\dot{K}_I^d / K_{Ic}^S$ (unit s^{-1}), as shown in Fig. 9.9, for the aluminum alloy 2024-T3, titanium alloy Ti–6Al–4V, polyester Homalite-100, and epoxy/graphite fiber composite, the dynamic initiation toughness K_{Id} is almost equal to the quasistatic plane strain fracture toughness K_{Ic} in the range of low loading rate of $\dot{K}_I^d / K_{Ic}^S \leq 10^4\ s^{-1}$, while increases rapidly in the range of high loading rate of $\dot{K}_I^d / K_{Ic}^S \geq 10^4 s^{-1}$.

For rock-like materials, the research results from different researchers have consistently shown that the dimensionless dynamic initiation toughness $K_{Id} / K_{I\,c}^S$ increases monotonically with the loading rate, as shown in Fig. 9.10 (Zhang and Zhao, 2014) In the figure, the symbol F, L, and Y represents the rock material of Fangshan marble, Laurentian granite and Ya'an marble separately. The results for Plexiglas PMMA, polyester (Homalite), and epoxy/graphite fiber composite (Rosakis et al., 1999) are also shown in same figure for comparison. It can be seen that rate dependence of rock-like materials and polymers has a similar trend, while the

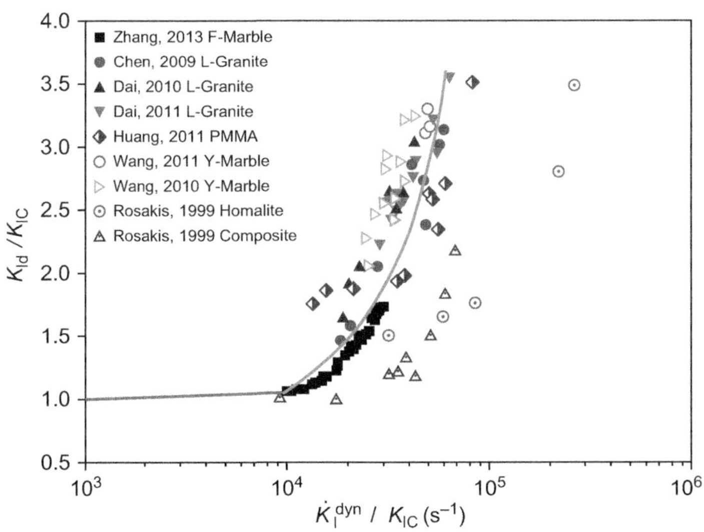

Figure 9.10 Dependence of dimensional dynamic initiation toughness K_{Id}/K_{Ic} on loading rate for some rock-like materials. *From Zhang Q.B., Zhao J., 2014. A review of dynamic experimental techniques and mechanical behaviour of rock materials, Rock Mech. Rock Eng. 47, 1411–1478, Fig. 40, p.1451. Reprinted with permission of the publisher.*

experimental data points of rocks are all above those of polymers, i.e., the K_{Id}/K_{Ic}^S for rocks is higher than that of polymers at the same loading rate.

However, there are also reports of K_{Id} remaining almost unchanged or even decreasing with increasing loading rate. Yokoyama (1993) reported that the dynamic initiation toughness K_{Id} for 7075-T6 aluminum alloy almost remains unchanged at the loading rate $\dot{K}_I$ in the range of $0.5–10^5$ $\text{MPa}\cdot\text{m}^{1/2}\cdot\text{s}^{-1}$.

Another example is the results of K_{Id} for epoxy resin Araldite B reported by Kalthoff (1986) as shown in Fig. 9.11. In the same figure, the data obtained from fast-tension tests and drop-weight tests, as well as the static fracture toughness K_{Ic} are also given. As can be seen, the results of dynamic initiation toughness K_{Id} for epoxy resin Araldite B do not show a significant dependence on loading rate. The data are scattered around a value which is identical with the static fracture toughness K_{Ic} of the material.

Klepaczko (1990) performed a series tests to aluminum alloys PA6 and DTD 502A; the results are shown in Fig. 9.12. As can be seen from the figure, the dynamic initiation toughness K_{Id} even decreases with increasing loading rate exhibiting a negative sensitivity of loading rate, which is different from the results on 7075-T6 aluminum alloy (rate insensitive) by Yokoyama (1993) and from the results on 2024-T3 aluminum alloy (a positive rate sensitivity) by Rosakis et al. (1999).

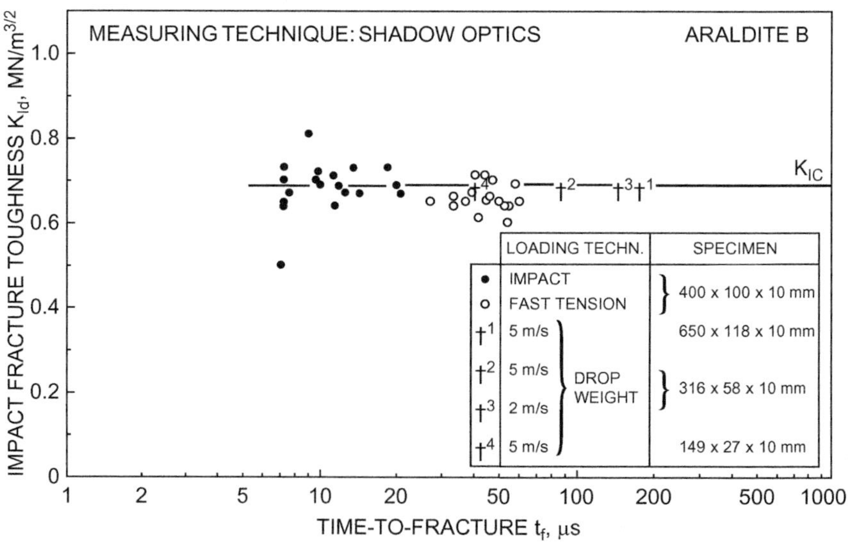

Figure 9.11 Change of dynamic initial fracture toughness of epoxy resin Araldite B with loading rate. *From Kalthoff J.F., 1986. Fracture behavior under high rates of loading, Eng. Fract. Mech. 23, 289–298, Fig. 8, p.296. Reprinted with permission of the publisher.*

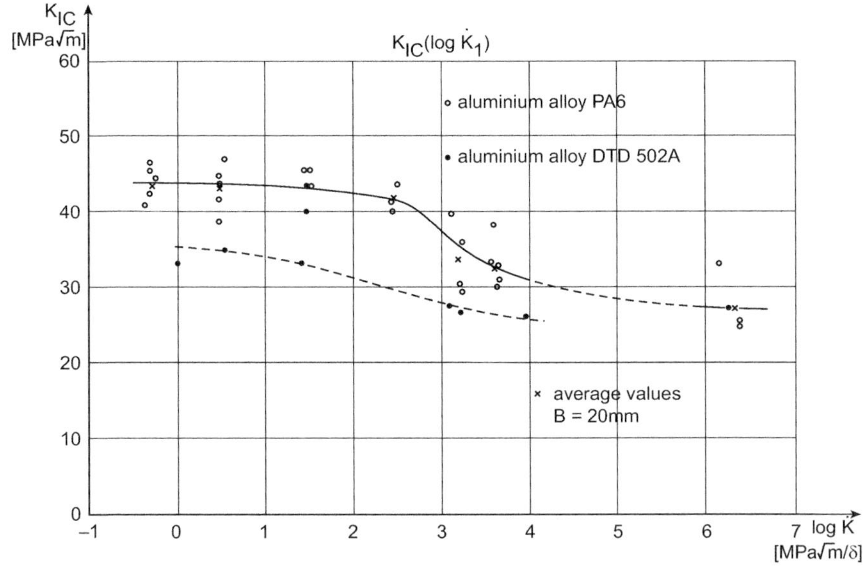

Figure 9.12 Dependence of dynamic initiation toughness of PA6 and DTD 502A aluminum alloys on loading rate. *From Klepaczko J.R., 1990. Dynamic crack initiation, some experimental methods and modeling, In: Klepaczko J.R. (Ed.), Crack Dynamics in Metallic Materials, Springer, Vienna, 255—454, Fig. 5.13, p.347. Reprinted with permission of the publisher.*

Kalthoff's (1986) results on a high strength steel X2NiCoMo18-9-5 are shown in Fig. 9.13. The loading rate used in those experiments was in the range of stress intensity factor rate $\dot{K}_I$ of 1×10^6 to 1×10^7 $\text{MPa}\cdot\text{m}^{1/2}\cdot\text{s}^{-1}$. In this range, the dynamic initiation toughness K_{Id} of the high strength steel does not vary monotonically with the loading rate (characterized by crack initiation time t_f, and a decrease in t_f corresponding to an increase in $\dot{K}_I$, as shown in Eq. (9.15b)), exhibits an initial drop in the dynamic initiation toughness then followed by a sharp increase at higher loading rate. The K_{Id} increasing once again under higher loading rate may be related to the stress wave effect (inertia effect) on stress intensity factor at crack-tip field for crack initiation under extremely short-duration pulse loading as well as the adiabatic temperature rise effect under high loading rate, or the coupling effect of the two.

In Chapter 5, we discussed the strain-rate effect of dynamic distortion law of materials. It has been pointed out that there is a certain rate—temperature equivalence between strain-rate effect and temperature effect, that is, lowering temperature is equivalent to increasing strain rate.

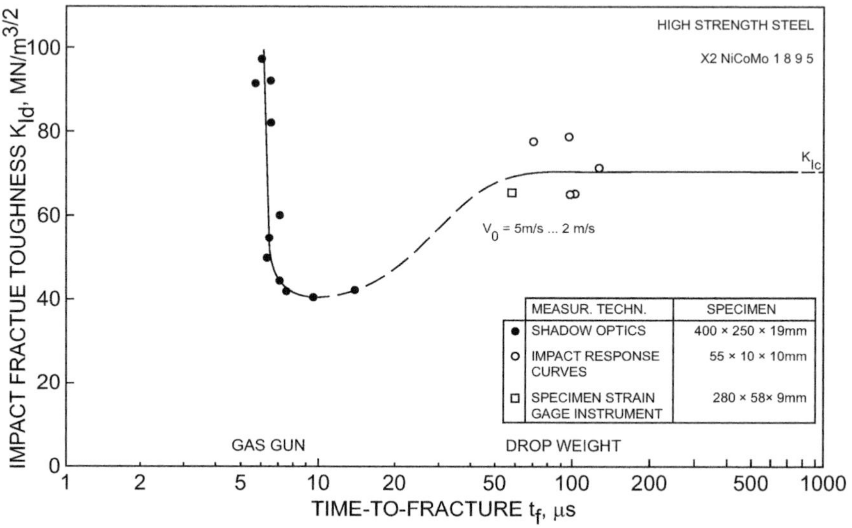

Figure 9.13 Change of dynamic initial fracture toughness of high strength steel X2NiCoMo18 9.5 with loading rate. *From Kalthoff J.F., 1986. Fracture behavior under high rates of loading, Eng. Fract. Mech. 23, 289–298, Fig. 9, p.296. Reprinted with permission of the publisher.*

Therefore, the researchers also pay attention to loading rate and temperature dependence on dynamic initiation toughness of materials.

Klepaczko (1984) has performed tests to determine the loading rate and temperature dependence on dynamic initiation toughness K_{Id} of A533B steel used in reactor, in which the Hopkinson bar apparatus is used to generate impact load on cracked specimens at a stress intensity factor rate $\dot{K}_I$ from 3×10^0 to 1×10^5 $MPa \cdot m^{1/2} \cdot s^{-1}$. The results are shown in Fig. 9.14. The dynamic initiation toughness K_{Id} was found to decrease with increasing loading rate (negative rate sensitive) while to increase with increasing temperature.

In contrast, Fig. 9.15 shows the experimental results of Wilson et al. (1980) on dynamic initial toughness and quasistatic fracture toughness for 1018 cold-rolled steel and 1020 hot-rolled steel under different temperatures. The loading rate for measuring dynamic initiation toughness K_{Id} is $\dot{K}_I = 2 \times 10^6$ $MPa \cdot m^{1/2} \cdot s^{-1}$, while for quasistatic fracture toughness K_{Ic} is $\dot{K}_I = 1$ $MPa \cdot m^{1/2} \cdot s^{-1}$.

The following two important features can be found in Fig. 9.15:

(1) Both steels at low temperature appear as brittle fracture shown as cleavage fracture by fracture microscopic analysis, and at the higher

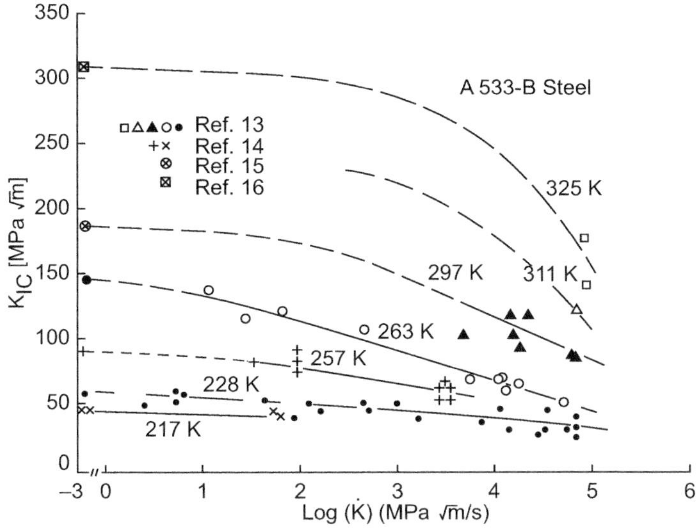

Figure 9.14 Change of dynamic initial fracture toughness K_{Id} of A533B reactor steel with temperature and loading rate. *From Klepaczko, 1984, Fig. 5, p.184. Reprinted with permission of the publisher.*

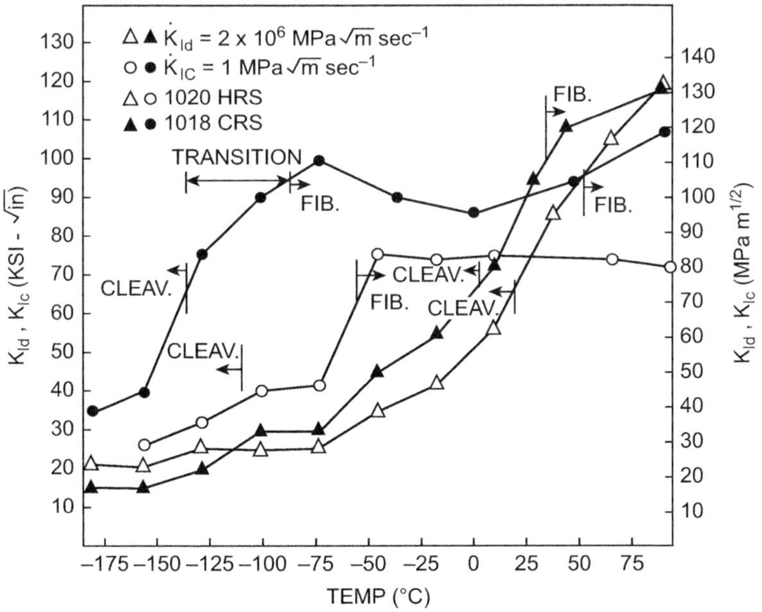

Figure 9.15 Dependence of K_I^d and K_{Ic} of 1018 cold-rolled steel and 1020 hot-rolled steel on temperature. *From Wilson M.L., Hawley R.H., Duffy J., 1980. The effect of loading rate and temperature on fracture initiation in 1020 hot-rolled steel. Eng. Fract. Mech. 13 (2), 371–385, Fig. 13, p.383. Reprinted with permission of the publisher.*

temperatures transformed into ductile fracture shown as fibrous fracture by fracture microscopic analysis. Thus, there is a ductile—brittle transition temperature zone, which the fracture mechanism changes from brittle cleavage to ductile fibrous fracture. The ductile—brittle transition temperature is often characterized by the upper limit temperature of brittle fracture, which is called nil ductility transition temperature (NDT). It can be found from the figure that the NDT increases with the increase of loading rate.

(2) In the brittle cleavage fracture zone, the dynamic initiation toughness K_{Id} is lower than the quasistatic fracture toughness K_{Ic} (negative rate sensitivity), while in the ductile fibrous fracture zone, the dynamic initiation fracture toughness K_{Id} is higher than the quasistatic fracture toughness K_{Ic} (negative rate sensitivity). It indicates whether the rate sensitivity of dynamic initiation toughness K_{Id} is positive or negative is closely related to the fracture mechanism.

Taking a comprehensive view of the above experimental results, we can find that the rate dependence of dynamic initiation toughness K_{Id} is much more complex than that of the material distortion law discussed in the part 2. This complexity is obviously related to the following factors:

(1) Firstly, the rate dependence of materials constitutive relation will directly affect the rate dependence of dynamic initiation fracture toughness K_{Id}. The higher the loading rate is, the more important the rate dependence of the constitutive relation (nonlinear viscoelastic, nonlinear visco elasto—plastics etc.) will play in the process zone of crack tip. For example, the constitutive relation of aluminum alloy is less sensitive to the strain rate, so its K_{Id} is also relatively insensitive to loading rate $\dot{K}_I$, but at higher $\dot{K}_I$, the K_{Id} of aluminum alloy also exhibits a certain rate sensitivity.

(2) The experimental determination of dynamic initiation toughness K_{Id} of materials is mainly by measuring the critical dynamic stress intensity factor $K_I^d(t_f)$ when $t = t_f$ (initiation time). The initiation time t_f is usually in the order of magnitude $10-10^2$ μs (even in the order of 10^0 μs); therefore, the dynamic stress intensity factor $K_I^d(t)$ at crack tip has not yet to reach the quasistatic and dynamic equilibrium state, and the stress wave effect (inertia effect) must be considered in the test. The coupling of strain-rate effect and stress wave effect make the research extremely complicated.

(3) The process of high loading rate is an approximate adiabatic process, and the thermomechanical coupling process due to adiabatic temperature rise will lead to the coupling of strain rate effect and temperature effect, and strengthen the positive effect of strain rate on dynamic initiation toughness, which is no longer a simple mechanical process.

(4) Considering the combined action of loading rate effect and temperature effect, the crack initiation mechanism covers the whole range from brittle cleavage fracture to ductile fibrous fracture. Due to the difference of initiation mechanism, different rate sensitivity types such as "positive rate sensitivity", "rate insensitivity" and "negative rate sensitivity" can be observed.

Anyhow, the rate dependence of dynamic initiation fracture toughness K_{Id} is a problem that remains to be further studied.

9.1.5 Kinetic energy and limiting propagating speed of moving crack

Once the critical condition for initial extension of a stationary crack subjected to dynamic loading (initiation criterion) is satisfied (see Eq. 9.11), i.e.,

$$K_{\mathrm{I}}^{d}(a, \sigma, t) = K_{\mathrm{Id}}(\dot{K})$$

the crack must initiated and subsequently grow at a certain speed $\left(\dot{a} = \frac{da}{dt}\right)$ and thus becomes a propagating/moving crack. Differing from the problems of stationary cracks, two new factors are introduced: firstly, moving cracks propagate with kinetic energy and the kinetic energy required for crack propagation must be taken into account in the total energy balance of the system; secondly, the mechanical field at the crack tip is now at a high loading rate caused by the crack propagating speed $\dot{a}$, and thus the dynamic stress intensity factor K_I^d and dynamic fracture toughness K_{Id} of propagating crack will both be a function of crack propagating speed $\dot{a}$.

The dynamic initial extension of stationary crack under dynamic load actually involves only one state point in the process of fracture, while the rapid propagating of crack is involved in a process, including the acceleration or deceleration of moving crack, the limiting propagating speed, the crack bifurcation (crack branching), the stop of propagation or the so-called crack arrest, and so on. To describe such a dynamic process, the complexity of the problem is imaginable.

Mathematically, the problem is a moving boundary problem. It is difficult to find its analytical solution except for some special cases. Therefore, we will focus here on the physical nature of this problem, such as the estimation of kinetic energy for propagating crack, the influence of kinetic energy on crack propagation, the effects of crack speed $\dot{a}$ on dynamic stress intensity factor K_I^d and dynamic fracture toughness K_{Id}, etc.

9.1.5.1 Kinetic energy of propagating cracks

The significant difference between propagating crack and stationary crack lies in the kinetic energy of the propagating crack, which as an expression of inertial effect is not negligible and must be considered first.

As we pointed out at the beginning of Section 9.1, from the perspective of development history of crack mechanics, researchers have explored the same crack problems from two distinct approaches, i.e., the energy approach and the mechanical field approach. The kinetic energy of propagating crack is also discussed, respectively, by these two approaches.

Mott (1948)[9.32] first studied the kinetic energy of propagating cracks based on dimensional analysis of crack-tip field, which is considered as a milestone in crack initiation dynamics.

Considered the extension of the central crack in the infinite elastic plate from the critical size of instability $2a_c$ to $2a$, as shown in Fig. 9.16. According to the solution of crack statics (Eq. 9.6), the components u and v of the displacement field near crack tip in the X-axis and the Y-axis, respectively, have the following forms:

$$\left.\begin{aligned} u &\propto \frac{K_1}{E}\sqrt{r}f_1(\theta) \\ v &\propto \frac{K_1}{E}\sqrt{r}f_2(\theta) \end{aligned}\right\} \tag{9.16a}$$

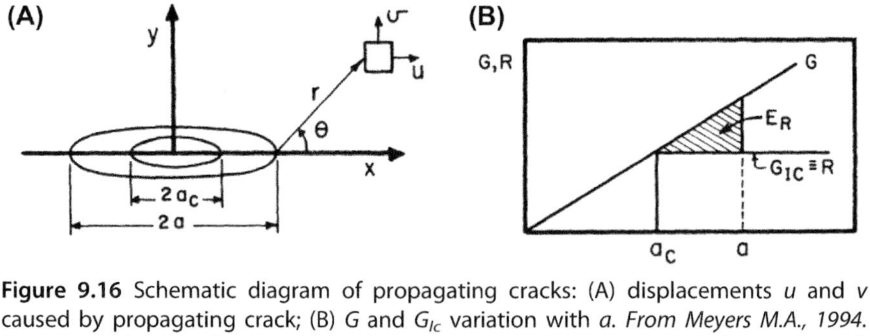

Figure 9.16 Schematic diagram of propagating cracks: (A) displacements u and v caused by propagating crack; (B) G and G_{Ic} variation with a. *From Meyers M.A., 1994. Dynamic Behavior of Materials, Wiley-Interscience, Fig. 16.7, p.498? Reprinted with permission of the publisher.*

where r and θ are polar coordinates measured from the crack tip (see Fig. 9.16a), and $f_1(\theta), f_2(\theta)$ are two functions of angular distribution.

Because of $K_I \propto \sigma\sqrt{\pi a}$, the above equations can also be written as follows:

$$\left.\begin{aligned} u &\propto \frac{\sigma}{E}\sqrt{ar}f_1(\theta) \\ v &\propto \frac{\sigma}{E}\sqrt{ar}f_2(\theta) \end{aligned}\right\} \tag{9.16b}$$

From the dimensional analysis, a is the only quantity with length dimension in this problem. Obviously, $r \propto a$, so the above equations can be expressed as:

$$\left.\begin{aligned} u &= c_1\sigma\ a/E \\ v &= c_2\sigma\ a/E \end{aligned}\right\} \tag{9.16c}$$

where c_1 and c_2 are proportionality constants and E is a material constant. If it is also assumed that the external load σ does not change with time when the crack propagates rapidly, then only the crack length $a(t)$ is a function of time t. Differentiate u and v in the above equation with respect to t.

$$\begin{aligned} \dot{u} &= c_1\sigma\ \dot{a}/E \\ \dot{v} &= c_2\sigma\ \dot{a}/E \end{aligned} \tag{9.17}$$

The total kinetic energy E_k of propagating crack is obtained by the integration of the square of velocity field:

$$E_k = \frac{1}{2}\rho\int\int_{\Omega}\left(\dot{u}^2 + \dot{v}^2\right)dxdy \tag{9.18}$$

where ρ is the density of the material and Ω represents the area of integrated areas. Substituting Eq. (9.17) into Eq. (9.18) to obtain:

$$E_k = \frac{1}{2}\rho\dot{a}^2\frac{\sigma^2}{E^2}\int\int_{\Omega}\left(c_1^2 + c_2^2\right)dxdy \tag{9.19}$$

In the case of the infinite plate discussed, the crack size a is the only length-dependent parameter, so that the area Ω has the same dimensions as a^2. The integral at the right hand of Eq. (9.19) has the same dimension as a^2, so it can be obtained:

$$E_k = \frac{1}{2}k\rho a^2\dot{a}^2\frac{\sigma^2}{E^2} \tag{9.20}$$

where k is an unknown undetermined constant.

On the other hand, the kinetic energy of propagating crack can also be studied from the energy approach, that is, the Griffith energy release rate criterion of crack statics (Eq. 9.5b) can be extended to propagating crack by adding kinetic energy term in the energy balance equation.

If we assume that the crack extension resistance $R =$ constant ($=G_{Ic}$), and the strain energy release rate G is always greater than the R at crack tip, then the crack will have sufficient driving force to extend at a certain propagating speed $\dot{a}$. This excess energy (G-R) determines how much energy can be converted into kinetic energy, which determines the propagation speed of crack.

Since G and R represent the energy per unit crack extension, when the crack extends from the critical size of instability $2a_c$ to $2a$, i.e., extends a value of Δa (see Fig. 9.16a), the two corresponding energy values are $G\Delta a$ and $R\Delta a$, respectively, so

$$\Delta E_k = G\Delta a - R\Delta a \tag{9.21a}$$

The integral of Eq. (9.21a), that corresponds to the shaded area shown in Fig. 9.16b, can be expressed as:

$$E_k = \int_{a_c}^{a} (G - R)da \tag{9.21b}$$

The external load σ can be assumed to be unchanged. According to Eq. (9.5b), the strain energy release rate G and the crack extension resistance $R(=G_{Ic})$ as material characteristic constants are respectively:

$$\left.\begin{aligned} G &= \pi\sigma^2 a/E \\ R &= \pi\sigma^2 a_c/E \end{aligned}\right\} \tag{9.5b★}$$

Thus, (Eq. 9.21b) is turned into:

$$E_k = -R(a - a_c) + \int_{a_c}^{a} \frac{\pi\sigma^2 a}{E} da \tag{9.22}$$

Substituting R in Eq. (9.5b★) into Eq. (9.22):

$$E_k = \frac{\pi\sigma^2}{2E}(a - a_c)^2 \tag{9.23}$$

Eqs. (9.20) and (9.23) are the total kinetic energy of a propagating crack derived by mechanical field approach and energy approach, respectively. Both of them are consistently proportional to $\sigma^2 a^2$. Under the given external load σ, the kinetic energy increases with the crack extension according to the square relation of crack size a. Eq. (9.20) is also proportional to $\dot{a}^2$, but independent of the critical size of instability a_c, while Eq. (9.23) is proportional to $\left(1-\frac{a_c}{a}\right)^2$, but independent of the crack propagating speed $\dot{a}$.

9.1.5.2 Limiting crack propagating speed

Comparing the equivalent kinetic energy expressions Eqs. (9.20) and (9.23), we have:

$$\dot{a} = \sqrt{\frac{\pi}{k}}\sqrt{\frac{E}{\rho}}\left(1-\frac{a_c}{a}\right) \tag{9.24}$$

where $\sqrt{E/\rho} = C_0$ is the velocity of longitudinal elastic wave in one-dimensional stress state, which has the following relationship with the longitudinal wave velocity C_l, shear wave velocity C_s, and Rayleigh surface wave velocity C_R under the three-dimensional stress state:

$$\begin{aligned} C_0 &= \sqrt{\frac{(1+\nu)(1-2\nu)}{(1-\nu)}}C_l = \sqrt{2(1+\nu)} \\ C_s &= \sqrt{2(1+\nu)}\left(\frac{1+\nu}{0.862+1.14\nu}\right)C_R \end{aligned} \tag{9.25}$$

where ν is Poisson's ratio. Therefore, Eq. (9.24) can be rewritten as a dimensionless form as follows:

$$\frac{\dot{a}}{C_w} = k_w\left(1-\frac{a_c}{a}\right) \tag{9.26}$$

In the above equation, C_w can be any one of C_0, C_l, C_s, and C_R, just change the corresponding k_w on the right side of the equation. It can be seen that the key is the determination of the value of k_w, which determines the limit crack propagating speed.

The k_w can be estimated by the value at $a_c/a \to 0$. For example, in case of the one-dimensional stress longitudinal elastic wave velocity C_0, as shown in Fig. 9.17, $k_0 \to 0.38$, this yields (Broek, 1982; Meyers, 1994):

$$\frac{\dot{a}}{C_0} = 0.38\left(1-\frac{a_c}{a}\right) \tag{9.27a}$$

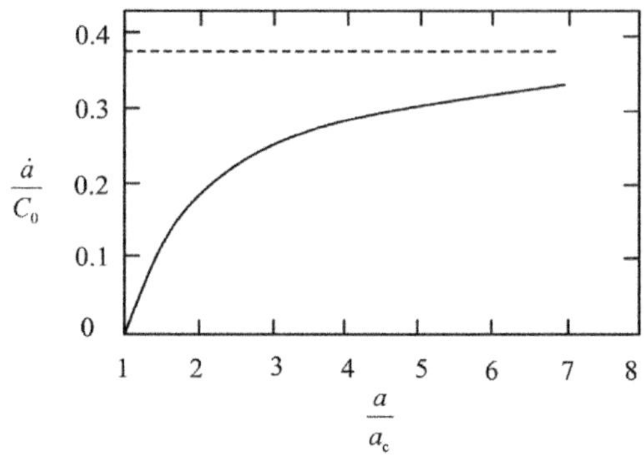

Figure 9.17 Variation of crack propagating velocity $\dot{a}$ with crack extension length a according to Eq. (9.27a).

This means that when the crack extends from the instability critical size a_c to a enough distance to make $a >> a_c$, the crack propagating velocity will increase from its initial speed of 0 to its limit speed $0.38C_0$.

On the other side, the researches by Broberg (1960) and Freund (1972) show that the crack limiting propagating speed is the Rayleigh surface wave speed CR, i.e.,

$$\frac{\dot{a}}{C_R} = \left(1 - \frac{a_c}{a}\right) \tag{9.27b}$$

If it is considered that the energy required to generate a new crack surface in the process of dynamic crack propagation is transmitted by surface waves, it is not difficult to understand why the crack-limiting propagating speed does not exceed the surface wave velocity C_R. However, there are also studies on "intersonic" crack propagation, that is, the crack propagating speed is between the transverse wave velocity C_s and the longitudinal wave velocity C_l: $C_s < \dot{a} < C_l$ (Ravi-Chandar, 2004).

The measured crack propagation velocity of brittle crack is much lower than the theoretical value calculated by Eq. (9.27), as shown in Table 9.3 (Meyers, 1994). Ravi-chandar and Knauss (1984a) believe that it is related to the formation of microcracks in the "process zone" ahead of the brittle crack tip. For ductile crack propagation, the measured crack propagating velocity is also considerably lower than the theoretical value because of additional energy dissipation due to the plastic deformation and the nucleation-growth-coalescence process of microvoids in the "process zone" ahead of the crack tip.

Table 9.3 The measured crack propagating velocity for several brittle materials (Meyers, 1994).

Materials	Crack velocity/ longitudinal wave velocity $\dot{a}/C_l$	Crack velocity/ surface wave velocity $\dot{a}/C_R$	Crack velocity (m/s)
Glass	0.29	—	1500
Steel	0.20	—	1000
Glass/Steel	0.28	—	1400
Cellulose acetate	0.37	—	400
Glass	0.29	0.51	
	0.28	0.47	
	0.30	0.52	
	0.39	0.66	
Plexiglas	0.33	0.58	
	0.36	0.62	
	0.36	0.62	
Homalite-100	0.19	0.33	357
	0.22	0.38	411
	0.25	0.41	444
	0.27	0.45	487
AISI 4340 steel	0.21	0.30	1100

9.1.6 Crack mechanical field near crack tip of propagating crack

As mentioned earlier, once the following critical condition of dynamic initial extension of stationary crack under dynamic loading (the initial extension criterion) is satisfied, i.e.,

$$K_I^d(t) = K_{Id}(\dot{K}), \quad t = t_f \tag{9.11$\star$}$$

the crack will extend at a speed $\dot{a} = \frac{da}{dt}$ and thus become a propagating/ moving crack. It is generally considered that when the crack propagation speed $\dot{a}$ is fast enough, for example, $\dot{a} > 0.2C_R$ (Freund, 1972), on the one hand, $\dot{a}$ will affect the dynamic stress intensity factor of crack by inertia effect (for example, the stress wave radiated continuously from the crack tip of a moving crack will affect the mechanics field at the crack tip, etc.), and on the other hand, the dynamic fracture toughness K_{ID} of material will be affected due to the loading rate effect. Therefore, the dynamic initial extension criterion (Eq. 9.11★) is no longer applicable to the moving cracks and must be replaced by the following **dynamic crack growth criterion** (Ravi-Chandar and Knauss, 1984; Freund, 1972):

$$K_I^d(\sigma(t), a(t), \dot{a}(t)) = K_{ID}(\dot{a}(t)), \quad t > t_f \tag{9.28}$$

The above formula is a nonlinear first-order differential equation for $a(t)$, also called the motion equation for the crack tip by Freund (1990). The dynamic stress intensity factor $K_I^d(\sigma, a, \dot{a})$ on the left side of the equation characterizes the dynamic structural response of the crack-tip mechanical field of the body with moving crack, while the dynamic crack growth toughness $K_{ID}(\dot{a}(t))$ on the right side of the equation characterizes the dynamic response of the material against the growth of moving crack. Here, the subscript of dynamic crack growth toughness $K_{ID}(\dot{a}(t))$ is capital D in order to distinguish it from the dynamic crack initiation toughness K_{Id}, of which the subscript is lowercase d.

In this section, the influence of crack propagation speed on the crack-tip mechanical field and the stress intensity factor will be first discussed. The influence of crack propagation speed on the dynamic crack growth toughness will be discussed in the next section.

Firstly, let us consider a simple case, in which a plane strain crack propagates at a constant speed. For such a stationary propagating crack problem, there are analytical solutions. Suppose that the crack tip of a semiinfinite crack at the origin O_1 in the fixed coordinate system (x_1, y_1) propagates in the x_1-direction at a constant speed $\dot{a}$, as shown in Fig. 9.18. The relation between the coordinate system (x, y) moving with crack propagation and the fixed coordinate system (x_1, y_1) is as follows:

$$x = x_1 - a(t) = x_1 - \dot{a}t, y = y_1 \tag{9.29a}$$

The following polar coordinates (r_1, θ_1) and (r_2, θ_2) are introduced, which are related to the longitudinal wave velocity C_l and the transverse wave velocity C_s, respectively, to account for the stress wave effect.

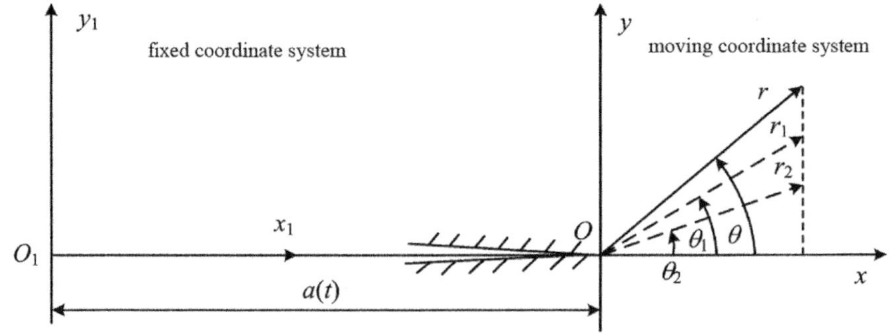

Figure 9.18 Fixed coordinate system and moving coordinate system for propagating crack.

$$\left.\begin{aligned}
&x = x_1 - a(t) = r\cos\theta = r_1\cos\theta_1 = r_2\cos\theta_2 \\
&y = y_1 = \frac{r_1}{\alpha_1}\sin\theta_1 = \frac{r_2}{\alpha_2}\sin\theta_2 \\
&r_1 = \sqrt{x^2 + \alpha_1^2 y^2},\ \gamma_2 = \sqrt{x^2 + \alpha_2^2 y^2} \\
&\theta_1 = arctg\left(\alpha_1 \frac{y}{x}\right),\ \theta_2 = arctg\left(\alpha_2 \frac{y}{x}\right) \\
&\alpha_1^2 = 1 - \frac{\dot{a}^2}{C_1^2},\ \alpha_2^2 = 1 - \frac{\dot{a}^2}{C_2^2} \\
&C_1 = C_l = \sqrt{\frac{\lambda + 2\mu}{\rho}} = \sqrt{\frac{E}{\rho}}\sqrt{\frac{1-v}{(1+v)(1-2v)}} \\
&C_2 = C_s = \sqrt{\frac{\mu}{\rho}} = \sqrt{\frac{E}{\rho}}\sqrt{\frac{1}{2(1+v)}}
\end{aligned}\right\} \tag{9.29b}$$

Note that C_1 here is the longitudinal wave (expansion wave) velocity C_l in the three-dimensional elastic body mentioned previously; C_2 is the transverse wave (shear wave) velocity C_s; λ, μ, v, E are known as the material elastic constants; and ρ is the material density.

Based on the above polar coordinate, Rice (1968) derived the solutions of crack-tip fields under the plane strain state by using complex function method similar to that of crack statics (Eq. 9.6); the principal terms of stress field and displacement field near crack tip as:

$$\left.\begin{aligned}
\sigma_{xx} &= \frac{K_I^d(\dot{a})}{\sqrt{2\pi}} \frac{1+\alpha_2^2}{4\alpha_1\alpha_2 - \left(1+\alpha_2^2\right)^2} \times \left\{\left(1 + 2\alpha_1^2 - \alpha_2^2\right)\frac{\cos\frac{\theta_1}{2}}{\sqrt{r_1}} - \frac{4\alpha_1\alpha_2}{1+\alpha_2^2}\frac{\cos\frac{\theta_2}{2}}{\sqrt{r_2}}\right)\right\} \\
\sigma_{yy} &= \frac{K_I^d(\dot{a})}{\sqrt{2\pi}} \frac{1+\alpha_2^2}{4\alpha_1\alpha_2 - \left(1+\alpha_2^2\right)^2} \times \left\{-\left(1+\alpha_2^2\right)\frac{\cos\frac{\theta_1}{2}}{\sqrt{r_1}} + \frac{4\alpha_1\alpha_2}{1+\alpha_2^2}\frac{\cos\frac{\theta_2}{2}}{\sqrt{r_2}}\right)\right\} \\
\sigma_{xy} &= \frac{K_I^d(\dot{a})}{\sqrt{2\pi}} \frac{1+\alpha_2^2}{4\alpha_1\alpha_2 - \left(1+\alpha_2^2\right)^2} \times \left\{2\alpha_1\left(\frac{\sin\frac{\theta_1}{2}}{\sqrt{r_1}} - \frac{\sin\frac{\theta_2}{2}}{\sqrt{r_2}}\right)\right\}
\end{aligned}\right\} \tag{9.30a}$$

and

$$\left.\begin{aligned}
u_x &= \frac{K_I^d(\dot{a})}{\sqrt{2\pi}} \frac{4(1+\nu)\left(1+\alpha_2^2\right)^2}{E\left[4\alpha_1\alpha_2-\left(1+\alpha_2^2\right)^2\right]} \times \left\{\left(\sqrt{r_1}\cos\frac{\theta_1}{2} - \sqrt{r_2}\ \frac{2\alpha_1\alpha_2}{1+a_2^2}\cos\frac{\theta_2}{2}\right\}\right. \\
u_y &= \frac{K_I^d(\dot{a})}{\sqrt{2\pi}} \frac{4(1+\nu)\left(1+\alpha_2^2\right)^2}{E\left[4\alpha_1\alpha_2-\left(1+\alpha_2^2\right)^2\right]} \times \left\{\left(-\alpha_1\sqrt{r_1}\sin\frac{\theta_1}{2} + \sqrt{r_2}\ \frac{2\alpha_1}{1+\alpha_2^2}\sin\frac{\theta_2}{2}\right\}\right.
\end{aligned}\right\}$$

(9.30b)

For Eq. (9.30), the following three points are worth noting:

(1) The mechanical field near the crack tip for moving cracks is dominated by K_I^d factor, just like stationary cracks. It can be seen that $\sigma_{yy} = K_I^d(\dot{a})/\sqrt{2\pi r}$ in front of the moving crack when taking $\theta_1 = \theta_2 = 0$, which is consistent with the conclusion of crack statics (Eq. 9.6a).

(2) The above stress fields and displacement fields are functions of α_1 and α_2, and thus of r_1 and r_2, which, respectively, include the expansion wave velocity C_l and the shear wave velocity C_s, reflecting the effects of stress waves with different wave velocities on the mechanical field near the crack tip. Moreover, due to $C_l > C_s$, both of them could affect the crack-tip mechanical field in the form of individual terms (e.g., α_1 or α_2, respectively) and multiplied coupling terms (e.g., $\alpha_1\alpha_2$).

(3) The ratio of σ_{yy} to σ_{xx} directly in front of the propagating crack ($\theta = 0$) characterizes the degree of stress triaxiality (there is $\sigma_{zz} = \nu(\sigma_{xx}+\sigma_{yy})$ in the current plane strain state). From formula (9.30), when $\theta = 0$ there is (Rosakis and Ravichandran, 2000):

$$\left(\frac{\sigma_{yy}}{\sigma_{xx}}\right)_{\theta=0} = \frac{4\alpha_1\alpha_2 - \left(1+\alpha_2^2\right)^2}{\left(1+2\alpha_1^2-\alpha_2^2\right)\left(1+\alpha_2^2\right) - 4\alpha_1\alpha_2} \tag{9.31}$$

As can be seen that the numerator term on the right side of above equation is exactly the so-called Rayleigh function $R(C)$ which is familiar to us when we solve for Rayleigh surface waves. When $R(C) = 0$, $\dot{a} = C_R$, or when $\dot{a} = C_R$, $R(C) = 0$. Eq. (9.31) shows that the value of $(\sigma_{yy}/\sigma_{xx})_{\theta=0}$ decreases with the increase of crack propagating speed from 1 (corresponding to a stationary crack, $\dot{a} = 0$) to 0 (corresponding to $\dot{a} = C_R$), that is, the stress triaxiality decreases with increasing crack propagating speed until it drops to 0 when $\dot{a} = C_R$ as shown in 9.19. This will have an important influence on whether the dynamic fracture is brittle or ductile.

After the asymptotic solution of the crack-tip mechanical field for moving cracks is given by Eq. (9.30), the key of the problem now is how to determine the change of the dynamic stress intensity factor $K_I^d(\dot{a})$ with the propagating velocity $\dot{a}$. Many researchers have contributed to this problem (Ravi-Chandar, 2004). Among them, Freund (1990) obtained a key result (Freund, 1973, 1990) by treating the crack propagating with a nonuniform velocity as a crack propagating with a series of different constant velocity. That is, if the dynamic stress intensity factor for a moving crack subjected to a generalized load $\sigma(t)$ and propagating at a transient velocity $\dot{a}(t)$ is expressed as $K_\mathrm{I}^d(\sigma(t), a(t), \dot{a}(t))$, and the stress intensity factor for a corresponding stationary crack ($\dot{a}(t) = 0$) having the same crack length and subjected to the same loading $\sigma(t)$ is expressed as $K_I^0(\sigma(t), a(t), 0)$, then the relationship of $K_\mathrm{I}^d(\sigma(t), a(t), \dot{a}(t))$ with the propagating velocity $\dot{a}(t)$ and $K_I^0(\sigma(t), a(t), 0)$ is established:

$$K_I^d(\sigma(t), a(t), \dot{a}(t)) = k(\dot{a}(t))K_I^0(\sigma(t), a(t), 0), \quad k(\dot{a}) \approx \frac{(1 - \dot{a}/C_R)}{\sqrt{1 - \dot{a}/C_l}} \tag{9.32}$$

where $k(\dot{a})$ is a universal function of crack propagating velocity $\dot{a}$. As the crack propagating velocity increases from 0 to C_R (Rayleigh wave velocity), the $k(\dot{a})$ decreases from 1 to 0, as shown in Fig. 9.20. This means that the kinetic energy of moving cracks increases with the increase of crack propagating velocity $\dot{a}$, but the dominant factor of crack-tip field $K_I^d(t)$ and the corresponding strain energy decrease with increasing crack velocity. Once the crack propagating velocity $\dot{a}$ tends to C_R, the dynamic stress intensity factor of moving crack $K_I^d(t)$ tends to 0. Thus, the crack propagating velocity obviously cannot exceed C_R.

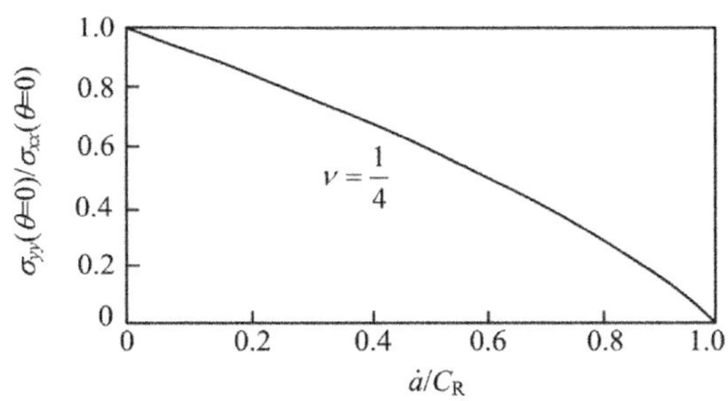

Figure 9.19 The influence of crack propagating speed on the stress triaxiality (Poisson's ratio = 0.25).

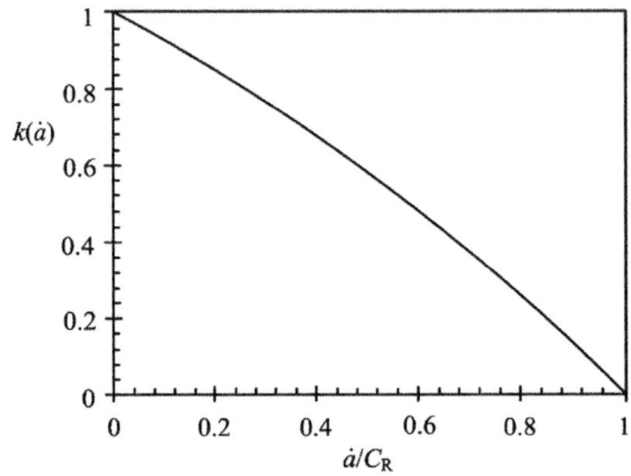

Figure 9.20 Change of universal function $k(\dot{a})$ with crack propagation velocity.

Eq. (9.32) separates the influence of instantaneous crack propagation velocity $\dot{a}(t)$ in the form of separated variable $k(\dot{a})$. In this way, the dynamic stress intensity factors of the crack with different propagating velocity under arbitrary loading can be determined by the stress intensity factor of the equivalent stationary cracks, which is relatively easy to be determined. This is an extremely important development, although it is based on the theory of linear elastodynamics.

9.1.7 Dynamic crack growth toughness

As discussed earlier, for propagating cracks, the dynamic crack growth criteria is described by Eq. (9.28).

$$K_I^d(\sigma(t), a(t), \dot{a}(t)) = K_{ID}(\dot{a}(t)), \quad t > t_f$$

The stress intensity factor K_I^d for propagating cracks on the left side has been discussed in the previous section. Now, we will discuss how the dynamic crack growth toughness on the right side $K_{ID}(\dot{a}(t))$ changes with the crack propagation velocity $\dot{a}$. The $K_{ID} - \dot{a}$ relationship is mainly determined experimentally.

Dally (1979) examined the $K_{ID} - \dot{a}$ relationship for polyester Homalite-100 through direct experimental measurements, and the results are shown in Fig. 9.21. The experiments were performed in a dynamic photoelastic apparatus coupled to a high-speed camera. Because Homalite-100 is suitable for the direct study of $K_{ID} - \dot{a}$ relationship by optical methods (dynamic photoelastic, caustics) combined with high-speed photography technology; it is the most studied material. Fig. 9.21a shows the relationship between the

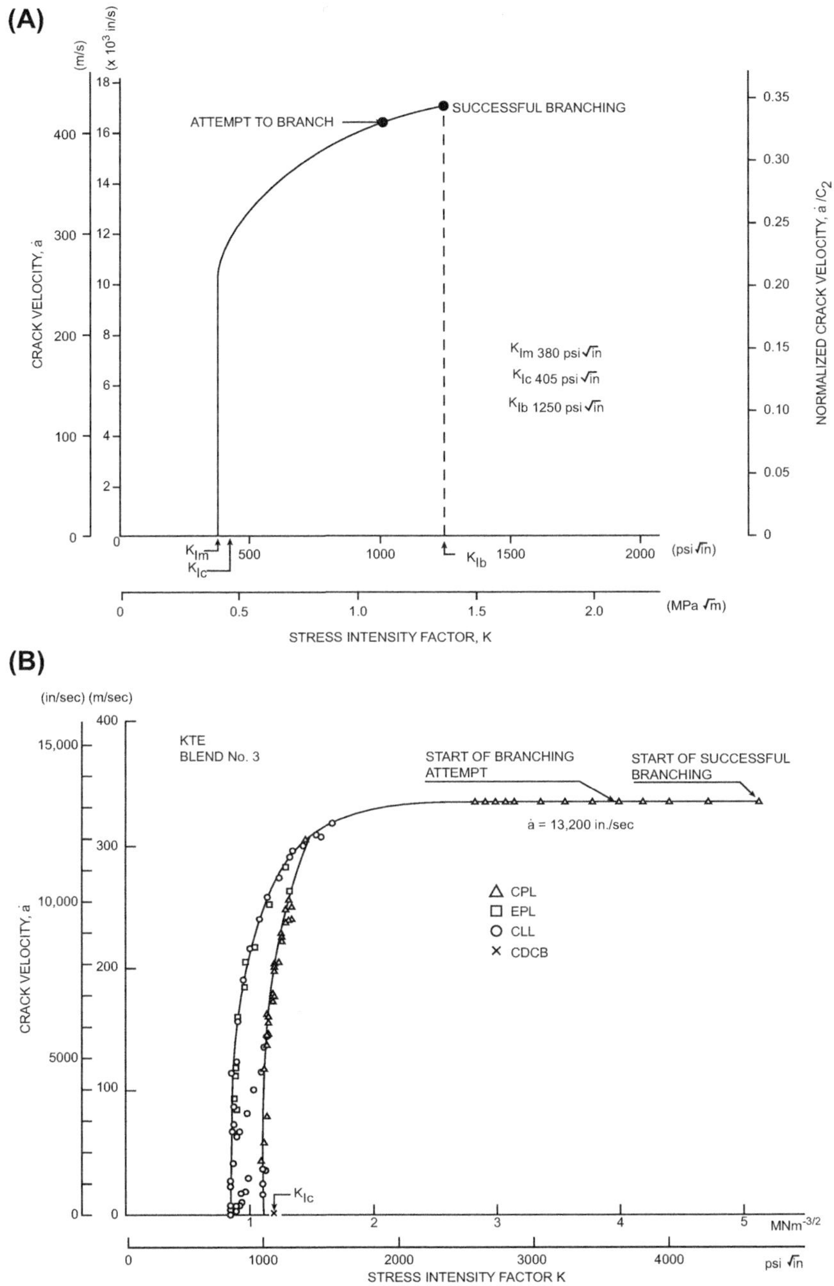

Figure 9.21 $K_{ID} - \dot{a}$ relationship for (A) Homalite-100 (B) epoxy resin. *From Dally J.W., 1979. Dynamic photoelastic studies of fracture, Exp. Mech. 19, 349–361, Fig. 7, p.354. Reprinted with permission of the publisher.*

dynamic stress intensity factor K_I^d and the crack propagation velocity $\dot{a}$, which is in fact the experimental characterization of the relationship of $K_{ID} - \dot{a}$ according to Eq. (9.28). In addition, please note the point marked "attempt to branch" and the point marked "successful branching", and note the K_{Im}, which is in fact equivalent to the dynamic crack arrest toughness K_{Ia}; all of them will be further discussed in the next Section 9.1.8.

Dally (1979) also studied the $K_{ID} - \dot{a}$ relationship for epoxy resin KTE, as shown in Fig. 9.21b, in which CPL, EPL, and CLL refer to different loading schemes for single-edge-notched (SEN) specimens, respectively, and CDCB refers to the contoured double-cantilever beam specimens.

As can be seen from both Fig. 9.21A and B, as K_I^d increases above the initiation toughness K_{Ic}, there is a rather sharp increase in the crack propagation velocity $\dot{a}$, which changes from 0 to about 300 m/s (roughly 30% of the shear wave speed) when K_I^d increases only about 20%. Beyond this, the crack propagation velocity increases much more slowly between 300 and 400 m/s, when K_I^d increases more than doubles, and then tends to propagate at constant speed. It seems that there is no one-to-one correspondence between the K_{ID} and $\dot{a}$. If K_I^d increases further, as shown in Fig. 9.21b, the crack begins to branch (as marked as "attempt to branch" and "successful branching"), which will be discussed in the next section.

Another noteworthy phenomenon from Fig. 9.21b is that the $K_{ID} - \dot{a}$ curve of the acceleration process of crack propagation and the $K_{ID} - \dot{a}$ curve of the deceleration process form a hysteretic loop, of which the right branch represents the acceleration curve, while the left branch represents the deceleration curve. This means that the acceleration process of propagating crack occurs at a higher scale of K_{ID}. However, Arakawa and Takahashi (1991) reported contrary experimental results that the deceleration process occurred at higher K_{ID}. Anyway, these experimental results at least indicate that the $K_{ID} - \dot{a}$ relationship depends not only on crack propagation velocity $\dot{a}$, but also on other factors, such as whether the crack was accelerating or decelerating, and the related mechanism needs further study.

Zehnder and Rosakis (1990) studied the $K_{ID} - \dot{a}$ relationship of AISI 4340 high strength steel, a ductile material, using the method of reflection caustics and high-speed photography; the results are shown in Fig. 9.23. When K_{ID} increases from approximately 60 $\text{MPa}\cdot\text{m}^{1/2}$ at the time of crack unstable extension to approximately 200 $\text{MPa}\cdot\text{m}^{1/2}$, the crack propagation velocity increases from 0 to 1000 m/s. It is worth noting that, unlike the case of brittle materials in Fig. 9.22, the relationship $K_{ID} - \dot{a}$ of 4340 steel has no difference between accelerating or decelerating cracks, which means that

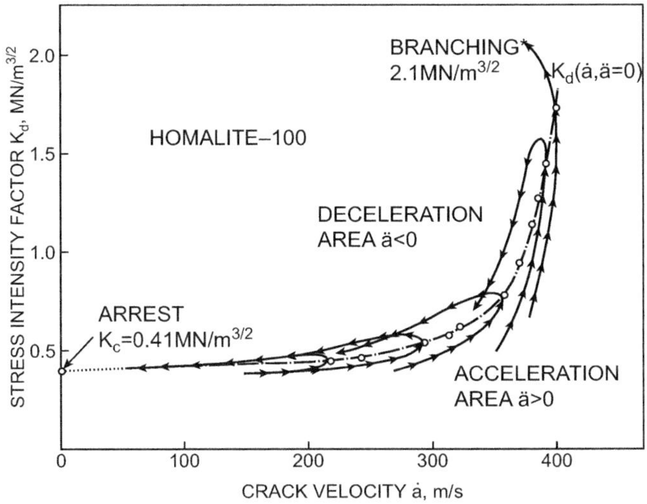

Figure 9.22 Relation of $K_{ID} - \dot{a}$ for Homalite-100. *From Arakawa K., Takahashi K., 1991. Relationship between fracture parameters and surface roughness of brittle polymers. Int. J. Fract. 48, 103–114, Fig. 5, p.107. Reprinted with permission of the publisher.*

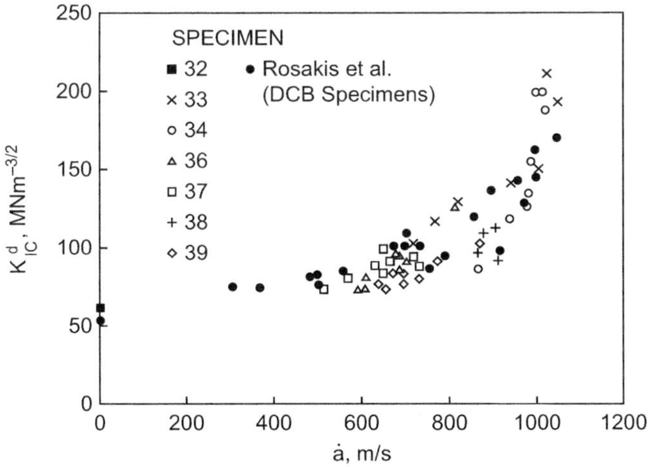

Figure 9.23 $K_{ID} - \dot{a}$ relationship for 4340 high strength steel. *From Zehnder A.T., Rosakis A.J., 1990. Dynamic fracture initiation and propagation in 4340 steel under impact loading. Int. J. Fract. 43, 271–285, Fig. 11, p.282. Reprinted with permission of the publisher.*

there is a single $K_{ID} - \dot{a}$ relationship corresponding to one to one for ductile metals. It might be attributed to the deformation and failure mechanism in the fracture process zone ahead of the crack tip. For ductile materials, the energy dissipation associated with plastic deformation in the process zone is a significant fraction of overall energy expended by the propagating crack and must be taken into account in dynamic fracture model.

9.1.8 Crack branching and arrest

Fig. 9.21 shows that crack propagation velocity $\dot{a}$ first increases rapidly with increasing K_{ID} (i.e., critical dynamic stress intensity factor of propagating crack) and is in the accelerating process. However, when at a certain critical value, the crack propagation velocity $\dot{a}$ will increases much more slowly with the increase of K_{ID}, and if K_{ID} continues to increase, then the crack branching will occur. On the contrary, no matter whether the crack is in the process of acceleration or deceleration, the crack arrest may occur when certain conditions are satisfied.

9.1.8.1 Branching/bifurcation

The typical pattern of crack branching observed in the experiments is schematically shown in Fig. 9.24; the start of successful branching often occurs after the branching attempted.

The crack branching will be discussed below firstly starting from the energy approach. Recalling that when discussing the kinetic energy of propagating cracks, it was pointed out that the kinetic energy of propagating crack comes from the difference between the strain energy release rate G and the crack extension resistance R ($=G_{Ic}=$ constant), as shown in Eq. (9.21) and Fig. 9.16b.

Obviously, when the excess energy represented by the difference between the G-Δa curve and the R-Δa curve is large enough to drive two cracks, as shown in Fig. 9.25, the crack will bifurcate into two cracks. As long as the difference of (G-R) is high enough, further bifurcation can be carried out.

Yoffe (1951) attempted to explain crack branching by means of an analysis of crack-tip mechanical field as early as 1951. She examined the crack-

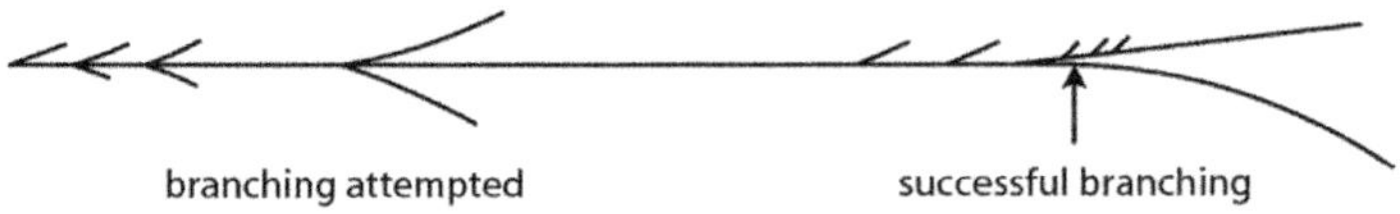

Figure 9.24 The successful branching usually occurs after the branching attempted.

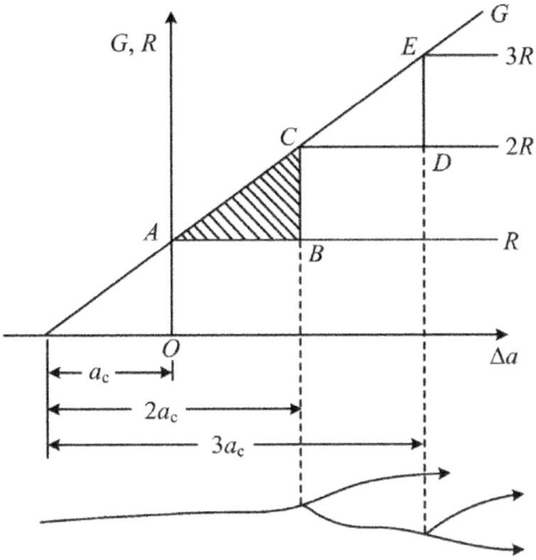

Figure 9.25 The difference between G and R is large enough for cracks to branch.

tip stress field for a Mode-I crack propagating at a constant velocity $\dot{a}$ under the action of uniform load σ_{xx} in infinite medium and emphatically investigated the hoop stress component $\sigma_{\theta\theta}$ that has important influence on the crack extension and branching. Under different constant velocities $\dot{a}$, that is, when $\dot{a}/C_s = 0$, 0.5, 0.8, 0.9, the change of hoop stress $\sigma_{\theta\theta}$ with θ is shown in Fig. 9.26. It can be found from Fig. 9.26 that the maximum of $\sigma_{\theta\theta}$ occurs at $\theta = 0$ (right in front of the crack tip) when the crack propagates at low velocity. However, as $\dot{a}$ increases, the maximum of $\sigma_{\theta\theta}$ occurs in $\theta > 0$ (deviating from the crack front). When

$$\dot{a} = 0.6C_s \tag{9.33}$$

the maximum of $\sigma_{\theta\theta}$ occurs at $\theta = 60°$. It means that the crack will change its extension direction, leading to crack curving or crack branching. Eq. (9.33) can be regarded as the crack branching criterion based on the maximum $\sigma_{\theta\theta}$.

Yoffe's theory, while concise and appealing, is not supported by experimental facts. Based on the viewpoint that the crack-tip mechanical field is dominated by stress intensity factor K, later researchers (e.g., Kobayashi and Ramalu, 1985) proposed the following K-based crack branching criterion:

$$K_I^d \geq K_{Ib} \tag{9.34}$$

where K_{Ib} is the so-called **crack branching onset toughness**.

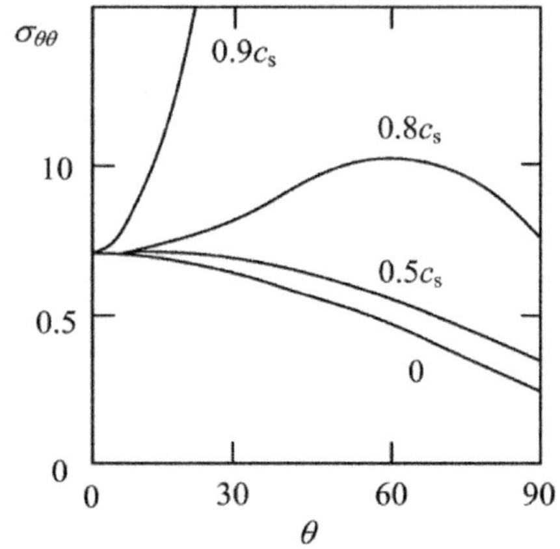

Figure 9.26 The change of hoop stress $\sigma_{\theta\theta}$ at crack tip with θ at different crack propagating velocities $\dot{a}$. *From Yoffe E., 1951. The moving Griffith crack, Philos. Mag. A 42, 739–750, Fig. 1, p.749. Reprinted with permission of the publisher.*

The crack branching velocities C_b and crack branching onset toughness K_{Ib} measured by different researchers for some materials are summarized, respectively, in Tables 9.4 and 9.5 (Kobayashi and Ramalu, 1985). In the tables, C_0 is the wave velocity of one-dimensional elastic bar, that is $C_0 = (2(1 + \nu))^{1/2} C_s$. According to Yoffe's crack branching criterion, $C_b = 0.6C_s$, which is equivalent to $C_b = 0.38C_0$ (taking $\nu = 0.25$). The crack branching onset toughness K_{Ib} listed in Table 9.5 were measured in plane strain state.

It can be seen that both crack branching criteria, Eq. (9.33) and Eq. (9.34), are not consistently supported by the experimental results shown in Tables 9.4 and 9.5. By summarizing various experimental results on crack branching, Brandon (1987) concluded that typical crack branching speed C_b is in the order of $0.3C_0$, K_{Ib} is in the order of $4K_{Ic}$, and the bifurcation angle of cracks is distributed in a wide range of 20–120 degrees.

The experimental fact that the branching onset toughness K_{Ib} could be as large as four times the dynamic initiation toughness K_{Ic}, which obviously means that the branching of crack consumes much more energy than the elastodynamics analysis. Ravichandar and Knauss (1984b) pointed out that this is closely related to the evolution of mesodamage in the process zone ahead of branching crack tip, as shown in Fig. 9.27. Note that among a number of microbranches ("attempted branching") in the process zone, three of them are of "successful branching". The mechanism of how the damages in the process zone evolve to successful branching is illustrated in

Table 9.4 Comparison of measured crack branching velocity C_b with theoretical prediction (Kobayashi and Ramalu, 1985, Table 1, p.C5-198).

Source	Material	C_b/C_o
Theoretical prediction		
Yoffe (1951)		0.38
Experimental measurements		
Anthony et al. (1968)	Glass	0.39
Bowden et al. (1967)	Glass	0.29
Congleton (1973)	Tool steel	0.26
Doll (1975)	Plate glass	0.28
	FK-52 glass	0.30
Hahn et al. (1977)	A533B steel	0.10
Irwin et al. (1977)	Homalite-100	0.24
Kobayashi et al. (1974, 1981)	Homalite-100	0.22
Paxson et al. (1973)	Plexiglass	0.36
Schardin (1959)	Glass	0.30

Table 9.5 Comparison of measured branching onset toughness K_{Ib} with fracture toughness K_{Ic} (Kobayashi and Ramalu, 1985, Table 2, p.C5-198).

Source	Material	K_{Ib}/K_{Ic}
Congleton et al. (1973)	Tool steel	2.36
Dally et al. (1977, 1979)	Homalite-100	3.80
Doll (1975)	Glass	4.23
Irwin et al. (1966)	Homalite-100	4.65
Hahn et al. (1977)	A533B steel	1.00
Kirchner and Kirchner (1979)	Flint glass	3.66
Kirchner et al. (1981)	Ti–ZrO	4.74
Kobayashi et al. (1974, 1981)	Homalite-100	3.60
Weimer and Rogers (1979)	HF-steel	4.55
	FS-01 steel	4.55

Fig. 9.28 (Ravichandar and Knauss, 1984c).This is a complex problem to be further studied.

9.1.8.2 Dynamic crack arrest

When discussing the kinetic energy of propagating crack (Fig. 9.16) and the kinetic energy of crack branching (Fig. 9.25) by energy approach, two basic assumptions have been made, that is: (i) the energy release rate $G = \pi\sigma^2 a/E$,

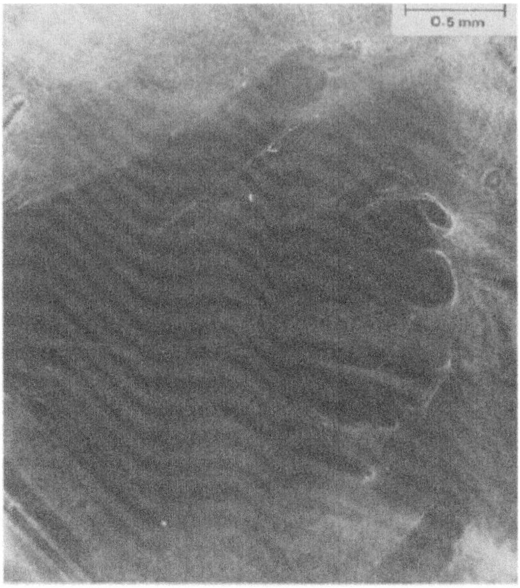

Figure 9.27 Real-time micrograph of crack branching in Homalite-100 (the field of view is 3 mm). *From Ravi-Chandar K., Knauss W.G., 1984b. An experimental investigation into dynamic fracture-II. Microstructural aspects, Int. J. Fract. 26, 65–80, Fig. 14, p.78. Reprinted with permission of the publisher.*

so that G increases proportionally with the crack size a, and (ii) the crack extension resistance R ($=G_{Ic}$) is a constant. In more practical cases, G is not a linear function of crack length a, even G may decrease with a instead of increasing because both the load $\sigma(t)$ and the crack length $a(t)$ are functions of time t. In some cases (see e.g., Broek, 1982), as shown in Fig. 9.29, the energy release rate G first increases nonlinearly with crack extension $a(t)$, and after reaching the maximum point B, it begins to decrease. When the crack extends to point C, the energy release rate G is equal to the crack extension resistance R again, then the crack arrest occurs ($a = a_a^*$), but because the influence of kinetic energy is not considered, so it is called static crack arrest analysis. If the effect of kinetic energy is taken into account, the kinetic energy of propagating crack at point C can be equal to area ABC in number, and this part of kinetic energy can also be used for further propagation of crack. Therefore, although the energy release rate $G < R$ after that, the crack can still extend with the support of kinetic energy. Finally, the crack stops at point E, making the area $CDE = ABC$, that is, the crack stops only after the crack expands to point E and reduces the total kinetic energy to zero ($a = a_a^{**}$).

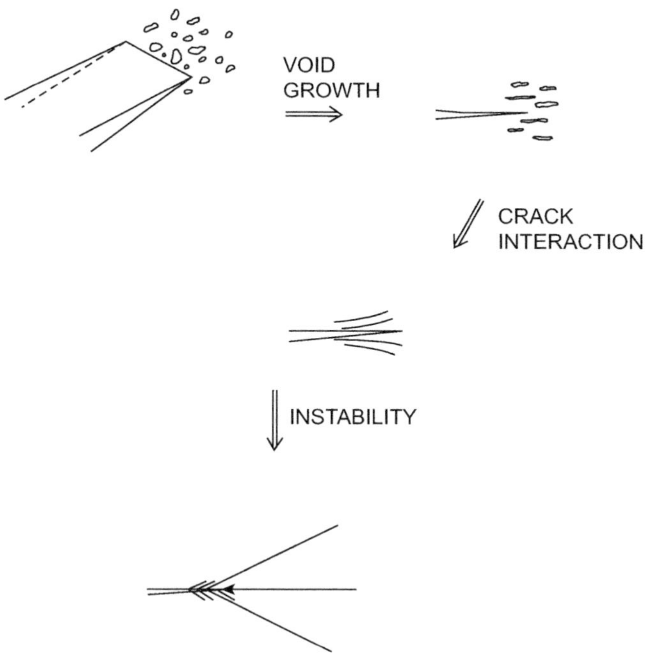

Figure 9.28 Schematic diagram for the mechanism of crack branching. *From Ravi-Chandar K., Knauss W.G., 1984c. An experimental investigation into dynamic fracture-111. On steady state crack propagation and branching. Int. J. Fract., 26, 141–154, Fig. 6, p.151. Reprinted with permission of the publisher.*

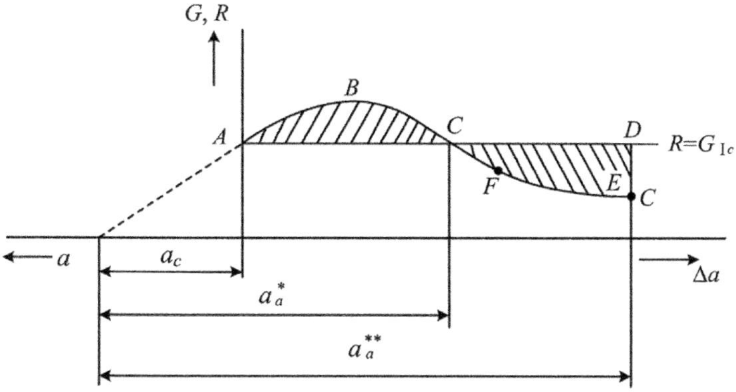

Figure 9.29 Crack arrest due to the reduction of *G*.

The above analysis implies a basic assumption that the critical energy release rate is equal to the static crack extension resistance. In other words, the crack arrest toughness of the material is equal to the static fracture toughness of the material. However, this assumption has not been confirmed by a large number of experiments conducted by subsequent researchers.

Considering the dynamic effect, a large number of studies (see e.g., Ravi-Chandar, 2004) show that the dynamic crack arrest toughness K_{Ia} of materials is different from the dynamic crack initiation toughness $K_{Id}(\dot{K})$. Specifically, if the dynamic stress intensity factor $K_I^d(t)$ first increases with time, once it reaches the dynamic crack initiation toughness $K_{Id}(\dot{K})$, the crack will initiate to extend. After that, if the dynamic stress intensity factor $K_I^d(t)$ decreases with time, the crack extension process will not be arrested when the $K_I^d(t)$ is reduced to the dynamic crack initiation toughness $K_{Id}(\dot{K})$, but will be arrested when the $K_I^d(t)$ is reduced to the dynamic crack arrest onset toughness K_{Ia} of the material, and generally there is $K_{Ia} < K_{Id}(\dot{K})$. In fact, a specific example has been given in Fig. 9.21a (Dally, 1979), where the dynamic crack arrest toughness K_{Ia} of the material is expressed as $K_{Im}(<K_{Ic})$ instead.

Thus, the criterion for dynamic crack arrest at the moment of $t = t_a$ is expressed as:

$$K_I^d(t) < K_{Ia}(T), \quad t > t_a \tag{9.35}$$

where the dynamic crack arrest toughness K_{Ia} is defined as the smallest value of the dynamic stress intensity factor for which the extension of crack cannot be maintained and T is the temperature, reflecting the experimental phenomenon that $K_{Ia}(T)$ increases with increase of temperature.

Dynamic crack arrest toughness should be determined by the experimental investigations. The crack arrest experiments designed by different researchers have one of the following characteristics: (i) The experiment contains the process of $K_I^d(t)$ as the decreasing function of time t, so as to obtain crack arrest information; (ii) The sample is designed to have a nonuniform temperature field, making it a nonuniform temperature-dependent crack arrest toughness field $K_{Ia}(T)$. When the crack starts from the specimen lower temperature end with low crack arrest toughness and extends to the specimen higher temperature end with higher crack arrest toughness, once the propagating crack arrests at some point, the corresponding $K_{Ia}(T)$ can be determined by measuring the dynamic stress intensity factor and temperature at the crack arrest point; (iii) The specimen is composed of low crack arrest toughness material and high crack arrest toughness material. The crack is initiated from low crack arrest toughness material and extends into the high crack arrest toughness material and hence arrests.

In crack arrest experiments, it is necessary to avoid the interference caused by the stress wave reflected from the specimen boundary onto the crack tip.

Ravi-Chandar and Knauss (1984a) used an electromagnetic loading device with adjustable load amplitude and duration to apply pressure pulse on the inner surface of the crack. In the same experiment, the dynamic crack initiation toughness K_{Id} and the dynamic crack arrest toughness K_{Ia} can be studied simultaneously by combining the high-speed camera (200,000 frames/second) with the optical caustics method. They used successive double-trapezoidal pulse loading with each pulse lasting about 70 μs. The duration on the one hand is long enough for crack initiation and extension, and on the other hand is short enough for unloading at the crack tip after a few microns of crack extension, so $K_I^d(t)$ that decreases, leading to crack arrest. The second trapezoidal pulse can reload the crack, causing a "reinitiation". Thus, in such experiments, the interference of stress wave reflected from the specimen boundary can be avoided. The results from this series tests on Homalite-100 are shown in Fig. 9.30 and Fig. 9.31.

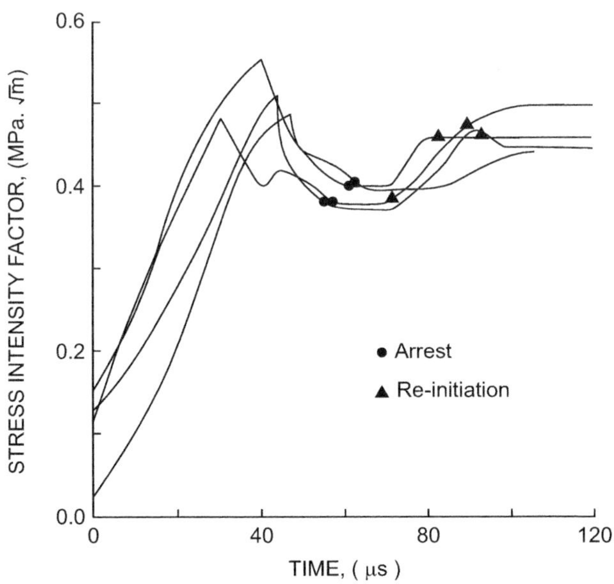

Figure 9.30 Variation of dynamic stress intensity factor $K_I^d(t)$ with time t for the crack arrest experiments. *From Ravi-Chandar K., Knauss W.G., 1984a. An experimental investigation into dynamic fracture-I. Crack initiation and crack arrest. Int. J. Fract. 25, 247–262, Fig. 8, p.259. Reprinted with permission of the publisher.*

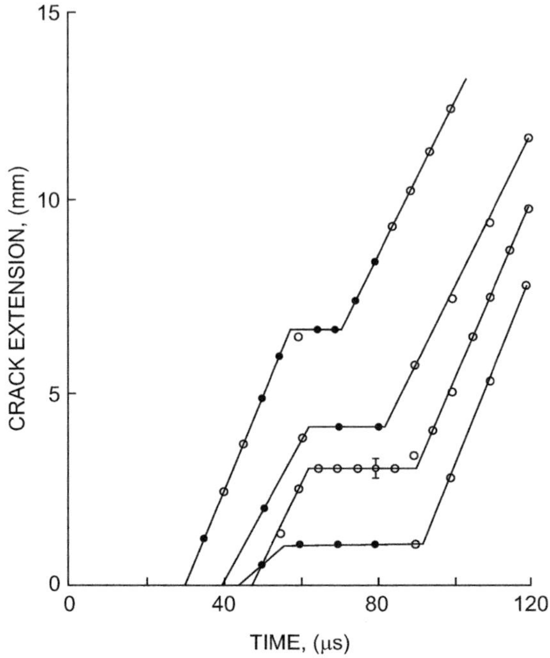

Figure 9.31 Change of extension distance of crack tip with time for the crack arrest experiments. *From Ravi-Chandar K., Knauss W.G., 1984a. An experimental investigation into dynamic fracture-I. Crack initiation and crack arrest. Int. J. Fract. 25, 247–262, Fig. 9, p.260. Reprinted with permission of the publisher.*

Fig. 9.30 shows the variation of dynamic stress intensity factor $K_I^d(t)$ with time t, while Fig. 9.31 shows the variation of the crack-tip extension with time t, of which the slope represents the crack-tip propagating velocity. As can be seen from Fig. 9.30, when the dynamic stress intensity factor $K_I^d(t)$ reaches the dynamic crack initiation toughness K_{Id}, the crack initiates and propagates at a constant velocity $\dot{a}$, which could be estimated by the slope of the first oblique line in Fig. 9.31. However, for propagating crack, as shown in Eq. (9.32) and Fig. 9.20, its dynamic stress intensity factor $K_I^d(t)$ will decrease. Once it is unloaded to the dynamic crack arrest toughness K_{Ia} of the material, the crack arrest takes place, which is expressed as a horizontal segment with the zero slope (i.e., zero propagating velocity) in Fig. 9.31 and is corresponding to the lowest value of $K_I^d(t)$ in Fig. 9.30. From this, the dynamic crack arrest toughness $K_{Ia} = 0.4\ \text{MPa}\cdot\text{m}^{1/2}$ can be determined, which is about 11% lower than K_{Ic} for the material (ref. Fig. 9.21). The "reinitiation" of the crack under the action of the second trapezoidal pulse can be also clearly seen in the two figures.

In summary, the static fracture toughness K_{Ic}, dynamic crack initiation toughness K_{Id}, dynamic crack growth toughness K_{ID}, and dynamic crack arrest toughness K_{Ia} of materials are independent and different material toughness indexes.

A schematic diagram of the relative independent relation for the three dynamic toughness in crack dynamics is shown in Fig. 9.32 (see e.g., Ravi-Chandar K, 2004). When dynamic stress intensity factor $K_I^d(t)$ reaches the dynamic crack initiation toughness K_{Id}, the crack will initiate to unstable extension and propagates at a certain velocity. After that, as indicated by the right arrow in Fig. 9.32, whether the propagating crack can continue to extend depends on whether $K_I^d(t)$ can reach the dynamic crack growth toughness $K_I^d(\dot{a})$. While whether the propagating crack can be stopped depends on whether $K^d{}_I$ (t) can be reduced to less than the crack arrest toughness K_{Ia} as indicated by the left arrow in Fig. 9.32.

The dynamic crack arrest criteria shown in Eq. (5.13) cannot only be used to judge whether the crack is arrested or not, but also provides a guiding principle for the design of crack arrest in engineering applications. The left-hand side and right-hand side of the inequality sign of Eq. (9.35) give the design principles of structure and material, respectively. From the point of view of structural crack arrest design, the dynamic stress intensity factor at crack-tip field $K_I^d(t)$ should be reduced as far as possible. From the perspective of material crack arrest design, the dynamic toughness of material should be improved as far as possible. In engineering practice, the following measures for crack arrest are often adopted.

Assume that a crack exists on the steel plate A with dynamic crack arrest toughness K_{Ia}^A, as schematically shown in Fig. 9.33; an arrester strip B with dynamic crack arrest toughness $K_{Ia}^B (> K_{Ia}^A)$ can be welded (or riveted) in

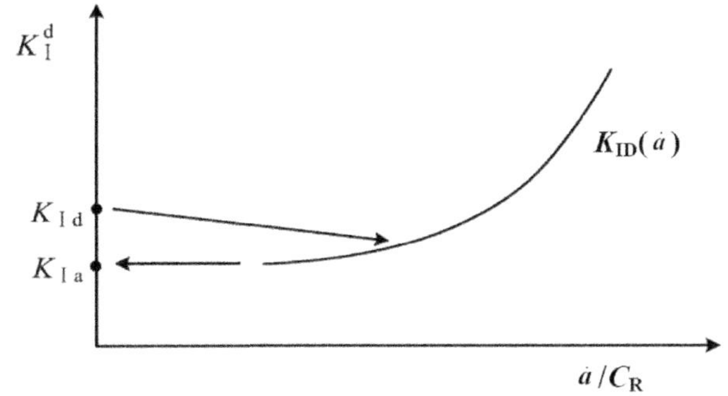

Figure 9.32 Independence of dynamic crack initiation toughness K_{Id}, dynamic crack growth toughness K_{ID}, and dynamic crack arrest toughness K_{Ia}.

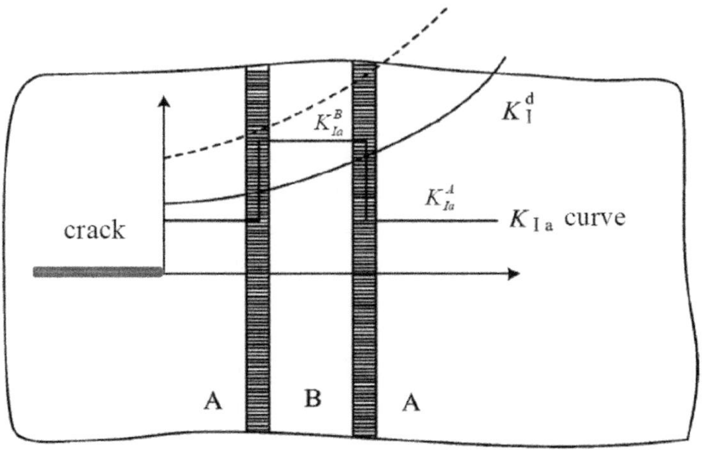

Figure 9.33 Schematic of crack arrest using arrester strip with high crack arrest toughness K_{Ia}.

front of the crack to achieve crack arrest. Now the distribution of K_{Ia} is shown by the convex curve in Fig. 9.33. If the dynamic stress intensity factor $K_I^d(t)$ increases monotonically with crack extension, as indicated by the solid line in Fig. 9.33, the crack will continue to propagate in steel plate A and not be arrested as long as $K_I^d > K_{Ia}^A$. However, as crack propagates into the strip B, it suddenly meets the material B with much larger crack arrest toughness K_{Ia}^B; as long as $K_I^d < K_{Ia}^B$, the condition of Eq. (9.35) is satisfied and the moving crack will be arrested within material B. Note that if another higher loading is applied on the crack making the dynamic stress intensity factor $K_I^d(t)$ increases with the crack extension as shown by the dotted line in Fig. 9.33, the moving crack will no longer be arrested because the crack arrest toughness K_{Ia}^B of material B is not high enough.

Fig. 9.34 illustrates a method for improving the crack arrest resistance K_{Ia} of a brittle material by adding high performance fibers.

In Fig. 9.35, a crack arrest hole is drilled at the crack tip, which makes the sharp crack tip become "passivized" by increasing the curvature radius of crack tip and can greatly reduce its stress intensity factor. In addition, a tool with a diameter slightly larger than the diameter of crack arrest hole can be used to perform cold working on the inner surface of the crack arrest hole, forming a plastic-hardening layer with residual compressive stress, or even a pin made of high-toughness material can be inserted directly in the hole in the form of tight fit, so as to improve the material crack arrest toughness at the surface of crack arrest hole.

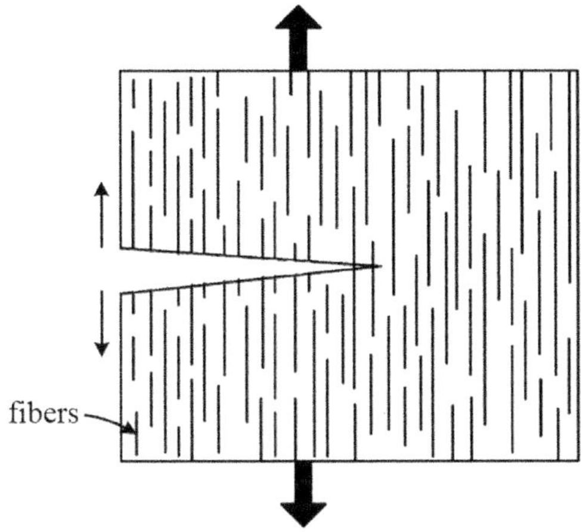

Figure 9.34 Improve crack arrest toughness K_{Ia} of brittle materials by adding high-performance fibers.

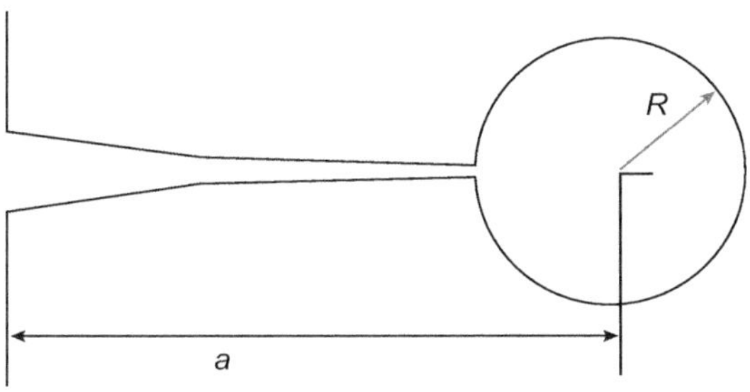

Figure 9.35 Drilling a crack arrest hole at the crack tip.

9.1.9 Experimental techniques for crack dynamics

Crack dynamics involves the study of the dynamic stress intensity factor near the crack tip (structural dynamic response) and the fracture toughness of the material against crack unstable extension under high loading rate (material dynamic response). As far as material dynamics is concerned, the latter is of primary concern, but it is also closely related to the former. The study of various dynamic fracture toughness (dynamic crack initiation toughness, dynamic crack growth toughness, dynamic crack arrest toughness, etc.) of materials mainly relies on experiments, which is essentially an experimental science. Although it also requires a lot of mathematics, it is not a deductive science or applied mathematics.

The implementation of dynamic experimental research on the dynamic fracture toughness of materials consists of two aspects of the key technology:

(1) dynamic loading technology for high loading rate (e.g., $\dot{K}_I^d(t) \geq 10^5$ MPa$\sqrt{m}$ /s) and (2) dynamic/instantaneous diagnostic technology to measure the critical value of dynamic stress intensity factor (e.g., $\dot{K}_I^d(t) \geq 10^5$ MPa$\sqrt{m}$ /s) and the corresponding critical time t_f (e.g., on the order of 10^0-10^2 μs) under high loading rate. Since the determination of dynamic fracture toughness of materials (for example, the dynamic crack initiation toughness) is achieved through dynamic measurement of $K_I^d(a, \sigma, t)\big|_{cr}$ and t_f. These two aspects are discussed separately below.

9.1.9.1 Loading technique

Early studies on dynamic fracture toughness of materials used pendulum impact test (the so-called Charpy impact test) or drop-weight test with notched specimens. It should be noted that such loading technology, on the one hand, is difficult to achieve high loading rate; on the other hand, more importantly, it is difficult to quantitatively analyze the stress wave effect in the specimen and loading device. Therefore, it has gradually been eliminated by other more perfect loading technology (projectile impact, explosion, and electromagnetic loading) as shown in Table 9.6 (see e.g., Ravi-Chandar, 2004), which have already covered in the previous sections of this chapter and in the previous chapters of this.

Recall Fig. 9.8 when discussing the dynamic crack initiation toughness of materials, and Fig. 9.30 and Fig. 9.31 when discussing the dynamic crack arrest toughness of materials; those results were obtained by Ravi-Chandar and Knauss (1984a) by using the **electromagnetic loading technology** with adjustable load amplitude and duration. The principle of this electromagnetic loading technology is schematically illustrated in Fig. 9.36.

A flat copper strip is folded back on itself and the space between the two layers is filled with an insulating strip. This assembly is then introduced into

Table 9.6 Range of loading rate $\dot{K}_I^d(t, \dot{a})$ and crack initiation time t_f for different loading techniques.

	Quasistatic loading	**Drop-weight**	**Projectile impact**	**Explosion**	**Electromagnetic loading**
Loading rate $\dot{K}_I^d(t, \dot{a})$, $MPa \cdot m^{1/2} \cdot s^{-1}$	1	10^4	10^4-10^8	10^5	10^5
Time to fracture (μs)	$>10^6$	~100	1–100	1–20	10–100

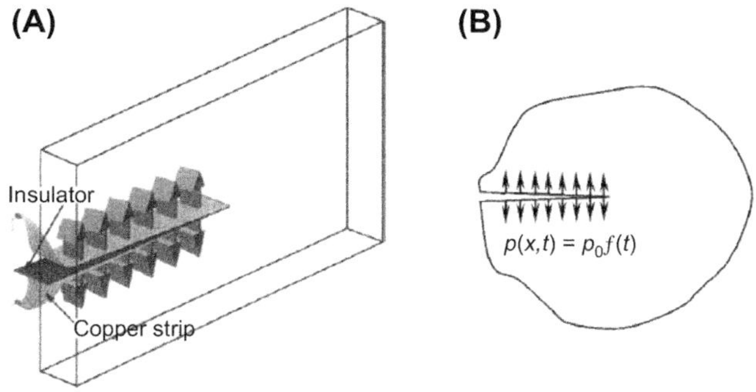

Figure 9.36 Principle of crack loaded by electromagnetic pressure pulse.

the crack of single-edge crack specimen (Fig. 9.36a). When a pulse current flows through the copper loop, each leg generates a magnetic field. The current vector in each leg interacts with the magnetic field of the other leg to produce an electromagnetic repulsion that forces the conductors apart. The two legs press upon the top and bottom surfaces of the crack with a uniform pressure (Fig. 9.36b). The trapezoidal pulse duration can be adjusted within the range of 150μs with a rising time in about 25μs. The pressure on the crack surface can be adjusted in the range of 1−20 MPa, and the loading rate is in the order of $\dot{K}_I^d(t) = 10^5 \text{ MPa}\sqrt{\text{m}}/\text{s}$. The major advantages of electromagnetic loading are: (i) it provides good repeatable loading that makes experiments easy; (ii) the experiment has been finished before the stress wave reflected from the specimen boundary back to the crack tip and can be modeled as a pressurized semiinfinite crack in an unbounded medium, so that is easy to analyze. Moreover, the dynamic crack initiation toughness of the material (Fig. 9.8) and the dynamic crack arrest toughness of the material (Fig. 9.30 and Fig. 9.31) can be carried out in the same experiment.

Projectile impact is a typical method to generate a high loading rate. Recall the plate impact experiment discussed in Chapter 4 "dynamic experimental study of state equation of solids under high pressure", in which a flyer launched by gas gun impacts onto a target (specimen), and recall the Hopkinson bar technique discussed in Chapter 7 "dynamic experimental study of material distortion law", both belong to the **projectile impact technology** list in Table 9.6. Different from the drop-weight technique which directly impacts specimen by falling mass, the projectile impact loading technique is essentially loaded by stress waves.

A typical example of the dynamic fracture toughness experiment of materials adopting plate impact technique is shown in Fig. 9.37 (Ravichandran and Clifton, 1989). The 4340 steel circular plate specimen with the thickness of h and a circumferential precrack across half of its cross-section is impacted at velocity v_0 by a flyer with the thickness of $h/2$ driven by a gas gun.

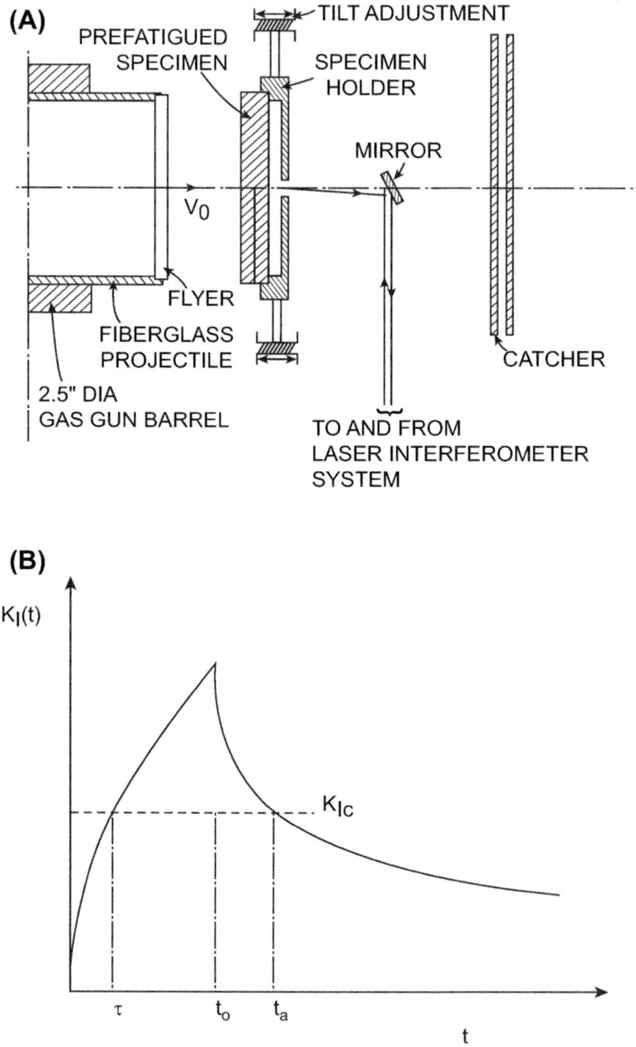

Figure 9.37 Dynamic fracture toughness experiment for cracked plate specimen impacted by flyer. *From Ravichandran G., Clifton R.J., 1989. Dynamic fracture under plane wave loading. Int. J. Fract. 40, 157–201, Fig. 2, p.163, Fig. 15, p.182. Reprinted with permission of the publisher.*

An incident compressive trapezoidal pulse is first applied to the specimen (without effect on crack extension), and after reflected from the free surface of the specimen, a tensile pulse load required by the experiment is formed at the crack of half thickness of the specimen. By varying v_0 and h, the load magnitude and duration (in the order of μs) can be adjusted. After analyzing the particle velocity history on the back surface of the specimen measured by the laser interferometer (VISAR), the dynamic stress intensity factor $K_I^d(t)$ (in the order of 100 $\text{MPa}\cdot\text{m}^{1/2}$) can be obtained. The corresponding loading rate $\dot{K}_I^d(t)$ achieved in these experiments is as high as 10^8 $\text{MPa}\cdot\text{m}^{1/2}\ \text{s}^{-1}$. $K_I^d(t)$ first increases with time t and is proportional to $t^{1/2}$, which is consistent with the analytical solution as shown in Fig. 9.6. After crack initiation and propagation, $K_I^d(t)$ then decreases with time, as shown in Fig. 9.37B.

Hopkinson bar technique has been adopted to conduct experimental research on dynamic fracture toughness of materials by more and more researchers (e.g., Kalthoff, 1986; Jiang and Vecchio, 2009), which has become more and more popular and mature.

The Hopkinson bar experiment technique has already introduced in detail in chapter 7 of this book. Here, we focus on its application in the study of dynamic fracture toughness of materials.

Costin et al. (1977) first used the Hopkinson bar principle to study the dynamic crack initiation toughness of materials, as schematically shown in Fig. 9.38. The specimen is a long round rod with a circumferential precrack. An explosive charge is detonated at one end of the specimen, generating a tensile pulse applied on the specimen. Two strain gauges are mounted to both sides of the precrack of the long rod specimen. The measured incident waves, reflected waves, and transmitted wave are used to determine the dynamic crack initiation toughness of the material. Crack initiation occurs in the range of 20–25μs, and the loading rate $\dot{K}_I^d(t)$ reaches 10^6 $\text{MPa}\cdot\text{m}^{1/2}\cdot\text{s}^{-1}$. However, because the long rod specimen itself plays the role of both

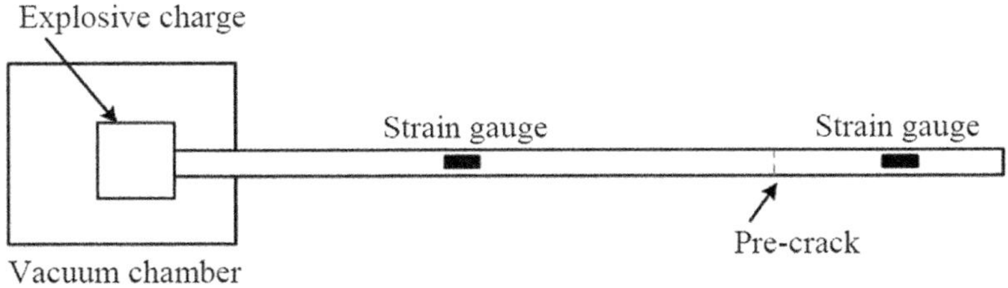

Figure 9.38 Circumferential precracked rod specimen subjected to a tensile pulse generated by an explosive charge.

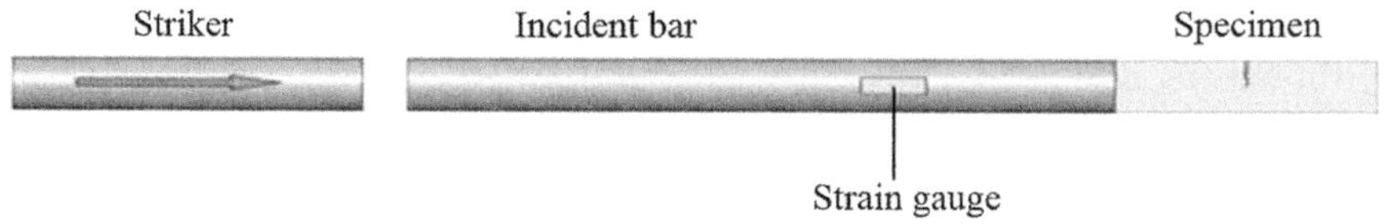

Figure 9.39 Cracked specimen loaded by the reflected tensile pulse using Hopkinson pressure bar.

Hopkinson incident bar and transmission bar at the same time, it consumes a lot of material and requires high processing technology.

In addition to directly apply the tensile stress wave loading to the pre-cracked specimen, the specimen can also be loaded through the Hopkinson compressive bar. In such case, a tensile stress wave can be generated by reflecting the incident compressive stress wave from the free surface of the short specimen (Stroppe et al., 1992), as schematically shown in Fig. 9.39. It is assumed that the compressive wave first passing through the specimen has no effect on the crack. Some researchers (e.g., Lee et al., 2002) added a free sleeve to the short-cracked specimen, which was fixed between the Hopkinson incident bar and the transmitting bar (see Fig. 7.7 in Chapter 7), so that the incident compressive wave in the incident bar propagates into the transmitted bar through the sleeve, while the tensile wave reflected at the free end of the transmitted bar then loads the cracked specimen.

The above is to load the cracked specimen by tensile stress. On the other side, based on the split Hopkinson pressure bar (SHPB) experimental technique, researchers have also developed a variety of ways to use compressive stress wave to load cracked samples.

Fig. 9.40 schematically shows that a compact compression (CC) specimen is directly compressively loaded by compressive pulse in an SHPB apparatu-s (see e.g., Rittel et al., 1992).

A wedge-loaded compact tension specimen (WLCT specimen) is loaded by compressive pulse using an SHPB apparatus (see e.g., Klepaczko, 1979), as schematically illustrated in Fig. 9.41. The WLCT specimen is placed between the loading wedge and the transmitted bar.

The SHPB apparatus is also especially suitable for dynamic three-point bending test to various cracked specimen by using compression pulses, which is convenient for the measurement of dynamic fracture toughness of materials.

Fig. 9.42 schematically shows a single-edge cracked specimen loaded in the one-point bend (1 PB) mode by compression pulse propagating through the Hopkinson incident bar (see e.g., Ruiz and Mines, 1985). If the

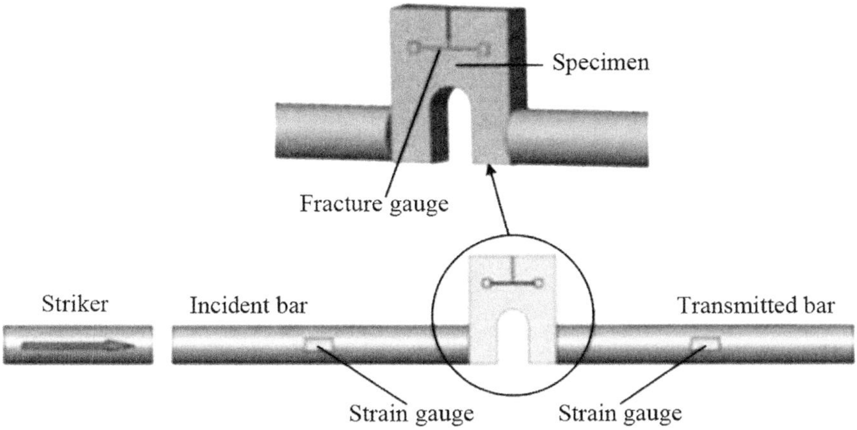

Figure 9.40 Compact compressed specimen loaded by compression pulse.

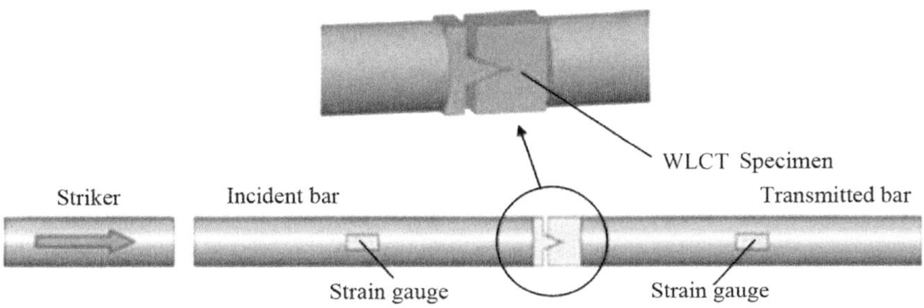

Figure 9.41 WLCT specimen loaded by compression pulse.

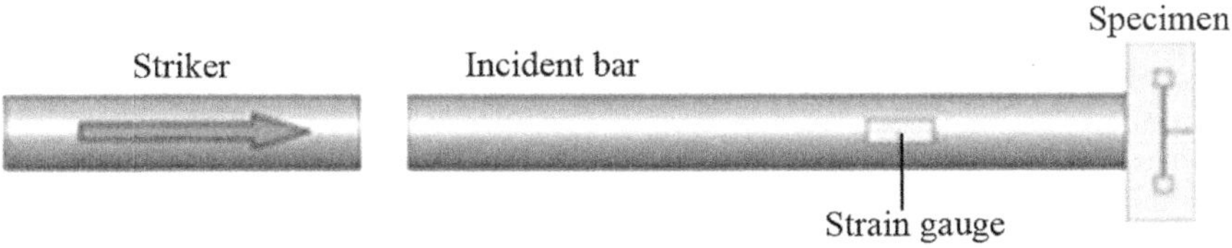

Figure 9.42 Single-edge cracked specimen loaded by compression pulse.

single-edge cracked specimen and the loading mode are the same as that of the usual Charpy test, it can be regarded as an improved Charpy test that can take into account the stress wave effect and thus can reliably determine the dynamic initiation toughness of materials (Weisbrod and Rittel, 2000).

In addition to the one-point bend (1 PB) test mentioned above, researchers have also developed a variety of dynamic experimental techniques for three-point bend (3 PB) specimens using Hopkinson pressure bar (Jiang

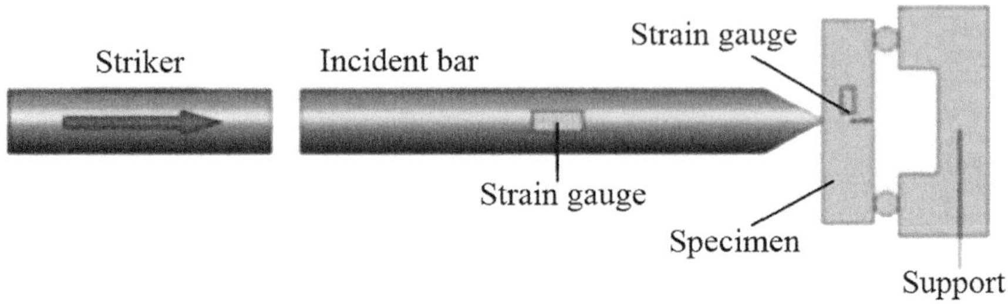

Figure 9.43 Three-point bend (3 PB) cracked specimen loaded by a single incident bar (1bar/3 PB).

and Vecchio, 2009) to study the dynamic fracture toughness of materials, as shown in Fig. 9.43 (see e.g., Mines and Ruiz, 1985), Fig. 9.44 (see e.g., Jiang and Vecchio, 2007), and Fig. 9.45 (see e.g., Yokoyama and Kishida, 1989).

Fig. 9.43 shows that compression pulse loading is directly applied to the three-point bending crack sample by using a single incident bar.

Fig. 9.44 shows that the compression pulse loading is applied to the three-point bending cracked specimens sandwiched between the incident bar and the transmission bar.

Fig. 9.45 shows that the three-point bending cracked specimen sandwiched between the incident bar and the double transmission bar is subjected to compression pulse loading.

Mainly the Mode-I cracked specimens are discussed above. Similarly, the Hopkinson pressure bar technique can also be developed to study dynamic fracture toughness on Mode-II cracked specimen. As shown in Fig. 9.46, Dong et al. (1998) adopted Hopkinson pressure bar apparatus to load a

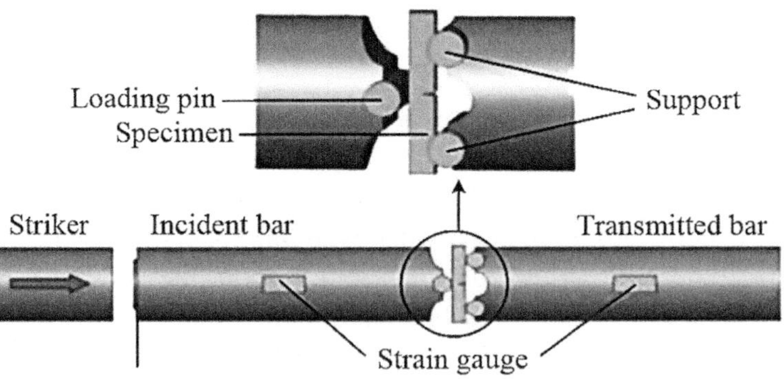

Figure 9.44 Three-point bend (3 PB) cracked specimen placed between the incident bar and the transmission bar (2bar/3 PB).

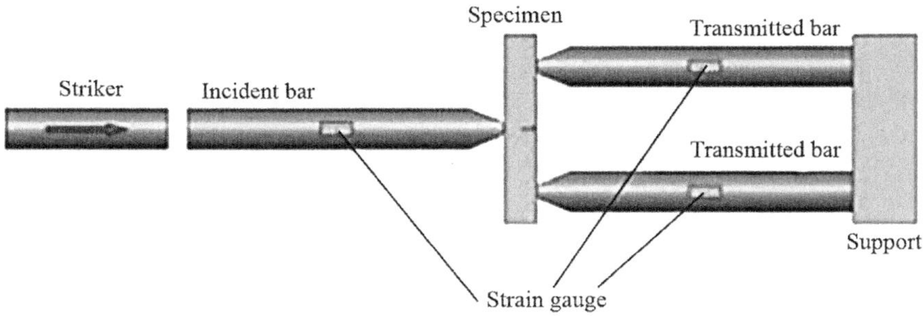

Figure 9.45 Three-point bend (3 PB) cracked specimen placed between the incident bar and double transmission bars (3bar/3 PB).

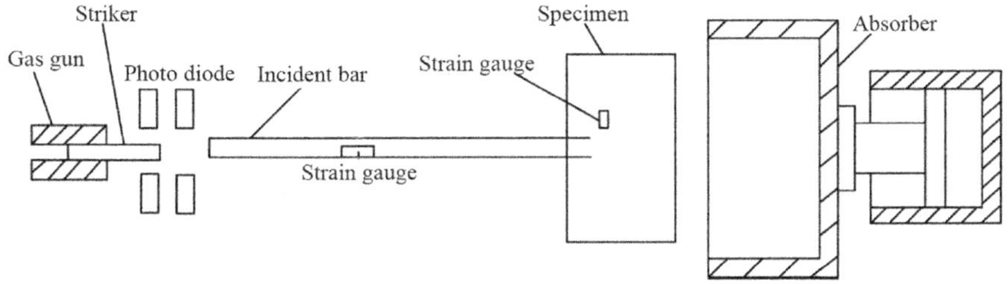

Figure 9.46 Experimental study on dynamic fracture toughness of Mode-II crack by Hopkinson pressure bar (Dong et al., 1998).

single-edge parallel double-cracks specimen to study the interaction between crack initiation and adiabatic shear for Mode-II crack. In the test, the incident bar is arranged in contact with the specimen at the position between the parallel double cracks; a compression pulse is transmitted to the specimen through the incident bar, resulting in a Mode-II fracture (see Chapter 10.1.5 "Interaction between adiabatic shear bands and cracks").

9.1.9.2 Measurement technique

As mentioned previously, the experimental determination of the dynamic fracture toughness of materials is achieved by measuring the dynamic stress intensity factor $K_I^d(a, \sigma, t)$ varying with time under high loading rate and the corresponding critical time t_f(in the order of 10^0—10^2μs) when some specific critical condition is satisfied. When using Hopkinson pressure bar technique, if the basic assumption of "uniformity" could be satisfied (see Chapter 7.1 "Experimental technique of split Hopkinson pressure bar (SHPB)"), i.e., the stress between the interfaces of the specimen-incident bar and the specimen-transmitted bar, can keep the dynamic equilibrium, it can be treated as a quasistatic process under high loading rate. However, for most experiments concerning crack dynamics, it is difficult to satisfy the basic assumption of "uniformity" since the critical time t_f is often of the same order as the rising time of incident wave. So it is necessary to directly measure the dynamic stress intensity factor $K_I^d(a, \sigma, t)$ and the critical time t_f.

At present, there are two main methods to measure dynamic stress intensity factor (dsif) at the crack tip: optical methods and electrical methods, which are discussed as follows.

(1) Optical method

The optical methods that have been used to measure dynamic stress intensity factor include dynamic photoelasticity, dynamic moire interferometry, and dynamic caustics, etc. Among them, the dynamic caustics method combined with high-speed camera has been successfully used not only in direct measurement of dynamic initiation toughness K_{Id} (see Fig. 9.8 for example), but also in the dynamic crack growth toughness K_{ID} and the dynamic crack arrest toughness K_{Ia} for propagating cracks (see Fig. 9.30 and Fig. 9.31 for example), which become the mainstream in the experimental technique of crack dynamics.

The method of dynamic caustics is mainly discussed below.

The method of caustics developed by Mannogg (1964), Theocaris (1970), Kalthoff (1987), and Rosakis (1980) in the 1970s has been particularly popular due to high accuracy in measurement and relatively simple to use. In addition to a high-speed camera, the required equipment is only a point light source. The physical principle of the caustics method is illustrated in Fig. 9.47.

The parallel light beam generated by the point light source irradiates on the surface of the cracked plate specimen which is subjected to a uniform tensile stress σ. If the specimen is optically isotropic transparent material, the stress concentration on the region surrounding the crack tip causes such changes as the uneven reduction of the specimen thickness (poisson's ratio effect) and the reduction of the refractive index of the material (Maxwell-Neumann stress-optical law). Thus, the concave surface around the crack tip due to the nonuniform reduction in thickness acts like an optical diverging lens, coupled with changes in the material's refractive index; the light passing through the specimen is deflected outward. Therefore, on the image plane behind the specimen at a distance of z_0, the image of the crack shows that the crack tip is surrounded by a black spot, as shown in Fig. 9.47B,C. This black spot is bounded by a very bright light, which is a singular curve formed by the focus of many light rays. Different researchers

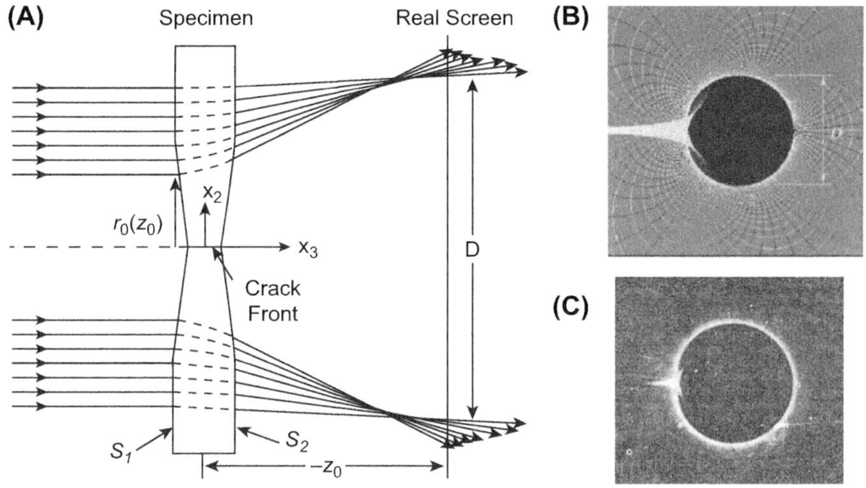

Figure 9.47 Schematic diagram of the principle of caustics formation for transparent specimen of Mode-I crack tip.

have different names for this, such as "stress corona," "shadow spot," or "caustic" etc. It is referred to as caustics hereafter. Fig. 9.47B shows the theoretical simulation pattern of caustics; the "caustic" and the "caustic curve" are clearly seen in this simulation. Fig. 9.47C shows the pattern measured experimentally. Both show the same clear dark shadow spots and bright caustics curves surrounding them. Note that the maximum diameter D of the caustics curve appears in the transverse direction, i.e., the direction of x_2 axis of Fig. 9.47A, as indicated in Fig. 9.47B.

The curve on the specimen plane (object plane) that maps with the caustics curve on the image plane is called the initial curve, with r_0 as the radius, see Fig. 9.47A. The light outside the initial curve falls on the outside of the caustics curve, the light inside the initial curve falls on or the outside of the caustics curve, and the light passing through the initial curve falls on the caustics curve, making the caustics curve a bright curve with many rays focused on it.

Since the caustic curve of the cracked body is originated from the singular mechanical field at the crack tip, the stress intensity factor of the singular mechanical field at the crack tip can be derived from the information contained in the caustic curve.

Unlike crack statics, crack dynamics focuses on dynamic stress intensity factors. For a Mode-I propagating crack extended at a steady speed $\dot{a}$, from the crack-tip mechanical field analysis (see Section 9.1.6), combined with the geometrical optics, after some mathematical calculations, the relationship between the dynamic stress intensity factor K_I^d and the maximum diameter D of the caustics curve can be deduced as follows (ref. e.g., Fan, 2006; Ravi-Chandar, 2004).

$$K_I^d = \frac{2\sqrt{2\pi}F(\dot{a})}{3m^{3/2}z_0 h c_t}\left(\frac{D}{f}\right)^{5/2} \tag{9.36}$$

where

Z_0—distance between object plane and image plane (Fig. 9.47a)

D—maximum diameter of caustics (Fig. 9.47b)

h—initial thickness of object

f—shadow optical constant (ratio of D to r_0)

c_t—material optical constant

m—scalar factor (ratio of any length on the image plane to its corresponding length on the object plane) $\left\{\begin{array}{l} = 1, \ \textit{for parallel light beam} \\ < 1, \ \textit{for convergent light beam} \\ > 1, \ \textit{for divergent light beam} \end{array}\right\}$

$F(\dot{a})$—correction factor of the crack propagating velocity (dynamic correction factor)

K_I^d—dynamic stress intensity factor

According to Ravi-chandar and Knauss (1984c), the crack propagating velocity correction factor $F(\dot{a})$:

$$F(\dot{a}) = \frac{4\alpha_1\alpha_2 - \left(1+\alpha_2^2\right)^2}{\left(\alpha_1^2 - \alpha_2^2\right)\left(1+\alpha_2^2\right)} \tag{9.37a}$$

The definitions of α_1 and α_2 in the above equation are the same as those in Eq. (9.29b), which are, respectively, related to the longitudinal wave velocity C_l and the shear wave velocity C_s, so:

$$\alpha_1^2 = 1 - \frac{\dot{a}^2}{C_1^2}, \qquad \alpha_2^2 = 1 - \frac{\dot{a}^2}{C_2^2} \tag{9.37b}$$

$$C_1 = C_l = \sqrt{\frac{\lambda + 2\mu}{\rho}} \tag{9.37c}$$

$$C_2 = C_s = \sqrt{\frac{\mu}{\rho}} \tag{9.37d}$$

When $\dot{a}/C_R$ increases from 0 to 1, $F(\dot{a})$ decreases from 1 to 0. For static cracks ($\dot{a}$=0), $F(\dot{a}) = 1$, Eq. (9.36) reduces to the corresponding solution for static cracks. In other words, the dynamic stress intensity factor K_I^d can be obtained by the static caustics solution K_I multiplying the dynamic correction factor $F(\dot{a})$.

The material optical constant c_t for optically isotropic transparent materials consists of two parts, i.e.,

$$c_t = c + \frac{(n_0 - 1)\nu}{E} \tag{9.38}$$

where c is the stress-optical coefficient for the material, n_0 is the refractive index of the material under unstrained state, ν is Poisson's ratio, and E is Young's modulus.

The above equation indicates that the caustic curve formed by the crack specimen is caused by the changes in two aspects: the first term on the right side of the formula represents the influence of the change of material refractive index, and the second term is the influence of the change of specimen thickness under the plane stress state.

If the specimen is made of an opaque material, the specimen shall have a reflective surface. It can be proved that the same caustic image as the transparent specimen can be obtained in principle on the virtual image plane behind the specimen due to the light deflection from the reflective surface, as shown in Fig. 9.48.

However, the light deflection in the reflected case is only caused by the thickness change in specimen; therefore, the c_t in Eq. (9.38) should be replaced by $c_n = -\nu/E$, and the initial thickness h of the specimen in Eq. (9.36) should be replaced by $h/2$ because whether or not the specimen is transparent, the object plane should be half of the specimen thickness.

The above dynamic caustics method combined with high-speed photography can be used to study crack propagation and subsequent crack arrest. Fig. 9.49 shows a sequence of caustics of rectangular double-cantilever beam (DCB) specimens made of Araldite B epoxy resin under wedging loading (Kalthoff et al., 1977). By analyzing the changes of shadow size and crack-tip position recorded in each photo, the variations of the dynamic stress intensity factor and at same time the crack velocity $\dot{a}$ can be obtained. Then, the dynamic crack growth toughness and the dynamic crack arrest toughness of the propagating crack are determined. It can be seen from Fig. 9.49 that with the growth of the crack length, the size of shadow spot decreases, which indicates that the dynamic stress intensity factor of

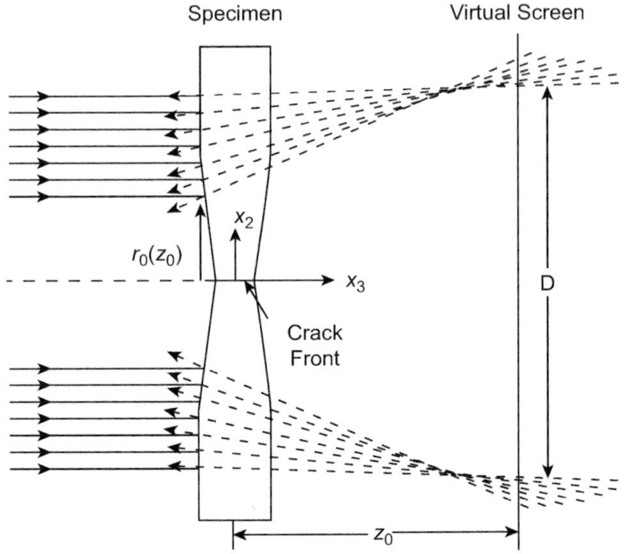

Figure 9.48 Schematic illustration of caustics formation for refection arrangement of opaque specimen of Mode-I crack tip.

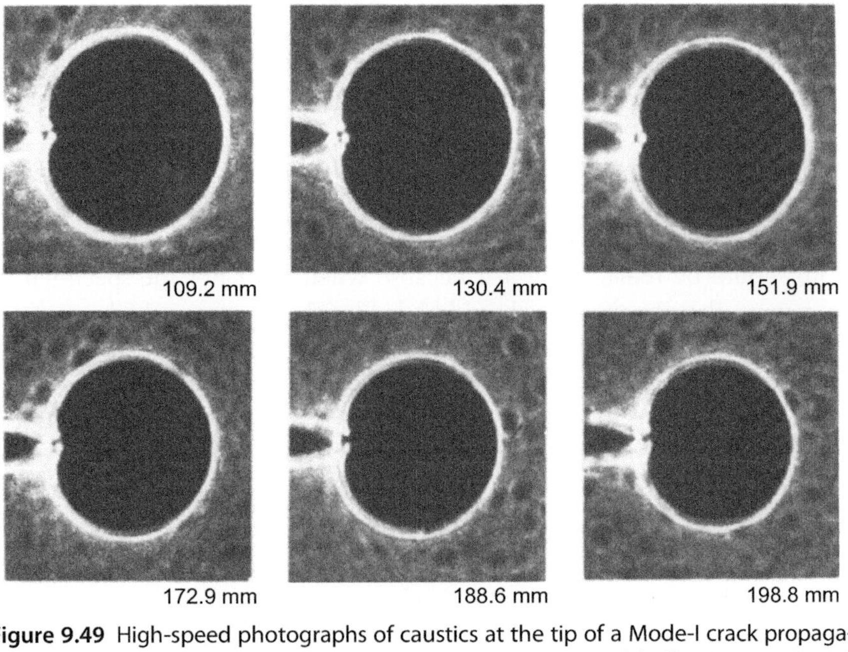

Figure 9.49 High-speed photographs of caustics at the tip of a Mode-I crack propagation and subsequent crack for epoxy resin. *From Beinert, J., Kalthoff, J.F., 1981. Experimental determination of dynamic stress intensity factors by shadow patterns. In: Experimental Evaluation of Stress Concentration and Intensity Factors (pp. 281–330). Springer, Dordrecht, Fig.5.17, p.323. Reprinted with permission of the publisher.*

propagating crack decreases with the increase of crack length, which is consistent with the result shown in Fig. 9.30.

The experimental results of the dynamic stress intensity factor K_I^d obtained from the caustic measurements versus the crack length a, $K_I^d - a$, by Kalthoff et al. (1977) are shown in Fig. 9.50, represented by experimental symbols. The cracked specimens with different crack notch radii were used so that the critical crack initiation stress intensity factor (marked as K_{Iq} in the figure) and, hence, the crack velocity are different

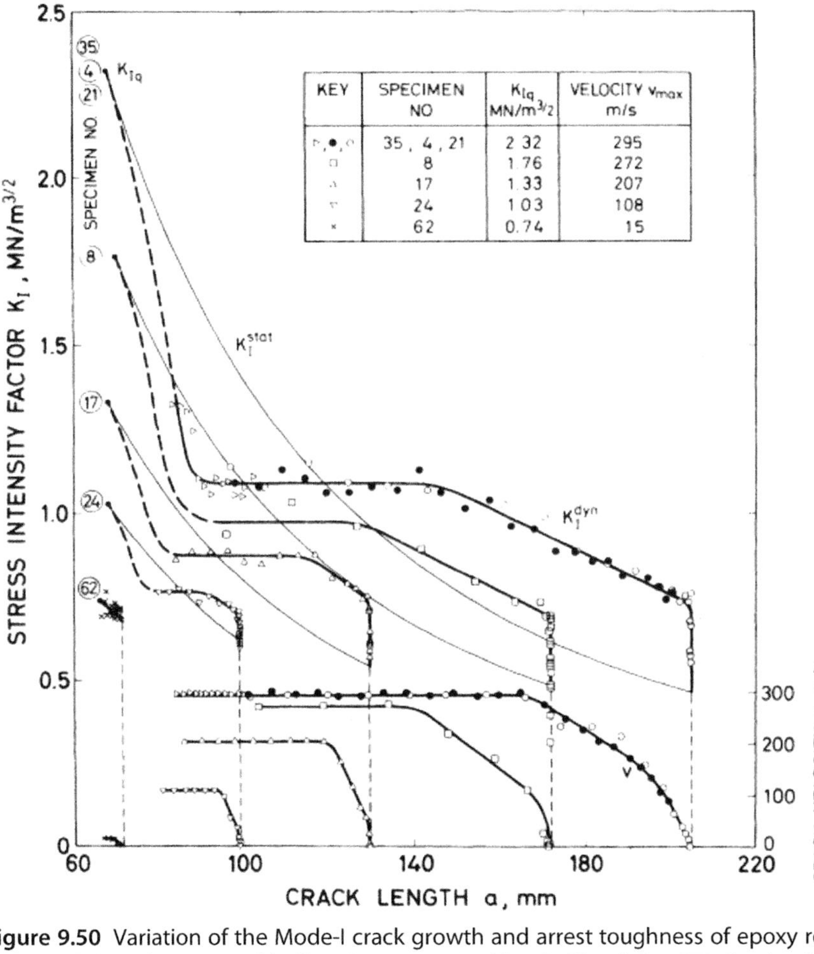

Figure 9.50 Variation of the Mode-I crack growth and arrest toughness of epoxy resin with crack length. *From Kalthoff, J., Beinert, J., Winkler, S., Blauel, J.. 1977. On the determination of the crack arrest toughness. In: ICF4, Waterloo (Canada) 1977, Fig. 4, p.755. Reprinted with permission of the publisher.*

in each experiment. The calculated results of the static stress intensity factor K_I^s versus the crack length a, $K_I^s - a$, are also given in the same figure represented by solid lines. In addition, the results of the crack propagating velocity measured versus the crack growth length a, $\dot{a} - a$, are shown at the bottom of the figure.

From these results, it can be inferred that the process of crack propagation and crack arrest has the following characteristics:

First, at the beginning of crack propagation, different from the $K_I^s - a$ curve that declines continuously, the dynamic stress intensity factor curve $K_I^d - a$ measured by caustics first decreases rapidly and then maintains at a constant value during the long crack-growth period and finally slowly drops down to the crack arrest value. Thus, at the beginning of crack propagation, the dynamic stress intensity factor measured by caustics K_I^d is smaller than the corresponding static stress intensity factor K_I^s, but in the constant velocity propagation stage, the K_I^d is larger than the corresponding K_I^s.

Second, when the dynamic stress intensity factor K_I^d remains constant, the crack propagation velocity $\dot{a}$ also remains constant, and K_I^d when decreases, the crack propagation velocity $\dot{a}$ also decreases too, although with a slight lag.

Third, during crack arrest, the dynamic stress intensity factor K_I^d decreases to crack arrest toughness K_{Ia}. The dynamic stress intensity factor K_I^d of different $K_I^d - a$ curves tends to a common value, of about $K_{Ia} = 0.7\ \text{MPa}\cdot\text{m}^{1/2}$, indicating that it is a material constant.

However, it should be pointed out that the method of dynamic caustics is based on the analysis of linear elastic crack dynamics, which is the premise of its wider application. In addition, the derived Eq. (9.36) based on the plane stress assumption does not fully reflect the three-dimensional stress effect caused by different specimen thicknesses nor the dynamic effect of stress wave on the K_I^s-dominated crack-tip mechanical field. In principle, in order to reduce the three-dimensional effect, it is better to increase the ratio of the initial-curve radius r_0 to the specimen thickness h, (r_0/h) (see Fig. 9.47A); however, in order to reduce the stress wave effect, it is better to decrease the ratio (r_0/h), which is exactly contradictory and should be coordinated in the design of experiments.

(2) Electrical measurement method

The electrical measurement in the study of crack dynamics mainly refers to the resistance strain gauge method. Because of its simplicity and low cost, it has been widely used in experimental stress–strain analysis, is well known to many researchers, and needs no detailed introduction.

In the following, the discussion will be focused on the application of resistance strain gauges to the determination of dynamic stress intensity factors.

The earliest application of resistance strain gauge in dynamic fracture experiment was in weigh-drop test, in which a resistance strain gauge is attached on the falling hammer head in order to measure the time variation of the load applied on the specimen. For this purpose, dynamic calibration is required to determine the relationship between the strain signal $\varepsilon(t)$ and the load $P(t)$.

However, the measured signal oscillates too much due to the effect of stress wave propagating in the weigh-drop test. In addition, the time when the load curve $P(t)$ measured on the hammer head reaches its maximum value, $P_{\max}$ is inconsistent with the time that the dynamic stress intensity factor $K_I^d(t)$ measured on the specimen reaches its critical value $K_I^d(t_f)$ at crack initiation time t_f. Thus, this method is less reliable and has a fundamental problem. Furthermore, people consider mounting stain gauge near the crack tip, so that the dynamic stress intensity factor $K_I^d(t)$ can be directly determined by dynamic strain signal $\varepsilon(t)$, and the crack initiation time t_f can be determined more accurately.

There are two methods to determine the dynamic stress intensity factor $K_I^d(t)$ from the measured dynamic strain gauge signals. One is to determine the dynamic stress field $\sigma_{ij}(t)$ near the crack tip by the measured strain signals, so that Eq. (9.30) can be used to determine $K_I^d(t)$. The second is to directly determine $K_I^d(t)$ by the measured dynamic strain field $\varepsilon_{ij}(t)$ near the crack tip. In this case, similar to Eq. (9.30), there is the following relationship between $\varepsilon_{ij}(t)$ and $K_I^d(t)$ (Ravi-Chandar, 2004):

$$\varepsilon_{ij}(r,\theta)=\frac{K_I^d(t,\dot{a})}{E\sqrt{2\pi r}}\Phi_{ij}^I(\theta;\dot{a},\nu)+\frac{K_{II}^d(t,\dot{a})}{E\sqrt{2\pi r}}\Phi_{ij}^{II}(\theta;\dot{a},\nu)+\cdots \tag{9.39}$$

where ν is Poisson's ratio of the material, and $\Phi_{ij}^I(\theta;\dot{a},\nu)$ and $\Phi_{ij}^{II}(\theta;\dot{a},\nu)$ are the known angular distribution functions for Mode-I crack and Mode-II crack, respectively (Ravi-Chandar, 2004).

Let the strain gauge be mounted at the position as shown in Fig. 9.51 near the crack tip, where (x, y) represents the coordinates of the crack tip and (x', y') represents the coordinates of the strain gauge.

For stationary ($\dot{a}=0$) Mode-I crack, the relationship between the tensile strain $\varepsilon_{x'x'}(t)$ in the direction x' of strain gauge coordinate indicated in

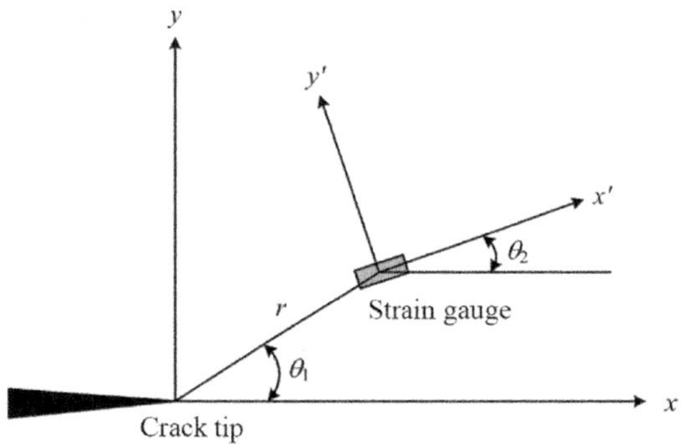

Figure 9.51 Position and orientation of strain gauge with respect to crack tip.

Fig. 9.51 and the dynamic stress intensity factor $K_I^d(t)$ can be derived according to Eq. (9.39) as follows.

$$\varepsilon_{x'x'}(t) = \frac{2(1+\nu)}{E}\frac{K_I^d(t)}{\sqrt{2\pi r}} \times \left(k\cos\frac{\theta_1}{2} - \frac{1}{2}\sin\theta_1\sin\frac{3\theta_1}{2}\cos 2\theta_2 + \frac{1}{2}\sin\theta_1\cos\frac{3\theta_1}{2}\sin 2\theta_2\right) + \mathrm{O}\left(r^{1/2}\right) + \cdots \tag{9.40}$$

where k=$(1-\nu)/(1+\nu)$. If θ_1 and θ_2 satisfy the following relation, the second term, the third term, etc. in the right-hand side of the equal sign in Eq. 9.30 are zero.

$$\cos 2\theta_1 = -k = -(1-\nu)/(1+\nu), \quad \tan(\theta_1/2) = -\cot(2\theta_2) \tag{9.41}$$

At this case, it means that the relationship between θ_1 and θ_2 only depends on the Poisson's ratio of the material.

For a material with $\nu = 1/3$, obviously $k = 1/3$ and $\theta_1 = \theta_2 = \pi/3$. Substituting them into Eq. (9.40), there is (Dally and Sanford, 1987):

$$K_I^d(t) = E\sqrt{\frac{8}{3}\pi r}\varepsilon_{x'x'}(t, r, \pi/3) \tag{9.42}$$

This means that if a single strain gauge is mounted according to the condition of $\theta_1 = \theta_2 = \pi/3$, then the dynamic stress intensity factor $K_I^d(t)$ can be directly determined from the measured strain $\varepsilon_{x'x'}(t)$.

Dong et al. determined the dynamic stress intensity factor for Mode-II crack $K_{II}^d(t)$ and the crack initiation time t_f by a similar method (Wang, et al., 1994; Dong et al., 1998). The dynamic experiments were carried out on the Hopkinson pressure bar apparatus shown in Fig. 9.46.

The strain gauge with small gauge length is mounted near the crack tip of the Mode-II crack specimen ($x = 1$ mm and $y = 2$ mm in the coordinate of crack tip shown in Fig. 9.51) (see Fig. 9.52A). The dynamic stress intensity factor $K_{II}^{d}(t)$ converted from the measured dynamic strain history is shown by the experimental data (circle symbol) in Fig. 9.52B. At the same time, the dynamic load applied on the crack specimen can be measured by the strain

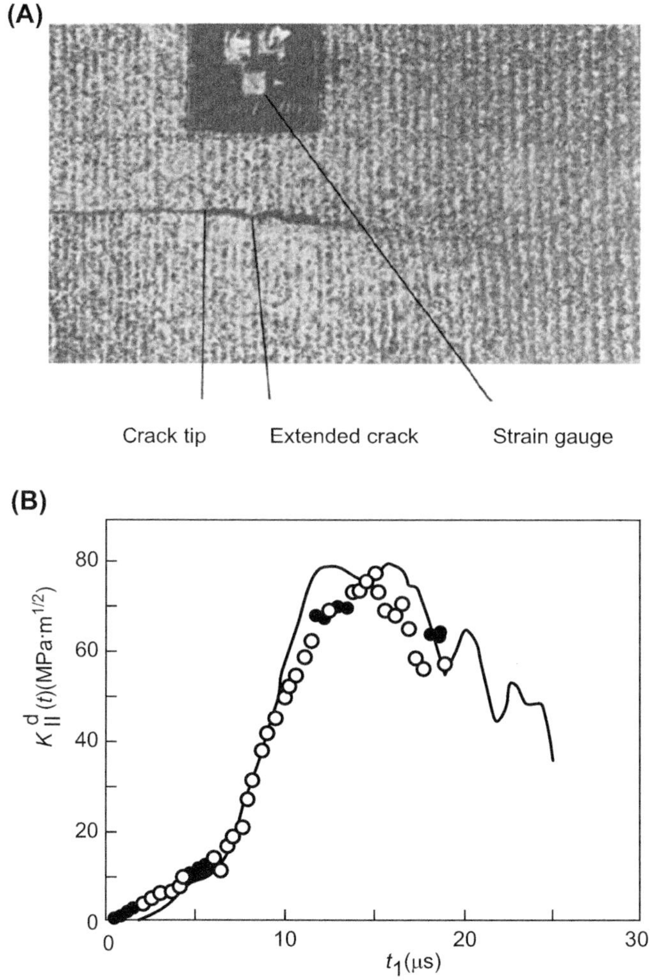

Figure 9.52 Measurement of dynamic stress intensity factor $K_{II}^{d}(t)$for Mode-II crack and time t_f of crack initiation by strain gauge method (Dong et al., 1998).

gauge on the Hopkinson pressure bar. On this basis, the corresponding dynamic stress intensity factor $K_{II}^{d}(t)$ can be calculated by the dynamic finite element program, which is shown in Fig. 9.52B by a solid line. It can be seen that the two coincide quite well before the crack initiation. The departure point between the measured value and the simulated value describes the crack initiation time.

9.2 Dynamic fragmentation

In the preface of the part III, we have emphasized that the dynamic failure of materials is essentially a time/rate-dependent process, which is characterized by the intrinsic failure characteristic time t_F of materials. In the case of quasistatic loading, t_F is negligible compared with the time scale t_L used to characterize the variation of quasistatic external load with times. Thus, the influence of intrinsic failure characteristic time t_F and the effect of stress wave can often be ignored in the case of quasistatic failure, which is then regarded as time/rate independent. However, under the short duration and high strain-rate loading such as explosion/impact loading, t_F is not negligible compared with the time scale t_L characterized the short duration of dynamic loading in the case of dynamic failure, which must be considered as time/rate dependent.

Take a one-dimensional tensile test as an example: under the quasistatic load, for the "crack-free body" that does not need to consider the macroscopic crack, as soon as the macroscopic failure criterion is satisfied in a certain weakest link, the instantaneous failure will occur and the sample will be divided into two parts. For the "crack body" that needs to take into account the macroscopic crack, once the fracture toughness failure criterion of crack mechanics is satisfied at the main crack with the largest crack size, the crack will initiate and extend until the failure, and the specimen is also divided into two parts. Thinking about those failure phenomena more carefully and deeply, there are actually two hidden facts in this description of quasistatic load failure. Firstly, no structure is perfectly made of homogeneous media, but rather always has randomly distributed defects, such as the "weak link" in crack-free body or the cracks in cracked body; otherwise the specimen will be mysteriously broken into an infinite number of infinitesimal fragments. Secondly, when the failure starts from the weakest link of the crack-free body or the largest crack of the crack body, the load of the adjacent subweak link or crack will be relaxed or reduced simultaneously (generalized unloading effect), and the failure in the adjacent area will no longer occur; otherwise, it is possible to break into an infinite number of infinitesimal fragments.

Unlike the case in the quasistatic test where the specimen is usually divided into two parts during failure, under explosion/impact loading characterized by short duration and high strain rate, the object is characterized by breaking into many pieces, that is, forming the so-called **dynamic fragmentation** in which many cracks expand simultaneously and break into many fragments. In fact, since the characteristic time required for the evolution of mesodamage (mesocracks, mesovoids, mesodeformation localization, etc.) in materials can be compared with the loading time and the characteristic time of stress wave propagation, the dynamic failure of an object is often accompanied by the generation and evolution process of multiple damages (multisource damages). For brittle objects, multiple cracks are usually developed simultaneously due to internal multisource damage. On the one hand, the apparent dynamic strength increases with increasing strain rate, while the randomness of material strength decreases with the increase of the strain rate (Grady and Kipp, 1980; Lankford and Blanchard, 1991; Hild et al., 2003; Zhou and Molinari, 2004). On the other hand, the object breaks into many fragments under impact load (Grady, 1982; Grady and Kipp, 1985; Kipp and Grady, 1985). However, for ductile objects, fracture mainly occurs in the form of tensile fracture. The main processes include that with the development of deformation/flow, material constitutive softening occurs due to the evolution of mesodamages (void growth, strain localization, and adiabatic shear banding, etc.). Furthermore, multiple necking occurs in multiple places in the object at the same time due to geometric softening, resulting in multiple fragments (Grady and Benson, 1983; Grady et al., 1984; Grady and Kipp, 1995, 1997).

In the following, the reasons, mechanism, and fragment sizes distribution of the fragmentation are discussed from the mechanical analysis approach and the energy balance approach.

9.2.1 Dynamic fragmentation phenomenon

In this section, we first take the simplest phenomena of expansion and tensile fracture of metal rings as an example to understand the difference between dynamic fragmentation and static failure process of objects. Altynova et al. (1996) used the electromagnetic expansion ring device to study the expansion—tensile fragmentation process of solutionized 6061 aluminum alloy ring. By applying different discharge voltages to the driving coil, the maximum radial expansion velocity range $V_r^{\max}$ of the metal ring specimen was obtained to be 50—300 m/s. The higher the expansion velocity is, the higher the tensile strain rate in the ring specimen is. The fragmentation process of 6061 aluminum alloy is frequently initiated by multiple necking as illustrated in Fig. 9.53, in which the photos of case 1 to case 5 present different pictures of one-dimensional stress tensile fracture of the specimen ring with the increase of the maximum expansion velocity, and the fracture under quasistatic loading is given in the photo of case 6 for comparison. Fig. 9.54 shows the variation of the number of fragments and necking with the maximum expansion velocity of ring specimen. As can be seen from those figures:

(1) Only one necking and the consequent fracture occur under quasistatic loading;
(2) During dynamic expansion and tensile failure, many necking and fragments are generated, and the number increases with the increase of expansion speed or strain rate;
(3) The development of some necks was suppressed before fracture, as shown in Fig. 9.55 (Zhang and Ravi-Chandar, 2006). In fact, it can

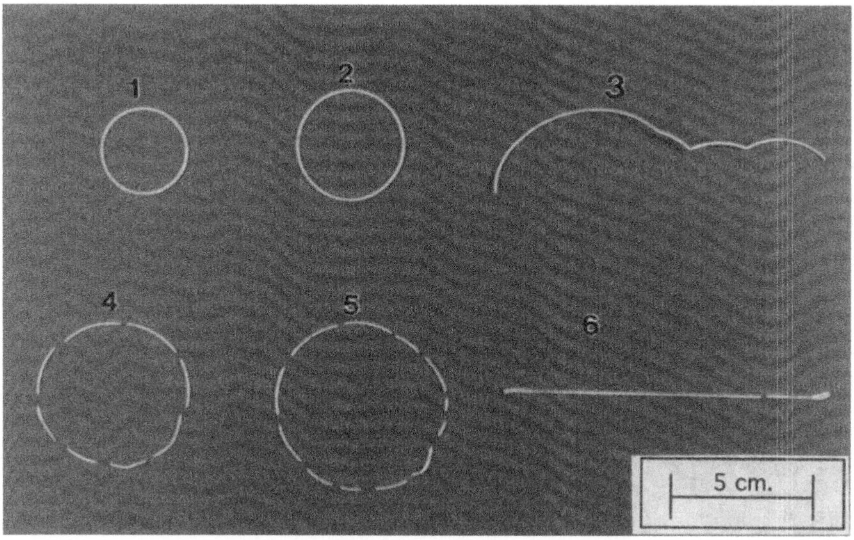

Figure 9.53 Necking and fragmentation of ring of solutionized 6061 aluminum alloy subject to electromagnetic expansion: case 1: original; case 2: 0.94 kJ; case 3:1.38 kJ; case 4: 2.06 kJ; case 5: 2.38 kJ, and case 6: tensile test. *From Altynova M., Hu X., Daehn G.S., 1996. Increased ductility in high velocity electromagnetic ring expansion, Metall. Trans. A 27, 1837—1844, Fig. 4 on page 1840. Reprinted with permission of the publisher.*

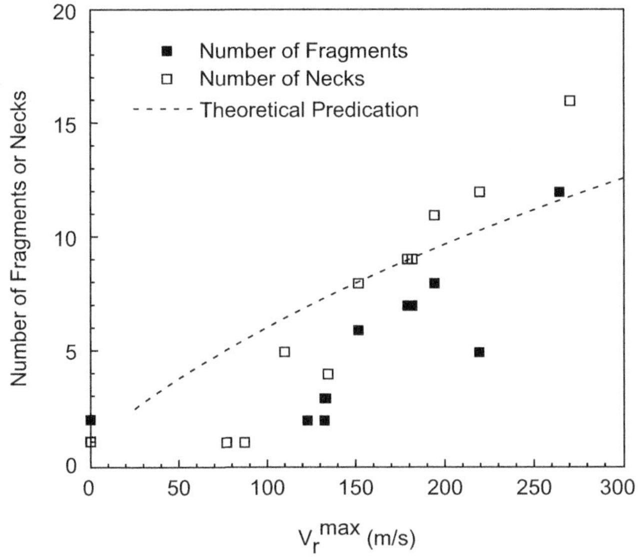

Figure 9.54 Variation of fragment number and necking number with maximum expansion speed. *From Altynova M., Hu X., Daehn G.S., 1996. Increased ductility in high velocity electromagnetic ring expansion, Metall. Trans. A 27, 1837–1844, Fig. 5 on page 1840. Reprinted with permission of the publisher.*

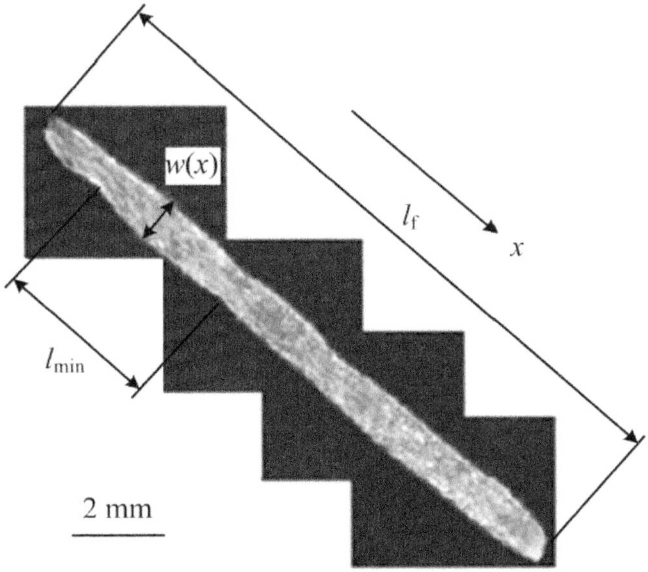

Figure 9.55 Three arrested necks are observed between the ends of the fragment. *From Zhang, H., Ravi-Chandar, K., 2006. On the dynamics of necking and fragmentation—I. Real-time and post-mortem observations in Al 6061-O. Int. J. Fract. 142 (3–4), 183–217, Fig. 25, p.213. Reprinted with permission of the publisher.*

also be seen in Fig. 9.54 that at a given expansion velocity, the number of necking is greater than the number of fragments.

(4) The fracture strain increases with the increase of expansion velocity (loading rate), and the numerical simulation also shows that these phenomena are closely related to the inertial effect under dynamic loading, but not to the strain-rate effect of the material constitutive relationship.

The fragmentation of several steel cylinders (the ratio of diameter to thickness is about 24 for thin-wall cylinders and about seven for thick-wall cylinders) under explosive expansion was investigated experimentally by Hiroe et al. (2008). The results show that the fragment morphology is mostly long and narrow, with the length being about 3–6 times of the width. Under the same explosion load, the thin-wall cylinder has a higher expansion speed, resulting in more fragments and smaller size as shown in Fig. 9.56.

Another typical example related to dynamic fragmentation is the hypervelocity impact effect. In order to protect spacecraft against hypervelocity impact of space debris, a layer or multiple layers of thin-plate protective screen (bumper sheet) are currently installed outside the spacecraft bulkhead at a certain distance, which is used to break space debris, consume and disperse its impact energy, so as to effectively protect the safety of spacecraft. Under the condition of hypervelocity impact, the projectile-bumper structure will be fragmented, spattered, penetrated, even melted, or gasified, forming a "debris cloud". By using the flash X-ray photography system, Piekutowski (1996) obtained the shadow photographs of debris clouds, of which two-represented photographs are shown in Fig. 9.57, which are produced by the hypervelocity impact of aluminum alloy spherical projectile onto an aluminum alloy sheet.

The debris cloud consists of a large number of fragments, which can be divided into three parts, as shown in Fig. 9.58 (Piekutowski, 1996, 2001): (1) "ejecta veil" composed of almost entirely of bumper fragments and a small number of projectile fragments, which was ejected in the opposite direction of impact; (2) "external bubble" composed mainly of fragments of bumper sheet; and (3) "internal structure" consisting mainly of projectile fragments and a small number of bumper fragments. The main body of the debris cloud is the internal structure, which is composed of three elements, as shown in Fig. 9.59 (Piekutowski, 1995, 2001, 2003), i.e., (a) the "front element" composed of bumper fragments and projectile fragments, (b) the "center element" composed mostly of projectile fragments, and (c) the "rear

(A)

Smooth thin-walled cylinder, t = 1.65 mm (filled with PETN)

(B)

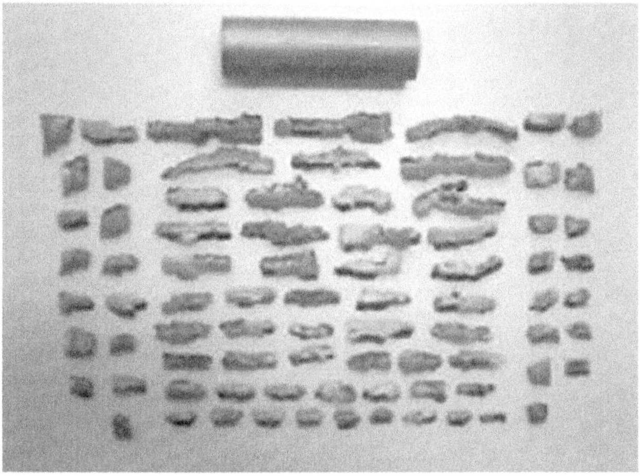

Smooth thick-walled cylinder, t = 6 mm (filled with PETN)

Figure 9.56 Photos of recovered fragments of exploded cylinders of 304 SS stainless steel. (A) Smooth thin-walled cylinder, t = 1.65 mm (filled with PETN); (B) Smooth thick-walled cylinder, t = 6 mm (filled with PETN). *From Hiroe T., Fujiwara K., Hata H., et al., 2008. Deformation and fragmentation behaviour of exploded metal cylinders and the effects of wall materials, configuration, explosive energy and initiated locations. Int. J. Impact Eng. 35 (12), 1578–1586, Fig. 7, p.1583. Reprinted with permission of the publisher.*

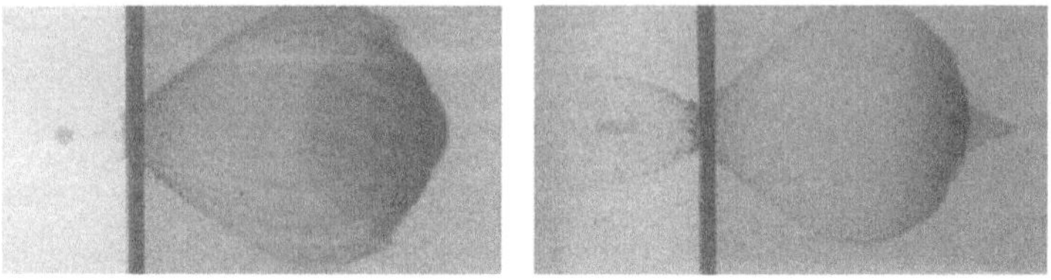

Figure 9.57 X-ray shadowgraph of typical debris cloud (Piekutowski, 1996).

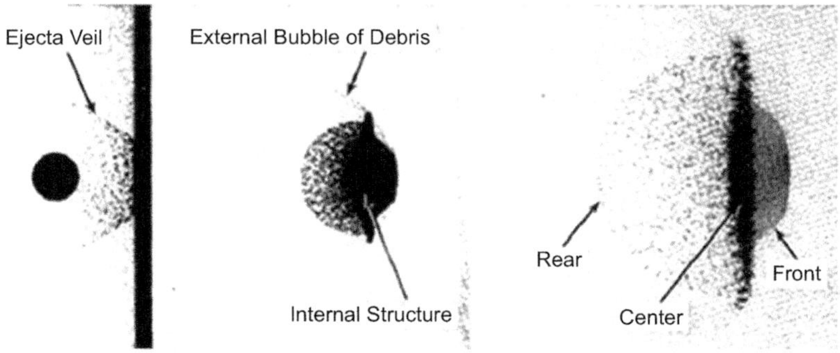

Figure 9.58 Basic composition of debris cloud. *From Piekutowski, A.J., 2001. Debris clouds produced by the hypervelocity impact of nonspherical projectiles. Int. J. Impact Eng. 26 (1–10), 613–624, Fig. 1, p.615. Reprinted with permission of the publisher.*

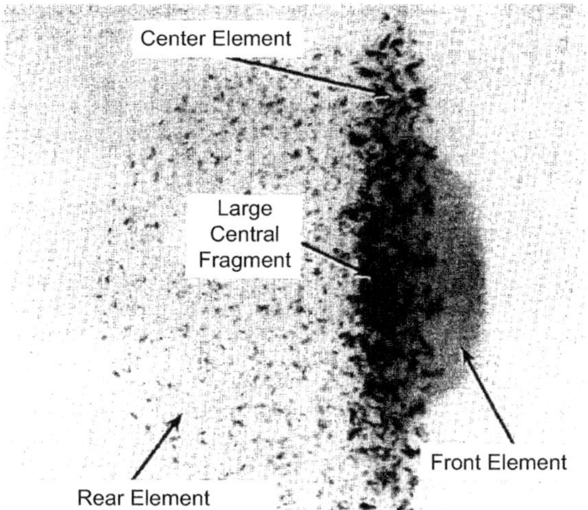

Figure 9.59 Basic composition of the "internal structure" of debris cloud body. *From Piekutowski, A.J., 2001. Debris clouds produced by the hypervelocity impact of nonspherical projectiles. Int. J. Impact Eng. 26 (1–10), 613–624, Fig. 1, p.615. Reprinted with permission of the publisher.*

element" composed of shell of spall fragments of the projectile rear. With the development of space activities, the problem of space debris has attracted more and more attention. The research on the formation mechanism, structural characteristics, movement law, and its mutilate features injuring characteristics of debris clouds has become a hot topic in recent years, and the study on dynamic fragmentation is the basis.

Although hypervelocity impact on the mechanism is related to dynamic fragmentation, it is more involved in the structural dynamic response, like penetration, plugging, and other projectile–target interaction. Those problems are classified into structural impact dynamics or structural hypervelocity impact dynamics. This book mainly focuses on the topic of material dynamics, and these issues will not be discussed further below. Readers who are interested can refer to relevant monographs (e.g., Zhang and Huang, 2000; Kinslow, 2012).

9.2.2 Dynamic fragmentation theory

The study on the fragmentation phenomenon of solid was first carried out by N.F. Mott (1943, 1947) during the World War II [9.78, 9.79].

- Mott Model

Mott proposed an analysis method to fragmentation problem while studying the dynamic fragmentation failure of artillery shells (Mott, 1943, 1947; Grady, 2006). Firstly, the uniform expansion problem of artillery cylindrical shell under the action of internal explosive detonation wave is assumed to be idealized as an uniform expansion problem of a series of overlapping rings with equal diameter, simplifying the two-dimensional expansion problem of cylindrical shell to a one-dimensional expansion problem of ring, that is, the expansion ring expanded outward at a constant speed u and subjected to uniform circumferential tensile load, as shown in Fig. 9.60. Such a ring subjected to uniform circumferential tensile load is equivalent to an infinite one-dimensional rod subjected to uniform tensile load (without the boundary constraint at the rod end) and is also equivalent to a finite one-dimensional rod with the same boundary condition at the two ends. This simple and clever treatment was first used by Born and Karman (1912) when discussing lattice vibration (recall Eq. (3.44) in Chapter 3).

Next, the governing equations of Mott problem will be established, which are composed of kinematics equation, dynamics equation, and material constitutive equation.

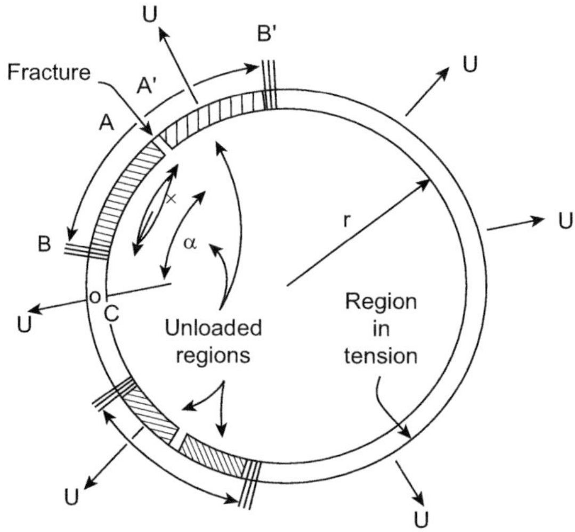

Figure 9.60 Schematic diagram of one-dimensional Mott ring.

Mott looked at the ring as a rigid-perfectly-plastic body, which deforms/flows under the tension action of a constant flow stress σ_c. Once the strain ε reaches the fracture strain ε_c of the material ($\varepsilon = \varepsilon_c$), the fracture occurs and the tensile stress is instantaneously released to zero (i.e., rigid unloading). The corresponding "rigid-perfectly-plastic-rigid unloading model" is shown in Fig. 9.61, and the corresponding constitutive relation of material can be expressed as:

$$\begin{cases} \sigma = \sigma_c, & \text{if } \varepsilon < \varepsilon_c \\ \sigma = 0, & \text{if } \varepsilon \geq \varepsilon_c \end{cases} \tag{9.43}$$

The ring is subjected to uniform tensile deformation before fracture occurs. Under constant radial velocity u outward expansion, the change of radius r at the moment of dt is $dr = udt$, and the corresponding strain increment of circumferential infinitesimal element $rd\theta$ can be expressed as:

$$d\varepsilon = \frac{(r + udt)d\theta - rd\theta}{rd\theta} = \frac{udt}{r}$$

So, the tensile strain rate $\dot{\varepsilon}$ is:

$$\dot{\varepsilon} = u/r \tag{9.44a}$$

In the expansion ring problem, $\dot{\varepsilon}$ is sometimes called the expansion strain rate. On the other hand, in order to ensure the single value and continuity of the circumferential displacement, the following compatibility relationship

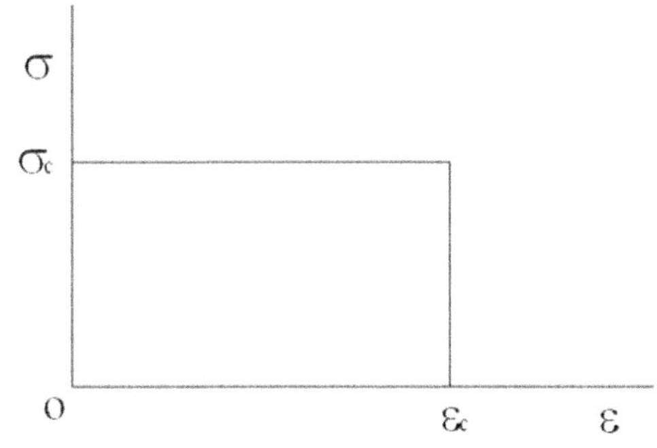

Figure 9.61 Ideal rigid–plastic unloading model.

should be satisfied between the circumferential particle velocity v and the circumferential strain rate $\dot{\varepsilon}$:

$$dv = \dot{\varepsilon}dx \tag{9.44b}$$

That is, the circumferential strain rate is equal to the velocity gradient of the circumferential particle. Eq. (9.44) gives the relation between kinematical quantities of expansion ring and constitutes the kinematic equation of the Mott problem.

Mott assumes that the location of instantaneous fracture was random in the stretching ring; at the moment of fracture, unloading wave was generated at the fracture point at the same time, and the unloading wave propagated to the surrounding region at a limited velocity, so as to relieve the tensile stress in the adjacent region; while other fracture related to the strain only occurred in the region where the unloading wave did not arrive. From this, Mott established a theoretical model to analyze the propagation process of unloading wave induced by the fracture once the expansion ring (or the corresponding one-dimensional rod) is fractured. If fracture occurs instantaneously at the circumferential point A, as shown in Fig. 9.60, once fracture initiates, an unloading wave (Mott wave) propagates outward from the point of fracture. The wave front as a propagating boundary divides the adjacent region into two different parts: a) The adjacent regions not yet reached by the Mott wave front are still subjected to uniform tensile stress σ_c; b) In the region where Mott wave passes, rigid unloading occurs, which can be regarded as rigid body. Thus, the front of Mott unloading wave is actually the boundary of the region of perfectly plastic flow and the region of rigid unloading. The velocity field in the front region of the Mott wave can be determined by the integral of Eq. (9.44) and the region behind the Mott wave is, as a rigid body, moving at the same velocity.

Apply the Newton's second law (momentum conservation equation) for the rigid unloading region (the segment AB with unit sectional area as shown in Fig. 9.60), we have:

$$\sigma_c = \rho x \frac{dv}{dt} \tag{9.45}$$

This is the dynamic equation of a rigid body undergoing rigid unloading in the Mott problem. Eq. (9.43) to (9.45) jointly constitute the governing equations of the Mott problem.

Substituting Eq. (9.44) into Eq. (9.45), we have:

$$\sigma_c = \rho \dot{\varepsilon} x \frac{dx(t)}{dt} = \frac{\rho \dot{\varepsilon}}{2} \frac{dx^2(t)}{dt} \tag{9.46}$$

To solve this ordinary differential equation, assuming that $\dot{\varepsilon}(= u/r)$ is approximately a constant, and by using the initial condition $x(0) = 0$, the position $x(t)$ of the Mott wave propagation at any time t, i.e., the width of unloading segment AB related to the fragment size (see Fig. 9.60), can be obtained.

$$x(t) = \sqrt{\frac{2\sigma_c}{\rho \dot{\varepsilon}}} \cdot t^{1/2} = \sqrt{\frac{2r\sigma_c}{\rho u}} \cdot t^{1/2} \tag{9.47}$$

It can be seen that the width of AB (i.e., the fragment size) increases in a manner proportional to $t^{1/2}$ and decreases with increasing strain rate $\dot{\varepsilon}$. By differentiating Eq. (9.47), the wave velocity of Mott wave at any time C_{Mott} can be obtained.

$$C_{Mott} = \frac{dx}{dt} = \sqrt{\frac{\sigma_c}{2\rho \dot{\varepsilon}}} \cdot t^{-1/2} = \sqrt{\frac{r\sigma_c}{2\rho u}} \cdot t^{-1/2} \tag{9.48}$$

It can be seen that the Mott wave velocity C_{Mott} decreases inversely with $t^{1/2}$.

Eq. (9.47) indicates that the propagation process of Mott wave is related to material parameters and kinematics parameters. Within the regions passed by the unloading wave, no further fracture occurs. Subsequent fracture will only occur in regions not yet reached by the unloading waves, and these regions continue to be subjected to tensile yield stress σ_c under strain rate $\dot{\varepsilon}$.

On the other hand, Mott considered that the fracture strain is randomly distributed, and when the strain increases $d\varepsilon$, let the fracture probability of nonfractured one-dimensional rod per unit length be:

$$dp = Ce^{\gamma \varepsilon} d\varepsilon$$

where C and γ are constants. Obviously, for those regions that have reached strain ε before fracture, the fracture probability when strain increases $d\varepsilon$ is:

$$dp = (1 - p)Ce^{\gamma\varepsilon}d\varepsilon$$

Integrating the above equation yields:

$$p = 1 - \exp\left(-\frac{C}{\gamma}e^{\gamma\varepsilon}\right) \tag{9.49}$$

Mott determined the fragment length distributions by Eqs. (9.47) and (9.49), although it brings some difficulties in engineering applications because it involves the processing of probability function by using graphing method.

Starting from formula (9.47), it can also be understood that the fragment size is controlled by the propagation distance of Mott waves at a certain characteristic time T_c, that is, the mean fragment size s_M is:

$$s_M(\varepsilon_c) = 2x(T_c) = \sqrt{\frac{8\sigma_c T_c}{\rho\dot{\varepsilon}}} = \sqrt{\frac{8\sigma_c\varepsilon_c}{\rho\dot{\varepsilon}^2}} \tag{9.50}$$

where $\varepsilon_c = \dot{\varepsilon}T_c$ is the apparent strain (fracture strain) of a one-dimensional rod (or the expanding ring) reached within the characteristic time T_c, which is distributed according to the probability function of Eq. (9.49).

Noted that $\sigma_c\varepsilon_c$ represents the strain energy density in the rigid unloading region (the area enclosed by the perfectly plastic stress–strain curve in Fig. 9.61), while $\dot{\varepsilon}^2$ is proportional to u^2 (see Eq. 9.44) which represents the kinetic energy density of expanding ring, so Eq. (9.50) indicates that the average fragment size s_M depends on the ratio of the strain energy density to the kinetic energy density.

The key contributions of the Mott model in the pioneering study of dynamic fragmentation are: first, the Mott unloading wave was proposed, and furthermore the whole fracture process of the expansion ring was determined by some random function. However, in the classical Mott theory, the occurrence of each fracture is an instantaneous phenomenon, the energy dissipation at the moment of fracture is ignored, and the material fracture resistance at the fracture site and the corresponding fracture energy are not considered, which still belongs to the classical mechanics category of "crack-free body" and has not yet involved the crack mechanics discussed earlier in this chapter.

In fact, fragmentation is a process of microcrack evolution (brittle fragmentation) or a process of microvoid evolution (ductile fragmentation), which does not occur instantaneously, and there is energy dissipation due to the formation of new crack surfaces. Therefore, Grady et al. further developed Mott model from the perspective of mechanical analysis (Grady and Kipp, 1980) and energy analysis (Grady, 1982). The following focuses mainly on the Grady-Kipp cohesive fracture model (Kipp and Grady, 1985; Grady, 2006).

9.2.2.1 Grady—Kipp cohesive fracture model

Based on the Mott's idea that the fragment size is dominated by the propagating distance of unloading wave from the fracture site, Grady and Kipp thought that the formation and separation of the crack surface is a process to overcome the cohesion force of crack, i.e., the so-called cohesive fracture process (Kipp and Grady, 1985; Grady, 2006), rather than occurs instantaneously. This is similar to what we discussed on the bonds between grains in the Section 3.4 in chapter 3.

Assume that the cohesive stress σ_{coh} is reduced linearly from σ_c to 0 with crack-opening displacement δ_{coh}, i.e., obeying the so-called linear cohesive fracture model, as expressed in Eq. (9.51a).

$$\sigma_{coh} = \sigma_c\left(1 - \frac{\delta_{coh}}{\delta_c}\right) \tag{9.51a}$$

where δ_c is the crack opening displacement when crack is broken completely and σ_{coh} is reduced to 0.

At the same time, Grady and Kipp also introduced the parameter of fracture energy per unit crack area (including two fracture surfaces) G_c to characterize the dissipated energy during the fracture process. Noting the linear relation (9.51a), the work done by the cohesive stress is $\sigma_c\delta_c/2$, let it equal to the dissipation energy of fracture G_c, then Eq. (9.51b) can be written as:

$$\delta_c = \frac{2G_c}{\sigma_c} \tag{9.51b}$$

which is illustrated in Fig. 9.62. Recall the discussion of Griffith energy release rate criterion in Section 9.1.1, the G_c here is exactly the fracture energy consumed by the critical unstable extension of the Mode-I crack G_{Ic}.

Compared with the governing equations of Mott model, i.e., Eqs. (9.43) to (9.45), the governing equations of Grady—Kipp's linear cohesive fracture model have the following improvements: (a) Eq. (9.51) is added as a

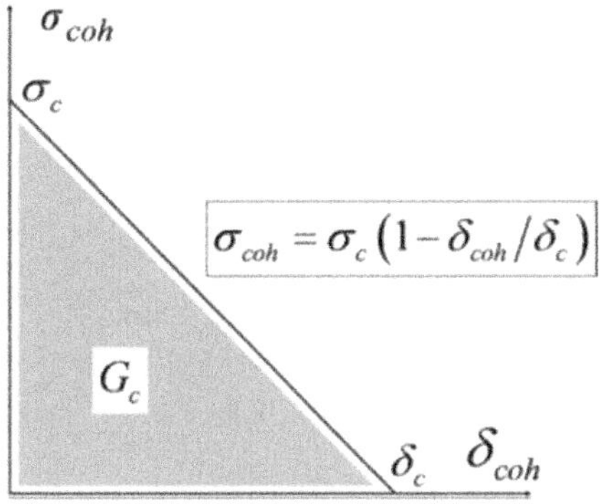

Figure 9.62 The cohesive stress as function of crack opening displacement in the Grady–Kipp model.

supplementary constitutive relation of the material to describe the characteristics of linear relation between the cohesive force and the crack opening displacement. (b) In terms of kinematics equation, besides the basic relation of expansion ring, Eq. (9.43), considering that the rigid unloading region is a rigid body moving as a whole, the particle velocities at both ends of the rigid region should be equal: $v(\delta_{coh}) = v(x(t))$, then the following equation should be added to reflect that the opening velocity of the crack end of the rigid body is equal to the movement velocity of the unloading wave end.

$$\frac{d}{dt}\left[\frac{\delta_{coh}(t)}{2}\right] = \dot{\varepsilon}x(t) \tag{9.52}$$

(c) In terms of the kinetic equation, when Eq. (9.45) was derived, the unloading rigid segment is only affected by σ_c acted on the one end adjacent to a fracture point, and now both ends are affected by σ_c and σ_{coh}, respectively, so the σ_c in Eq. (9.45) should be replaced by $(\sigma_c - \sigma_{coh})$, then we have:

$$\sigma_c - \sigma_{coh} = \rho x \frac{dv}{dt} \tag{9.53a}$$

Substituting Eq. (9.44) and Eq. (9.51) into Eq. (9.53a), and setting $\dot{\varepsilon}$ remains constant, the following momentum balance relation can be obtained:

$$\rho\dot{\varepsilon}x(t)\frac{dx(t)}{dt} = \frac{\sigma_c^2}{2G_c}\delta_{coh}(t) \tag{9.53b}$$

Eq. (9.52) and Eq. (9.53b) constitute a set of ordinary differential equations about $x(t)$ and $\delta_{coh}(t)$. By using the initial condition $x(0) =$

$\delta_{coh}(0) = 0$ at time $t = 0$, the propagation distance $x(t)$ of the unloading wave and the crack opening displacement $\delta_{coh}(t)$ can be solved:

$$\begin{cases} x(t) = \dfrac{\sigma_c^2}{6\rho G_c} t^2 \\ \delta_{coh}(t) = \dfrac{\sigma_c^2 \dot{\varepsilon}}{9\rho G_c} t^3 \end{cases} \tag{9.54a}$$

It can be seen that the propagation distance $x(t)$ of unloading wave and the crack opening displacement $\delta_{coh}(t)$ depend on the material properties such as density ρ, flow stress σ_c, and fracture energy G_c, as well as the expansion strain rate $\dot{\varepsilon}$. Among these three material properties, G_c/σ_c has a length dimension (see Eq. 9.51) and σ_c/ρ has a dimension of the square of wave speed. If introducing the characteristic fragment size s_0 and the characteristic strain rate $\dot{\varepsilon}_0$ defined as below,

$$s_0 = \frac{G_c}{\sigma_c}, \quad \dot{\varepsilon}_0 = \sqrt{\frac{\sigma_c}{\rho}}\frac{\sigma_c}{G_c} = \frac{\sigma_c^{3/2}}{\rho^{1/2} G_c} \tag{9.55a}$$

and on this basis, the corresponding dimensionless unloading wave propagation distance $\overline{x}$, the dimensionless crack opening displacement $\overline{\delta}_{coh}$, the dimensionless strain rate $\overline{\dot{\varepsilon}}$, the dimensionless time $\overline{t}$, and the dimensionless fragment size $\overline{s}$are defined as:

$$\overline{x} = \frac{x}{s_0}, \quad \overline{\delta}_{coh} = \frac{\delta_{coh}}{s_0}, \quad \overline{\dot{\varepsilon}} = \frac{\dot{\varepsilon}}{\dot{\varepsilon}_0}, \quad \overline{t} = t\dot{\varepsilon}_0, \quad \overline{s} = \frac{s}{s_0} \tag{9.55b}$$

then Eq. (9.54a) can be rewritten as the following simple dimensionless form:

$$\begin{cases} \overline{x}(\overline{t}) = \dfrac{\overline{t}^2}{6}, \\ \overline{\delta}_{coh}(\overline{t}) = \dfrac{\overline{\dot{\varepsilon}}}{9}\overline{t}^3 \end{cases} \tag{9.54b}$$

Let t_c represents the time required for the fracture from starting to completing, according to Eq. (9.51b), $\delta_{coh}(t_c) = \delta_c = 2G_c/\sigma_c$ at time t_c. By substituting it into the second equation of Eq. (9.54a), the time t_c required for the crack's complete development can be obtained:

$$t_c = \left(\frac{18\rho G_c^2}{\sigma_c^3 \dot{\varepsilon}}\right)^{1/3} \tag{9.55c}$$

It can be seen that with the introduction of the linear cohesive fracture model, the characteristic time T_c in the Mott problem has been clearly defined. Therefore, Grady et al. obtained a fragment size relationship characterized by fracture energy (referred as Grady—Kipp fragment formula).

$$s_{G-K} = 2x(t_c) = \left(\frac{12G_c}{\rho\dot{\varepsilon}^2}\right)^{1/3} \tag{9.56a}$$

The above formula shows that the mean size of fragments depends on the ratio of the fracture energy to the expansion kinetic energy density $\frac{G_c}{\rho\dot{\varepsilon}^2}$. Compared with Mott's Eq. (9.50), the energy ratio $\frac{8\sigma_c\varepsilon_c}{\rho\dot{\varepsilon}^2}$ in the Mott formula is replaced by the $\frac{12G_c}{\rho\dot{\varepsilon}^2}$ in the Grady—Kipp fragment formula, which takes into account the fracture energy of crack mechanics, thus the Grady—Kipp fragment formula is obviously more reasonable. If the dimensionless parameters defined in Eq. (9.54) are introduced, the above formula can be rewritten as a simple dimensionless form:

$$\bar{s}_{G-K} = \left(\frac{12}{\bar{\dot{\varepsilon}}^2}\right)^{1/3} \tag{9.56b}$$

This is displayed as a straight line with a slope of (−2/3) on the $\log\bar{s} - \log\bar{\dot{\varepsilon}}$ graph, which indicates that the dimensionless mean fragments size depends only on the dimensionless strain rate, and the mean fragment size decreases with increasing strain rate. Note that G_c in the above equation is referred to the fracture energy of two fracture surfaces. If G_c is defined as the fracture energy of one fracture surface as in some literature, the coefficient of 12 in Eq. (9.56) should be replaced by 24.

Grady also studied the dynamic fragmentation of brittle objects based on the perspective of local energy conservation (Grady, 1982). He proposed that if a rapidly expanding object is broken into many fragments which will fly away at a certain velocity, the kinetic energy of the mass of each fragments does not contribute to the generation of new surface. He proposed that surface area created in the fragmentation process is governed by an equilibrium between the fracture energy of fragments and the component of kinetic energy with respect to the mass center before fragmentation (called "local kinetic energy"). Based on this principle, the expression of the mean fragment size of brittle materials derived by Grady is exactly consistent with Eq. (9.56).

9.2.2.2 Glenn–Chudnovsky model

Glenn and Chudnovsky (1986) improved the Grady's local energy conservation model (kinetic energy model), taking into account the elastic strain energy accumulated during the tensile dilation. Note that the material with strength of σ_c has stored elastic strain energy $\sigma_c^2/2E$ (E is Young's modulus) per unit volume at the moment of fracture. By taking into account this elastic tensile strain energy, the following mean fragment size relation was derived by Glenn and Chudnovsky:

$$s_{G\&C} = 4\sqrt{\frac{\alpha}{3}}\sinh\left(\frac{\phi}{3}\right),$$
$$\phi = \sinh^{-1}\left[\beta\left(\frac{3}{\alpha}\right)^{3/2}\right], \quad \alpha = \frac{3\sigma_c^2}{\rho E\dot{\varepsilon}^2}, \quad \beta = \frac{3}{2}\frac{G_c}{\rho\dot{\varepsilon}^2} \tag{9.57a}$$

Note that the physical significance of α is just the ratio of the elastic strain energy density to the expansion kinetic energy density, and the physical significance of β is the ratio of the fracture energy to the expansion kinetic energy density as contained in the Grady–Kipp model of Eq. (9.56). At the high strain rate, the expansion kinetic energy density of the material is far greater than the elastic strain energy density; thus the elastic strain energy becomes insignificant. The Glenn–Chudnovsky model (G–C) of Eq. (9.57) and Grady–Kipp model of Eq. (9.56) give the same result, showing as a coincident straight lines with a slope of $-2/3$ on the $\log(s) - \log(\dot{\varepsilon})$ diagram. At the quasistatic condition ($\dot{\varepsilon} \le 10^0\ \text{s}^{-1}$), the $-C$ model gives a constant fragment size, $(s_{G-C})_{qs} = \frac{G_c}{\sigma_c^2/2E}$, which is strain-rate independent, but determined by the ratio of the fracture energy to the elastic strain energy density of the material. In other strain rates between the two, the G–C model accounts the contribution of elastic strain energy (modulus E and failure stress σ_c) and expansion kinetic energy ($\dot{\varepsilon}$).

Compared with the Grady model, the G–C model added a material parameter of E, so the dimensionless parameter defined in Eq. (9.55) is no longer applicable. If taking $(s_{G-C})_{qs}\Big/2 = \frac{EG_c}{\sigma_c^2}$ as a characteristic fragment size s_c, it is the ratio of the characteristic fragment size s_0 defined by Eq. (9.55a) to the dimensionless number σ_c/E.

$$s_c = \frac{s_0}{\sigma_c/E} = \frac{EG_c}{\sigma_c^2} \tag{9.58a}$$

Noting also that $\beta/\alpha^{3/2}$ in Eq. (9.57), after the calculus, can be expressed as:

$$\frac{\beta}{\alpha^{3/2}}=\frac{\frac{3}{2}\frac{G_c}{\rho\dot{\varepsilon}^2}}{\frac{3^{3/2}\sigma_c^3}{(\rho E)^{3/2}\dot{\varepsilon}^3}}=\frac{1}{2\sqrt{3}}\frac{\dot{\varepsilon}}{\left(\frac{c_0\sigma_c^3}{G_cE^2}\right)}=\frac{1}{2\sqrt{3}}\frac{\dot{\varepsilon}}{\left(\frac{c_0\sigma_c}{s_cE}\right)}$$

where $c_0=(E/\rho)^{1/2}$ is the velocity of elastic wave in the rod, and the denominator of above formula is a parameter with a dimension of strain rate containing relevant material parameters, which can be taken as a characteristic strain rate $\dot{\varepsilon}_c$, which is the product of the characteristic strain rate $\dot{\varepsilon}_0$ defined in Eq. (9.55a) and the dimensionless number $(\sigma_c/E)^{3/2}$.

$$\dot{\varepsilon}_c=\frac{c_0\sigma_c^3}{G_cE^2}=\frac{c_0\sigma_c}{s_cE}=\dot{\varepsilon}_0\bullet\left(\frac{\sigma_c}{E}\right)^{3/2} \tag{9.58b}$$

Then after introducing the dimensionless fragment length $\bar{s}_{G-C}$ and the dimensionless strain rate $\bar{\dot{\varepsilon}}$ defined below:

$$\bar{s}_{G-C}=\frac{s_{G-C}}{s_c},\quad \bar{\dot{\varepsilon}}=\frac{\dot{\varepsilon}}{\dot{\varepsilon}_c} \tag{9.58c}$$

The mean fragment size of G—C model can be reduced to the dimensionless form:

$$\bar{s}_{G-C}=\frac{4}{\bar{\dot{\varepsilon}}}\sinh\left[\frac{1}{3}\sinh^{-1}\left(\frac{3}{2}\bar{\dot{\varepsilon}}\right)\right],\quad (\bar{s}_{G-C})_{qs}=2 \tag{9.57b}$$

This implies that the dimensionless mean fragment size $\bar{s}_{G-C}$ for the G—C model still only depends on the dimensionless strain rate $\bar{\dot{\varepsilon}}$.

9.2.2.3 Zhou Fenghua et al. model

The Grady fragment model (Eq. 9.56) and G—C fragment model (Eq. 9.57) are derived from the energy conservation relationship. The physical images are simple and the expressions are concise. Both models reflect the basic physical phenomenon of fragmentation. In a certain range, they are also qualitatively consistent with the experimental results and have been widely adopted. However, compared with some numerical simulation and theoretical analysis in recent years, the formula significantly overestimated the fragment size (up to 5—10 times). The main reason is that dynamic fragmentation is a complicated process involving the dynamic mechanisms of crack nucleation, propagation, and multisource crack interaction, etc.

The above formula ignores the details of these mechanisms, the physical process of fragmentation is oversimplified, and the propagation of unloading interface is not an unloading wave in the strict sense.

Zhou Fenghua et al. established a model that takes into account the mechanism of fragmentation (Zhou et al., 2005, 2006a), taking into account the random generation of cracks in the process of fracture of brittle objects, the extension and fracture process of cracks under the action of irreversible linear cohesive force, and the complex interaction of elastic unloading waves between multiple cracks. In fact, in a process of dynamic fragmentation, a large number of cracks are nucleated due to rapid loading, and most of them will stop expanding due to mutual unloading of cracks, becoming internal damages of the fragments. In order to comprehensively analyze the interaction process of multiple cracks, it seemed that numerical simulation is the only way. Based on this model, Zhou Fenghua et al. simulated the dynamic fragmentation process of 11 brittle materials at different strain rates by using the characteristic line method to deal with the stress wave propagation and obtained the curves of the average fragment size s versus the strain rate $\dot{\varepsilon}$ for each material. These s-$\dot{\varepsilon}$ curves are different. However, when the fragment size s and the strain rate $\dot{\varepsilon}$are normalized by the corresponding characteristic quantities (Eq. 9.58), all data points for the different materials converge to a master curve on the dimensionless coordinate $\bar{s} - \bar{\dot{\varepsilon}}$, which can be approximately expressed by the following fitting equation (Zhou—Molinari—Ramesh formula, or ZMR formula for short):

$$\bar{s}_{ZMR} = \frac{4.5}{1 + 6.0\bar{\dot{\varepsilon}}^{2/3}}, \quad (\bar{s}_{ZMR})_{qs} = 4.5 \tag{9.59}$$

Fig. 9.63 (Zhou and Wang, 2008; Wang Li-li et al., 2013) shows the comparison of dimensionless fragment size curves predicted by the Grady model (Eq. 9.56), G—C model (Eq. 9.57) and ZMR model (Eq. 9.59) on $\bar{s} - \bar{\dot{\varepsilon}}$ dimensionless coordinates, and the dynamic numerical calculation results of other researchers are also given in this figure.

It can be seen from the figure that in the high strain-rate region of $\bar{\dot{\varepsilon}} >> 1$, the fragment size versus strain-rate curves of the three models are basically parallel, indicating that they have the same scaling law. However, in terms of absolute values, the fragment size predicted by the ZMR model is only 1/4 to 1/5 of that predicted by the Grady model and G—C model, although the ZMR model is closer to the results of dynamic numerical calculation. It means that the Grady and G—C models overestimate the

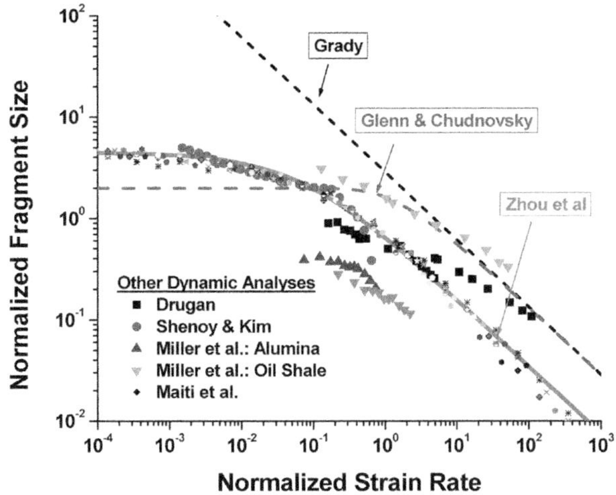

Figure 9.63 Dimensionless $\bar{s} - \bar{\dot{\varepsilon}}$ curves for various models (Zhou and Wang, 2008).

fragment size in this strain-rates region, showing that the assumption of local kinetic energy of fragments is oversimplified. This can be understood as the fact that no matter how quickly dynamic fragmentation occurs, it is always an evolution process that develops over time. In this process, the system can extract far more energy from the overall kinetic energy than from the "local kinetic energy", which is used to provide the fracture energy, thus producing more and finer fragments. In the low strain-rate region of $\bar{\dot{\varepsilon}} << 1$, the dimensionless average fragment size $\bar{s}$ given by the ZMR model is 4.5 (Eq. 9.59), which is more than twice as large as the value of 2.0 given by the G-C model 2.0 (Eq. 9.57b), and note that the $\bar{s}$ given by the ZMR model is closer to the dynamic numerical calculation result. This can be understood as that in the quasistatic region, once the fragmentation process occurs, the rapid energy release within the object will induce stress waves, which will consume more than half of the strain energy, and the rest will be converted into fracture energy. Thus Glenn and Chudnovsky's prediction was too low, and the actual quasistatic fragment size is twice as larger as that of G—C's prediction.

Zhou Fenghua et al. further studied the interaction process of crack-expansion fracture and stress wave for the crack array composed of several equidistant virtual cracks in a one-dimensional brittle solid under uniform expansion strain rate (Zhou et al., 2010; Zhou and Wang, 2010). The governing equations are composed of the linear elastic wave equations and the linear cohesive fracture model of the crack. After solving by Laplace transform, the curve of the cohesive stress changing with time in the process of crack expansion, the critical time of full fracture, and the critical expansion displacement of unit crack body (fragment) are obtained. It is found that for any given strain rate, there is always an optimal crack spacing, so that the time of full fracture is the shortest. Accordingly, Zhou Fenghua et al. put forward a "rapidest unloading principle", i.e., the rapidest full unloading of unit crack body corresponds to optimal crack spacing. The fragment size estimated by this optimal crack spacing is the average fragment size of brittle solids in a natural fragmentation. The calculated results of dimensionless fragment size versus dimensionless strain rate $\bar{s} - \bar{\dot{\varepsilon}}$ based on this "rapidest unloading principle" are given with solid dots in Fig. 9.64. For comparison, the predictions by the Grady model (Eq. 9.56), G–C model (Eq. 9.57) and ZMR model (Eq. 9.59) are also given in the same figure. It can be seen that the calculation results based on the "rapidest unloading principle" are in the best agreement with the ZMR model (Eq. 9.59) using completely random numerical simulation.

9.2.2.4 Fragment size distribution law

The dynamic fragmentation process is actually a complicated process involving multiple mechanisms, and the fracture location is random. Therefore, the fragment size discussed above should be an average characteristic

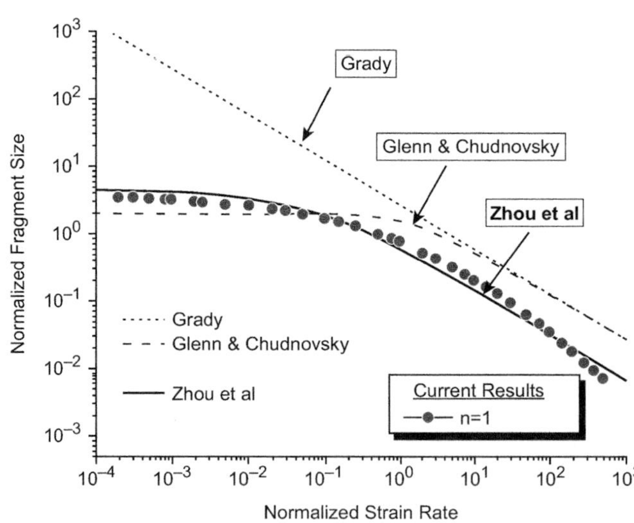

Figure 9.64 Comparison of calculation results of $\bar{s} - \bar{\dot{\varepsilon}}$ based on the "rapidest unloading principle" with other models (Zhou et al., 2010, Fig. 8, p.292).

size with a statistical distribution. This is already covered by Eq. (9.55) when introducing the Mott model.

As for the study on the statistical distribution law of fragments, the geometric statistical distribution model was adopted in the early stage. The representative ones are lognormal distribution (Epstein, 1947; Ishii, and Matsushita, 1992), Lienau (1936) distribution, Mott-Linfood (1943) distribution, and Weibull distribution (Brown and Wohletz, 1995) etc. Later, the probability distribution based on energy model was developed (Oddershede et al., 1993; Marsili and Zhang, 1996; Meibom and Balslev, 1996; Inaoka et al., 1997; Kadono, 1997).

Lienau (1936) proposed an exponential distribution model in 1936. He thinks that the one-dimensional fragmentation problem can be described with an appropriate probability distribution function of random fracture. When the fragment size l is much smaller than the total length L ($l \ll L$), it is shown that the random distribution of one-dimensional fracture points can be described by Poisson distribution, and the probability density distribution function of the fragment size can be obtained:

$$f(l) = (1/\lambda)e^{-l/\lambda} \tag{9.60}$$

where λ is the size scale parameter of the distribution (the average size of fragment). However, when the fragment size l is not less than the total length L (i.e., the number of fragments is very small), binomial distribution should be used to describe the random distribution of one-dimensional fracture points, so as the probability density distribution function of the fragment size is obtained as:

$$f(l) = \frac{n-1}{L}\left(1 - \frac{l}{L}\right)^{n-2} \tag{9.61}$$

where $n = L/\lambda$ is the total number of fragments. When $\lambda \ll L$ and $l \ll L$, the binomial probability density function is equivalent to the exponential probability density function suggested by Lienau.

Mott and Linfoot (1943) extended the Lienau random statistics to the fragmentation problem in multidimensional problem. Assume that the fragment size is proportional to $a^{1/2}$, $l \sim \sqrt{a}$, i.e., (where a is the fragment area), which also conforms to the Poisson distribution, and for the fragmentation of thin-walled shell, the fragment area a is proportional to its mass m, so Mott and Linfoot obtained the probability density distribution function of fragment mass:

$$f(l) = \frac{1}{2\mu}\left(\frac{m}{\mu}\right)^{-1/2} e^{-(m/\mu)^{1/2}} \tag{9.62}$$

where μ is the characteristic distribution mass of fragments.

The above-mentioned fragment size distribution model mainly depends on the use of an appropriate random function to describe the distribution of the fracture point, and there is no reasonable explanation on the physical mechanism.

Based on the average fragment size theory on the basis of the fracture strain distribution function, Grady (1981, 2006) studied the fragment distribution law of impact tensile fragmentation. In the one-dimensional tensile fracturing process, it is assumed that the cracks are randomly distributed in the impact process, and the development and formation of each crack can unload the surrounding area, thus preventing the new cracks in the unloading area. Grady then used the following probability density distribution function to describe the distribution characteristics of fragment size l:

$$\begin{cases} f(l) = \dfrac{4}{\sqrt{\pi}}\dfrac{1}{l_0}e^{-(l/l_0)^2}\operatorname{erf}(l/l_0) & \text{brittle fragmentation} \\ f(l) = \dfrac{\beta^2}{4}\dfrac{1}{l_0}\left(\dfrac{l}{l_0}\right)^3 e^{-\frac{1}{4}(l/l_0)^3}\displaystyle\int_0^1 \left(1-y^2\right)e^{-\frac{3}{4}(l/l_0)^3 y^2}\,dy & \text{ductile fragmentation} \end{cases} \tag{9.63}$$

where l_0 is the characteristic scale of fragment distribution; erf(x) is the error function; and $\beta = 3\Gamma(2/3)$ and $\Gamma(x)$ is the gamma function.

Based on the idea of Mott unloading wave, Zhou Fenghua et al. (2006b) used the characteristic line method to study the fragmentation process of ceramic rings. The results showed that the following Weibull distribution function can better describe the distribution law of fragments:

$$N(>l) = N_0 \exp\left[-\left(\frac{l-l_{\min}}{l_0}\right)^n\right] \quad (l > l_{\min}) \tag{9.64}$$

where N is the number of fragments with size larger than l accumulated per unit length, N_0 is the average number of fragments per unit length, $l_{\min}$ is the effective minimum fragment size, l_0 is the scale parameter, and n is the shape

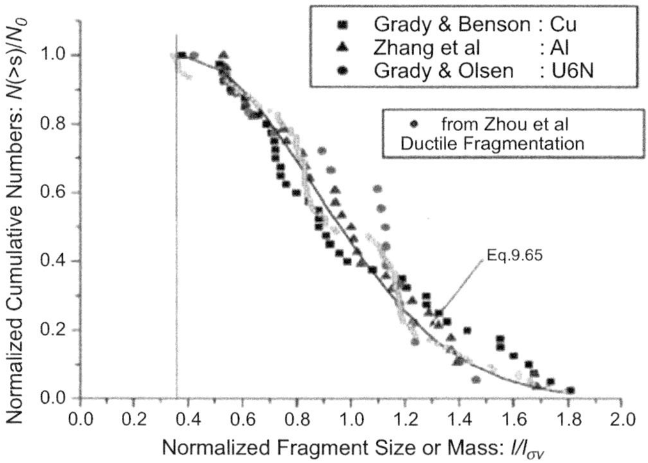

Figure 9.65 Comparison of Rayleigh distribution function (Eq. 9.65) with various experimental data (Zhou et al., 2006b, Characteristic fragment size distributions in dynamic fragmentation, Appl Phys Letters, 88(26), Fig. 3, p.201918–2).

parameter of distribution function. Furthermore, by comparing with the existing experimental data, it was found that when the shape parameter is $n = 2$, the Weibull distribution of Eq. (9.64) can be well fitted with the experimental data (see Fig. 9.65). In other words, the Weibull distribution of Eq. (9.64) can be converted into its specific case—the Rayleigh distribution function—as follows:

$$N(>l) = N_0 \exp\left[-\left(\frac{l - l_{\min}}{l_0}\right)^2\right] \quad (l > l_{\min}) \tag{9.65}$$

9.2.3 Experimental study on fragmentation of ring and cylinder shell

In the early stage, the experimental research on the impact tensile fragmentation was carried out mainly for the ring specimens of ductile metal. Since the thin-walled ring specimen is in a one-dimensional tensile stress state along the circumference in the process of rapid radial expansion, and there was no circumferential boundary condition, the expansion ring experiment became an important means to study impact tensile fragmentation.

The experimental technology of explosive expansion ring was first reported by Johnson et al. (1963) in 1963. As shown in Fig. 9.66, the driver expands outward under the action of high pressure of detonation product

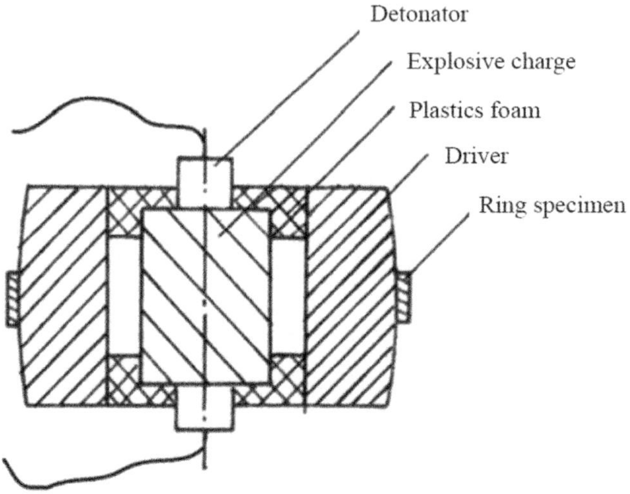

Figure 9.66 Schematic diagram of experimental arrangement of explosive expansion ring.

and drives the ring specimen to expand together. When the stress wave is reflected from the outer surface of the ring specimen and back to the interface between the ring and the driver, the ring specimen will disengage from the driver and enter the stage of free expanding. By measuring the radial displacement data of expanding ring, Johnson et al. calculated the relationship between tensile flow stress, plastic strain, and strain rate. Subsequently, Hoggatt and Recht (1969) also obtained the dynamic stress—strain relationships for various engineering materials by explosive expansion ring experiments. However, at that stage, the explosive expansion ring experiment was mainly focused on the study of dynamic constitutive relation of materials, and since the test techniques could only measure the expanding displacement of ring, the stress and strain of the ring specimen could only be calculated by the second derivative of the radial displacement, resulting in a large error. Thus, the explosive expansion ring test was not widely adopted in the following 10 years until the rapid development of the laser velocity interferometer (VISAR). Warner et al. (1981) directly measured the radial velocity of the explosive expansion ring by VISAR (see Fig. 9.67), which overcame the difficulty of second-order differential of displacement in the calculation of stress and strain.

During the same period, Niordson (1965) developed an electromagnetic expansion ring apparatus, as schematically shown in Fig. 9.68. In the experiment, the metal ring specimen is placed concentrically outside a solenoid.

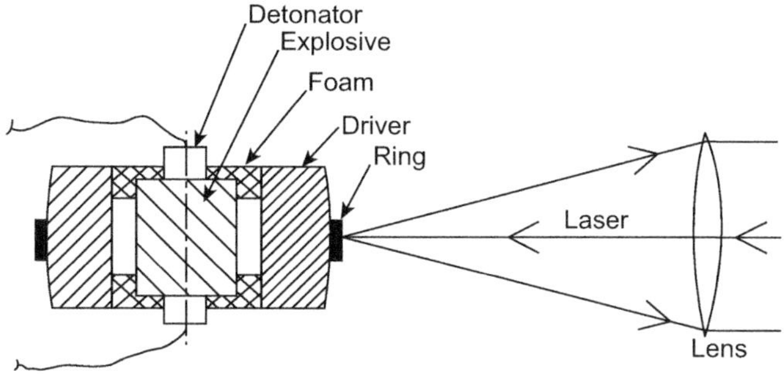

Figure 9.67 Direct measurement of radial velocity of expanding ring by VISAR.

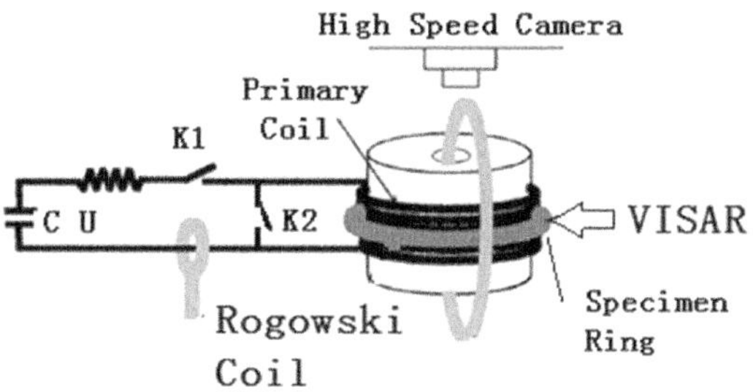

Figure 9.68 Schematic diagram of electromagnetic expansion ring apparatus.

A capacitor is discharged through the solenoid, generating a strong transient magnetic field, which induces a current inside the specimen. The magnetic field of the ring itself interacts with the magnetic field of the solenoid, producing a large outward radial electromagnetic force and causing the ring to expand outward. A series of dynamic fragmentation experiments were performed by Niordson on copper rings and aluminum alloy rings.

The electromagnetic expansion ring technology was also adopted by Grady and Benson (1983) to study the fragmentation characteristic of 1100-0 aluminum and OFHC copper at a strain rate of 10^2 s^{-1} to 10^4 s^{-1}, corresponding to the initial expansion velocity about 18–220 m/s for Al alloy and 6–138 m/s for OFHC. The experimental results of the fragment number versus the expansion velocity are shown in Fig. 9.69a. It is shown that the experimental fragment number is approximately linear dependence on expansion velocity, which is different from the predictions by the Grady–Kipp model (Eq. 9.56) and G–C model (Eqn. 9.57). Grady et al. believed that the deviation was caused by the thermal softening effect of residual current on the metal ring. After that, Grady and Olsen (2003) carried out a series of fragmentation experiments for U6Nb uranium–niobium

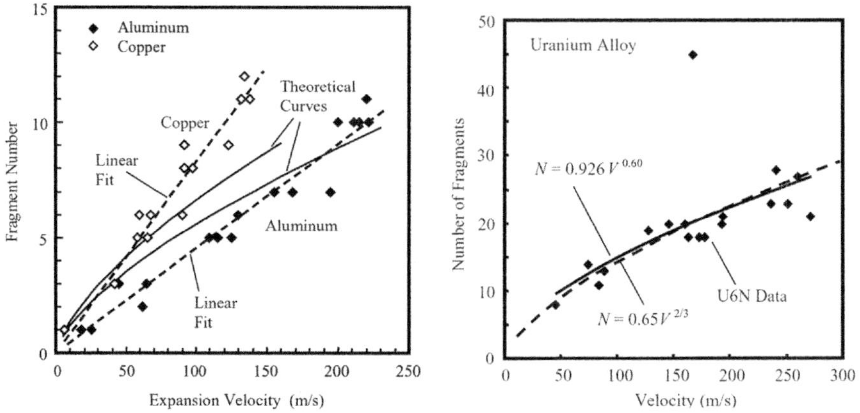

Figure 9.69 Relationship between the number of fragments of the metal expansion ring and the initial expansion velocity. *From Grady D.E., Benson D.A., 1983. Fragmentation of metal rings by electromagnetic loading, Exp. Mech. 23 (4), 393–400, Fig. 8, p.399; Grady, D.E., Olsen, M.L., 2003. A statistics and energy based theory of dynamic fragmentation, Int. J. Impact Eng. 29 (1–10), 293–306, Fig. 3, p.299. Reprinted with permission of the publisher.*

alloy rings with initial expansion velocity of 50–300 m/s. As shown in Fig. 9.69b, the number of fragments of U6Nb ring data has a power relationship of 2/3 with the strain rate, which is consistent with the trend predicted by the Grady model and G–C model.

Zhang and Ravi-Chandar (2006) used the electromagnetic expansion ring technology combined with a rotating mirror high-speed camera to accurately obtain high-speed expansion photos of 6061-0 Al alloy ring, as shown in Fig. 9.70. The distribution of the spacing between the initial necks was observed to conform to the Weibull distribution. If the fragment size is normalized according to the characteristic size, the normalized fragment size distribution generated by all experimental data also conforms to the Weibull distribution.

Explosive expansion ring experiment and electromagnetic expansion ring experiment have their own advantages and disadvantages. In the former, the ring specimen is driven by shock wave, enabling it to expand freely after obtaining the initial kinetic energy. Shock wave can enable the specimen to obtain greater kinetic energy and reach a higher strain rate. The material of the specimen is not limited, and the temperature rise is not noticeable. However, the initial shock wave generated by the explosion will affect the one-dimensional tensile state of the specimen, and the randomness of the explosion also makes it difficult to adjust the strain rate

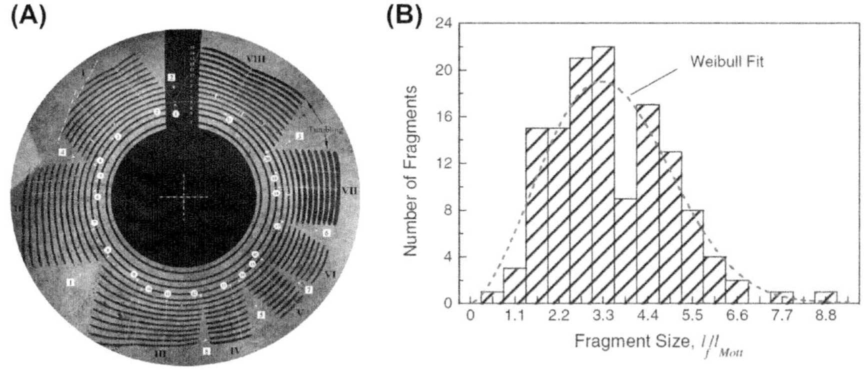

Figure 9.70 (A) Composite high-speed image for 1100-O aluminum ring expanding test; (B) normalized fragment size distribution. *From Zhang, H., Ravi-Chandar, K., 2006. On the dynamics of necking and fragmentation—I. Real-time and post-mortem observations in Al 6061-O, Int. J. Fract. 142 (3–4), 183–217, Fig. 11, p.198; Fig. 18, p.208. Reprinted with permission of the publisher.*

of the experiments, and the cost is high. In the electromagnetic expansion ring experiments, the expansion of the ring specimen is derived through the electromagnetic force, which can control the expansion strain rate of the ring more accurately without the influence of the initial stress wave effect. However, the strain rate that can be achieved is relatively low, and the induced large current in the specimen will bring inevitable temperature rise, which will affect the experimental results.

In addition to explosive expansion ring and electromagnetic expansion ring, Zhou Fenghua et al. (Wang and Zhou, 2008; Zheng et al., 2014) developed the experimental device of impact expanding ring based on the Hopkinson pressure bar technique, which have been widely used to study the dynamic constitutive relation of materials (see §7.1 "Split Hopkinson Pressure Bar (SPPB) Experimental Technique" in Chapter 7) as well as the dynamic fracture toughness (see Section 9.1.9 in the present Chapter). For brittle fragmentation study, a specially designed additional expanding device is used, as shown in Fig. 9.71. The ring specimen (on the upper left of Fig. 9.71) is set on the driving collar (in the middle of Fig. 9.71A) with a cylinderical outer surface and an inside conical surface, and then the driving collar is tightly mounted on the outer taper surface of the loading cone rod (on the lower right of Fig. 9.71). The driving collar is cut out three to four slots in advance by wire-electrode cutting; so that it will separate in sliding forward. The assembly of the expanding device with the ring specimen is placed between the incident bar and the transmission bar of

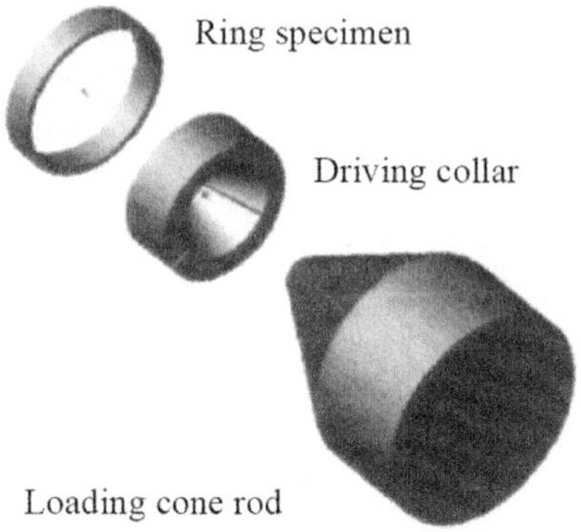

Figure 9.71 The expanding device consists of a ring specimen, a driving collar, and a loading cone rod.

SHPB. During impact loading through the incident bar, the forward motion of the loading cone rod is converted to the radial motion of the driving collar, resulting in radial expansion of the ring specimen. The fragmentation experiments of Al_2O_3 ceramic rings under different impact velocities were carried out. The fragment number of the ceramic ring increases with the impact velocity and the apparent dynamic tensile strength increases significantly compared to the corresponding static strength.

For the study of ductile fragmentation, the loading device as shown in Fig. 9.72 is used (Zheng et al., 2014). When the impact piston is derived by the Hopkinson incident bar, the liquid in the cavity enclosed by the impact piston and the sleeve will drive the thin-walled ring specimen to radially expand. The low-speed axial impact of the piston can be converted into the high-speed radial expansion of the ring specimen due to the

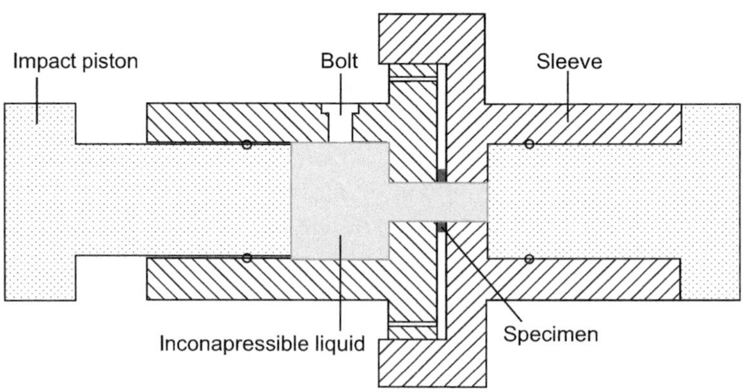

Figure 9.72 Schematics of the loading device for impact expending ring experiments based on the SHPB technology.

approximate incompressible property of the liquid and the step-like reduction of the large ratio of the cross-sectional area of the hydraulic cavity. In the process of high-speed expansion, the ring specimen undergoes rapid tensile deformation in the circumferential direction until it is fragmented. The impact expansion fragmentation experiment was carried out on 14 groups of aluminum alloy rings. The impact velocity of the strike bar is in the range from 10 to 35 m/s. The experimental results showed that with increase of the impact velocity, the apparent fracture strain of ring specimens increased, while the average fragment size decreased.

All of the above are experimental studies of one-dimensional fracturing problems. Due to military requirements, the experimental study of two-dimensional fragmentation problem actually started much earlier than that of the one-dimensional expansion ring fragmentation problem. The typical experiment is the expansion tube (shell) experiment, which started in the World War II. At that time, mathematical mechanics modeling and prediction as well as macro-experimental observation were mainly used, but the failure mechanism, stress wave effect, and loading—unloading effect in the fragmentation formation process were not paid attention to. In the 1970s, research gradually shifted to the study of material properties, fragmentation mechanism, and fragments distribution.

Hoggatt and Recht (1968) conducted different experiments on explosion expanding fragmentation of thick-walled cylinders and observed both the tensile fragmentation mode mainly in the form of radial cracks and the shear fragmentation modes mainly in the form of tangential cracks. The analysis indicates that under low detonation pressure, the radial crack forms on the outer wall and expands rapidly to the inner wall, so the tensile fracture is the main failure mode. However, under high detonation pressure, in the inner wall area, there exists a triaxial stress zone of circumferential compressive stress caused by stress wave effect, and most of the fragments are in the shear failure mode controlled by the maximum shear stress.

Grady and Hightower (1992) conducted fragmentation experiments on 4410 steel thick-walled cylinder (the ratio of diameter to thickness is about 3) driven by explosion. They also found that in addition to tensile fracture, shear fracture is another important failure mode, as shown in Fig. 9.74. In addition, the concept of local energy conservation was extended to the fragmentation problem of two-dimensional metal cylinder. It is believed that the fragment size of the metal cylinder could still be characterized by the fracture energy G_c at the crack interface, and the specific expression is still $s = \left(24G_c/\rho\dot{\varepsilon}^2\right)^{1/3}$, i.e., the Grady—Kipp model (see Eq.(9.56), note that

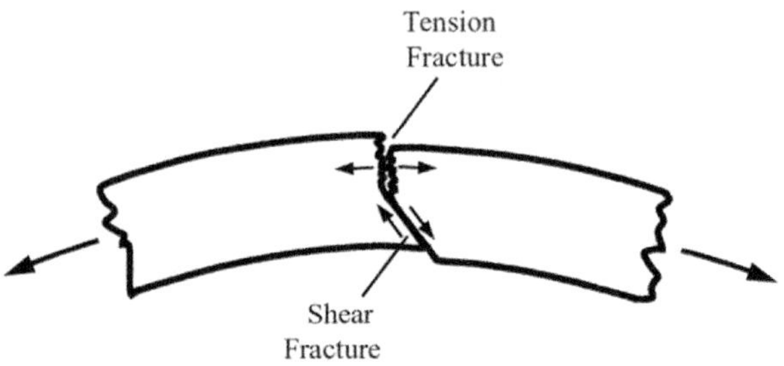

Figure 9.73 Tensile fracture and shear fracture during 4140 steel cylinder fragmentation.

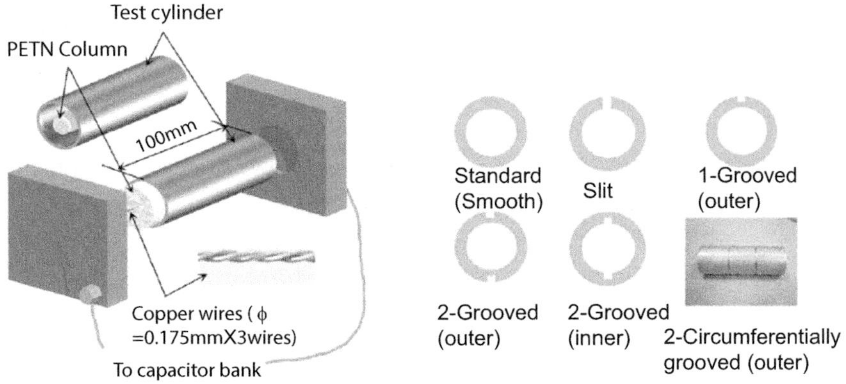

Figure 9.74 Experimental devices and test specimens of different types of prefabricated grooves. *From Hiroe T., Fujiwara K., Hata H., et al., 2008. Deformation and fragmentation behaviour of exploded metal cylinders and the effects of wall materials, configuration, explosive energy and initiated locations. Int. J. Impact Eng. 35 (12), 1578–1586, Fig. 1, p.1579. Reprinted with permission of the publisher.*

G_c here is defined as the fracture energy of one crack surface and exactly half of that of Eq. (9.56)). However, the fracture surface here is dominated by the failure modes of tensile fracture and shear fracture; so the fracture energy here can be divided into the contributions of tensile fracture and shear fracture, and the correspondingly fracture energy can be expressed as (Grady and Kipp, 1987):

$$\begin{cases} G_c = \dfrac{K_c^2}{2E} & \text{Tensile failure} \\ G_c = \dfrac{\rho c}{\alpha}\left(\dfrac{9\rho^3 c^3 \kappa^3}{Y^3 \alpha^2 \dot{\gamma}}\right)^{1/4} & \text{Shear failure} \end{cases} \tag{9.66}$$

where K_c is the fracture toughness of the material, E the Young's modulus, ρ the material density, c the specific heat of the material, κ the thermal conductivity coefficient, α the thermal softening coefficient, Y the yield strength, and $\dot{\gamma}$ the shear strain rate (close to the tensile strain rate of the metal cylinder). However, how to accurately determine the proportion of tensile failure and shear failure remains to be solved.

Hiroe et al. (2008, 2010) studied the fragmentation of various metal cylinders under explosive expanding; the ratio of diameter to thickness for thin-walled cylinder is about 24 and for thick-walled cylinder is about 7. Representative photos of fragments have been given in Fig. 9.56 previously. Based on the analysis of multiple experimental results of the same material, it is found that the material fracture energy $Gc = s^3\rho\dot{\varepsilon}^2/24$ derived inversely from the Grady's energy model covers a wide area varying with the strain rate and is not a constant. Hiroe et al. believed that this was related to the change of material properties caused by preshock waves. That is, with the increase of strain rate, the yield stress of the material increases and the fracture strain decreases, thereby affecting the fracture energy of the material. By metallographic analysis of the fragments, it is found that the shear failure is the main failure mode of metal cylinder fragmentation. Hiroe et al. also studied the influence of various types of initial defects (prefabricated grooves with different number, locations, and depths) on the fragmentation of cylinder, as shown in Fig. 9.74. The experimental results showed that the grooves had little influence on the fragmentation process at high strain rate, but significant influence at low strain rate.

CHAPTER TEN

Adiabatic shearing and dynamic evolution of meso-damage

10.1 Adiabatic shearing

Adiabatic shearing is a typical phenomenon characterizing the dynamic mechanical behavior, particularly the dynamic failure, of materials under impact loading. Such phenomenon exists in various high-speed deformation processes, such as high-speed impact, penetration, punching, high-speed forming, machining, erosion and so on, where the explosive/impact loading is involved, regardless materials are metals, plastics, or rock. The reason why it is called adiabatic shear is because this phenomenon generally has the following three most basic characteristics:

(1) Microscopically, it is characterized by observation of the so-called shear band, namely the shear deformation highly localized band-like zone with width in the order of about $10-10^2$ μm magnitude, as shown in Fig. 10.1 (Wingrove, 1973; Rogers, 1979). Correspondingly, the structure generally will fail in the mode of shear failure dominated by shear stress. Some examples of formation of shear bands in military applications are shown in Fig. 10.2, while some examples of formation of shear bands in civil applications (plastic-forming processes) are shown in Fig. 10.3 (Meyers, 1994).

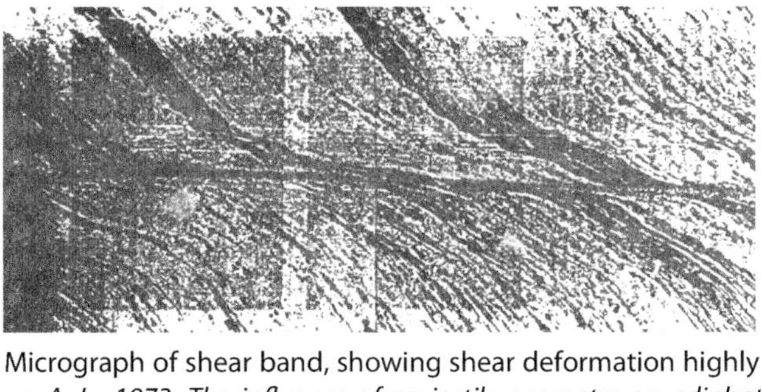

Figure 10.1 Micrograph of shear band, showing shear deformation highly localization. *From Wingrove, A. L., 1973. The influence of projectile geometry on adiabatic shear and target failure. Metallurgical Transactions 4 (8), 1829–1833., Fig. 9, p.1832. Reprinted with permission of the publisher.*

Dynamics of Materials
ISBN: 978-0-12-817321-3
https://doi.org/10.1016/B978-0-12-817321-3.00010-3

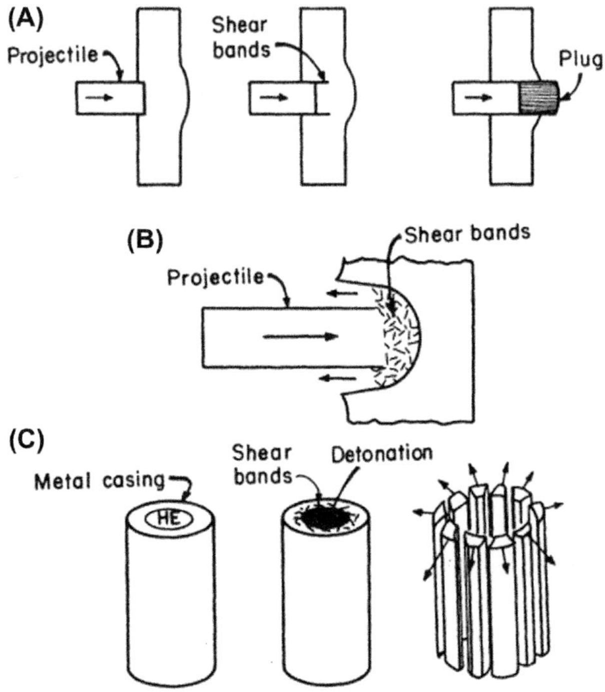

Figure 10.2 Formation of shear bands in military applications: (A) defeat of armor by plugging; (B) shear bands breaking up projectile; (C) shear bands determining fracture in exploding cylinders. *From Meyers, M.A., 1994. Dynamic Behavior of Materials. Wiley-Interscience., Fig. 15.1, p.449. Reprinted with permission of the publisher.*

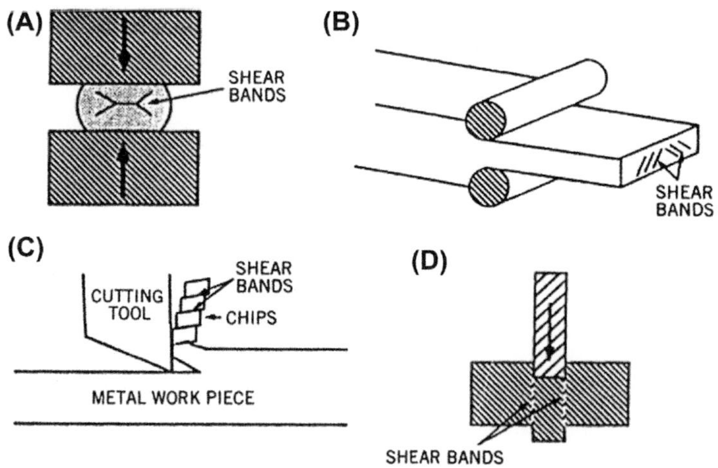

Figure 10.3 Formation of shear bands in plastic-forming processes: (A) upset forging; (B) rolling; (C) machining; (D) punching and shearing. *From Meyers, M.A., 1994. Dynamic Behavior of Materials. Wiley-Interscience., Fig. 15.2, p.450. Reprinted with permission of the publisher.*

(2) From the thermodynamic point of view, the high-speed deformation of the material under the impact load is close to that of the adiabatic process, so it is called adiabatic shear. In such case, the heat converted from the irreversible inelastic work ($\sigma d\varepsilon$ or $\tau d\gamma$) in the deformation process will cause adiabatic temperature rise.

$$dT = \frac{\beta \tau d\gamma}{\rho C_v} \tag{10.1}$$

which consequently introduce the corresponding thermal softening $\frac{\partial \tau}{\partial T}$ (<0) mechanism, where ρ is the material density, C_v is the specific heat, and β is called Taylor–Quinney coefficient which characterizes the ratio of the converted heat into the total inelastic work and is often assumed 0.9 according to Taylor–Quinney (1934). It is usually recognized that the formation and development of adiabatic shear bands are inseparable from this feature.

(3) Macroscopically, in the adiabatic stress–strain curve (compression or shear) of the material, the corresponding instability phenomenon can be always observed, namely the apparent strain hardening ($d\sigma/d\varepsilon > 0$) is changed to the apparent strain softening ($d\sigma/d\varepsilon < 0$), as shown by the solid line in Fig. 10.4.

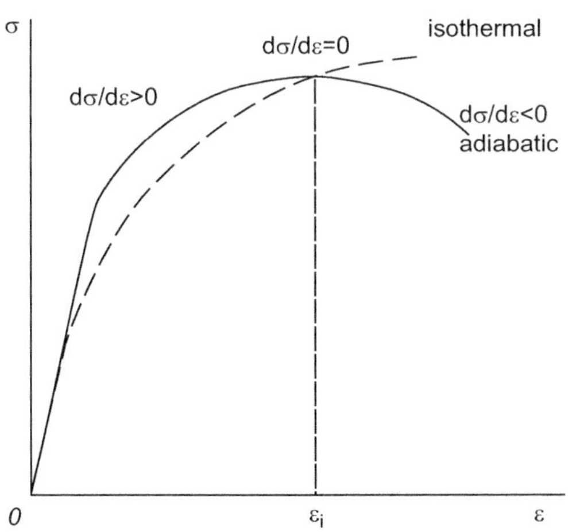

Figure 10.4 The macroinstability phenomenon in the adiabatic stress–strain curve, corresponding to the microadiabatic shear banding.

It is well known that under the experimental condition of quasistatic isothermal uniaxial compression, the stress generally increases monotonically with the increase of strain, as shown by the isothermal stress—strain curve (dotted line) in Fig. 10.4. However, under the impact adiabatic condition, the instability phenomenon ($d\sigma/d\varepsilon = 0$) appears on the adiabatic stress—strain curve (solid line) as shown in Fig. 10.4. The adiabatic stress—strain curve under impact condition is a comprehensive expression of strain hardening effect, strain rate hardening effect, and thermal softening effect. Under the impact of high strain rate condition, due to the strain rate hardening effect, the stress of the adiabatic stress—strain curve for a given strain is higher than that of the isothermal stress—strain curve. But simultaneously due to the thermal softening effect induced by the adiabatic temperature rise, the adiabatic stress—strain curve appears a transform from the apparent strain hardening ($d\sigma/d\varepsilon > 0$) to the apparent strain softening ($d\sigma/d\varepsilon < 0$), namely appears the so-called constitutive instability ($d\sigma/d\varepsilon = 0$). Once the strain reaches a big enough value, e.g., ε_i in Fig. 10.4, the adiabatic curve intersects with the isothermal curve. After that, the stress of the adiabatic stress—strain curve for a given strain is lower than that of the isothermal curve. Thus, it can be understood why the impact force and the impact energy for the high-speed impact punch is less than that for the low-speed impact punch.

Zener and Hollomon (1944) firstly attributed the adiabatic shearing, i.e., the constitutive instability of the material to the equilibrium of the plastic strain hardening and the thermal softening caused by the adiabatic temperature rise. When the plastic strain hardening exceeds the thermal softening caused by the adiabatic temperature rise, the material is in the apparent strain hardening phase, or the stable plastic deformation stage ($d\sigma/d\varepsilon > 0$ or $d\tau/d\gamma > 0$) as shown in Fig. 10.4; while when the thermal softening caused by the adiabatic temperature rise exceeds the plastic strain hardening, the material is in the apparent strain softening phase, or the unstable plastic deformation stage ($d\sigma/d\varepsilon < 0$ or $d\tau/d\gamma < 0$). Further plastic deformation occurs under lower and lower stress, which corresponds to an instability process, until it is failed. Zener and Hollomon's view, which explains the microadiabatic shear as the macrothermoplastic instability, is a milestone in the studies on adiabatic shearing and established a bridge between microcosmic research and macroscopic study of materials. The critical point of stable plastic deformation and unstable plastic deformation, namely the maximum stress condition ($d\sigma/d\varepsilon = 0$ or $d\tau/d\gamma = 0$), represents the critical condition of this thermoplastic instability.

After understanding those three interrelated basic features of adiabatic shearing, in the following, the microstructure of adiabatic shear bands, the strain rate effect, the temperature effect, the macroscopic instability criterion, the interaction of adiabatic shear band with crack, and so on will be further discussed, respectively.

10.1.1 Microstructure of adiabatic shear band—deformed band and transformed band

The microscopic observation of adiabatic shear band shows that adiabatic shear band has two basic types: the **deformed band** and the **transformed band**. The former is characterized by highly localization of shear deformation, severe distortion and fragmentation of grains, and is usually observed in nonferrous metals, whereas the latter is characterized by phase transformation or recrystallization and is usually observed in steels and titanium alloys. Because the transformed band in steel specimen after corrosion treatment is characterized by its shiny appearance and high hardness, so is called "white band."

Further experimental studies show that deformation band and transformed band are just different morphologies describing different stages of an adiabatic shearing process. The evolution of deformation band to transformed band can be observed for the same material under suitable conditions.

Take β-titanium alloy Ti−8Cr−5Mo−5V−3Al (called TB2 for short) impact compression tests as example (Lu and Lu, 1985; Wang et al., 1985, 1987; Lu et al., 1986), it differs from that of quasistatic deformation process where the specimen is mainly to achieve plastic deformation in the conventional uniform sliding way (Fig. 10.5), the specimen in the process of high-speed deformation is mainly to achieve plastic deformation in the way of adiabatic shear deformation localization. Fig. 10.6 shows a typical metallographic photograph of the adiabatic deformed band. As can be clearly seen that the deformed shear band is a zone with high strain localization, and the grains in such zone experience severe shear deformation along the band direction, and even fragment, whereas the grains outside the shear band experience much less deformation, even there is no visible slip line. The apparent impact deformation of the specimen is 30%, but in fact the deformation is mainly localized in the band, and the localized deformation within the band is up to the order of 1000% magnitude.

When the high-speed deformation is larger and the strain rate is higher, the transformed band is formed, as shown in Fig. 10.7, the grains within the

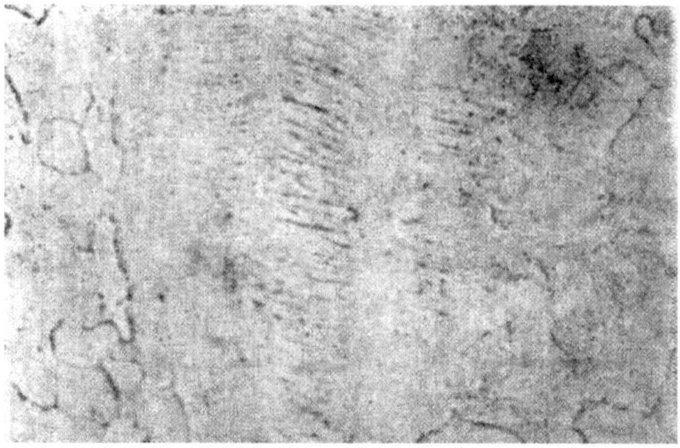

Figure 10.5 Metallographic photograph of the TB2 specimen after quasistatic compression deformation 10%. *From Lu, Wei-xian., Lu, Zai-qing., 1985. The microscopic analysis of adiabatic shear for a titanium alloy in the stress-wave-riveting. Explosion and Shock Waves 5 (1), 67–72 (in Chinese)., Fig. 6, p.70.*

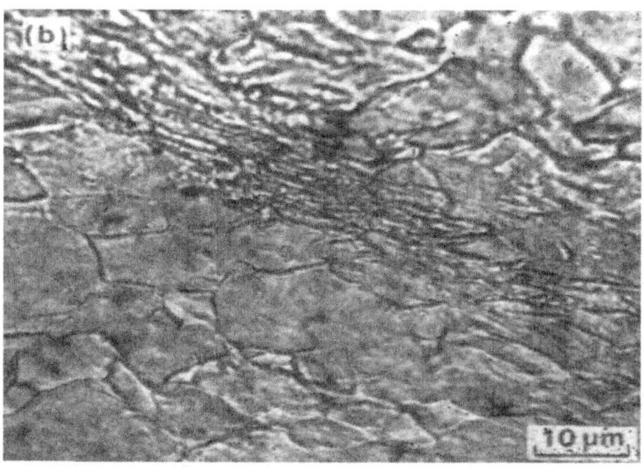

Figure 10.6 Metallographic photograph of the TB2 specimen after high-speed compression deformation 30%, showing adiabatic deformed band. *From Lu, Wei-Xian. Wang, Li-li., Lu, Zai-Qing., 1986. The adiabatic shear of a β-titanium alloy under high strain rates. Acta Mitallurgica Sinica 22 (4), A317–320., pA41–42 (in Chinese)., Fig. 1, p.A41.*

band become very small, characterizing that the microstructure of the material within the band has been changed. The boundary of the transformed band with the surrounding grains is clear. The deformation of the outer grains is very small, and almost there is no visible slip line.

Whether the deformed band or the transformed band, its width and length both have a growth and development process along the band and

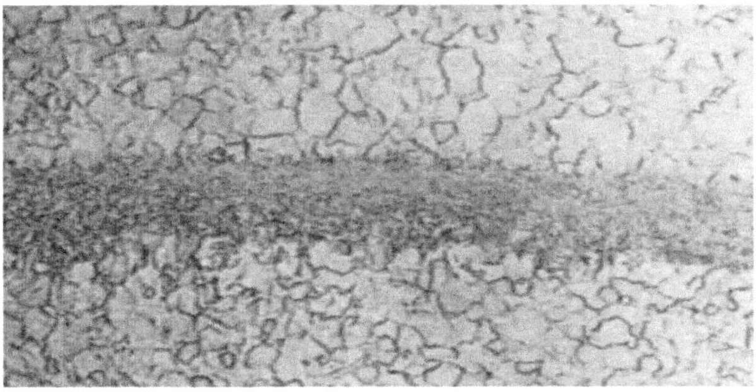

Figure 10.7 Metallographic photograph of the TB2 specimen after larger high-speed compression deformation and higher strain rate, showing adiabatic transformed band. *From Wang et al., 1985, 1987, Fig. 7, p.401. Reprinted with permission of the publisher.*

even can be developed to the intersection of two shear bands, as shown in Fig. 10.8A (Lu et al., 1986). By the scanning electron microscope at 3500 times, it was observed that the narrower shear band on the left shows a fine strip-like microstructure with a slight directivity (Fig. 10.8B), while the wider shear band on the right side shows an equiaxed very fine micro-crystalline structure (Fig. 10.8C). This indicates that in a sufficiently high strain rate and sufficiently large strain, the adiabatic temperature rise is high enough to cause the transformation of the microstructure in the shear band, and also indicates that the transformed band is an inevitable consequence of the further development of the deformed band. In fact, the so-called "mixed shear band" with microstructure of both deformed band and transformed band was observed in TB2 (Wang et al., 1985).

The adiabatic shear deformation, once develops to a certain extent, can lead to adiabatic shear failure. The microscopic observation shows that the crack is generally extended along the adiabatic shear band and is basically in the center of the shear band, as shown in Fig. 10.9. Moreover, in the shear band, voids are visible in front of the main crack tip (Fig. 10.10). The main crack will extend through those voids.

Of course, for all materials experienced adiabatic shearing, it does not have to be possible that both deformed band and transformed band, as well as their evolution and development process, can be all observed, it depends on the properties of the material itself and the external loading conditions. Dormeval et al. (Stelly et al., 1981; Dormeval, 1987) have observed both deformed band and transformed band in uranium alloy U−Mo1.5, as shown in Fig. 10.11.

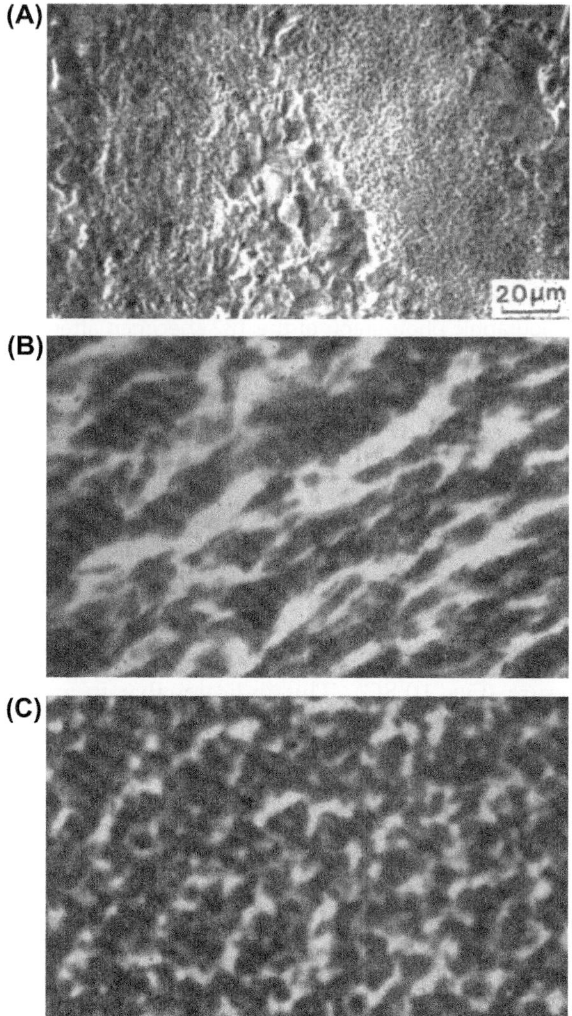

Figure 10.8 SEM micrograph of transformed shear band in specimen TB2 after high-speed compression deformation 46%: (A) the intersection of two shear bands, (B) the narrower shear band on the left showing a fine strip-like microstructure with a slight directivity, (C) the wider shear band on the right side showing an equiaxied very fine microcrystalline structure. *From Lu, Wei-Xian., Wang, Li-li., Lu, Zai-Qing., 1986. The adiabatic shear of a β-titanium alloy under high strain rates. Acta Mitallurgica Sinica 22 (4), A317–320, pA41–42 (in Chinese)., Fig. 3, p.A41 and Fig. 4, p.A42.*

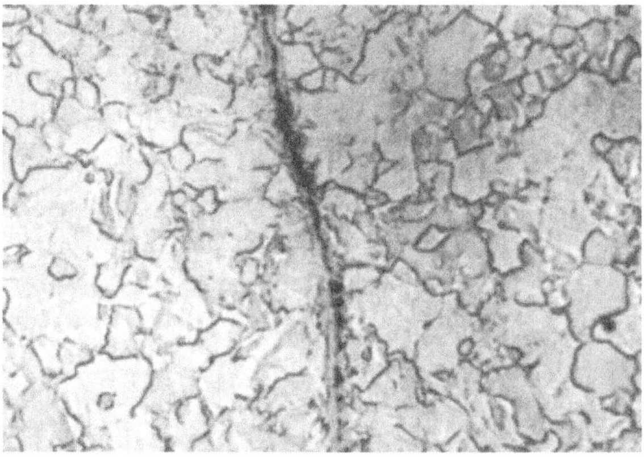

Figure 10.9 Micrograph of adiabatic shear band in specimen TB2, showing crack extended along the shear band. *From Wang et al., 1985, 1987, Fig. 8, p.402. Reprinted with permission of the publisher.*

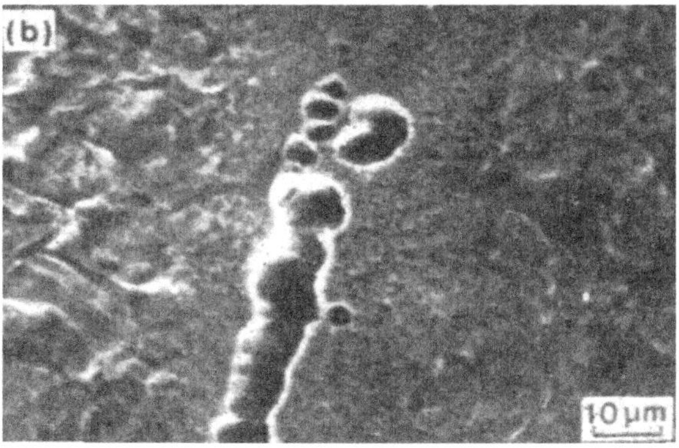

Figure 10.10 Micrograph of adiabatic shear band in specimen TB2, showing voids in front of crack tip. *From Lu, Wei-Xian., Wang, Li-li., Lu, Zai-Qing., 1986. The adiabatic shear of a β-titanium alloy under high strain rates. Acta Mitallurgica Sinica 22 (4), A317–320, pA41–42 (in Chinese)., Fig. 5, p.A42.*

10.1.2 Strain and strain rate relativity of adiabatic shearing

As indicated above, adiabatic shearing is a series of strain and strain rate–dependent process that includes the initiation and development of deformed band, the transformation from deformed band to transformed band (development of mixed zone), the development of the transformed band, until the crack propagating along the shear band and leading to failure. Regardless of the material, this process occurs under specific external loading conditions, mainly the sufficient high strain rates and sufficient large strain

(A)

(B)

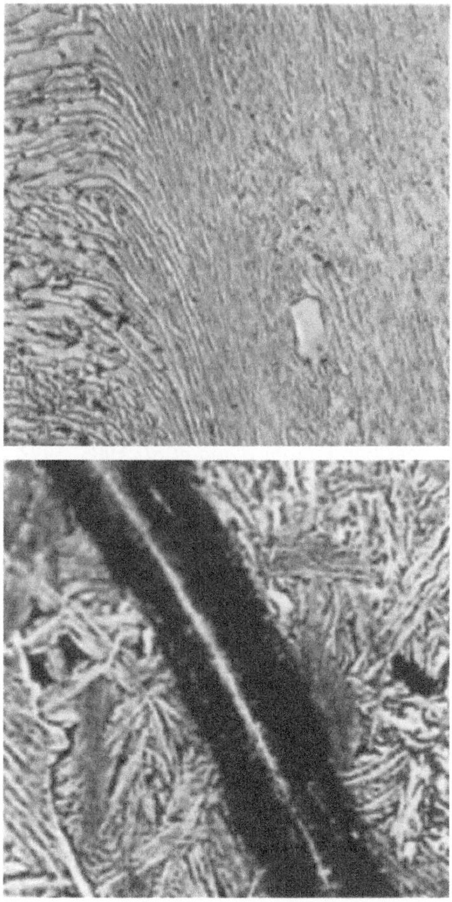

Figure 10.11 Micrograph of adiabatic shear band in uranium alloy U–Mo1.5: (A) deformed band and (B) transformed band. *From Stelly, M., Legrand, J., Dormeval, R., 1981. Some metallurgical aspects of the dynamic expansion of shells, In: Murr, L.E., Meyers, M.A. (Eds.), Shock Waves and High-Strain-Rate Phenomena in Metals, 113–125., Fig. D and Fig. E, page 120. Reprinted with permission of the publisher.*

conditions. Obviously, the strain rate and strain should be two equally important factors affecting this process at a given environmental temperature.

In fact, the experimental study and microscopic observation of titanium alloy TB2 (Wang et al., 1988) showed that under a given high enough strain rate ($1.5 \times 10^3\ s^{-1}$), with the increase of strain, the following results were observed in sequence: (A) trace of localized shear deformation, although there is not yet a continuous shear band (Fig. 10.12A), (B) the formation

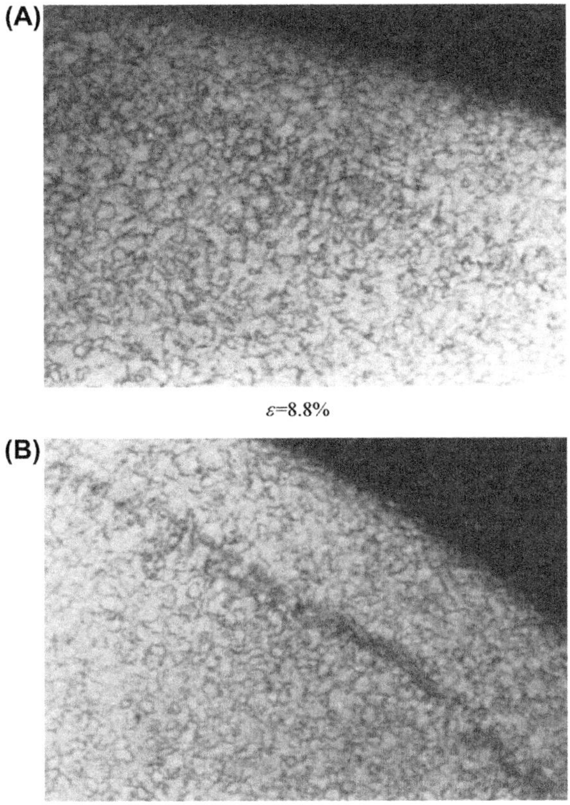

Figure 10.12 Micrographs of adiabatic shear band in TB2, showing evolution of adiabatic shear band with strain, under strain rate $1.5 \times 10^3\ s^{-1}$ and temperature 20°C. (A) $\varepsilon = 8.8\%$, (B) $\varepsilon = 16.3\%$, (C) $\varepsilon = 27.4\%$, (D) $\varepsilon = 42.3\%$. *From Wang, L.L., Bao, H.S., Lu, W.X., 1988. The dependence of adiabatic shear banding on strain-rate, strain and temperature. Journal de Physique 49 (9), C3 207–214., Fig. 2, p.C3–209. Reprinted with permission of the publisher.*

of discontinue or continue deformed band (Fig. 10.12B), (C) transformation of deformed band to transformed band (Fig. 10.12C), and (D) crack propagating along the intersected transformed band (Fig. 10.12D).

In the same way, for a given large enough strain ($\varepsilon = 16\%$), with the increase of strain rate, the following results were observed in sequence (Wang et al., 1988), as shown in Fig. 10.13, (A) initiation of deformed band (Fig. 10.13A), (B) development of deformed band (Fig. 10.13B), and (C) transformation of deformed band to transformed band (Fig. 10.13C).

(C)

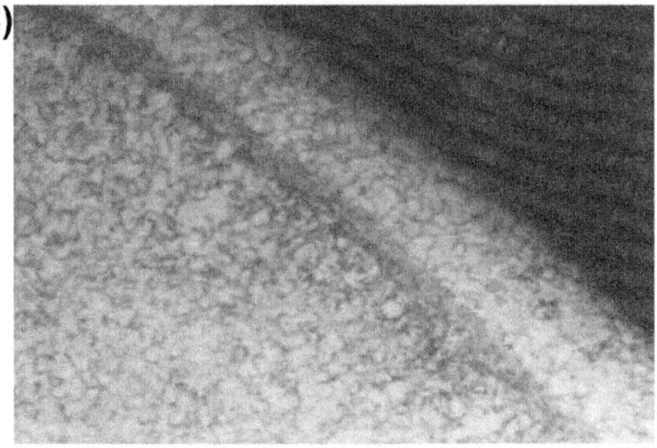

ε=27.4%

(D)

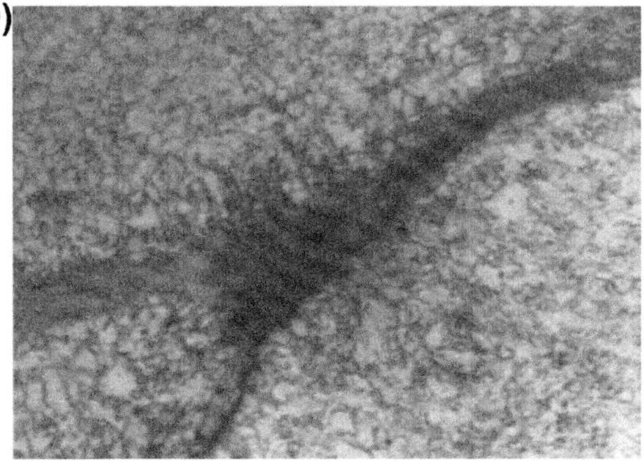

ε=42.3%

Figure 10.12 (*continued*).

For titanium alloy TB2, a large number of experimental observation are summarized in the strain rate-strain ($\dot{\varepsilon} - \varepsilon$) coordinates, as shown in Fig. 10.14 (Wang et al., 1985), showing whether and what type of adiabatic shear band is observed. The facts provided by these large quantities of experiments and microscopic observations convincingly show that both the strain effect and the strain rate effect all cannot be ignored in adiabatic shearing analysis.

10.1.3 Temperature relativity of adiabatic shearing

Recall the contents about the dynamic constitutive relations of materials under impact loading in Chapter 5, certain equivalence between the strain rate effect and the temperature effect has been discussed, namely the decrease of environment temperature is often equivalent to the increase of strain rate

(A)

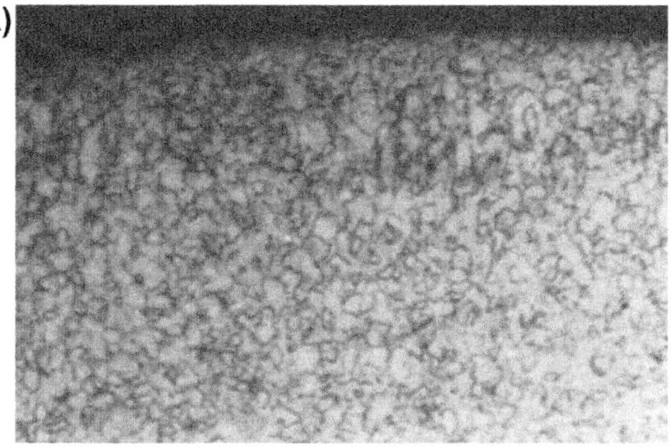

$\dot{\varepsilon}$=0.88x10^3 s^{-1}.

(B)

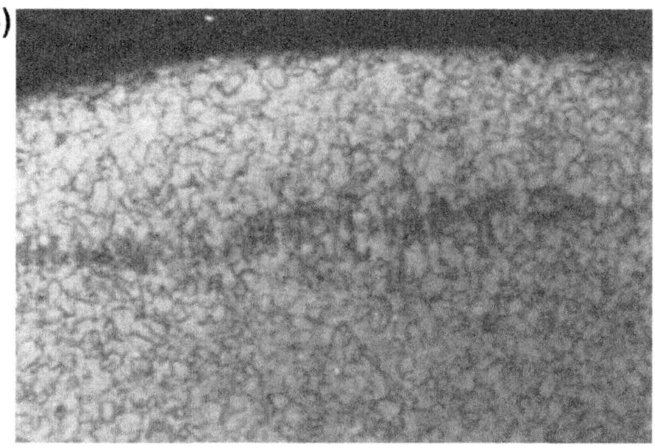

$\dot{\varepsilon}$=1.45x10^3 s^{-1}.

(C)

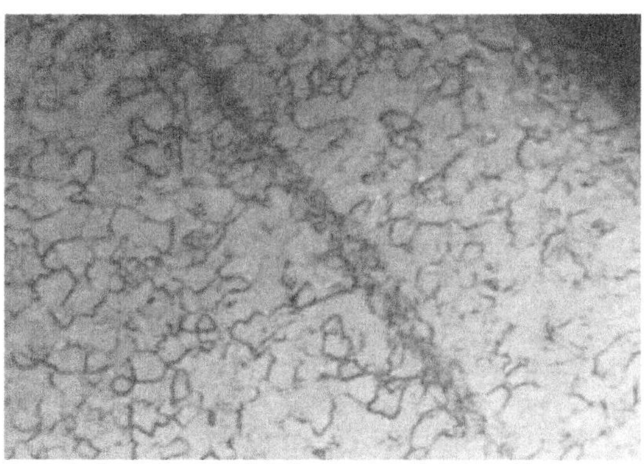

$\dot{\varepsilon}$=2.21x10^3 s^{-1}.

Figure 10.13 Micrographs of adiabatic shear band in TB2, showing evolution of adiabatic shear band with strain rate, under given strain $\varepsilon = 16\%$ and temperature 20°C. (A) $\dot{\varepsilon} = 0.88 \times 10^3\ s^{-1}$, (B) $\dot{\varepsilon} = 1.45 \times 10^3\ s^{-1}$, (C) $\dot{\varepsilon} = 2.21 \times 10^3\ s^{-1}$. *From Wang, L.L., Bao, H.S., Lu, W.X., 1988. The dependence of adiabatic shear banding on strain-rate, strain and temperature. Journal de Physique 49 (9), C3 207–214., Fig. 3, p.C3–210. Reprinted with permission of the publisher.*

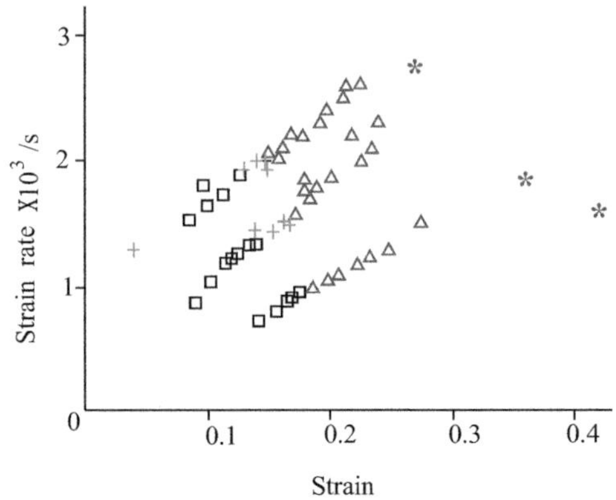

Figure 10.14 Dependence of evolution of adiabatic shear band on both strain$\varepsilon$and strain rate$\dot{\varepsilon}$. Symbols: □—no shear band, +—deformed band, △—transformed or mixed band, *—cracking band. *(From Wang et al.,* 1985, *1987, Fig. 10, p.402. Reprinted with permission of the publisher).*

(rate-temperature equivalency). Therefore, considering that the adiabatic shearing is a thermomechanics coupling and rate-sensitive phenomenon, the environmental temperature effect should be considered as an important factor too and should not be neglected.

In fact, the following results are provided by the experimental study of titanium alloy TB2 (Wang et al., 1988; Bao et al., 1989).

(1) With the decrease of environment temperature, adiabatic shear band is observed under lower critical strain rate-strain condition. Fig. 10.15 shows the micrographs of adiabatic shear band in impact compression tests for titanium alloy TB2 at environment temperature of $-90°$C, $-110°$C, and $-190°$C, respectively. It is worthwhile to note that when the environment temperature is as low as $-190°$C, although the strain rate is only $3.4 \times 10^2\ s^{-1}$ and strain is only 3.5%, a bright and wide transformed band with clear boundary has been observed (Fig. 10.15C).

(2) At low temperature, once the adiabatic shear band is observed, it is usually the type of transformed band, or at least the type of mixed band. This fact indicates that with the decrease of environment temperature, the initiate and development process of shear band is intensified, and the evolution of microstructure of shear band is accelerated.

It thus can be seen that there is also certain rate-temperature equivalence in adiabatic shearing, namely reducing the environment temperature is often equivalent to increasing the strain rate. In comparison with the situation at

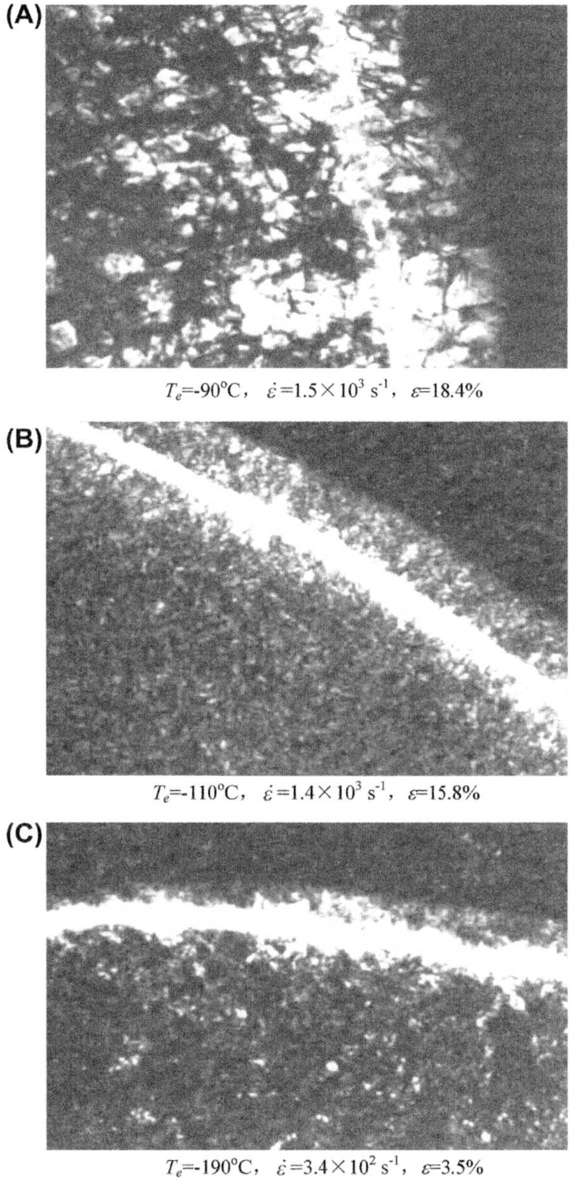

Figure 10.15 Micrographs of adiabatic shear band in TB2 at different environment temperature T_e. (A) $T_e = -90°C$, $\dot{\varepsilon} = 1.5 \times 10^3\ s^{-1}$, $\varepsilon = 18.4\%$, (B) $T_e = -110°C$, $\dot{\varepsilon} = 1.4 \times 10^3\ s^{-1}$, $\varepsilon = 15.8\%$, (C) $T_e = -190°C$, $\dot{\varepsilon} = 3.4 \times 10^2\ s^{-1}$, $\varepsilon = 3.5\%$. *From Wang, L.L, Bao, H.S, Lu, W.X, 1988. The dependence of adiabatic shear banding on strain-rate, strain and temperature. Journal de Physique 49 (9), C3 207–214., Fig. 4, Fig. 5 and Fig. 6, p.C3–212. Reprinted with permission of the publisher.*

the normal temperature, at low enough temperature, adiabatic shear band can be formed under a strain rate of an order of magnitude less as well as under a strain of an order of magnitude less.

10.1.4 Macroscopic constitutive instability criteria for adiabatic shearing

In the research on plastic deformation in plastic mechanics, on the one hand, it is needed to study the physical mechanism of plastic deformation at the microlevel, on the other hand, to satisfy practical engineering applications, it is needed to establish a criterion at the macrolevel for judging the critical condition from the elastic deformation to plastic deformation, namely the so-called yield criterion.

Similarly, in the previous three sections, the material microstructure characteristics of adiabatic shear band and its main influence factors were discussed from the microscopic level, providing the physical mechanism of the occurrence and development of adiabatic shearing. It is now needed to establish the adiabatic shearing criterion at the macrolevel, namely how to determine the critical condition of adiabatic shearing, for practical engineering application.

In this regard, the thermoplastic instability criterion proposed by Zener and Hollomon (1944) has the milestone significance. According to this criterion, the adiabatic shearing can be attributed to the balance between the plastic strain hardening and the thermal softening caused by the adiabatic temperature rise. Since then, the researchers have proposed various adiabatic shearing criteria, but the basic idea can be roughly summarized as the type of thermoplastic instability criteria, or the type of thermo-viscoplastic instability criteria taking account of strain rate effect. The following will be discussed mainly in terms of the representative criterions of these two types.

- **Thermoplastic Constitutive Instability Criterion**

According to Zener and Hollomon's view of thermoplastic instability, the delimitation of steady plastic deformation ($d\tau/d\gamma > 0$) and unsteady plastic deformation ($d\tau/d\gamma < 0$), namely the maximum stress condition, is the critical condition characterizing the adiabatic shearing

$$\frac{d\tau}{d\gamma} = 0 \tag{10.2}$$

where the stress τ is regarded as a function of strain γ and temperature T, while the temperature T is also in dependent on strain γ:

$$\tau = \tau(\gamma, T(\gamma)) \tag{10.3}$$

Substitute it into Eq. (10.2), we have

$$\frac{d\tau}{d\gamma} = \frac{\partial \tau}{\partial \gamma} + \frac{\partial \tau}{\partial T}\frac{dT}{d\gamma} = 0 \tag{10.4a}$$

Or, it can be rewritten as

$$\frac{\dfrac{\partial \tau}{\partial \gamma}}{-\dfrac{\partial \tau}{\partial T}\dfrac{dT}{d\gamma}} = 1 \tag{10.4b}$$

Eq. (10.4) provides a general form describing thermoplastic instability criterion, where $\frac{\partial \tau}{\partial \gamma}(>0)$ characterizes the strain hardening feature of material, $\frac{\partial \tau}{\partial T}(<0)$ characterizes the thermal softening feature of material, both can be determined by experimental measurements, while the term $\frac{dT}{d\gamma}$ is related to different temperature rise models, which is one of the keys in the establishment of different adiabatic shearing criterion.

Below, we will discuss two representative thermoplastic instability criteria based on Eq. (10.4).

(1) Recht's critical strain rate criterion

Based on Eq. (10.4), Recht R F proposed a **critical strain rate criterion** for adiabatic shearing (Recht, 1964; Kou, 1979).

To establish how the rate of temperature changes with the increase of strain $\frac{dT}{d\gamma}$, he proposed the following analysis model. Consider a specimen with length L and unit area, suffered stress τ, as shown in Fig. 10.16. The specimen contains a strain localized adiabatic shear plane A with a thickness h, which is negligible in comparison with L. In the analysis, this plane can be considered as a plane of uniform heat generation. The heat generation is originated from the irreversible inelastic work ($\tau d\gamma$), so the heat generation rate per unit area $\dot{Q}$ is

$$\dot{Q} = \frac{dQ}{dt} = \frac{L}{J}\frac{\tau d\gamma}{dt} = \frac{L\tau\dot{\gamma}}{J} \tag{10.5}$$

where J is the heat equivalent of work.

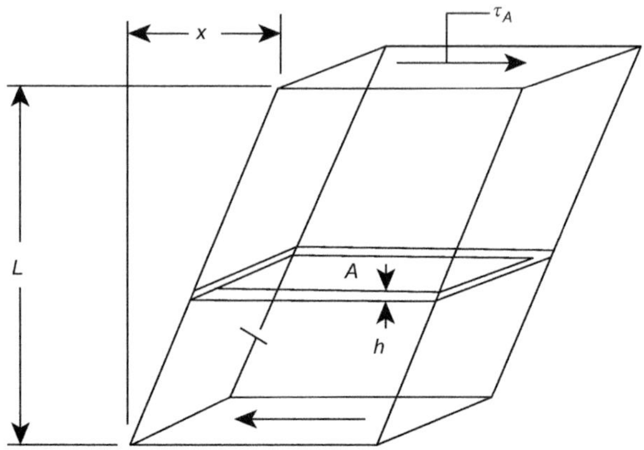

Figure 10.16 Recht's model for determination of $dT/d\gamma$.

According to Carslaw and Jaeger (1947), when the heat generation rate $\dot{Q}$ is constant, the solution of how temperature T of A plane changes with time t was obtained as

$$T = \dot{Q}\left(\frac{t}{\pi k \rho_w C_v}\right)^{1/2} = \frac{L\tau\dot{\gamma}}{J}\left(\frac{t}{\pi k \rho_w C_v}\right)^{1/2} \tag{10.6a}$$

where k denotes the coefficient of heat conduction, ρ is the specific weight, C_v is the specific heat; the second equal sign of the above equation has taken into consideration of Eq. (10.5). Thus, there is

$$dT = \frac{L\tau\dot{\gamma}}{J}\left(\frac{1}{\pi k \rho_w C_v t}\right)^{1/2} dt \tag{10.6b}$$

In the above equation, it has taken into consideration that the stress τ can be taken as a constant under the critical condition $\frac{d\tau}{d\gamma} = 0$ and assume that the average strain rate of specimen is constant, that is (see Fig. 10.16)

$$\dot{\gamma} = \frac{\dot{X}}{L} = const., \quad d\gamma = \dot{\gamma}dt, \quad \gamma = \dot{\gamma}t + \gamma_y$$

where γ_y is the initial yield strain. Substitute the above equation into Eq. (10.6b), after put in order, the change rate of temperature with strain $\frac{dT}{d\gamma}$ can be obtained

$$\frac{dT}{d\gamma} = \frac{1}{2}\frac{L\tau}{J}\left(\frac{\dot{\gamma}}{\pi k \rho_w C_v (\gamma - \gamma_y)}\right)^{1/2} \tag{10.6c}$$

Substitute the above equation into Eq. (10.4), the general form of thermoplastic instability criterion is reduced to

$$\frac{\partial \tau}{\partial \gamma} = -\frac{1}{2}\frac{L\tau}{J}\left(\frac{\dot{\gamma}_c}{\pi k\rho C_v(\gamma - \gamma_y)}\right)^{1/2}\frac{\partial \tau}{\partial T} \tag{10.7a}$$

where $\dot{\gamma}_c$ is the minimum value of the average strain rate which satisfies the thermoplastic critical condition for adiabatic shearing. Thus, from Eq. (10.7a), the $\dot{\gamma}_c$ can be expressed as

$$\dot{\gamma}_c = 4\pi k\rho_w C_v(\gamma - \gamma_y)\left(\frac{\partial \tau/\partial \gamma}{\partial \tau/\partial T}\right)^2\frac{J^2}{L^2\tau^2} \tag{10.7b}$$

Recht focused on the critical strain rate condition of thermoplastic instability, Eq. (10.7) is called Recht's **critical strain rate criterion** for adiabatic shearing. According to Eq. (10.7), obviously, the lower the material specific weight ρ_w, the heat conduction coefficient k, the specific heat C_v, and the strain hardening $(\partial\tau/\partial\gamma)$, as well as the higher the thermal softening $(\partial\tau/\partial T)$, then the smaller the $\dot{\gamma}_c$ is, namely more sensitive to adiabatic shearing. In the same plastic deformation condition, the critical strain rate ratio of two materials is

$$\frac{\dot{\gamma}_{c1}}{\dot{\gamma}_{c2}} = \frac{(k\rho_w C_v)_1}{(k\rho_w C_v)_2}\left[\frac{\left(\frac{\partial \tau}{\partial \gamma}\right)_1\left(\frac{\partial \tau}{\partial T}\right)_2}{\left(\frac{\partial \tau}{\partial \gamma}\right)_2\left(\frac{\partial \tau}{\partial T}\right)_1}\right]^2\left(\frac{\tau_2}{\tau_1}\right)^2 \tag{10.7c}$$

where subscript 1 and 2 denote two kinds of materials, respectively. By Eq. (10.7c),the sensitivity to adiabatic shearing of two materials can be compared. For example, the ratio of the adiabatic critical strain rate of titanium alloy to medium carbon steel is about 1:1400, indicating that titanium alloy is much more sensitive to adiabatic shearing than medium carbon steel.

(2) Culver's critical strain criterion

Culver (1973) established the temperature rise model directly from Eq. (10.1), namely, let $\frac{dT}{d\gamma}$ as

$$\frac{dT}{d\gamma} = \frac{0.9\tau_D}{\rho C_v} \tag{10.1b}$$

where shear stress τ indexed subscript D is to emphasize the dynamic stress under the impact adiabatic condition. On the other side, to quantitatively analyze the strain hardening term $\frac{\partial \tau}{\partial \gamma}$, he assumed that the isothermal quasistatic stress—strain curve can be expressed as

$$\tau_T = B_T \gamma^{n_T} \tag{10.8a}$$

where n_T is the strain hardening index, the subscript T under shear stress τ denotes the isothermal quasistatic stress. Then we have

$$\frac{\partial \tau}{\partial \gamma} = n_T B_T \gamma^{(n_T - 1)} = \frac{n_T \tau_T}{\gamma} \tag{10.8b}$$

Substitute both Eqs. 10.1b and 10.8b into the general form of thermoplastic instability criterion (Eq. 10.4), the critical strain of adiabatic shearing thermoplastic instability γ_c can be obtained

$$\gamma_c = \frac{n_T \rho C_v}{0.9\left(-\dfrac{\partial \sigma}{\partial T}\right)} \cdot \frac{\tau_T}{\tau_D} \tag{10.9}$$

It is called Culver's **critical strain criterion** for adiabatic shearing. Eq. (10.9) also indicates that the lower the material density ρ, the specific heat C_v and the strain hardening index n_T, as well as the higher the thermal softening $(\partial \tau / \partial T)$, then the smaller the critical strain γ_c is, namely more sensitive to adiabatic shearing.

Other criteria based on Zener and Hollomon thermoplastic instability, or can be classified as critical strain rate type, or can be classified as critical strain type, will not be discussed here. For readers who are interested in it can refer to, for example, Table 2.2 in the paper of Dormeval (1987).

However, the role of the Recht's critical strain rate criterion in fact is rather to compare the relative sensitivity on the adiabatic shearing of different materials, but not a strict critical criterion for predicting the occurrence of adiabatic shearing. On the other side, the criteria of critical strain type based on thermoplastic instability all ignore the important strain rate effect, which should be taken into account in adiabatic shear instability.

Furthermore, whether the critical strain rate criterion or the critical strain criterion, it is considered that the occurrence of adiabatic shearing instability only depends on one control variable, although the experimental researches and microobservations given previously all show that it seems more reasonable to think of both strain and strain rate as control variables.

In the following, the thermo-viscoplastic constitutive instability criterion which takes into account simultaneously of both the strain rate effect and strain effect will be briefly discussed.

- **Thermo-Viscoplastic Constitutive Instability Criterion**

The experimental researches and microobservations given in Section 10.1.1—10.1.3 all show that under impact high strain rate loading, material displays obvious strain hardening effect and strain rate hardening effect as well as thermal softening effect, which should be treated as thermo-viscoplastic material. The related adiabatic shearing constitutive instability should be discussed on the basis of the thermo-viscoplastic constitutive equation, namely to generalize the Zener—Hollomon's thermoplastic instability criterion to thermo-viscoplastic instability criterion, taking into account of the balance between all three effects, namely the strain hardening $\left(\frac{\partial\tau}{\partial\gamma}>0\right)$, the strain rate hardening $\left(\frac{\partial\tau}{\partial\dot{\gamma}}>0\right)$, and the thermal softening $\left(\frac{\partial\tau}{\partial T}<0\right)$.

In the following, we will discuss the thermo-viscoplastic constitutive instability criterion of adiabatic shearing from two analysis approaches, namely the model based on the apparent uniform deformation and the model based on nonuniform deformation (localization), respectively.

(1) Thermo-viscoplastic constitutive instability analysis based on the apparent uniform deformation model

Firstly, it is worthwhile to notice that for the material with microadiabatic shear bands, when its macroscopic stress—strain curve is measured in macrolevel, a macroscopic "averaging" approach is implicitly adopted, namely the macrostress and macrostrain are all measured on a macroapparent uniform deformation model regardless whether shear bands exists in microlevel. Below, we will analyze the thermo-viscoplastic instability criterion on the basis of such macroapparent uniform deformation model. This is what the same approach adopted in Zener and Hollomon's thermoplastic instability analysis, but we will consider the strain rate effect.

The general form of thermo-viscoplastic constitutive relation, in case of shear deformation, can be expressed as

$$\tau=f(\gamma,\dot{\gamma},T) \tag{10.10}$$

Thus, the critical condition for constitutive instability is

$$\frac{d\tau}{d\gamma}=\frac{\partial\tau}{\partial\gamma}+\frac{\partial\tau}{\partial\dot{\gamma}}\frac{d\dot{\gamma}}{d\gamma}+\frac{\partial\tau}{\partial T}\frac{dT}{d\gamma}=0 \tag{10.11}$$

where $\frac{\partial\tau}{\partial\gamma}>0$, $\frac{\partial\tau}{\partial\dot{\gamma}}>0$, and $\frac{\partial\tau}{\partial T}<0$ characterize the strain hardening, the strain rate hardening, and the thermal softening, respectively, but notice that the $\frac{d\dot{\gamma}}{d\gamma}$ can be positive or negative. Eq. (10.11) indicates that whether the constitutive instability occurs depends on the balance between the strain hardening effect, the strain rate hardening effect, and the thermal softening effect.

For adiabatic process, according to Eq. (10.1), there is

$$\frac{dT}{d\gamma}=\frac{\beta\tau}{\rho C_v} \tag{10.1c}$$

where β is the ratio coefficient of how many percent of the viscoplastic work converted to the heat, which is usually taken as 0.9 according to Taylor and Quinney (1934).

Note that the stress τ in the above equation is generally a function of strain γ, strain rate $\dot{\gamma}$, and temperature T according to Eq. (10.10), while from Eq. (10.1b) the temperature T can further be solved as a function of γ and $\dot{\gamma}$, so that there is $\tau=f(\gamma,\dot{\gamma},\Psi(\gamma,\dot{\gamma}))=\Phi(\gamma,\dot{\gamma})$. Substitute it into Eq. (10.11), the critical condition of constitutive instability is reduced to a differential equation only containing γ and $\dot{\gamma}$, which is not hard to be solved.

For example, if Eq. (10.10) is expressed as the following thermo-viscoplastic constitutive equation in Johnson–Cook model

$$\tau=\tau_0\gamma^n\left(1+g\ln\frac{\dot{\gamma}}{\dot{\gamma}_0}\right)\left(1-\alpha\frac{T}{T_e}\right) \tag{10.12}$$

where n, g, and α characterize the strain hardening, strain rate hardening, and thermal softening, respectively, τ_0, $\dot{\gamma}_0$, and T_e are characteristic stress, characteristic strain rate, and characteristic environmental temperature at quasistatic experiment, respectively. Substitute Eqs. 10.12 and 10.1b into Eq. (10.11), the following equation can be obtained

$$\frac{n}{\gamma}+\frac{1}{1+g\ln\frac{\dot{\gamma}}{\dot{\gamma}_o}}\frac{g}{\dot{\gamma}}\frac{d\dot{\gamma}}{d\gamma}-\frac{\alpha\beta\tau_o}{T_e\rho C_v}\gamma^n\left(1+g\ln\frac{\dot{\gamma}}{\dot{\gamma}_o}\right)=0 \tag{10.13}$$

The above equation is a first-order ordinary differential equation for $\dot{\gamma}$ with respect to γ, it is not hard to put it into a standard form of Bernoulli differential equation, and its general solution can be obtained as

$$\gamma^n\left(1+g\ln\frac{\dot{\gamma}}{\dot{\gamma}_o}\right)\left(A-\frac{\alpha\beta\tau_o}{T_e\rho C_v}\gamma\right)=1 \tag{10.14}$$

where A is an integral constant characterizing a certain state of adiabatic shearing process, such as the initiation of shear band, the transformation of deformed band to transformed band, and so on. Thus, on the $\dot{\gamma}$-γ plot, the solution (Eq. 10.14) corresponds to a set of $\dot{\gamma}$-γ curves, each of them corresponds to a critical condition for a specific state of adiabatic shearing process, the whole set of curves corresponds to the whole adiabatic shearing instability process composed of a series of stages.

Similarly, if Eq. (10.10) is expressed in the following form, respectively,

$$\tau=(\tau_o+E_1\gamma)\left(1+g\ln\frac{\dot{\gamma}}{\dot{\gamma}_o}\right)\left(1-\alpha\frac{T}{T_e}\right) \tag{10.15}$$

$$\tau=\tau_o\gamma^n\left(\frac{\dot{\gamma}}{\dot{\gamma}_o}\right)^m\left(1-\alpha\frac{T}{T_e}\right) \tag{10.16}$$

$$\tau=(1+E_1\gamma)\left(\frac{\dot{\gamma}}{\dot{\gamma}_o}\right)^m\left(1-\alpha\frac{T}{T_e}\right) \tag{10.17}$$

where E_1 is the linear strain hardening modulus, and the exponent m is another form of the parameter that characterizes strain rate hardening, then the corresponding critical conditions of thermo-viscoplastic instability are, respectively,

$$\left(\frac{\tau_o}{E_1}+\gamma\right)\left(1+g\ln\frac{\dot{\gamma}}{\dot{\gamma}_o}\right)\left(A-\frac{\alpha\beta E_1}{T_e\rho C_v}\gamma\right)=1 \tag{10.18}$$

$$(\gamma^n)\frac{\dot{\gamma}}{\dot{\gamma}_o}\left(A-\frac{\alpha\beta\tau_o m}{T_e\rho C_v}\right)=1 \tag{10.19}$$

$$\left(\frac{\tau_o}{E_1}+\gamma\right)\frac{\dot{\gamma}}{\dot{\gamma}_o}\left(A-\frac{\alpha\beta E_1 m}{T_e\rho C_v}\gamma\right)=1 \tag{10.20}$$

All of the above forms of thermo-viscoplastic instability criteria give a set of critical $\dot{\gamma}$-γ curves. Therefore, such criterion not only takes into account of the strain rate effect, which is ignored in the thermoplastic instability criterion, thus replaces the traditional univariable criterion with bivariable criterion but also can describe the whole adiabatic shearing process involving a series of stages, unlike the maximum stress criterion based on thermoplastic instability, which only describes the adiabatic shearing as a single state of critical instability.

Under special circumstances, such as for a given strain, the thermo-viscoplastic instability criterion will be reduced to a critical strain rate criterion; or for a given strain rate, it will be reduced to a critical strain criterion. Therefore, thermoplastic instability criterion can be regarded as a special case of thermo-viscoplastic instability criterion.

For titanium alloy TB2, the SHPB experimental results (Wang, 1986) show that its thermo-viscoplastic constitutive equation can be expressed by Eq. (10.15), and the corresponding thermo-viscoplastic instability criterion is Eq. (10.18). In terms of normal stress σ, normal strain $\boldsymbol{\varepsilon}$, and normal strain rate $\dot{\boldsymbol{\varepsilon}}$, the material constants experimentally determined are $\rho = 4.5$ g/cm^3, $C_v = 0.527$ J/g/K, $\sigma_0 = 935$ MPa, $E_1 = 1.59 \times 10^3$ MPa, $\dot{\varepsilon}_0 = 1.4 \times 10^{-3}$ s^{-1}, $g = 2.45 \times 10^{-2}$, $\alpha / T_e = 1.37 \times 10^{-3}$ K^{-1}, and approximately take $\beta = 1$. The theoretical critical $\dot{\boldsymbol{\varepsilon}}$-$\boldsymbol{\varepsilon}$ curves for different A values are given in Fig. 10.17, where the theoretical curve for

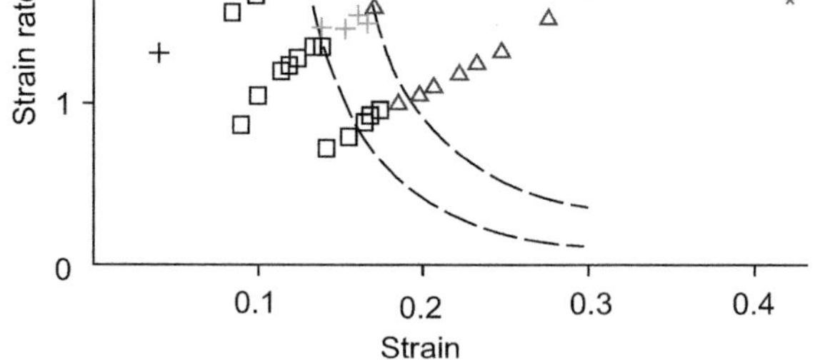

Figure 10.17 The theoretical curves of thermo-viscoplastic instability at room temperature compared with the experimental results (Fig. 10.14), theoretical curves: (A) $A = 1.1536$, (B) $A = 1.1382$, (C) $A = 1.1143$, experimental results: □—no shear band, +—deformed band, △—transformed or mixed band, *—cracking band. *From Wang, Lili., 1992. Adiabatic Shearing—the Constitutive Instability of Materials under Impact Loading, in Progress in Impact Dynamics. (Eds.), Wang, Lili., Yu, Tongxi., Li, Yongchi., the Press of University of Science and Technology of China, Hefei, 3–33., Fig. 14, p.20.*

$A = 1.1536$ corresponds to the initiation of deformed band of adiabatic shearing, the curve for $A = 1.1382$ corresponds to the initiation of mixed band, and the curve for $A = 1.1143$ corresponds to the fully transformed band. In the same figure, the experimental results shown in Fig. 10.14 are given too for comparison. As can be seen, the theoretical predictions accord with the experimental results pretty well.

For $\alpha+\beta$ titanium alloy TC4 (Ti−6Al−4V), the experimental results confirm that the same holds true for the thermo-viscoplastic instability criterion (Xu et al., 1987).

As can be seen from the theoretical critical curve in Fig. 10.17, along the critical curve, as the strain increases, the strain rate decreases and tends to a threshold strain rate $\dot{\varepsilon}_{\mathrm{th}}$, after that even if strain increases further, adiabatic shearing instability no more occurs. This threshold strain rate $\dot{\varepsilon}_{\mathrm{th}}$can be understood as the aforementioned critical strain rate criterion for the thermoplastic instability. On the contrary, along the critical curve, as the strain rate increases rapidly, the strain decreases slowly and tends to a threshold strain ε_{th}, after that even if strain rate increases further, adiabatic shearing instability is no more occurs. This threshold strain ε_{th} can be understood as the aforementioned critical strain rate criterion for the thermoplastic instability. Therefore, either the critical strain rate criterion or the critical strain criterion all can be regarded as certain special case of the thermo-viscoplastic instability criterion.

The Section 10.1.3 has shown that the adiabatic shearing is obviously dependent on the environmental temperature T_e too. In fact, as described by Eq. (10.14), Eqs. 10.18−10.20, all kinds of specific forms of thermo-viscoplastic instability criterion contain the multiplication factor T_e, reflecting the environmental temperature effect. In addition, as shown in Fig. 10.15, because the initiation of adiabatic shearing instability in different stages are very sensitive to the environmental temperature, the integrate constant A should be clearly dependent on T_e too.

For example, taking account of the environmental temperature effect, Eq. (10.18) will be expressed in the following form (in terms of normal stress, normal strain, and normal strain rate)

$$\left(\frac{\sigma_o}{E_1} + \gamma\right)\left(1 + g \ln\frac{\dot{\varepsilon}}{\dot{\varepsilon}_o}\right)\left(A(T_e) - \frac{\alpha\beta E_1}{T_e \rho C_v}\varepsilon\right) = 1 \tag{10.21}$$

Here, it has been approximately assumed that compared with the influence of T_e on A, the influence of Te on α, ρ, and C_v can be preliminary ignored. Eq. (10.21) represents a critical surface in three-dimensional spaces $\varepsilon - \dot{\varepsilon} - T_e$. It means that the thermo-viscoplastic instability criterion for adiabatic shearing depends on three variables $\dot{\gamma}$, γ, and T_e, called trivariable thermo-viscoplastic instability criterion.

The $A(T_e)$ is usually determined by fitting with experimental results. For titanium alloy TB2, the theoretical predictions for the initiation of adiabatic shear band under the environmental temperature of 20, -90, and −110°C, calculated by the thermo-viscoplastic instability criterion (Eq. 10.21), are shown in Fig. 10.18; and the experimental results are given in the same figure too for comparison (Wang et al., 1988; Bao et al., 1989). From a practical point of view, the theoretical predictions accord with the experimental results pretty well (Fig. 10.18).

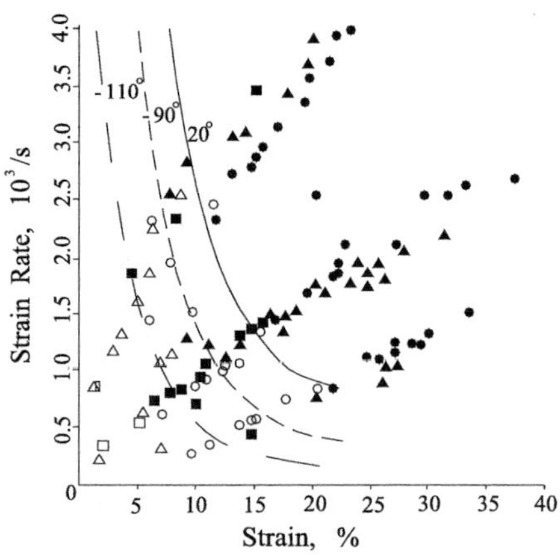

Figure 10.18 Initiation of adiabatic shear band for titanium alloy TB2 depends on strain, strain rate, and environmental temperature. Comparison of theoretical curves with experimental data

T_e	No shear band	With shear band
20°C	○	●
−90°C	△	▲
−110°C	□	■

From Wang, Lili., 1992. Adiabatic Shearing—the Constitutive Instability of Materials under Impact Loading, in Progress in Impact Dynamics. (Eds.), Wang, Lili., Yu, Tongxi., Li, Yongchi., the Press of University of Science and Technology of China, Hefei, 3—33, Fig. 21, p.28.

(2) Thermo-viscoplastic constitutive instability analysis based on the nonuniform deformation model

Because the adiabatic shear band is a highly localized strain region, people may assume that there exists a certain initial nonuniformity in material (the nonuniformity of such as material, geometry, deformation, temperature, etc.), and to examine that under what conditions such initial nonuniformity will further develop and lead to adiabatic shearing instability. This is a kind of constitutive instability analysis based on nonuniform (localized) deformation model, similar to micromechanics analysis, and is another kind of analytical method worthy of attention and introduction.

Semiatin et al. (Semiatin et al., 1983, 1984) developed such a localized analysis and indicated that the occurrence of adiabatic shearing could be predicted more accurately. But their analysis is still in the category of thermoplastic instability. Below taking titanium alloy TB2 as an example, a localization analysis model based on thermo-viscoplastic instability is presented (Bao and Wang, 1990; Wang and Bao, 1991), and then the effects of strain hardening, strain rate hardening, and thermal softening are taken into account simultaneously.

Assume that this initial nonuniform deformation body contains a defect region with volume percentage of V_d and a nondefect region (matrix) with volume percentage of V_u. Obviously, they should satisfy the following condition at any time:

$$V_u + V_d = 1 \tag{10.22}$$

Generally, the volume of the defect region is much smaller than that of the nondefect region, and in the following the V_d is assumed to be 0.1%.

The deformation within two regions, although has different magnitudes but is respectively uniform and all follow the thermo-viscoplastic constitutive equation described by Eq. (10.15), namely

$$\sigma = (\sigma_o + E_1 \varepsilon)\left(1 + g \ln \frac{\dot{\varepsilon}}{\dot{\varepsilon}_o}\right)\left(1 - \alpha \frac{T}{T_e}\right) \tag{10.23}$$

only that the material parameter σ_0 is different for two regions, namely $\sigma_{0u} \neq \sigma_{0d}$ (the subscript u and d correspond to nondefect region and defect region, respectively), other material parameters E_1 ($=1.59 \times 10^3$ MPa), g ($=2.45 \times 10^{-2}$), $\dot{\varepsilon}_0$($=1.4 \times 10^{-3}\ \mathrm{s}^{-1}$), and T_e($=293\ ^\circ$K) are all the same for two regions. At any time, the apparent average strain ε_{av} and the apparent average strain rate $\dot{\varepsilon}_{av}$ of the whole deformed body are defined as the geometric average value of the two regions, namely

$$\varepsilon_{av} = V_u\varepsilon_u + V_d\varepsilon_d \tag{10.24a}$$

$$\dot{\varepsilon}_{av} = V_u\dot{\varepsilon}_u + V_d\dot{\varepsilon}_d \tag{10.24b}$$

The temperature rise dT induced by the heat, which is converted from the irreversible inelastic work, is calculated according to Eq. (10.1). In the present localization analysis, considering that the strain rate is different in two regions, Eq. (10.1) is rewritten as

$$dT = \frac{\beta(\dot{\varepsilon})\tau(\varepsilon, \dot{\varepsilon}, T)}{\rho C_v(T)} d\gamma \tag{10.25a}$$

where $\rho = 4.5\ \mathrm{g/cm^3}$, the β will be not taken as a constant (e.g.,0.9) as in the adiabatic approximation, while should be regarded as a function of strain rate $\dot{\varepsilon}$. Because in the localization analysis, the difference of the strain rate between the defect region and the nondefect region may be as large as orders of magnitude, so the adiabatic approximation does not simultaneously hold for those two regions. Obviously, the lower limit of $\beta(\dot{\gamma})$ is zero, corresponding to the situation that all the heat converted from inelastic deformation work is lost, so that there is no temperature rise. The upper limit of $\beta(\dot{\gamma})$ is 1, corresponding to the situation that the inelastic deformation work is all converted into heat, so that there is no heat loss. The former is equivalent to the isothermal process at very low strain rate, whereas the latter is equivalent to the adiabatic process at very high strain rate. In the present analysis, the $\beta(\dot{\gamma})$ is described by the following form

$$\beta(\dot{\varepsilon}) = \frac{1}{\pi} arctg\left(\frac{1}{3}\left(tg\left(\frac{2\pi}{5}\right)\lg\dot{\varepsilon}\right)\right) + \frac{1}{2} \tag{10.25b}$$

The above equation can meet the following conditions: $\lim_{\dot{\varepsilon}\to 0}\beta(\dot{\varepsilon}) = 0$, $\lim_{\dot{\varepsilon}\to\infty}\beta(\dot{\varepsilon}) = 1$, $\beta = 0.9$ at $\dot{\varepsilon} = 10^3\ \mathrm{s^{-1}}$, and $\beta = 0.1$ at $\dot{\varepsilon} = 10^{-3}\ \mathrm{s^{-1}}$. In fact, this is in a particular form to take into account the different effects of heat conduction in different strain rates.

In the same way, the difference of temperature in the defect region and the nondefect region cannot be ignored too. The specific heat C_v in Eq. (10.25a) cannot be regarded as a constant, but a function of temperature $C_v(T)$. In the specific analysis, $C_v(T)$ can be calculated according to the Einstein model (see Chapter 3, Eq. 3.73).

$$C_v(T) = 3Nk_B\left(\frac{\theta_E}{T}\right)^2 \exp\left(\frac{\theta_E}{T}\right) \Big/ \left[\exp\left(\frac{\theta_E}{T}\right) - 1\right]^2 \tag{10.25c}$$

where k_B is the Boltzmann constant (1.38×10^{-23} J/K), N is the Avogadro constant (6.02×10^{23}), and θ_E is the Einstein characteristic temperature (taking 242.9K in the present calculation).

From Eq. (10.25), the T can be solved as a function of $\boldsymbol{\varepsilon}$ and $\dot{\boldsymbol{\varepsilon}}$, and then $\sigma(\boldsymbol{\varepsilon}, \dot{\boldsymbol{\varepsilon}}, T)$ can be expressed as a composite function of $\boldsymbol{\varepsilon}$ and $\dot{\boldsymbol{\varepsilon}}$.

Therefore, the problem comes down to Eqs. (10.23) and (10.25) hold in the two regions (but $\sigma_{0u} \neq \sigma_{0d}$), while on the boundary of the two regions the following conditions should be satisfied, namely the displacement continuity conditions, the following stress continuity conditions.

$$(\sigma_{0u} + E_1\varepsilon_u)\left(1 + g\ln\frac{\dot{\gamma}_u}{\dot{\gamma}_0}\right)\left(1 - \alpha\frac{T_u}{T_e}\right) = (\sigma_{od} + E_1\varepsilon_d)\left(1 + g\ln\frac{\dot{\gamma}_d}{\dot{\gamma}_o}\right)\left(1 - \alpha\frac{T_d}{T_e}\right) \tag{10.26}$$

as well as the following initial condition

$$t = 0: \quad \varepsilon_u = \varepsilon_d = 0, \quad T_u = T_d = T_e \tag{10.27}$$

For a given constant $\dot{\boldsymbol{\varepsilon}}_{av}$ and a given $\Delta\sigma_o = \sigma_{ou} - \sigma_{od}$, by the finite difference method it is not difficult to calculate the strain, strain rate, and temperature at any time of the defect region and the nondefect region.

Take $\dot{\boldsymbol{\varepsilon}}_{av} = 1.0 \times 10^3\ \mathrm{s}^{-1}$ and $\Delta\sigma_o = 0.05\sigma_{0u}$, with increasing the apparent average strain $\boldsymbol{\varepsilon}_{av}$, the variation of strain $\boldsymbol{\varepsilon}_d$, strain rate $\dot{\boldsymbol{\varepsilon}}_d$, and temperature T_d in the defect region as well as the variation of strain $\boldsymbol{\varepsilon}_u$, strain rate $\dot{\boldsymbol{\varepsilon}}_u$, and temperature T_u in the nondefect region can be calculated and are shown in Fig. 10.19.

From the above figure, it can be seen that the localization process experiences three stages of development.

(1) In the first stage, due to the defect-weakening effect, the shear strain $\boldsymbol{\varepsilon}_d$ and the temperature T_d within the defect region all increase more rapidly than the shear strain $\boldsymbol{\varepsilon}_u$ and the temperature T_u within the nondefect matrix. However, the strain rate $\dot{\boldsymbol{\varepsilon}}_d$ within the defect region decreases and gradually approaches the value of the strain rate $\dot{\boldsymbol{\varepsilon}}_u$ within the nondefect matrix. It implies that the strain hardening plays a dominant role when deformation is not large enough, actually showing stabilization process which tends to strain uniformization.

(2) However, with increasing the strain $\boldsymbol{\varepsilon}_d$, the thermal softening induced by the adiabatic temperature rise gradually enhances. When the temperature rise reaches to a certain value (as shown by the green dash-

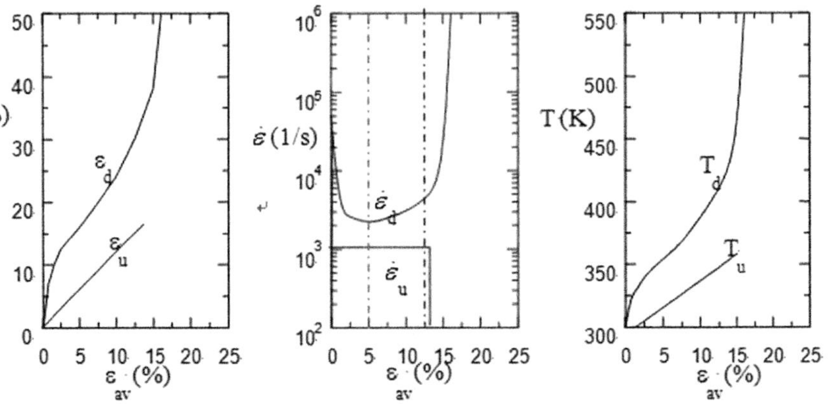

Figure 10.19 Evolution of adiabatic shearing by localization analysis for TB2, with increasing the apparent average strain ε_{av}, the variation of strain, strain rate, and temperature in both the defect region and the nondefect region. *From Bao, Hesheng., Wang, Li-li., 1990. An analysis of strain localization for adiabatic shearing. Journal of China University of Science and Technology 20 (71), Explosive Mechanics Issue, 18–24 (in Chinese)., Fig. 2, p.22.*

point line in Fig. 10.19), a transition appears; that is, the strain rate in defect region $\dot{\varepsilon}_d$ begins to increase with the average strain ε_{av}, and correspondingly both ε_d and T_d accelerated increase. It means that the thermal softening begins to play a dominant role in the second stage, showing an accelerated development of strain localization.

(3) As can be expected, the accelerated strain localization process finally leads to a dramatically increase of all the ε_d, T_d, and $\dot{\varepsilon}_d$, but an abrupt decrease of $\dot{\varepsilon}_u$ at a turning point (as shown by the red dash-point line in Fig. 10.19). This turning point corresponds to the "instability strain" ε_{in} under the given average strain rate $\dot{\varepsilon}_{av}$. Thus, in this stage, thermal softening plays a leading role and finally under the joint action of strain hardening, strain rate hardening, and thermal softening leads to the so-called thermo-viscoplastic instability or adiabatic shearing failure.

For different constant average strain rates at room temperature similar calculations can be made, the results are shown in Fig. 10.20. As can be seen, with the increase of average strain rate$\dot{\varepsilon}_{av}$, the critical strain ε_{in} which corresponds to the strain localization instability, decreases.

At different temperatures, similar calculations can be made, and the results can be plotted in a $\dot{\varepsilon}_{av}$–ε_{in}diagram, as shown in Fig. 10.21.

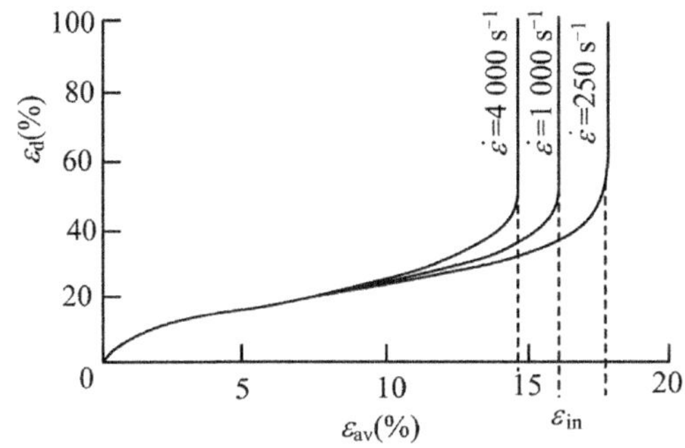

Figure 10.20 The ε_d-ε_{av} curves at different average strain rate $\dot{\varepsilon}_{av}$. *From Bao, Hesheng., Wang, Li-li., 1990. An analysis of strain localization for adiabatic shearing. Journal of China University of Science and Technology 20 (71), Explosive Mechanics Issue, 18–24 (in Chinese)., Fig. 3, p.22.*

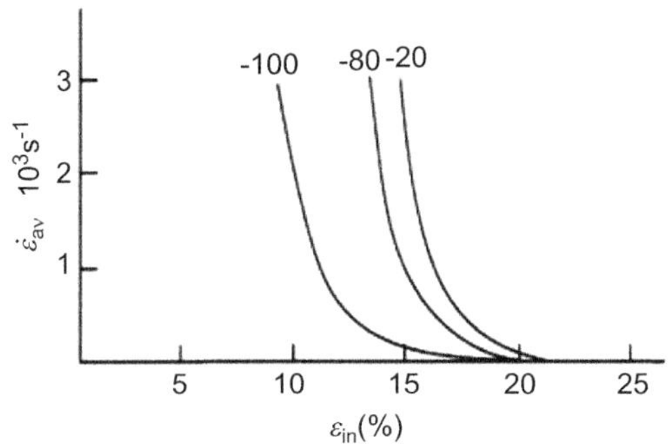

Figure 10.21 The $\dot{\varepsilon}_{av}$-ε_{in} curves at different temperature. *From Bao, Hesheng., Wang, Li-li., 1990. An analysis of strain localization for adiabatic shearing. Journal of China University of Science and Technology 20 (71), Explosive Mechanics Issue, 18–24 (in Chinese)., Fig. 4, p.22.*

Comparing the results given in Fig. 10.21 with the results given in Fig. 10.18, it can be seen that they are consistent qualitatively. This suggests that, the adiabatic shearing constitutive instability analysis based on the nonuniform deformation (localized) model after considering the strain rate effect is consistent qualitatively with the analysis based on the apparent uniform deformation model. Two approaches lead to a consistent conclusion: the occurrence and development of adiabatic shearing is dependent on the strain, strain rate, and temperature, it obeys the macroscopic criterion of the trivariable thermo-viscoplastic constitutive instability.

In retrospect of the previous discussions, it is necessary to emphasize the following two points.

(1) Comparing with the thermo-viscoplastic criterion, the adiabatic shearing thermoplastic instability criterion mainly ignores the strain rate effect, namely the second term in Eq. (10.11)

$$\frac{d\tau}{d\gamma}=\frac{\partial\tau}{\partial\gamma}+\frac{\partial\tau}{\partial\dot{\gamma}}\frac{d\dot{\gamma}}{d\gamma}+\frac{\partial\tau}{\partial T}\frac{dT}{d\gamma}=0 \tag{10.11b}$$

However, it should be noticed that strain rate effect plays an important role with respect to the adiabatic shearing, which is shown as

(a) In fact, considering the strain rate hardening effect, the dynamic flow stress at high strain rates is higher than the quasistatic flow stress at low strain rate. Because the adiabatic temperature rise dT is caused by the stress work $\tau d\gamma$ (see Eq. 10.1), so the temperature rise at high strain rates is higher than that at low strain rate. In this sense, the strain rate hardening effect will be transformed into a softening effect through adiabatic temperature rise and material thermal softening effect to promote the occurrence of instability.

(b) Under the adiabatic condition, the $\frac{d\dot{\gamma}}{d\gamma}$ in the second item of Eq. (10.11) actually is not independent of the strain hardening, strain rate hardening, and thermal softening, it can be positive and also can be negative in the evolution process of adiabatic shearing. In fact, as shown in Fig. 10.19, in the first stage there is $\frac{d\dot{\gamma}}{d\gamma}<0$, then the second item in Eq. (10.11) will be negative, namely showing a softening effect to balance the strain hardening, which will help achieve the critical instability condition. On the contrary, in the second stage, there is $\frac{d\dot{\gamma}}{d\gamma}>0$, then the second item in Eq. (10.11) will show a hardening effect to balance the thermal softening effect, thus will delay the occurrence of instability. Therefore, even if the value of the $\frac{d\dot{\gamma}}{d\gamma}$ related item in Eq. (10.11) is smaller than the other items, it cannot be ignored if the problem of instability is concerned. Thus, the adiabatic shearing instability is not a sudden event controlled by a single variable, but it is a continuous development process controlled by complicated positive and negative factors competing with each other.

(2) The attention of the above analysis mainly focuses on the **macrocriterion** of the occurrence and development of adiabatic shearing started from the view of **material constitutive instability.** Two points are worth noting here.

(a) It is necessary to clarify the dialectical relationship between macroscopic criterion and microscopic experimental observation, having both connection and difference between them. Zener and Hollomon linked the microscopic observation of adiabatic shear band to the macroscopic thermoplastic constitutive instability and established the bridge between the microcosmic investigation in material field and the macroscopic study in mechanics field. However, it is important to notice that meeting the adiabatic shearing criterion in the macroscopic level although means that the adiabatic shear band could be observed in the microlevel; but it is not strictly corresponding to the initiation of adiabatic shear band in the microlevel. This is like that the yield criterion in plastic mechanics is a macroscopic criterion, which represents the macroscopic critical condition for the transformation of elastic deformation to plastic deformation. However, meeting the plastic yield criterion in the macroscopic level although means that as a whole the crystalline slip line and dislocation motion that characterize plastic deformation could be observed in the microlevel; but it is not strictly corresponding to the initiation of crystalline slip line and dislocation motion in the microlevel. On the contrary, people have already realized from the "Kaiser's effect" in the acoustic emission experiment that before satisfying the macroyield condition, there exist already plastic slip lines in the crystalline grains with different orientations. This shows that although physical mechanisms of macrocriterion are based on the microlevel but cannot confuse the macrocriterion with the microcriterion in the microlevel.

Therefore, in the macroscopic mechanical analysis of the continuous medium, the region that satisfies the adiabatic shearing critical criterion is the macroscopic "adiabatic shear zone" not the microshear band itself. It is important to distinguish between the macroscopic "adiabatic shear zone" and the microscopic "adiabatic shear band" which are the different objects discussed at different scales, should not be confused with each other.

(b) It is necessary to clarify the difference between the constitutive instability of materials and the structural instability. Structural instability refers to that structures (especially slender and thin structures) under small forces lose their ability of keeping stability and balance, resulting in excessive deformation (such as buckling, wrinkling, etc.) so that reducing or completely losing the structural bearing capacity. This is the instability analysis of structures. Material constitutive instability while refers to a load instability ($d\sigma/d\varepsilon < 0$) of the material constitutive response, such as the yield drop and adiabatic shearing instability of material. This is the instability analysis of material constitutive response (stress—strain relation). Compared with the static stability analysis, structural dynamic instability under impact loading in general should take into account the inertia effect or stress wave effect of the structure, while the material dynamic constitutive instability generally should take into account the strain rate effect of material constitutive response. As one of the contents of material dynamics, this chapter discusses adiabatic shearing from the angle of material constitutive instability.

It should be pointed out that regarding the stability analysis of adiabatic shearing, in addition to the above discussion based on material dynamic constitutive instability, there are many researchers such as Clifton (1979), Bai et al. (1980, 1982, 1986, 1990), Morinali and Clifton (1983, 1988), Burns (1985, 1986), Wright and Batra (1985) and Wright and Walter (1987) following another approach—something like a structural instability analysis approach considering inertia effect. They started from the momentum conservation (considering the inertia effect), the energy conservation (the first law of thermodynamics), and the Fourier heat conduction equation (considering the thermal conduction effect), using perturbation analysis, by investigating under what conditions a small perturbation (of temperature, stress, or strain) can accelerate its development, to analyze the adiabatic shear localized instability phenomenon, then a series of results are obtained on the dynamic evolution of adiabatic shear band. Burns (1986) particularly emphasizes the importance of the inertial effect in the instability analysis. This is of great significance for the adiabatic shearing failure analyze of structures under impact loading taking account of inertial effect. Further progress in these areas and related work can be seen in two books by Bai and Dodd (Bai and Dodd, 1992; Dodd and Bai, 2012).

10.1.5 Interaction between adiabatic shear band and crack

So far, this chapter is aimed at the discussion of crackless body, while the microscopic observation showed that the adiabatic shear deformation once develops to a certain extent can initiate crack extending along the central part of shear band (Fig. 10.9), and voids can be seen in front of the main crack tip in the shear band (Fig. 10.10). However, when adiabatic shearing relates to the crack body discussed in Chapter 9, the problem is obviously further complicated: how will adiabatic shear bands and cracks interact with each other?

The impact shear experiments by Kalthoff et al. (Kalthoff and Winkler, 1987; Kalthoff, 2000) for mode II crack of two kinds of steel first displayed that with the increase of impact velocity or strain rate, failure mode will be changed, namely the failure mode is changed from the conventional failure mode which is characterized by a crack extension direction of 70 degrees relative to the crack plane, into the adiabatic shear failure model which is characterized by the crack extension direction close to the original crack plane.

Dong Xinglong et al. (Wang et al., 1994; Dong et al., 1998) further measured the dynamic stress intensity factor $K_{II}(t)$ in their mode II crack impact shear experiments for titanium alloy Ti6Al4V and steel 40Cr. The results show that depending on the different sensibility of different materials there are different failure modes. For steel 40Cr, mode II crack mainly extends along a direction of around 80°relative to the crack plane, it is the conventional shear failure mode, as shown in Fig. 10.22A, whereas for titanium alloy Ti6Al4V, with the increase of impact velocity, mode II crack may extend along a direction of around −5°relative to the crack plane, it is the adiabatic shear failure model, as shown in Fig. 10.22B.

A careful microscopic observation on the adiabatic shearing failure specimen shows that before crack initiation of a mode II crack, adiabatic shear band has been formed at a certain distance r_c ahead of the crack tip, as shown in Fig. 10.23. The typical micrograph of mode II crack after adiabatic shearing extension is shown in Fig. 10.24. As can be seen, the deformed

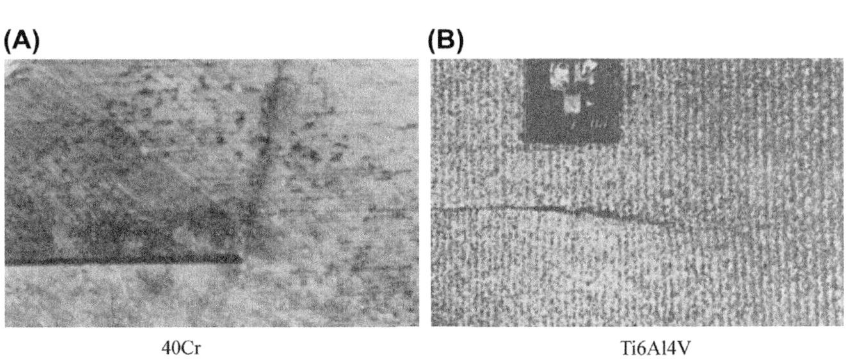

Figure 10.22 Two failure modes observed in mode II crack impact shear experiments. (A) 40Cr, (B) Ti6Al4V.

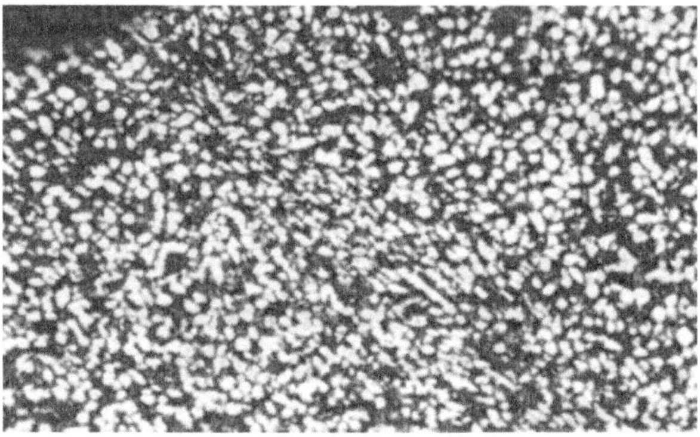

Figure 10.23 The formation of adiabatic shear band in front of crack tip (the upper left black part) before impact initiation of a mode II crack for Ti6Al4V.

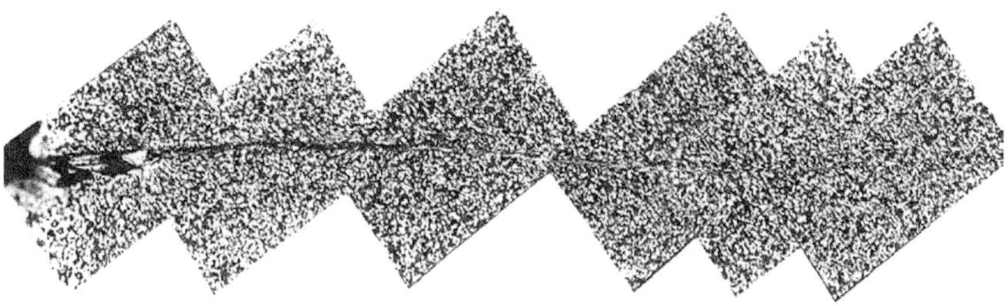

Figure 10.24 Micrograph of adiabatic shearing extension of mode II crack for Ti6Al4V (impact velocity 93 m/s).

band is observed in the front of the extended crack, while the transformed band is observed in the root part of the extended crack.

The above experimental results show that the adiabatic shearing extension of mode II crack includes the following three stages successively: (a) first, driven by the dynamic three-dimensional localized stress field, strain field, and strain rate field, in the front of the crack tip, a highly localized shear band is created along the original direction of the crack; (b) then, voids appear within the shear band ahead of the crack tip; (c) finally, by the void-coalescence mechanism, crack extends along the shear band.

Combining the adiabatic shearing thermal viscoplastic instability criterion discussed in the previous section and the mode II crack stress/strain field formula from the classical fracture mechanics, it is not difficult to deduce the critical criterion for adiabatic shearing crack initiation of mode II crack

(Wang et al., 1994). For example, when the thermo-viscoplastic constitutive equation is expressed by Eq. (10.15), namely in the following form,

$$\sigma_{eff} = (\sigma_o + E_1\varepsilon_{eff})\left(1 + g\ln\frac{\dot{\varepsilon}_{eff}}{\dot{\varepsilon}_o}\right)\left(1 - \alpha\frac{T}{T_e}\right) \tag{10.15★}$$

then the corresponding critical condition for thermo-viscoplastic constitutive instability is expressed by Eq. (10.13), namely in the following form

$$\left(\frac{\sigma_o}{E_1} + \varepsilon_{eff}\right)\left(1 + g\ln\frac{\dot{\varepsilon}_{eff}}{\dot{\varepsilon}_o}\right)\left(A - \frac{\alpha\beta E_1}{T_e\rho C_v}\varepsilon_{eff}\right) = 1 \tag{10.18★}$$

Notice that under the three-dimensional stress state discussed at present, the stress, strain, and strain rates in the above two formulas have been rewritten as the equivalent stress σ_{eff}, equivalent strain ε_{eff}, and equivalent strain rate $\dot{\varepsilon}_{eff}$, as introduced in Chapter 5 (Eq. 5.8).

On the other hand, by using the mode II crack stress/strain field formula from the classical fracture mechanics, it can be calculated (Wang et al., 1994) that the equivalent strain ε_{eff} and equivalent strain rate $\dot{\varepsilon}_{eff}$ along the crack direction ($\theta = 0$) at a distance r from the crack tip are, respectively

$$\varepsilon_{eff} = \frac{(1+\nu)K_{II}^2}{\pi E\sigma_s r}, \dot{\varepsilon}_{eff} = \frac{2(1+\nu)K_{II}\dot{K}_{II}}{\pi E\sigma_s r}, \text{for } \theta = 0 \tag{10.28}$$

where E is the Young's modulus, ν is the Poisson's ratio, σ_s (i.e., the σ_o in Eq. (10.15)★) is the yield stress, K_{II} is the stress intensity factor for mode II crack, and $\dot{K}_{II}$ is the rate of stress intensity factor for mode II crack. Substitute Eq. (10.28) into Eq. (10.18★), consider that the adiabatic shear band is formed at a distance r from the mode II crack tip (Fig. 10.23) so that the r in Eq. (10.28) should be taken as r_c, then obtain

$$\begin{gathered}\left(\frac{\sigma_0}{E_1} + \frac{(1+\nu)K_{II}^2}{\pi E\sigma_0 r_c}\right)\left(1 + g\ln\frac{2(1+\nu)K_{II}\dot{K}_{II}}{\pi E\sigma_0\dot{\varepsilon}_0 r_c}\right)\\ \times\left(A - \frac{\alpha\beta E_1}{T_e\rho C_v}\frac{(1+\nu)K_{II}^2}{\pi E\sigma_0 r_c}\right) = 1\end{gathered} \tag{10.29}$$

where the characteristic distance r_c and the integral constant A are material constants to be determined by experiments. Corresponding to the adiabatic shearing thermo-viscoplastic instability criterion (Eq. 10.18★) for crackless body, Eq. (10.29) gives the **bivariable thermo-viscoplastic instability criterion for adiabatic shearing crack initiation of mode II crack**. It shows that the adiabatic shearing crack initiation of mode II crack depends

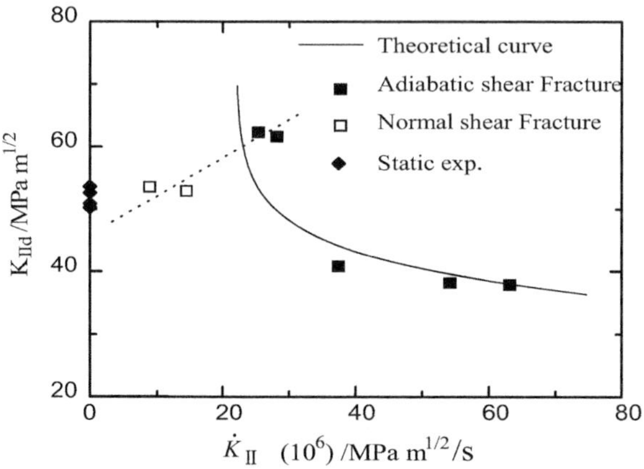

Figure 10.25 Comparison of the theoretical prediction with the experimental data of the adiabatic shearing crack initiation of mode II crack for Ti6Al4V. *From Dong, Xinlong., Wang, Lili., Yu, Jilin., 2000. An Investigation of Failure Behavior in Dynamic Shear Fracture, Jour. Ningbo Univ. (Natural Sci. & Engrg. Edition), 13, 174–180 (in Chinese)., Fig. 2, p.176.*

on both the stress intensity factor for mode II crack, K_{II}, and the rate of stress intensity factor for mode II crack, $\dot{K}_{II}$.

For Ti6Al4V mode II crack specimens, it is experimentally determined that $r_c = 0.8$ μm, $A = 0.86$. Fig. 10.25 shows in the K_{II}-$\dot{K}_{II}$ coordinates a comparison of the theoretical curve predicted by Eq. (10.29) with the experimental data. As can be seen, the theoretical predictions accord with the experimental data pretty well.

In the front of the crack tip, why it can cause a highly localized adiabatic shear band, of course, it is inseparable from the unique stress field (strain field and strain rate field) in front of the crack tip (see Chapter 9). It is mainly expressed in (a) a change from one-dimensional stress state of crackless body to three-dimensional stress states in front of crack tip and (b) the highly nonuniform localized mechanical field (the so-called concentrated stress/strain, strain rate). Therefore, it can be thought that these two factors should be also important to affect adiabatic shearing in crackless body.

In fact, once some appropriate changes are made to the specimen shape and consequently change the stress state and stress concentration, which then will also produce the conditions favorable for adiabatic shearing. It is sometimes called "geometric softening," while its essence is closely related to stress state and related stress/strain localization.

For example, tungsten heavy alloys (WHA), which is usually used to make armor-piercing bullet, is generally regarded as a material insensitive to the adiabatic shear deformation. However, from the studies by Yu Jilin et al. (Wei et al., 2001; Li et al., 2003), it has been found that although WHA did show insensitive to adiabatic shearing if using traditional cylindrical specimens, but adiabatic shearing localization will be observed if using step-cylindrical specimens and truncated-conic specimens which can cause the variation of stress state and the stress concentration, as shown in Figs. 10.26A and B, respectively. Further numerical simulation confirmed that for the traditional cylindrical specimen, the stress/strain distribution remains uniform even though the apparent compression strain is up to 50%, and consequently no shear localization was found, while for the truncated-conic specimen, the stress strain state changes from one-dimensional to three-dimensional, and the distribution of the corresponding effective strain is significantly ununiform, even if the apparent strain of the sample is only 10%, the shear localization has already observed. The maximum shear stress concentration zone in numerical simulation is consistent with the shear localization extension zone in experimental observation.

Yu Jilin et al. (Wei et al., 2001; Li et al., 2003) also found that when the stress state is properly changed by considering the microstructure and morphology of the material, it will be more advantageous to shear localization. They conducted SHPB experiments on pretwisted specimens of tungsten alloy WHA. If the traditional cylindrical specimen is adopted, adiabatic shear localization is not observed. However, if the compression-shear combined load is applied to the oblique cutting specimen with its axis inclined to

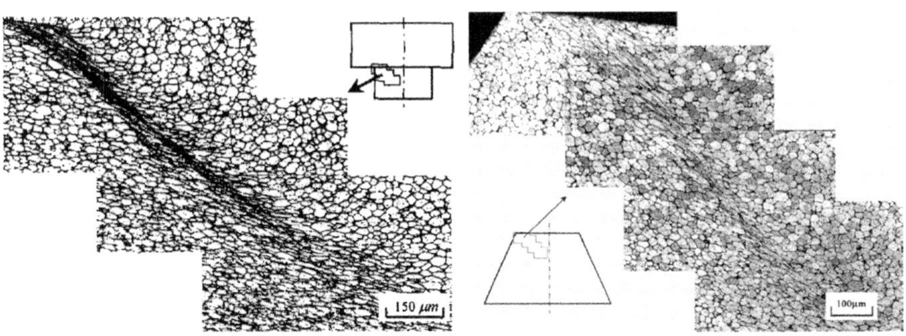

Figure 10.26 Adiabatic shear band observed in WHA specimens with different shapes. *From Wei, Z.G., Yu, J.L., Li, J.R., Li, Y.C., Hu, S.S., 2001. Influence of stress condition on adiabatic shear localization of tungsten heavy alloys. Int. J. Impact Engineering 26 (1—10)., 843—852., Fig. 3, p.846 and Fig. 4, p.847. Reprinted with permission of the publisher.*

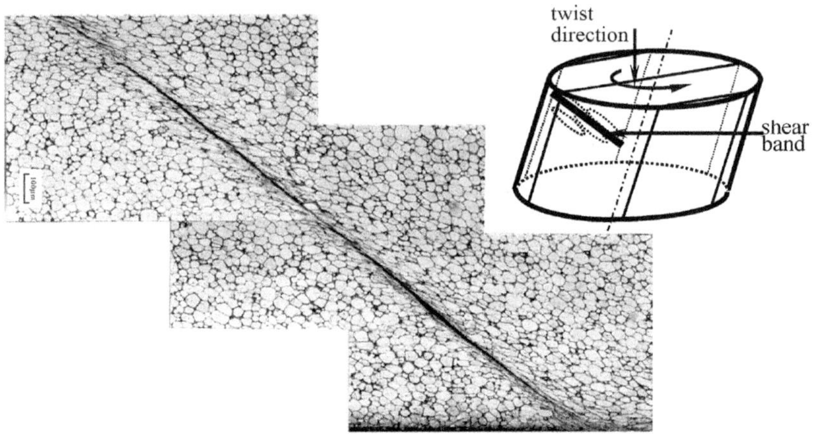

Figure 10.27 Adiabatic shear failure in oblique cutting specimen of WHA under compression-shear combined load. *From Wei, Z.G., Yu, J.L., Hu, S.S, Li, Y.C., 2000. Influence of microstructure on adiabatic shear localization of pre-twisted tungsten heavy alloys. Int. J. Impact Engineering 24 (6–7), 747–758., Fig. 4, p.750 and Fig. 10, p.754. Reprinted with permission of the publisher.*

the loading axis, the adiabatic shear failure occurs, as shown in Fig. 10.27. Microscopic observations show that the tungsten alloy WHA is a two-phase alloy with high strength tungsten particles distributed in low strength Ni–Fe matrix. After pretwisting, the aspect ratio of tungsten particles increases and is arranged in a certain orientation. Under the three-dimensional stress state of compression-shear combined load, when the direction of maximum shear stress is consistent with the elongated tungsten particle orientation, the shear localization will extend along the weaker Ni–Fe matrix (Fig. 10.27). These findings provide theoretical guidance for improving the performance of the tungsten alloy long rod penetrator and realizing "self-sharpening."

10.2 Dynamic evolution of damage

It was known from the discussion in Section 10.1 that the macroscopic adiabatic shear failure is due to, in the meso/microlevel, the accumulation of the formation and development of shear band. It was also pointed out in Chapter 8 that the spalling failure in the macroscopic level is not a simple and suddenly happening event but an accumulation process of the formation and development of damage in the meso/microlevel. In fact, look from the view of that any dynamic failure is essentially a dynamic process, various forms of macrodynamic failure all correspond to the evolution and accumulation process of different types of meso/microdamage, respectively.

In general, dynamic failure of materials under high strain rate can be classified into three types according to the damage type:

(a) Brittle failure, no significant plastic deformation at failure (corresponding to the term "fracture" in Fig. 5.15), which is attributed to the evolution of microcrack type damage, as shown in Fig. 10.28 (Curran, 1982). Eventually the crack extends along the cleavage plane of the grain, called a transgranular fracture, or extends along the grain boundary, known as intergranular fracture,

(b) Ductile failure, undergone a significant plastic deformation before the failure (corresponding to the term "rupture" in Fig. 5.15), which is attributed to the evolution of microvoid type damage, as shown in Fig. 10.29 (Curran, 1982).

(c) Adiabatic shearing failure, which is attributed to the evolution of the adiabatic shear band type damage as discussed in Section 10.1 of this chapter.

Therefore, the basic types of microdamage of dynamic failure are often attributed to microcracks, microvoids, and shear bands. The scale of such kind of damage is in the order of magnitude $10-10^2$ μm, which is in a larger scale relative to the material microscopic structure such as molecules, atoms,

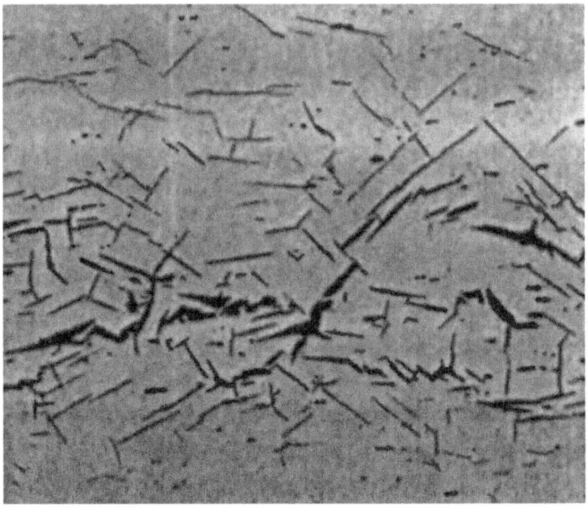

Figure 10.28 The meso-crack distribution in Armco iron plate specimen after brittle spalling. *From Curran, D.R., Seaman, L., Shockey, D.A., 1987. Dynamic failure of solids. Phys. Rep. 147, 253–388., Fig. 4.2, p.325. Reprinted with permission of the publisher.*

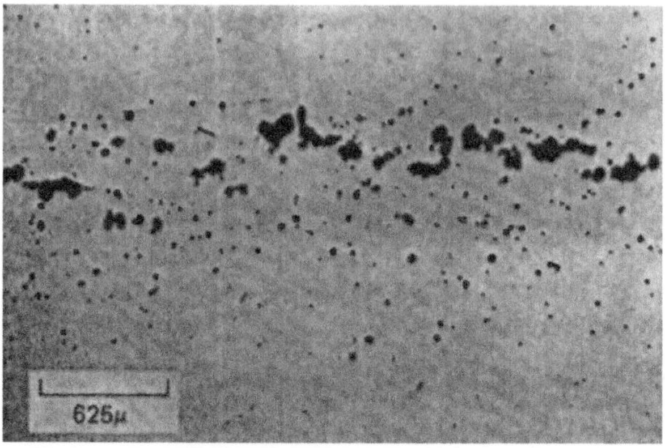

Figure 10.29 The meso-void distribution in 1145 aluminum plate specimen after ductile spalling. *From Curran, D.R., Seaman, L., Shockey, D.A., 1987. Dynamic failure of solids. Phys. Rep. 147, 253–388., Fig. 4.2, p.325. Reprinted with permission of the publisher.*

but it is in a smaller scale relative to the engineering macroscopic scale, so people often call them as meso-scale damage.

It is worth noting that under the impact loading, the meso-damage often appears in the form of multisource (as shown in Figs. 10.28 and 10.29). This is a significant difference from the microdamage evolution form in the quasistatic failure. It should also be pointed out that in the above discussion, we retain and follow the conventional terms such as microcrack and microvoid in the existing literature; in fact, they should be understood as meso-damages in mesoscopic level.

Because the number of these microdamages is as large as tens of thousands, it is not possible for people to deal with each crack in the same way as crack mechanics. For such kind of problem, it is usually dealt with by the following two types of statistical methods, namely (a) a mesoscopic statistical method is adopted, of which the typical is the model of Nucleation and Growth of meso-damages, referred to as NAG model. (b) A macroscopic continuous damage evolution method is adopted, of which the typical is the model of Thermal Activation Damage Evolution, referred to as TADE model. These two methods are discussed, respectively, below.

10.2.1 Statistical meso-damage model—NAG model

Seaman, Curran, and Shockey in 1973 proposed a model of nucleation and growth of meso-damages, referred to as NAG model (Curran et al., 1973, 1987; Seaman et al., 1976; Curran, and Seaman, 1996). This model is established on the basis of statistical processing of the experimental observation, that is, statistical processing of the meso-damage distribution on different sections of the recovery specimens after impact experiment is performed to obtain the surface distribution of the characteristic size of meso-damage. Then, it is transformed statistically into the corresponding body distribution, and the dynamic process of meso-damage nucleation, growth, and coalescence until the complete destruction is obtained. It is essentially a statistical meso-damage model.

- **Experimental observation of meso-damage evolution**

To analyze the evolution of mesoscopic damage with various sizes in the impacted materials, first of all, a series of experimental operations should be performed for the impacted specimens, such as dissecting, polishing, microscopic observation, and classification statistics. Moreover, to obtain the evolution law of the meso-damage in the process measured in the order of magnitude of μs and even of ns, it is necessary to "freeze" the various meso-damage evolution processes, thus a series of impact tests with different loading amplitude and different loading duration should be carried out. This is an experimental research with a lot of work and a lot of difficulty.

Usually the plate impact spalling test (see Chapter 8) is used to measure the evolution of meso-void and meso-crack. A relative thin flyer is driven by a gas gun to impact a relative thick flat target (the specimen), the compression shock waves first propagate in the target and the flyer, respectively, in opposite direction, after reflecting from their own free surface, two reflected release waves interacts in the target, eventually a tension zone is formed within the target and meso-damages in different forms are produced. Figs. 10.28 and 10.29 are obtained from such experiments and microscopic observation, showing the meso-crack distribution and meso-void distribution, respectively (the vertical axis corresponds to the impact direction). In a certain region near the back surface of the target, the duration of tensile stress is the longest, resulting in the most serious meso-damage, while with the increase of the distance either in front of or behind this region, the duration of tensile stress decreases gradually, then the number and size of meso-damage decreases too.

From a series of experiments, by changing the impact velocity and target plate thickness, as well as in each experiment to observe microscopically the meso-damage on a series of section with different distance from the target back surface, it is possible to change in a large range the tensile stress value and the pulse duration as required.

The statistical results of the meso-cracks of Armco iron specimen after the impact test is given in Fig. 10.30, expressed as a set of $N_{>R}$-R curves,

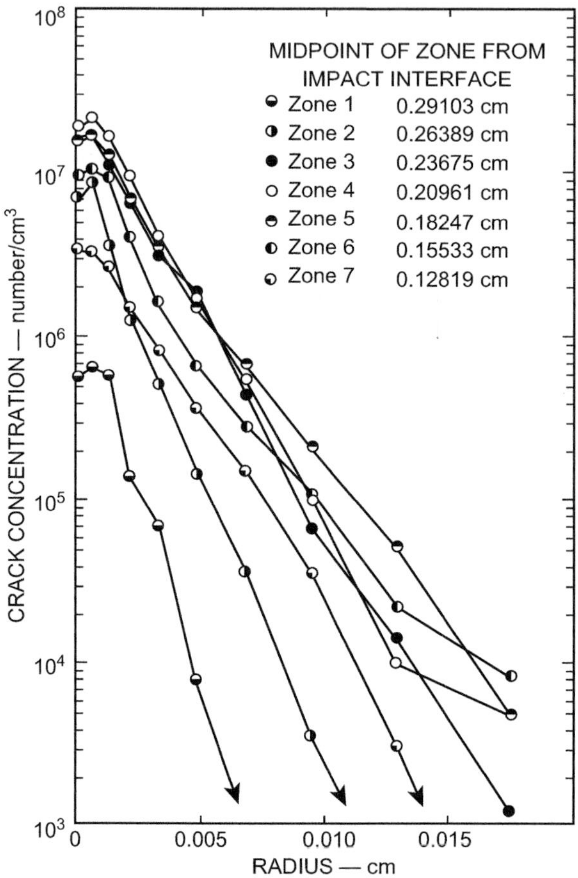

Figure 10.30 The relationship of the meso-crack concentration versus the crack radius measured on the sections with different distance from the free surface of specimen for Armco iron. *From Seaman, L., Barbee, T.W., Crewdson, R.C., 1972. Dynamic Fracture Criteria of Homogeneous Materials, AFWL-TR-71—156, Air Force Weapons Laboratory, Kirtland AFB, Albuquerque, New Mexico, February 1972., Fig. 55, p.151.*

where R denotes the crack radius which is actually the equivalent radius converted according to a given rule (Seaman et al., 1978), and the $N_{>R}$ denotes the cumulative number of cracks with radii larger than R, called the cumulative **meso-crack concentration**. In the figure, the $N_{>R}$-R distribution curves measured on the sections with different distance from the back surface (free surface) of specimen are displayed. Because the stress amplitude and duration are different on the different sections of specimen, it is possible to derive the relationship between the crack nucleation and growth rate with the stress amplitude and duration from those measured $N_{>R}$ -R distribution curves.

Similarly, it is possible to study the evolution of meso-void by the plate impact spalling test for ductile materials. Fig. 10.31 shows the $N_{>R}$-R distribution curves for 1145 aluminum (industrial pure aluminum) specimen measured after the plate impact spalling test (Meyers, 1994; Curran et al., 1987).

However, for the study of the evolution of shear band, it is difficult to produce shear band by using the plate impact spalling test, but instead the implosion test of metal cylinder can be used, as shown in Fig. 10.32 (Curran, 1982). When the specimen cylinder is expanded under the implosion load, the shear bands are nucleated and grown in the inner surface of the specimen cylinder. To avoid that the specimen cylinder may be broken by the implosion load, the outside of specimen cylinder is covered with a plastic (Plexiglas) tube and a massive steel containment annulus tube, which constitutes the so-called "Controlled Fragmenting Cylinder (CFC)," then it can stop the continuous expansion of CFC and "freeze" adiabatic shear

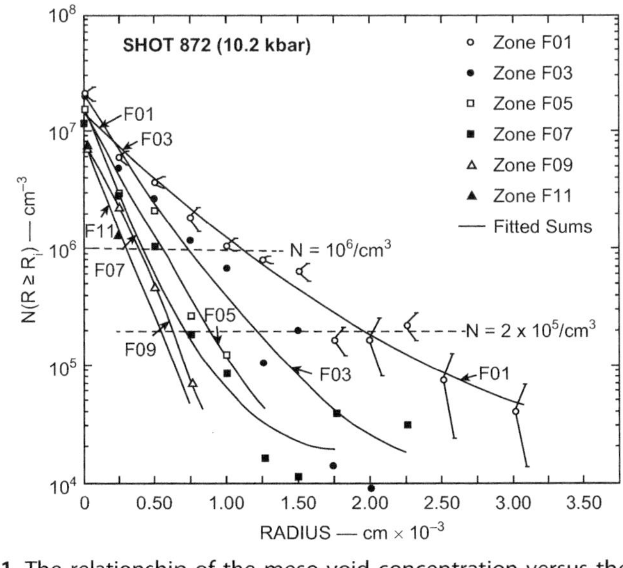

Figure 10.31 The relationship of the meso-void concentration versus the void radius measured on the sections with different distance from the free surface of specimen for 1145 aluminum. *(From Seaman, L., Barbee, T.W., Crewdson, R.C., 1972. Dynamic Fracture Criteria of Homogeneous Materials, AFWL-TR-71—156, Air Force Weapons Laboratory, Kirtland AFB, Albuquerque, New Mexico, February 1972., Fig. 3, p.16).*

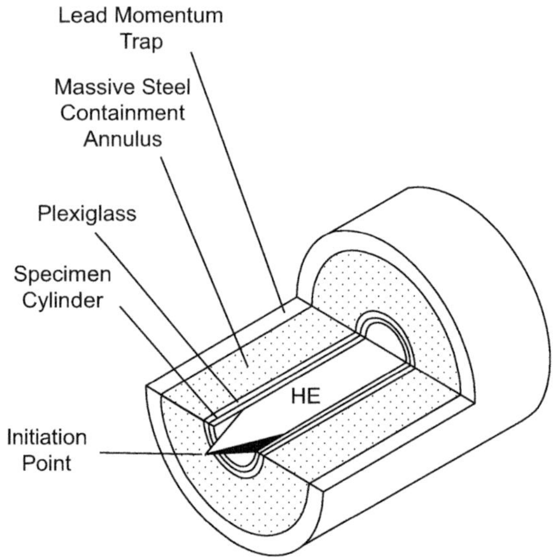

Figure 10.32 The implosion test of metal cylinder for adiabatic shear band research. *From Curran, D.R., Seaman, L., Shockey, D.A., 1987. Dynamic failure of solids. Phys. Rep. 147, 253–388., Fig. 2.13, p.273. Reprinted with permission of the publisher.*

damage at different loading stages. By measuring the adiabatic shear band distribution at each stage, the size distribution of shear band evolution can be obtained. Fig. 10.33 shows a set of $N_{>L}$ - L curves for 4340 steel, where L denotes the length of the shear band, and the $N_{>L}$ denotes the number of shear bands with length larger than L in the unit area, called the shear band surface density (Curran, 1982). The statistical properties of this shear band surface density $N_{>L}$ are very similar to those of the above meso-cracks (Fig. 10.30) and meso-voids (Fig. 10.31).

Curran et al. (Curran, 1982) found that the body distribution of meso-damage can be approximately expressed by the following equation

$$N_g(R) = N_0 \exp\left[-\left(\frac{R}{R_1}\right)^m\right] \tag{10.30}$$

where N_g is the number of meso-damage with radius greater than R per unit volume, N_0 is the total number of meso-damage per unit volume, R_1 is the distribution shape parameter, and m can be 1 (linear relation) or 2 (plane relation). In practice, when $m = 1$, it is found that when the nucleation size is in the range of several microns to a few millimeters, the size distribution of meso-damage observed in various materials can be well described by Eq. (10.30).

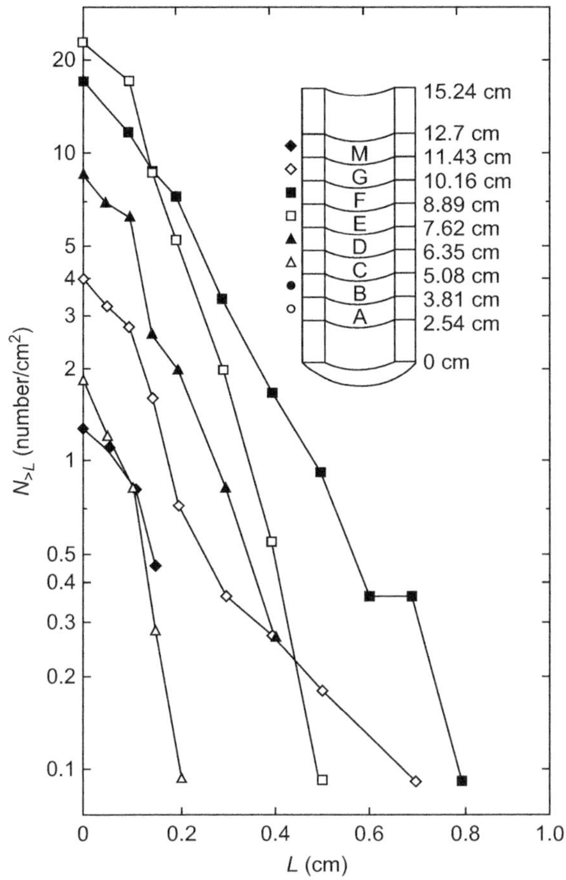

Figure 10.33 The $N_{>L}$ - L curves determined by implosion test of metal cylinder for 4340 steel. *From Shockey, D.A., Curran, D.R., Seaman, L. (1985). Development of Improved Dynamic Failure Models. Sri International Menlo Park CA., Fig. VI.5, p. VI-46.*

The above statistics data provide a basis for the establishment of the meso-damage dynamics model. Written this dynamic model into the constitutive relation of material, and apply it on a computational program, it is possible to calculate the process of nucleation, growth, and coalescence of meso-voids, meso-cracks, and shear bands that lead to final failure. In the following, the constitutive simulation of nucleation, growth, and coalescence of meso-damage will be discussed, respectively.

Cumulative evolution process of meso-damage consists of three stages: nucleation, growth, and coalescence into fragmentation:

- **Nucleation**

The meso-damage is usually nucleated in the inner nonuniform place of material, such as inclusion, grain boundary, and other flaws. First, nucleation must satisfy a certain threshold condition, e.g., the simplest threshold criterion is that the tensile stress σ is greater than or equal to the nuclear threshold stress σ_{n0}: $\sigma \geq \sigma_{n0}$. Once the threshold criterion is met, the nucleation will be developed at a certain rate in a certain nonuniform place having a certain size.

Let N be the number of meso-cracks in various sizes per unit volume

$$N = \widehat{N}(\underline{X}, t) \tag{10.31}$$

where $\underline{X}$ is the Lagrangian material coordinate vector and t is the time. Curran, Shockey, and Seaman () found that the nucleation rate $\dot{N}$ can be expressed by the following formula.

$$\dot{N} \equiv \left(\frac{\partial N}{\partial t}\right)_{\underline{X}} = \begin{cases} \dot{N}_0 \exp\left[\dfrac{(\sigma - \sigma_{n0})}{\sigma_1}\right], & \sigma > \sigma_{n0} \\ 0 & \sigma < \sigma_{n0} \end{cases} \tag{10.32}$$

where $\dot{N}_0$ is the material constant characterizing the threshold value of the nucleation rate, and σ_1 is the material constant characterizing the stress sensitivity of nucleation.

The above formula is derived from Zhurkov's thermal activated theory of the time dependent (or equivalently the rate dependent) strength of materials based on statistical theory (Zhurkov and Sanfirova, 1955; Zhurkov, 1965; Curran et al., 1977). It considers the nucleation process as a rate process caused by both stress and thermal activation; although temporary no explicit temperature term appears in the above formula. This is essentially similar to the theory of dislocation dynamics (see Chapter 6). As can be seen by comparing with the Seeger model in dislocation dynamics theory, it has been assumed here that the thermal activation energy of nucleation is proportional to the external stress. If the stress is less than the threshold value, namely $\sigma<\sigma_{n0}$, there will be no nucleation. Conversely, if the stress is greater than the threshold value, the nucleation rate increases exponentially with the stress.

- **Growth**

Growth is defined as the increase in the size of the meso-damage.

Large number of experiments found that under the dynamic load, for the meso-damage with a certain size, the relationship of the grow rate $\dot{R}$ with that the stress σ and the existing size R of the meso-damage is linear

$$\frac{\dot{R}}{R} = \left[\frac{(\sigma - \sigma_{g0})}{4\eta}\right] \tag{10.33}$$

where η is the viscosity coefficient (for crack, the crack tip viscosity), σ_{g0} the threshold stress of meso-void growth (often is equal to σ_{n0}). This kind of viscous growth law makes the meso-void growth maintain the exponential distribution characteristics as described by Eq. (10.30); in addition, it limits the growth rate of the meso-crack not to exceed the theoretical maximum, namely, the Rayleigh surface wave velocity C_R.

In this way, in the time of Δt, the growth radius of meso-damage can be obtained by the time integral of Eq. (10.33).

- **Coalescence and Fragmentation**

When the meso-damage grows to the point where its size can be compared with the average interval of meso-damage, it must start to coalesce.

In some of materials with better plasticity, the coalescence occurs through the direct contact of elliptical voids. However, in most plastic materials, there is first a certain type of plastic deformation localization between the voids. Studies by Curran et al. (1977) showed that the direct contact of voids occurs when the relative volume of voids is 50%—60%.

The coalescence mechanism for the brittle meso-cracks and the adiabatic shear bands is direct contact.

As the coalescence proceeds, the material splits into fragments; the final fragment size distribution is closely related to the meso-damage size that existed before the beginning of the coalescence. The mechanism of "coalescence" is the subject of further research.

- **Stress Relaxation**

With the growth of meso-damage, more stress-free surface appear in the material damage area, this increases the compliance of the damaged material and reduces the stress required to produce the same total deformation, in other words, the so-called stress relaxation phenomenon appears. This means that the formation and evolution of damage can lead to the weakening of materials, namely the so-called damage weakening effect.

The model with such effect is a kind of so-called active fracture model; with the growth of the damage, stress does not increase monotonically but decreases after it peaks.

When the meso-damage grows, stress relaxation can be contributed by the following two parts. (1) The contribution of opening displacement of meso-cracks or meso-voids under tensile stress. The open displacement will relieve some of the volume deformation applied, then corresponding to the elastic relaxation (elastic unloading) of volume deformation of matrix material, the related average tensile stress (tension spherical stress tensor) is relaxed. (2) The contribution of the reduction of material bearing area. Therefore, a correction factor υ should be introduced

$$\underline{\sigma} = \underline{\sigma}^{s}(1-\upsilon) \tag{10.34}$$

where $\underline{\sigma}^{s}$ is the stress tensor of matrix material taking account of the reduction of matrix bearing area, $\underline{\sigma}$ is the apparent continuum stress tensor and υ is the relative volume of void.

It is found from the study on the coalescence of meso-cracks and meso-voids that until the final stage of coalescence, the stress relaxation effect by the first mechanism is the main one. However, for the adiabatic shear band, the stress relaxation effect by the second mechanism is the main one, as there is no such situation that continuum medium subjected tensile stress.

By introducing the NAG model into the computer numerical program, the connection between the meso-damage evolution law and the macroscopic mechanical field is connected. On the one hand, the dynamical constitutive relation taking account of meso-damage evolution and the associated governing equations are established, and on the other hand, it is established the overall framework describing the development from the meso-damage evolution to the failure, thus a bridge is established between the physical evolution process of meso-damage and the macroscopic dynamic failure of the materials.

In the above, the basic ideas and principles of NAG model are briefly discussed. In fact, because the NAG model proposed, a lot of new progress has been made. Interested readers can refer to the review literature of Curran et al. (e.g., Curran et al., 1987).

It should also be pointed out that because the NAG model was proposed by Curran in the 1970s, a large number of new achievements have emerged in the field of statistical meso-damage mechanics (see e.g., Huang et al., 1993; Bai et al., 2005, 2006). Among those, the works of Bai et al. have attracted the attention of domestic and foreign scholars (Ke et al., 1990;

Bai et al., 1991; Xia et al., 1995a; 1995b; Bai et al., 2005, 2006). They emphasize that statistics meso-damage mechanics is a complex across-scale problem, a coupling problem of across macro-mesoscopic space and across time scale, so it is in need to solve a unified system of simultaneous equations consisting of the macroscopic mechanics equations and the meso-damage evolution equation. Considering that this is a problem of across macroscopic-mesoscopic levels, there are three kinds of coupling in the damage evolution, namely the coupling of macrodamage field and stress field, the coupling of mesoscopic level, as well as the across-scale coupling of macrocontinuous field and the meso-damage evolution. Start from this view point, Bai et al. first established the relationship between the macroscopic continuous damage D and the meso-damage number density n, then together with the three conservation equations of continuum mechanics constitute the cross-scale coupling framework, to solve such simultaneous equations. When the system is written into a dimensionless form, there appear two independent dimensionless numbers which control the cross-scale coupling, namely the intrinsic Deborah number $D^{\star}$ and the stress wave Deborah number D_e^*, as defined below:

$$D^* = \frac{n_N^*(c^*)^5}{V^*} \tag{10.35a}$$

$$D_e^* = \frac{ac^*}{LV^*} \tag{10.35b}$$

In Eqs. (10.35b), (L) is the characteristic size of specimen, a is the elastic wave speed, both are quantities in macroscopic scale, the ratio L/a represents the macroscopic characteristic time of stress wave loading, $t_{im}(= L/a)$, while $c^{\star}$ is the meso-damage size, $V^{\star}$ the extension speed of meso-damage, both are quantities in mesoscopic scale, the ratio $C^{\star}/V^{\star}$ represents the mesoscopic characteristic time of meso-damage extension, $t_v(= C^{\star}/V^{\star})$; then it is shown that the stress wave Deborah number D_e^* is an across-scale dimensionless parameter and can be expressed as $D_e^* = t_V / t_{im}$, namely the ratio of the mesoscopic characteristic time of meso-damage extension t_V and the macroscopic characteristic time of stress wave loading t_{im}, indicating the competition and coupling relationship between the stress wave loading and the meso-damage extension. As long as $D_e^* < 1$, it means that the meso-damage has enough time to extend.

In Eq. (10.35a), n_N^* is the nucleation rate of the meso-damage number density n, where the n is defined as the meso-damage number per unit volume and per unit meso-damage size (number/mm^3/μm), thus, the unit of n_N^* is "meso-damage number/mm^3/μm/t." So the dimensionless characteristic time of meso-damage nucleation can be expressed as $t_N = \left[n_N^*(c^*)^4\right]^{-1}$, and then the intrinsic Deborah number $D^\star$ can be expressed as $D^\star = t_V/t_N$, namely the ratio of the characteristic time of meso-damage extension t_V and the characteristic time of meso-damage nucleation t_N, characterizing the rate ratio of two meso-damage evolution processes, i.e., the ratio of the extension rate and the nucleation rate of meso-damage.

Thus, it can be seen that the macroscopic dynamic failure problem based on the microscopic damage evolution is an across-dimensional scale coupling problem, across the specimen size in macroscopic level and the meso-damage size in mesoscopic level; and on the other hand, it is an across-time scale coupling problem, across the loading time in macroscopic level and the time of meso-damage nucleation and extension in mesoscopic level. The Deborah number (Eq. 10.35) characterized by the ratio of multiple time scales is the key factor to control this cross-scale process. The reader who is interested in more specific and in-depth discussions, please refer to the relevant literature of Bai et al. (Ke et al., 1990; Bai et al., 1991; Xia et al., 1995a; 1995b; Bai et al., 2005, 2006).

10.2.2 Macroscopic continuum damage and thermal activated damage evolution model

In the field of continuum mechanics, Kachanov (1958) introduced the concept of macroscopic damage variable D, which attracted the attention of a lot of scholars and developed the continuum damage mechanics (Kachanov, 1986; Lemaitre, 1992).

Macroscopic damage D, as the internal variable of continuous thermodynamics, is defined as

$$D = \frac{\sigma_0 - \sigma}{\sigma_0}, \quad \text{or} \quad \sigma = (1 - D)\sigma_0, \quad (0 \le D \le 1) \tag{10.36a}$$

where σ_0 is the stress of the nondamaged material, and σ is the stress of the damaged material; when $D = 0$, there is no damage to the material, and when $D = 1$, the material completely loses its carrying capacity.

Under the condition of linear elastic deformation, Eq. (10.36a) is simplified to the continuum damage D_E measured by the reduction of elastic modulus

$$D_E = \frac{E_0 - E}{E_0}, \qquad \text{or} \quad E = (1 - D_E)E_0 \tag{10.36b}$$

where E_0 is the elastic modulus of the nondamaged material, E is the elastic modulus of the damaged material, and the subscript E is to show this continuum damage D_E is defined by the reduction of elastic modulus.

The damage variable D evolves with time (damage evolution law), and generally it is assumed to obey the following rate rule.

$$\dot{D} \equiv \frac{\partial D}{\partial t} = \frac{1}{(1-D)^{\nu}} \tag{10.37}$$

where the damage evolution rate $\dot{D}$ is defined as $\dot{D} \equiv \frac{\partial D}{\partial t}$, ν is a material parameter to be determined. Note that, for the sake of convenience, in the above, it has been assumed to be a one-dimensional problem, and the continuum damage D is thus simplified as a scalar.

Through the above approach, the macroscopic damage variable D and its evolution law are introduced into the constitutive equation, and the damage evolution can be described in the framework of continuum mechanics, thus provides an effective engineering method for studying the mechanical problems involving damage evolution in the macroscopic level. Here, the key to the problem is what kind of damage evolution law should be used and is there a kind of critical damage to predict the final failure.

It should be pointed out that the macroscopic damage evolution law described by Eq. (10.37) is mainly applicable to quasistatic load, while people know very little about the macroscopic damage evolution law under the impact high strain rates, which is still a question to be further studied.

In fact, taking the dynamic evolution of adiabatic shear bands discussed at the beginning of this chapter as an example, Fig. 10.12, Figs. 10.13 and 10.15 show that the evolution of the mesoscopic shear band depends on the strain, strain rate, and temperature simultaneously. The experimental observation and numerical simulation of the evolution of other mesodamages under the impact load also show that the evolution of all kinds of damage under impact load depends on both strain and strain rate (Wang et al., 1996; Wang et al., 1999; Huang et al., 2001). Considering that those mesoscopic facts should be reflected in the evolution of macroscopic continuum damage, the macrocontinuum damage evolution rate $\dot{D}$ should depend on the strain and strain rate too.

One of the difficulties in experimental research on the dynamic evolution of continuum damage is that so far, no dynamic test technique can be directly used to quantitatively measure the continuum damages themselves in real time during the short duration of the impact experiment. One of the possible approaches at present is to use the "damage freezing" method, namely it is assumed that the damage is irreversible, and it is considered that the damage has been frozen in the specimen experienced different impact loading conditions, then the frozen damage can be measured after the impact tests (e.g., to measure the D_E under quasistatic loading). Fig. 10.34 shows the D_E measured in the quasistatic test for PP/PA polymer blend specimens experienced different strain rate and strain in the SHPB experiments (Sun and Wang, 2005, 2008).

As can be seen from Fig. 10.34, (1) a threshold strain ε_{th} exists in the evolution of damage D, for PP/PA polymer blend ε_{th} depends on the strain rate, $4\% < \varepsilon_{th} < 6\%$; (2) the evolution of damage D depends on both strain and strain rate, $D_E = D_E\ (\varepsilon\dot{\varepsilon})$, so Eq. (10.37) should be replaced by a rate-dependent evolution law.

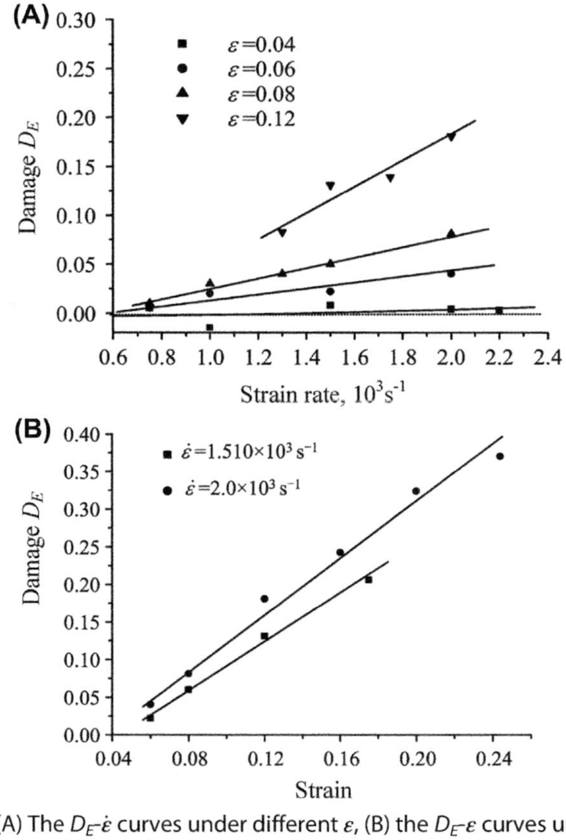

Figure 10.34 (A) The D_E-$\dot{\varepsilon}$ curves under different ε, (B) the D_E-ε curves under different $\dot{\varepsilon}$, for PP/PA polymer blend. *From Sun, Zi-jian., Wang, Li-li., 2005. Dynamic damage evolution described by modules change for PP /PA polymer blends with two different ompatibilizersm. Jour. Ningbo Univ. (Natural Sci. & Engrg. Edition) 18 (2), 159–162 (in Chinese)., Fig. 4 and Fig. 5, p.161–62.*

In the following, a **rate-dependent dynamic evolution law of continuum damage based on the thermal activation mechanism** will be discussed (Wang et al., 1993; Wang et al., 2006b; Wang et al., 2010).

Recall the rate-dependent constitutive relation of metallic materials (see Chapter 6), which is commonly explained by the thermal activation movement of dislocation in the microscopic mechanism, namely there is

$$\dot{\varepsilon}^p = \dot{\varepsilon}_0 \exp\left(-\frac{U_S(\sigma)}{kT}\right) \tag{10.38}$$

where $\dot{\varepsilon}^p$ is the viscoplastic strain rate, $\dot{\varepsilon}_0$ is the frequency factor, $U_S(\sigma)$ is the thermal activation energy related with the applied stress σ, k is the Boltzmann constant, and T is the absolute temperature. In the above equation, the key is the relationship between the thermal activation energy U_S and the applied stress σ. When there is a linear functional relationship between them, the above equation is reduced to the well-known Seeger model of the dislocation motion. In such case, there is a linear functional relationship between σ and $\log\dot{\varepsilon}$.

Noticing that, as shown in Fig. 10.33, the evolution of the macroscopic continuum damage D is rate-dependent, so should be characterized by a rate-dependent evolution law $D_E = D_E(\varepsilon,\dot{\varepsilon})$; furthermore, notice that, for the adiabatic shear band evolution, a rate-temperature equivalence relation exists between the strain rate effect and the temperature effect (ref. Section10.1.3), which is similar to the rate-temperature equivalence relation of plastic flow stress based on the thermal activation mechanism, then it is not difficult to consider the evolution of macroscopic continuum damage D as a statistical thermal activation process, namely similar to Eq. (10.38), the damage evolution law can be expressed as

$$\dot{D} = \frac{\partial D}{\partial t} = \dot{D}_0 \exp\left(-\frac{U_D}{kT}\right) \tag{10.39}$$

where $\dot{D}_0$ is the frequency factor of damage evolution, $U_D(\sigma)$ is the thermal activation energy of damage evolution related with the applied stress σ. Eq. (10.39) characterizes the so-called **Thermal Activated Damage Evolution** (**TADE**) model.

Recall the nucleation rate formula based on the thermal activated mechanism in the NAG model (Eq. 10.32), from the view of the basic mechanism, obviously the TADE model of macroscopic continuum damage is consistent with the NAG model of meso-damage.

For both Eqs. (10.38) and (10.39), the key lies in how to determine the relationship between the thermal activation energy and the applied stress.

For the same material, the $U_S(\sigma)$ and the $U_D(\sigma)$ should have an intrinsic link. Therefore, not specifically consider how $U_D(\sigma)$ depends on the applied stress σ, in substitute, assume that as a first approximation, there is a proportional relation between $U_S(\sigma)$ and $U_D(\sigma)$, namely

$$U_D = \lambda U_S \tag{10.40}$$

where λ is the material parameter. Substitute Eqs. (10.38) and (10.39) into Eq. (10.40), then after some calculations, obtain

$$\frac{\dot{D}}{\dot{D}_0} = \left(\frac{\dot{\varepsilon}}{\dot{\varepsilon}_0}\right)^{\lambda} \tag{10.41a}$$

or

$$\dot{D} = K_D\, \dot{\varepsilon}^{\lambda},\, K_D = \frac{\dot{D}_0}{\dot{\varepsilon}_0^{\lambda}} \tag{10.41b}$$

After integral, have

$$D = K_D \Psi_D(\dot{\varepsilon}),\, \Psi_D(\dot{\varepsilon}) = \int_{t_0}^{t} \dot{\varepsilon}^{\lambda} dt \tag{10.41c}$$

The above equation gives an explicit rate-dependent damage evolution law.

For the process of constant strain rate, and assume that there is a strain threshold ε_{th} in the damage evolution, then by integrating Eq. (10.41c), the following equation can be obtained.

$$D = K_D\, \dot{\varepsilon}^{\lambda-1}(\varepsilon - \varepsilon_{th}),\, (\varepsilon > \varepsilon_{th}) \tag{10.42a}$$

It is explicitly depicted by the above equation that the evolution of continuum damage depends on strain and strain rate. In the general situation, there may be a nonlinear relationship between D and ε, then the above equation can be generalized to the following generalized form

$$D = K_D\, \dot{\varepsilon}^{\lambda-1}(\varepsilon - \varepsilon_{th})^{\kappa},\, (\varepsilon > \varepsilon_{th}) \tag{10.42b}$$

where $\kappa(\geq 1)$ is the material constant describing the nonlinear function of D with ε.

Actually, the thermal softening induced by adiabatic temperature rise can be regarded as a kind of generalized damage (Zhou et al., 1992). In fact, according to the adiabatic temperature rise described by Eq. (10.1) and the definition of the stress decrease $d\sigma_T$ caused by the temperature rise, we have

$$dT = \frac{\beta \sigma d\varepsilon}{\rho C_v}, \quad d\sigma_T = \left(\frac{\partial \sigma}{\partial T}\right) dT = -\varphi dT \tag{10.43a}$$

where φ is the thermal softening coefficient of material, $\varphi = -\left(\frac{\partial \sigma}{\partial T}\right) \rangle 0$. Assume $\varphi =$ constant, then we have

$$dD = -\frac{d\sigma_T}{\sigma} = \frac{\varphi \beta}{\rho C_v} d\varepsilon, \quad \dot{D} = \frac{\varphi \beta}{\rho C_v} \dot{\varepsilon} \tag{10.43b}$$

The above equation can be regarded as a special case of Eq. (10.41b) namely when $\lambda = 1$ and $K_D = \frac{\varphi \beta}{\rho C_v}$. It means that the thermal softening induced by adiabatic temperature rise can be regarded as a kind of generalized damage that leads to weakening (softening). Similarly, the negative strain rate effect resulted from the strain-induced phase transition is derived from the weakening effect of phase transition under high strain rate and can also be attributed to another kind of generalized damage (Wang and Dong, 2001; 2004).

Based on Eq. (10.42), the dynamic failure criterion $D \geq D_c$ can be expressed in the following rate-dependent form

$$K_D \dot{\varepsilon}^{\lambda - 1} (\varepsilon - \varepsilon_{th})^{\kappa} \geq D_c \tag{10.44}$$

where D_c is the critical damage value, corresponding to the critical state of dynamic failure of material. The above formula shows that for each given critical damage D_c, there exists in $\dot{\varepsilon} - \varepsilon$ coordinates a critical curve, $(\dot{\varepsilon} - \varepsilon)_c$, so there will be different failure strains under different loading strain rates (two-variable failure criterion). Obviously, for different λ, these will be classified into three cases:

(1) If $\lambda > 1$, then with increase of strain rate, the failure strain decreases, presenting the so-called "impact embrittlement."

(2) If $\lambda < 1$, then with increase of strain rate, the failure strain increases, presenting the so-called "impact toughening."

(3) If $\lambda = 1$, then Eq. (10.44) is reduced to the critical strain criterion, $\varepsilon_c = \varepsilon_{th} + D_c / K_D$.

Therefore, in terms of the continuum damage evolution law of TADE model, the problem comes down to the point of how to determine ε_{th}, K_D, λ, and κ by experiments; while in terms of dynamic failure criteria, the question comes down to the point of how to determine D_c by experiments.

10.2.3 Coupling of macrocontinuum damage evolution with rate-dependent constitutive flow/deformation

How to determine ε_{th}, K_D, λ, and κ by experiments? Its complexity is that both the damage evolution process and the flow/deformation process are actually coupling and interacting with each other. On the one hand, the damage develops with the flow/deformation process, and the damage evolution depends on the stress, strain, and strain rate what the material experienced. On the other hand, the damage evolution will inevitably affect the mechanical behavior of the material, including the apparent constitutive relation and failure criterion. In other words, the research on the dynamic evolution of damage is usually only coupled with the research on the rate-dependent dynamic constitutive relationship. The problem became more complicated under the condition of short duration of the explosion/impact loading. It's not an exaggeration to say that under the condition of high strain rate, the studies on the dynamic constitutive relation and dynamic failure criterion coupling with the dynamic evolution of damage have become one of the leading research topics that physicists, mechanics, and materials scientists are concerned about.

For the material before damage, namely for the undamaged material, if it has been known the type of the dynamic constitutive relation, we can conduct the research on this basis, and the solution to the problem is relatively easy. Otherwise the problem is more complicated. In the following, these two kinds of situations will be discussed, respectively, through concrete examples.

- **Determination of damage evolution parameters, when the material parameters of rate-dependent constitutive relation of undamaged material are known**

For polymethylmethacrylate (PMMA), in the range of strain rate of 10^{-4} to $10^3\ \mathrm{s}^{-1}$ and the strain up to 6%–7%, the experimental results show that (Zhou et al., 1992) the rate-dependent constitutive relation for undamaged PMMA can be described by the ZWT nonlinear viscoelastic equation, namely Eq. (5.26) in Chapter 5 (the meaning of each parameter in the formula can be seen in Chapter 5)

$$\sigma = f_e(\varepsilon) + E_1 \int_0^t \dot{\varepsilon} \exp\left(-\frac{t-\tau}{\theta_1}\right) d\tau + E_2 \int_0^t \dot{\varepsilon} \exp\left(-\frac{t-\tau}{\theta_2}\right) d\tau \tag{5.26a$\star$}$$

$$f_e(\varepsilon) = E_0 \varepsilon + \alpha \varepsilon^2 + \beta \varepsilon^3 \tag{5.26b$\star$}$$

$$f_e(\varepsilon) = \sigma_m \left[1 - \exp\left(- \sum_{i=1}^{n} \frac{(m\varepsilon)^i}{i} \right) \right] \quad (5.26c\star)$$

The parameter values determined by the experiment are $\sigma_m = 91.8$ MPa, $n = 4$, $m = 22.3$, $E_1 = 0.897$ GPa, $\theta_1 = 15.3$ s, $E_2 = 3.07$ GPa, and $\theta_2 = 95.4$ μs

However, as the strain further increases, the theoretical prediction of undamaged PMMA deviates from the experimental curve, and on the experimental curve the relevant strain softening phenomenon ($\frac{d\sigma}{d\varepsilon} < 0$) appears, as shown in Fig. 10.34. Correspondingly, more and more tiny cracks are observed within the transparent PMMA specimen, so that such strain softening can be attributed to the constitutive instability caused by damage weakening. The difference between the experimental curve and the theoretical prediction of undamaged ZWT equation just represents the stress difference caused by the damage weakening.

The ZWT equation taking account of dynamic damage evolution can be composed of the ZWT nonlinear viscoelastic equation and the TADE continuum damage evolution law, called the damage-modified ZWT (DM-ZWT) model, namely in the following form

$$\sigma = (1 - D)\left[f_e(\varepsilon) + E_1 \int_0^t \dot{\varepsilon} \exp\left(-\frac{t-\tau}{\theta_1} \right) d\tau + E_2 \int_0^t \dot{\varepsilon} \exp\left(-\frac{t-\tau}{\theta_2} \right) d\tau \right],$$

$$D = K_D\, \dot{\varepsilon}^{\lambda-1} (\varepsilon - \varepsilon_{th})^{\kappa}, \text{ if } \varepsilon > \varepsilon_{th} \quad (10.45)$$

From the experimental results shown in Fig. 10.35, the damage evolution parameters for the PMMA specimen can be determined as $K_D = 1.82$, $\lambda = 1.17$, $\kappa = 1$, and $\varepsilon_{th} = 0.06$. It is worth to note, by comparison of the DM-ZWT equation (Eq.10.45) with the experimental curve, it can be seen that the theoretical prediction not only fits well with the experimental loading curve in a wide range of strain rate, reflecting the damage weakening induced constitutive instability under high strain rate, but also fits well with the experimental unloading curve, providing a fully support to the effectiveness of DM-ZWT model and TADE damage evolution law.

Once damage evolution parameters K_D, λ, κ, and ε_{th} have been determined, substitute the failure strain values of specimens measured under different strain rates into Eq. (10.44), the critical damage can be determined as $D_c = 10.2\%$. This indicates that the value of critical damage during failure

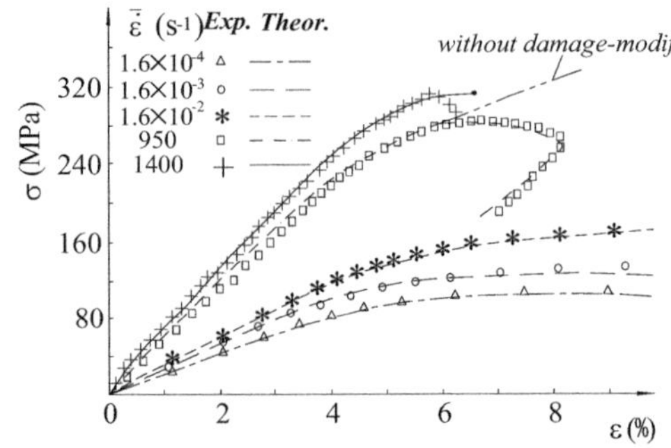

Figure 10.35 Comparison of the experimental data with the theoretical prediction by Eq. (10.45) under different strain rates. *From Zhou, Feng-hua., Wang, Li.li., Hu, Shi.sheng., 1992. A Damage-Modified Nonlinear Viscoelastic Constitutive Relation and Failure Criterion of PMMA at High Stain Rates. Explosion and Shock Waves 12 (4), 333–342 (in Chinese)., Fig. 4, p.338.*

is almost a constant, although the value of failure strain is different under different strain rate. In addition, although theoretically the material strength is completely lost when D = 1, actually before the theoretical strength is reduced to zero, the material has lost its carrying capacity, leading to failure.

Fig. 10.36 shows the theoretical loading curve determined by DM-ZWT model (Eq. 10.45) under different constant strain rates, as well as the theoretical failure point (curve endpoint) determined by Eq. (10.44) and their envelope. As a comparison, the measured failure points are also

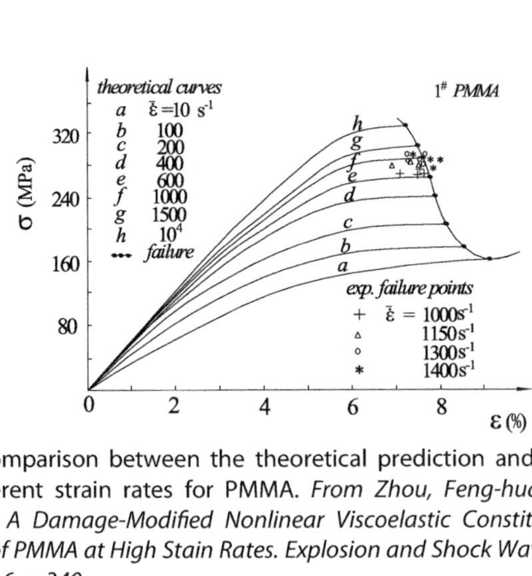

Figure 10.36 Comparison between the theoretical prediction and the experimental data under different strain rates for PMMA. *From Zhou, Feng-hua., Wang, Li.li., Hu, Shi.sheng., 1992. A Damage-Modified Nonlinear Viscoelastic Constitutive Relation and Failure Criterion of PMMA at High Stain Rates. Explosion and Shock Waves 12 (4), 333–342 (in Chinese)., Fig. 6, p.340.*

given in the same figure. Considering that the dispersion of experimental data is usually larger in the study of material failure characteristics, so in Fig. 10.36 the coincidence degree of the theoretical prediction and the measured result can be considered to be quite good.

- **Determination of damage evolution parameters, when the type of rate-dependent constitutive relation taking account of damage evolution is known**

  Further research shows that, even if material parameters of rate-dependent constitutive equation of undamaged material are unknown, as long as the type of rate-dependent constitutive equation taking account of damage evolution is known, and if experimental data are large enough, then we can find a way to determine the decoupling parameters of damage evolution.

  Take the DM-ZWT equation (Eq. 10.45) as an example, it can be applied to characterize the material constitutive relation after damage evolution ($\varepsilon > \varepsilon_{th}$) and can also be applied to characterize the material constitutive relation before damage evolution ($\varepsilon < \varepsilon_{th}$). Therefore, once the experimental data in a wide enough range of strain rate and in a sufficient large range of strain, including enough information of damage evolution, are obtained by experiments, then as long as the type of constitutive relation taking account of damage evolution is known, the optimized data fitting method, such as genetic algorithm (GA), can be used to perform global fitting of all material parameters, so finally can determine the decoupling parameters of damage evolution and constitutive relation (Wang et al., 2006; Sun and Wang, 2006).

  GA is a kind of optimization algorithm which is based on the idea of biological genetics and the natural law of survival of the fittest (Xing Wen-xun and Xie Jin-xing, 1999). Firstly, the numerical codes of each individual in the problem to be solved are generated by means of random method to form the initial group. Then, each individual is evaluated by the fitness function, eliminate the low-fitness individuals and select the individuals with high fitness to participate in genetic manipulation; the individual collection after the genetic manipulation forms the new group of next generation; this new group goes on the next evolution, so again and again, until the optimal individual in the group has not improved in successive generations, or the average fitness has not been improved in successive generations. That is the basic principle of GAs, and it provides a solution to the difficult problem of finding traditional mathematical models.

For polypropylene-polyamide (PP/PA) polymer blend in the strain rate range of 10^{-4}–10^{3} s^{-1} and in the strain range up to 24%, the related experimental results are shown in Fig. 10.37A in solid lines. Assume that its rate-dependent constitutive behavior taking account of damage evolution can be described by the DM-ZWT model (Eq. 10.45), then each material parameter in Eq. (10.45) can be determined by the GA (Wang et al., 2006; Sun and Wang, 2006), namely $\sigma_m = 44.1$ MPa, $n = 1$, $m = 39.4$, $E_1 = 0.278$ GPa, $\theta_1 = 7.22$ s, $E_2 = 2.29$ GPa, $\theta_2 = 0.107$ μs, $K_D = 0.639$, $\lambda = 1.17$,

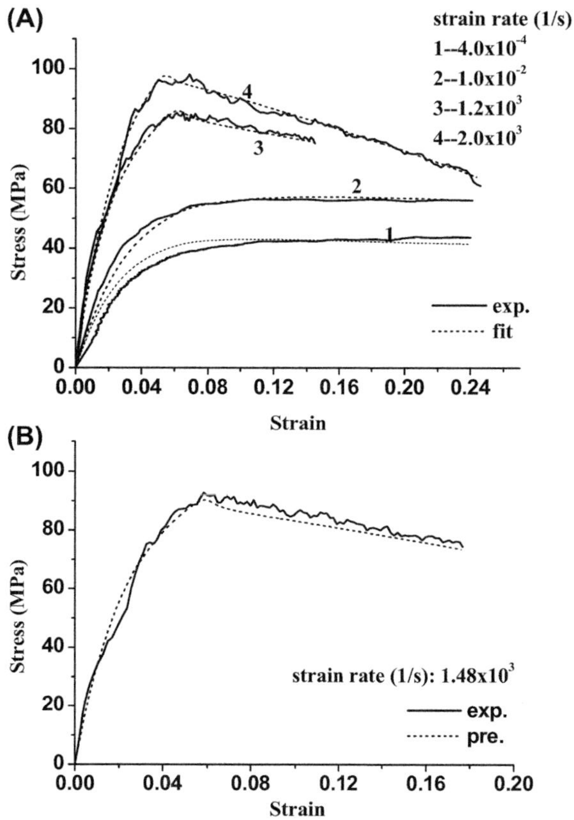

Figure 10.37 Comparison of (A) the theoretical prediction curves and (B) the experimental curves under different strain rates for PP/PA polymer blend. *From Sun, Z.J., Wang, L.L, 2006. Studies on impact constitutive behavior and damage evolution for PP/PA polymer blends at large deformation. Journal de Physique IV 134, 117–124., Fig. 1, p.120; Fig. 2, p.121.Reprinted with permission of the publisher.*

$\kappa = 0.617$, and the threshold strain ε_{th} depends on both strain and strain rate, $\varepsilon_{th} = 0.075 - 1.02 \times 10^{-5}\dot{\varepsilon}$. The theoretical curves drawn by these parameters are shown in Fig. 10.37A in dotted lines. It can be seen that, in the range of strain rate across eight orders of magnitudes, the theoretical predictions are in good agreement with the experimental data. As a test and verify, Fig.10.37B shows a comparison of the theoretical predicted curve (in dotted line) with the experimental curve (in solid line), which is tested under strain rate of $1.48 \times 10^3\ s^{-1}$ and not used as the fitting data. The validity of the model is verified by good agreement between them.

Once the decoupling parameters of damage evolution, K_D, λ, κ, and ε_{th}, are determined, the curves of damage evolution with the increase of strain, namely the $D-\varepsilon$ curves, under different strain rate can be drawn, as shown in Fig. 10.38. It can be seen that the strain rate plays a more important role in the damage evolution under impact loading. Compared with the damage evolution curves (curve 1 and 2) under the quasistatic load, the damage evolution curves (curve 3 to 6) under the high strain rate load show a much faster evolution rate.

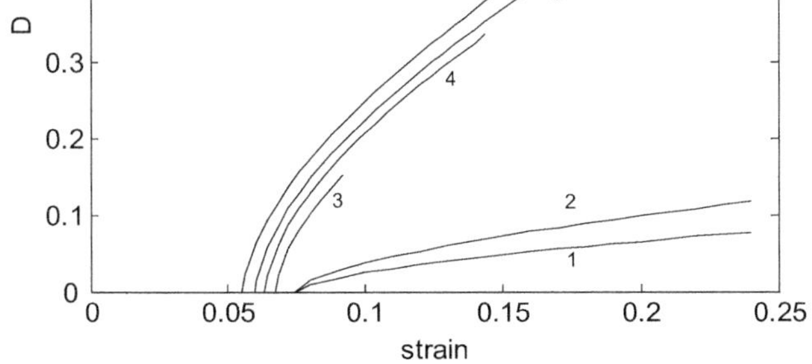

Figure 10.38 According to DM-ZWT model, the evolution of damage D with increasing strain and strain rate, for PP/PA polymer blend. *From Sun, Z.J., Wang, L.L, 2006. Studies on impact constitutive behavior and damage evolution for PP/PA polymer blends at large deformation. Journal de Physique IV 134, 117–124., Fig. 3, p.121. Reprinted with permission of the publisher.*

- **Determination of damage evolution parameters, when the rate-dependent constitutive relation taking account of damage evolution is an implicit function**

Further studies have found that the methods discussed in the previous section can be further developed. It is not necessary to restrict to an explicit type of the material rate-dependent constitutive equation which takes account of damage evolution, but as long as the material rate-dependent constitutive equation taking account of damage evolution does implicitly exist.

From the perspective of system science, the process of determining the interrelationship between the variables of damage evolution and the variables of material constitutive response is actually the process of system identification. That is, on the basis of the input and output data of causal relationship, a model equivalent to the measured system is determined from a given set of model classes. In other words, according to the known input and output data, a system model is selected and the system model is approximated to the real system by optimization method. Among the various methods of system identification, the artificial neural network (ANN) developed in recent years is especially suitable for nonlinear problems. In particular, Back-Propagation Neural network (BP Neural network) has the advantages of simple structure, stable working condition, and easy hardware implementation, so it is widely used (Li Xue-qiao and Ma Li, 1996). It has also been applied to determine the damage evolution parameters and so on without presupposition that the function form of material constitutive relation and damage evolution law (Wang et al., 2003; Wang et al., 2003; Xu and Wang, 2006; Wang et al., 2006; Sun and Wang, 2006).

The BP neural network in general consists of an input layer, one or more hidden layers (typically one layer), and an output layer. The output of the jth neural unit on the nth layer, $O_{n,j}$, is determined by the following formula

$$O_{n,j} = f\left(\Sigma w_{ji} O_{n-1,i} + \theta_j\right) \tag{10.46a}$$

where $O_{n-1,i}$ is the output of the ith neural unit in the preceding layer, w_{ji} is the weight between the jth neural unit of this layer and the ith neural unit of the preceding layer, θ_j is the threshold of the jth neural unit, and $f(x)$ is the so-called activation function, which is usually expressed by the following Sigmoid function

$$f(x) = \frac{1}{1 + \exp(-x)} \tag{10.46b}$$

During the learning or training steps of BP neural network, the weight w_{ji} is adjusted again and again by the "error-back-propagation algorithm," until the difference between the actual output and the expected output is small enough as required, then the network training is completed. The model final determined is the implicit relation between input parameters and output parameters.

When the BP neural network was used to determine the damage evolution parameters, a three-layer BP neural network is generally selected, as shown in Fig. 10.39. The material constitutive relation before and after the damage evolution (taking threshold strain ε_{th} as the boundary) can generally be expressed, respectively, as

$$\sigma(t) = f[\varepsilon(t),\ \dot{\varepsilon}(t)], \quad \varepsilon < \varepsilon_{th} \tag{10.47}$$

$$\sigma(t) = f[\varepsilon(t),\ \dot{\varepsilon}(t),\ D(t)] = f\left[\varepsilon(t),\ \dot{\varepsilon}(t),\ t^{-1}(D)\right], \quad \varepsilon > \varepsilon_{th} \tag{10.48}$$

The BP neural network is trained according to these two conditions, referred to as "Case 1" and "Case 2," respectively.

Specifically, in "Case 1", the $\varepsilon(t)$ and $\dot{\varepsilon}(t)$ measured in the SHPB experiment are taken as input, and the corresponding $\sigma(t)$ is taken as output.

In "Case 2," as shown in Eq. (10.48), originally the $\varepsilon(t)$, $\dot{\varepsilon}(t)$, and $D(t)$ should be taken as input. However, the technique of direct measurement of $D(t)$ has not been solved in the SHPB experiment. So instead of $D(t)$, taking a measurable $t^{-1}(D)$ which is the inverse function of $D(t)$, then the

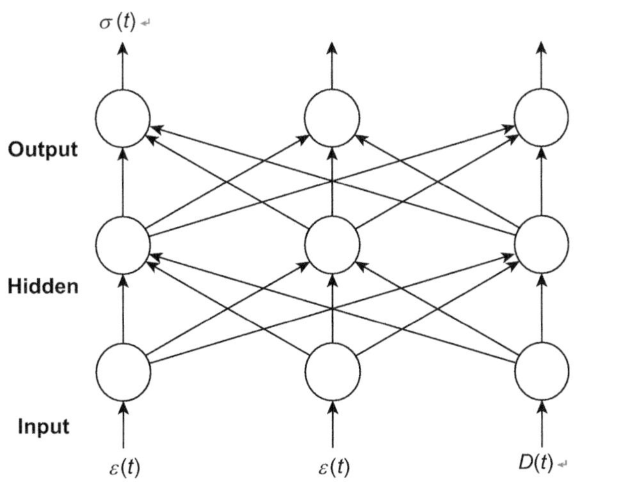

Figure 10.39 Schematic diagram of three-layer BP neural network.

$\varepsilon(t)$,$\dot{\varepsilon}(t)$, and $t^{-1}(D)$ been measurable in the SHPB experiment are taken as input, and the corresponding $\sigma(t)$ is taken as output.

Note that in "Case 1," the stress determined by BP neural network, as a function of strain and strain rate, is equivalent to the stress of undamaged material σ_0 in the definition equation of continuum damage (Eq. 10.36), whereas in "Case 2," the stress determined by BP neural network, as a function of strain, strain rate, and damage, is equivalent to the stress of damaged material σ in the definition equation of continuum damage (Eq. 10.36). Substitute the results of those two cases into the definition of continuum damage (Eq.10.36)

$$D = \frac{\sigma_0 - \sigma}{\sigma_0} \tag{10.36$\star$}$$

Thus, D and its dependence on strain and strain rate can be determined.

As a typical example, the BP neural network was used to analyze the damage evolution for PP/PA polymer blend (Sun and Wang, 2006). The damage evolution threshold strain ε_{th} has been measured by the "damage freezing method" (see Fig. 10.34). Both the constitutive relation and the damage evolution law are not presupposed. The representative results obtained by BP neural network in "Case 1" (strain rate of $1.22 \times 10^3\ \text{s}^{-1}$) are shown in Fig. 10.40 in dotted line. As a comparison, the experimental curve (in solid line) is also given. The ordinate in the graph is denoted by a dimensionless normalized stress $\overline{\sigma} = \sigma/\sigma_{\max}$, where $\sigma_{\max}$ is the maximum stress in all experiments, so that $\overline{\sigma} \leq 1$.

It can be seen from Fig. 10.40 that in the range of $\varepsilon<\varepsilon_{th}$, the predicted curve by BP neural network (actual output) is in good agreement with the experimental curve (expected output). In the range of $\varepsilon>\varepsilon_{th}$, the deviation of the two curves increases with strain, showing the weakening effect caused by damage evolution. However, the predicted curve by BP neural network in "Case 2" (strain rate of $1.22 \times 10^3\ \text{s}^{-1}$) is shown in Fig. 10.41 in dotted line, showing the effect of taking account of the damage evolution. For comparison, the experimental curve (in solid line) is also given. Obviously, the predicted curve by BP neural network (actual output) and the experimental curve (expected output) are in good agreement within the whole strain range.

Using the results identified by BP neural network under "Case 1" and "Case 2," the continuum damage D can be finally determined as a function of strain rate and strain $D = D(\varepsilon, \dot{\varepsilon})$ according to Eq. (10.36). Fig. 10.42 shows the evolution of D with strain at different constant strain rates.

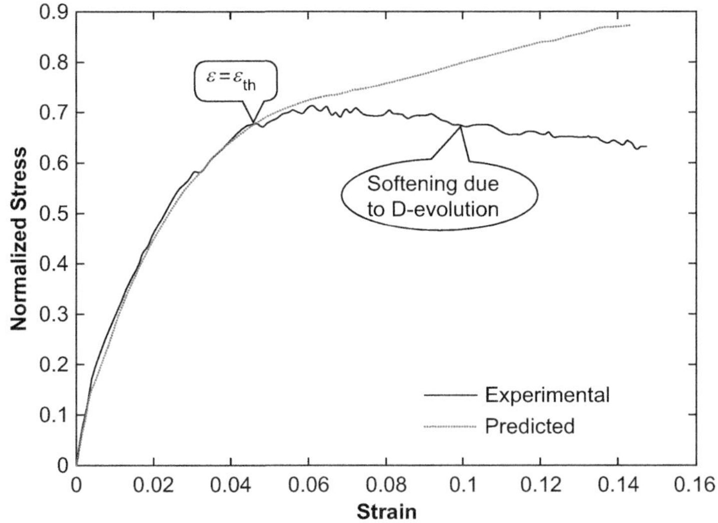

Figure 10.40 Comparison of experimental curve with the predicted curve obtained by BP neural network in "Case 1" for PP/PA polymer blend. *From Sun, Z.J., Wang, L.L, 2006. Studies on impact constitutive behavior and damage evolution for PP/PA polymer blends at large deformation. Journal de Physique IV 134, 117–124., Fig. 4, p.122. Reprinted with permission of the publisher.*

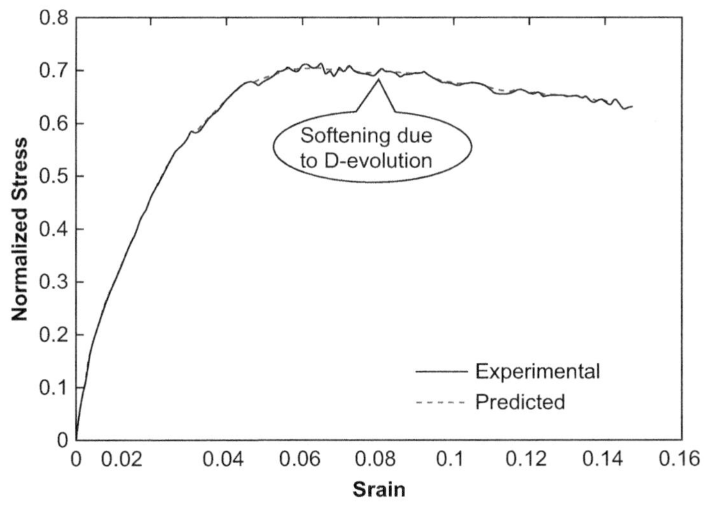

Figure 10.41 Comparison of experimental curve with the predicted curve obtained by BP neural network in "Case 2" for PP/PA polymer blend. *From Sun, Z.J., Wang, L.L, 2006. Studies on impact constitutive behavior and damage evolution for PP/PA polymer blends at large deformation. Journal de Physique IV 134, 117–124., Fig. 4, p.122. Reprinted with permission of the publisher.*

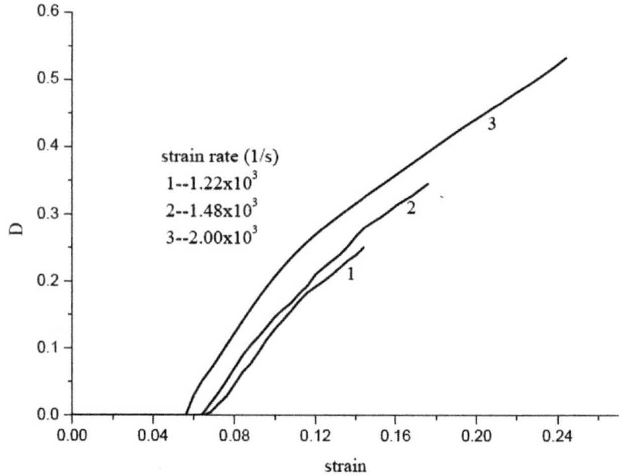

Figure 10.42 The $D = D(\varepsilon, \dot{\varepsilon})$ determined by BP neural network for PP/PA polymer blend. *From Sun, Z.J., Wang, L.L, 2006. Studies on impact constitutive behavior and damage evolution for PP/PA polymer blends at large deformation. Journal de Physique IV 134, 117–124., Fig. 5, p.123. Reprinted with permission of the publisher.*

It is noteworthy that the above $D = D(\boldsymbol{\varepsilon}, \dot{\boldsymbol{\varepsilon}})$ characteristics of PP/PA polymer blend identified by the BP neural network are very close to the $D = D(\boldsymbol{\varepsilon}, \dot{\boldsymbol{\varepsilon}})$ characteristics (Fig. 10.38) of the same PP/PA polymer blend determined by the DM-ZWT model using GA. Fig. 10.43 shows the comparison of D–$\boldsymbol{\varepsilon}$ curves determined by two methods under the same strain rate (2.0×10^3 s^{-1}). As can be seen from the figure, the D–$\boldsymbol{\varepsilon}$ curve (the curve II) determined by the BP neural network is slightly higher than the D–$\boldsymbol{\varepsilon}$ curve (the curve I) determined by the GA based on DM-ZWT model. This is understandable because curve I is limited by the DM-ZWT viscoelastic constitutive model, if PP/PA polymer blend possesses additional constitutive viscoplasticity, which may induce further related damage, then the curve I underestimates the possible damage evolution, whereas the curve II is determined without any preassumption on material constitutive model, thus can reflect more factors related to the damage evolution. This is right the advantage of BP neural network method.

In the second part of this chapter, the representative meso-damage evolution model and the representative continuum damage evolution model were discussed. It should be pointed out that the relation between continuum damage mechanics and meso-damage mechanics is not to evaluate which superior and which inferior, also not mutually exclusive, but is a

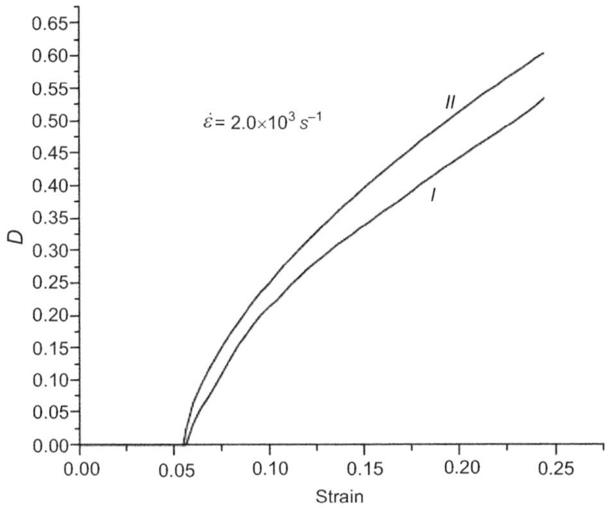

Figure 10.43 Comparison of the $D-\varepsilon$ curve determined by the BP neural network (the curve II) with the $D-\varepsilon$ curve determined by the GA based on DM-ZWT model (the curve I), under the strain rate of $2.0 \times 10^3\ s^{-1}$, for PP/PA polymer blend. *From Sun, Z.J., Wang, L.L, 2006. Studies on impact constitutive behavior and damage evolution for PP/PA polymer blends at large deformation. Journal de Physique IV 134, 117–124., Fig. 6, p.123. Reprinted with permission of the publisher.*

mutually complementary dialectical relation; just like that the macroscopic plasticity theory and the microscopic dislocation theory are mutually reinforcing. In the macroscopic continuum damage, on the macrosurface the details of the meso-damage evolution are ignored, but it is the macroscopic manifestation of the statistical average of the meso-damage evolution. On the other hand, more cognition of the meso-damage evolution law can promote to establish a macroscopic damage evolution law with more physical background. The meso-damage mechanics provides the physical mechanism for the macroscopic damage mechanics, and the macroscopic damage mechanics is used to solve the practical engineering problems.

References

Achenbach, J.D., 1972. Dynamic fracture effects in brittle materials. In: Nemat-Nasser, S. (Ed.), Mechanics Today, vol. 1. Pergamon Press, New York, pp. 1–57.

Albertini, C., Montagnani, M., 1994. Study of the true tensile stress-strain diagram of plain concrete with real size aggregate; need for and design of a large Hopkinson bar bundle. J. Phys. IV Colloq. 04 (C8), C8-113–C8-118. https://doi.org/10.1051/jp4:1994817 jpa-00253372.

Albertini, C., Boone, P.M., Montagnini, M., 1985. Development of the Hopkinson bar for testing large specimens intension. J. Phys. Fr. 46 (C5), 499–504.

Altynova, M., Hu, X., Daehn, G.S., 1996. Increased ductility in high velocity electromagnetic ring expansion. Metall. Trans. A 27, 1837–1844.

Antoun, T., Seaman, L., Curran, D.R., Kanel, G.I., Razorenov, S.V., Utkin, A.V., 2003. Spall Fracture. Springer-Verlag, New York.

Arakawa, K., Takahashi, K., 1991. Relationship between fracture parameters and surface roughness of brittle polymers. Int. J. Fract. 48, 103–114.

Bai, Y.L., 1990. Adiabatic shear banding. Res. Mech. 31, 133–203.

Bai, Y.L., 1982. Thermo-plastic instability in simple shear. J. Mech. Phys. Solids 30 (4), 195–207.

Bai, Y.L., 1980. A criterion for thermo-plastic shear instability. In: Meyers, M.A., Murr, L.E. (Eds.), Shock Waves and High Strain Rate Phenomena in Metals. Plenum, N.Y., pp. 277–284

Bai, Y.-L., Wang, H.-Y., Xia, M.-F., Ke, F.-J., 2006. Statistical mesomechanics of solid, linking coupled multiple space and time scales. Adv. Mech. 36 (2), 286–305 (in Chinese).

Bai, Y.L., Wang, H.Y., Xia, M.F., Ke, F.J., 2005. Statistical mesomechanics of solid, linking coupled multiple space and time scales. Appl. Mech. Rev. 58, 372–388.

Bai, Y.-L., Ke, F.-J., Xia, M.-F., 1991. Formulation of statistical evolution of microcracks in solids. Acta Mech. Sin. 22, 290–298 (in Chinese).

Bai, Y.L., Cheng, C.M., Yu, S.B., 1986. On evolution of thermo-plastic shear band. Acta Mech. Sin. 2, 1–7.

Bai, Y.L., Dodd, B., 1992. Adiabatic Shear Localization: Occurrence, Theories and Applications. Pergamon, Oxford.

Bailey, J.E., Hirsch, P.B., 1960. The dislocation distribution, flow stress, and stored energy in cold-worked polycrystalline silver. Philos. Mag. 5 (53), 485–497.

Baker, W.W., Yew, C.H., 1966. Strain rate effects in the propagation of torsional plastic waves. J. Appl. Mech. 33, 917–923.

Bancroft, D., Peterson, E.L., Minshall, S., 1956. Polymorphism of iron at high pressure. J. Appl. Phys. 27 (3), 291–298.

Bao, H., Wang, L.-L., 1990. An analysis of strain localization for adiabatic shearing. J. China Univ. Sci. Technol. 20 (71), 18–24. Explosive mechanics issue. (in Chinese).

Bao, H.-S., Wang, L.-L., Lu, W.-X., 1989. The high velocity deformation and adiabatic shearing of titanium alloy at low temperature. Explos. Shock Waves 9 (2), 109–119 (in Chinese).

Ващенко, В.Я., Зубарев, В.Н., 1963. Ф. Т. Т. 5, 886.

Beinert, J., Kalthoff, J.F., 1981. Experimental determination of dynamic stress intensity factors by shadow patterns. In: Experimental Evaluation of Stress Concentration and Intensity Factors. Springer, Dordrecht, pp. 281–330.

Bhat, H.S., Rosakis, A.J., Sammis, C.G., 2012. A micromechanics based constitutive model for brittle failure at high strain rates. J. Appl. Mech. 79, 1–12, 031016.

Bodner, S.R., Partom, Y., 1975. Constitutive equations for elastic-viscoplastic strain hardening materials. ASME J. Appl. Mech. 42, 385−389.
Bodner, S.R., Rubin, M.B., 1994. Modeling of hardening at very high strain rates. J. Appl. Phys. 76 (5), 2742−2747.
Bodner, S.R., Symonds, P.S., 1962. Experimental and theoretical investigation of the plastic deformation of cantilever beams subjected to impulsive loading. J. Appl. Mech. 29, 719−728.
Bodner, S.R., 1968. Constitutive equations for dynamic material behavior. In: Lindholm, U.S. (Ed.), Mechanical Behavior of Materials under Dynamic Loads. Springer-Verlag, New York, pp. 176−190.
Bodner, S.R., 2000. Unified Plasticity − An Engineering Approach. Faculty of Mechanical Engineering, Technion - Israel Institute of Technology, Haifa 32000, Israel.
Born, M., von Karman, T., 1912. Uber die Schwingungen in Raumgittern (The vibration of a lattice in space). Phyiks. Z. 13, 297.
Bowden, F.P., Field, J.E., 1964. The brittle fracture of solids by liquid impact, by solid impact and by shock. Proc. Roy. Soc. Lond. A282, 331−352.
Brandon, D.G., 1987. Dynamic loading and fracture. In: Blazynski, T.Z. (Ed.), Materials at High Strain Rates. Elsevier Applied Science Publishers, Essex, England, pp. 187−218.
Breed, R.B., Mader, C.L., Venable, D., 1967. Technique for the determination of dynamic tensile strength characteristics. J. Appl. Phys. 38 (8), 3271−3275.
Bridgman, P.W., 1949. The Physics of High Pressure. Bell and Sons, London.
Broberg, K.B., 1960. The propagation of a brittle crack. Arch fur Fysik 18 (1), 159−192.
Broek, D., 1982. Elementary Engineering Fracture Mechanics. Martinus Nijhoff, The Hague, Netherlands.
Brown, W.K., Wohletz, K.H., 1995. Derivation of the Weibull distribution based on physical principles and its connection to the Rosin-Rammler and lognormal distributions. J. Appl. Phys. 78 (4), 2758−2763.
Burgers, J.M., 1939. Some considerations on the fields of stress connected with dislocations in a regular crystal lattice. Proc. K. Neth. Akad. Sci. 42, 293−324.
Burns, T.J., 1985. Approximate linear stability analysis of a model of adiabatic shear band formation. Q. Appl. Math. 43, 65−84.
Burns, T.J., 1986. Influence of effective rate sensitivity on adiabatic shear instability. In: Murr, L.E., Staudhammer, K.P., Meyers, M.A. (Eds.), Metallurgical Applications of Shock-Wave and High-Strain-Rate Phenomena. Marcel Dekker, New York, pp. 741−747.
Cadoni, E., Dotta, M., Forni, D., Riganti, G., Albertini, C., 2015. First application of the 3D-MHB on dynamic compressive behavior of UHPC. In: EPJ Web of Conferences, vol. 94. EDP Sciences, p. 01031.
Campbell, J.D., Ferguson, W.G., 1970. The temperature and strain rate dependence of shear strength of mild steel. Philos. Mag. 21, 63−82.
Cao, J.-D., Li, Y.-C., 2006. Research of the damage evolution equation and spallation criterion. Eng. Sci. 8 (1), 40−45 (in Chinese).
Carslaw, H.S., Jaeger, J.C., 1947. Conduction of Heat in Solids. Clarendon Press, Oxford, p. 222.
Chen, Y.M., 1975. Numerical computation of dynamic stress intensity factors by a Lagrangian finite-difference method (the HEMP code). Eng. Fract. Mech. 7 (4), 653−660.
Chen, E.P., Sih, G.C., 1977. Transient response of cracks to impact loads. In: Sih, G.C. (Ed.), Mechanics of Fracture 4, Elastodynamic Crack Problem. Noordhoff International Publishing, Leyden, pp. 1−58.
Chen, J.-Y., Wang, L.-L., 1997. Damage-Modified non-linear constitutive relation for cement mortar at impact loading. Explos. Shock Waves (Suppl. 1), 282−286 (in Chinese).

Chen, J.-Y., Wang, L.-L., 2000. Rate-Dependent constitutive equation of cement mortar. J. Ningbo Univ. (Nat. Sci. Eng. Ed.) 13 (2), 1–5 (in Chinese).
Chen, W.N., Song, B., 2011. Split Hopkinson (Kolsky) Bar – Design, Testing and Applications. Springer.
China Encyclopedia, 1985. Mechanics. Encyclopedia of China Publishing House (in Chinese).
China Institute of Aeronautical Research, 1981. Stress Intensity Factor Manual. Science Press, Beijing (in Chinese).
Chou, P.C., Hopkins, A.K. (Eds.), 1972. Dynamic Response of Materials to Intense Impulsive Loading. U.S. Air Materials Lab., Wright Patterson AFB, Ohio.
Christensen, R.J., Swanson, S.R., Brown, W.S., 1972. Split-Hopkinson-bar tests on rocks under confining pressure. Exp. Mech. 12 (11), 508–513.
Chu, C., Wang, L., Xu, D., 1985. A nonlinear thermo-viscoelastic constitutive equation for thermoset plastics at high strain-rates. In: Chien Wei-Zang (Ed.), Proceedings of the International Conference on Nonlinear Mechanics (Oct. 28-31, 1985, Shanghai, China). Science Press, Beijing, pp. 92–97.
Clifton, R.J., 1979. Adiabatic shear banding. Report No. NMAB-356. In: Material Response to Ultra High Loading Rates. National Materials Advisory Board, NAS, Washington, DC (Chapter 8).
Соколовский, В.В., 1948. Распространение упруго-вязко-пластических волн в стержнях. прик. Mat. Mex. 12, 261.
Coleman, B.D., Noll, W., 1960. An approximation theorem for functional with applications in continuum mechanics. Arch. Ration. Mech. Anal. 6, 355.
Costin, L.S., Duffy, J., Freund, L.B., 1977. Fracture initiation in metals under stress wave loading conditions. In: Hahn, G.T., Kanninen, M.F. (Eds.), Fast Fracture and Crack Arrest, ASTM STP 627. American Society for Testing and Materials, Philadelphia, pp. 301–318.
Cowper, G.R., Symonds, P.S., 1957. Strain Hardening and Strain-Rate Effects in the Impact Loading of Cantilever Beams. Brown University Division of Applied Mathematics Report No. 28.
Cowperthwaite, M., Williams, R.F., 1971. Determination of constitutive relationships with multiple gauges in nondivergent wave. J. Appl. Phys. 42 (1), 456–462.
Cristescu, N., 1967. Dynamics Plasticity. North Holland Publishing Co., Amsterdam.
Culver, R.S., 1973. Thermal instability strain in dynamic plastic deformation. In: Rohde, R.W. (Ed.), Metallurgical Effects at High Strain Rates. Plenum Press, New York, pp. 519–530.
Curran, D.R., 1982. Chapter 9, dynamic fracture. In: Zukas, J.A., Nicholas, T., Swift, H.F., Gresczczuk, L.B., Curran, D.R. (Eds.), Impact Dynamics. John Wiley & Sons, Inc., New York, pp. 333–366.
Curran, D.R., Seaman, L., 1996. Simplified models of fracture and fragmentation. In: High-Pressure Shock Compression of Solids II. Springer-Verlag, New York.
Curran, D.R., Shockey, D.A., Seaman, L., 1973. Dynamic fracture criteria for a polycarbonate. J. Appl. Phys. 44 (9), 4025–4038.
Curran, D.R., Seaman, L., Shockey, D.A., 1987. Dynamic failure of solids. Phys. Rep. 147, 253–388.
Dally, J.W., 1979. Dynamic photoelastic studies of fracture. Exp. Mech. 19, 349–361.
Dally, J.W., Sanford, R.J., 1987. Strain gauge method for measuring the opening mode stress intensity factor. Exp. Mech. 27 (4), 381–388.
Dash, W.C., 1956. Copper precipitation on dislocations in silicon. J. Appl. Phys. 27 (10), 1193–1195.
Dash, W.C., 1957. The observation of dislocations in silicon. In: Fisher, J.C., Johnston, W.G., Thomson, R., Vreeland, T. (Eds.), Dislocations and Mechanical Properties of Crystals. John Wiley & Sons, Inc., New York.

Davidson, D.L., Lindholm, U.S., 1974. The Effect of Barrier Shape in the Rate Theory of Metal Plasticity (Based on Crystal Dislocations). Mechanical Properties at High Rates of Strain, pp. 124–137.

Davies, E.D.H., Hunter, S.C., 1963. The dynamic compression testing of solids by the method of the split Hopkinson pressure bar. J. Mech. Phys. Solids 11 (3), 155–179.

Davies, R.M., 1948. A critical study of the Hopkinson pressure bar. Phil. Trans. Roy. Soc. Lond. A240, 375–457.

Debye, P., 1942. Zur theorie der spezifischen wärmen. Ann. Physik 39, 789.

Ding, Y.Y., Yang, L.M., Wang, L.L., 2012. A further discussion on the Lagrangian analysis method based on particle velocity wave-profiles measurements. J. Ningbo Univ. (NSEE) 25 (4), 83–87 (in Chinese).

Dodd, B., Bai, Y., 2012. Adiabatic Shear Localization: Frontiers and Advances. Elsevier Science, Burlington.

Dong, X.-L., Yu, J.-L., Hu, S.-S., Wang, W., Wang, L.-L., 1998. Dynamic stress intensity factor and fracture behavior of mode-II cracked specimens at high loading rates. Explos. Shock Waves 18 (1), 62–68 (in Chinese).

Dong, X., Wang, L., Yu, J., 2000. An investigation of failure behavior in dynamic shear fracture. J. Ningbo Univ. (Nat. Sci. Eng. Ed.) 13, 174–180 (in Chinese).

Dong, X.-L., Chen, J.-Y., Gao, P.-Z., Qi, Z.-L., Wang, Y.-Z., Wang, Y.-G., Wang, L.-L., 2004. Experimental study on common and steel fiber reinforced concrete under tensile stress. J. Beijing Inst. Technol. 13 (3), 254–259.

Dong, X.L., Leung, M.Y., Yu, T.X., 2008. Characteristics method for viscoelastic analysis in a Hopkinson tensile bar. Int. J. Mod. Phys. B 22 (9–11), 1062–1067.

Dong, X., Gao, P., Chen, J., Hong, Z., Wang, L., 2009a. Study on the behavior of plain and steel fiber reinforced concrete under dynamic tensile stress. DYMAT 1, 633–637.

Dong, X.-L., Hong, Z.-Q., Gao, P.-Z., Chen, J.-Y., Shao, S.-M., Huang, D., Wang, L.-L., 2009b. Study on collapse of common and steel fiber reinforced concrete slabs subjected to contact detonation. Acta Armamentarii 30 (Suppl. 2), 280–283 (in Chinese).

Dong, X.L., Zhang, S.L., Yuan, H.L., Wang, L., 2011. Hopkinson bar bundle technique and experimental studies on dynamic behavior of concretes. In: Proceedings of 10th National Symposium on Impact Dynamics, July 26-31, 2011, Taiyuan, China (in Chinese).

Dormeval, R., 1987. The adiabatic shear phenomenon. In: Blazynski, T.Z. (Ed.), Materials at High Strain Rates. Elsevier Applied Science Publishers, Essex, England, pp. 47–70.

Dowling, A.R., Harding, J., Campbell, J.D., 1970. The dynamic punching of metals. J. Inst. Met. 98, 215–224.

Dremin, A.N., Molodets, A.M., Melkumov, A.I., Kolesmikov, A.V., 1992. On anomalous increase of steel spall strength and its relationship to martensitic transformation. In: Meyers, M.A., Murr, L.E., Staudhammer, K.P. (Eds.), Shock-Wave and High-Strain-Rate Phenomena in Materials. Marcel Dekker, New York, pp. 751–757.

Duffy, J., Campbell, J.D., Hawley, R.H., 1971. On the use of a torsional split Hopkinson bar to study rate effects in 1100-O aluminum. J. Appl. Mech. 38, 83–91.

Dugdale, J.S., McDonald, D.K.C., 1953. Phys. Rev. 89, 832.

Duvall, G.E., Graham, R.A., 1977. Phase transitions under shock-wave loading. Rev. Mod. Phys. 49, 523.

Duvall, G.E., 1971. Shock waves in condensed media. In: Caldirola, P., Knoepfel, H. (Eds.), Physics of High Energy Density. Academic Press, New York, p. 7.

Einstein, A., 1907. Die Plancksche theorie der strahlung und die theorie der spezifischen wärme. Annalen der Physik 327 (1), 180–190.

Eleiche, A.M., Campbell, J.D., 1976. The Influence of Strain-Rate History and Temperature on the Shear Strength of Copper, Titanium and Mild Steel. Air Force Materials Laboratory, Wright-Patterson Air Force Base, Ohio, USA. Tech. Rep. AFML-TR-76-90.

Epstein, B., 1947. The mathmatical description of certain breakage mechanisms leading to the logarithmico-normal distribution. J. Frankl. Inst. 244, 471–477.
Erkman, J.O., 1961. Smooth spalls and the polymorphism of iron. J. Appl. Phys. 32 (5), 939–944.
Fan, T.-Y., 2006. The Principle and Application of Fracture Dynamics. Beijing Institute of Technology Press, Beijing (in Chinese).
Feng, Z.-Z., Wang, X.-J., Wang, F.-S., Gao, H.-S., Yue, Z.-F., 2007. Implementation and its application in finite element analysis of constitutive model for ZWT nonlinear viscoelastic material. J. Mater. Sci. Eng. 25 (2), 269–272 (in Chinese).
Ferguson, W.G., Hauser, F.E., Dorn, J.E., 1967a. Dislocation damping in zinc single crystals. Br. J. Appl. Phys. 18, 411–417.
Ferguson, W.G., Kumar, A., Dorn, J.E., 1967b. Dislocation damping in Aluminum at high strain rates. J. Appl. Phys. 38, 1836–1869.
Field, J.E., Hutchings, I.M., 1987. Surface response to impact. In: Blazynski, T.Z. (Ed.), Materials at High Strain Rates. Elsevier Applied Science Publishers, Essex, England, pp. 243–293.
Field, J.E., Gorham, D.A., Rickerby, D.G., 1979. In: Adler, W.F. (Ed.), Erosion: Prevention and Useful Applications, ASTM, STP 664. ASTM, Philadelphia, PA, pp. 320–243.
Field, J.E., Sun, Q., Townsend, D., 1989. Ballistic impact of ceramics, mechanical properties of materials at high rates of strain. Inst. Phys. Conf. Ser. 102, 387–393.
Field, J.E., Walley, S.M., Proud, W.G., Goldrein, H.T., Siviour, C.R., 2004. Review of experimental techniques for high rate deformation and shock studies. Int. J. Impact Eng. 30 (7), 725–775.
Fleisher, R.L., 1962. Rapid solution hardening, dislocation mobility, and the flow stress of crystals. J. Appl. Phys. 33, 3504–3508.
Follansbee, P.S., Kocks, U.F., 1988. A constitutive description of the deformation of copper based on the use of the mechanical threshold stress as an internal state variable. Acta Metall. 36 (1), 81–93.
Follansbee, P.S., 1985. High Strain Rate Deformation in FCC Metals and Alloys. Los Alamos National Lab., NM, USA. No. LA-UR-85-3026; CONF-850770-8. https://www.osti.gov/servlets/purl/5276223.
Follansbee, P.S., 1986. High-strain-rate deformation of FCC metals and alloys. In: Murr, L.E., Staudhammer, K.P., Meyers, M.A. (Eds.), Metallurgical Applications of Shock-Wave and High-Strain Rate Phenomena. Dekker, New York, pp. 451–479.
Forquin, P., Gary, G., Gatuingt, F., 2008. A testing technique for concrete under confinement at high rates of strain. Int. J. Impact Eng. 35 (6), 425–446.
Forrestal, M.J., Sagartz, M.J., 1978. Elastic-plastic response of 304 stainless steel beams to impulse loads. J. Appl. Mech. 45, 685–687.
Fowles, R., 1970. Conservation relations for spherical and cylindrical stress wave. J. Appl. Phys. 41 (6), 2740–2741.
Fracture toughness – Wikipedia, the free encyclopedia, https://en.wikipedia.org/wiki.
Frank, F.C., Read Jr., W.T., 1950. Multiplication processes for slow moving dislocations. Phys. Rev. 79 (4), 722–723.
Freund, L.B., 1972. Crack propagation in an elastic solid subjected to general loading II. Nonuniform rate of extension. J. Mech. Phys. Solids 20 (3), 141–152.
Freund, L.B., 1973. Crack propagation in an elastic solid subjected to general loading, III stress wave loading. J. Mech. Phys. Solids 21, 47–61.
Freund, L.B., 1990. Dynamic Fracture Mechanics. Cambridge University Press, Cambridge.
Gary, G., Bailly, P., 1998. Behavior of quasi-brittle material at high strain rate. Experiment and modeling. Eur. J. Mech. A Solid. 17 (3), 403–420.
Gilman, J.J., 1960. The plastic resistance of crystals. Aust. J. Phys. 13, 327–348.
Gilman, J.J., 1965. Dislocation mobility in crystals. J. Appl. Phys. 36 (10), 3195–3206.

Glenn, L.A., Chudnovsky, A., 1986. Strain-energy effects on dynamic fragmentation. J. Appl. Phys. 59, 1379–1380.
Gong, J.C., Malvern, L.E., 1990. Passively confined tests of axial dynamic compressive strength of concrete. Exp. Mech. 30 (1), 55–59.
Gou, Q., 1978. Concise Course of Solid Physics. People's Education Press (in Chinese).
Grady, D.E., Benson, D.A., 1983. Fragmentation of metal rings by electromagnetic loading. Exp. Mech. 23 (4), 393–400.
Grady, D.E., Hightower, M.M., 1992. Natural fragmentation of exploding cylinders. In: Meyers, M.A., Murr, L.E., Staudhammer, K.P. (Eds.), Shock Wave and High-Strain Rate Phenomena in Materials. Marcel Dekker, Inc, New York, pp. 713–721.
Grady, D.E., Kipp, M.E., 1980. Continuum modeling of explosive fracture in oil shale. Int. J. Rock Mech. Min. Sci. 17 (3), 147–157.
Grady, D.E., Kipp, M.E., 1985. Geometric statistics and dynamic fragmentation. J. Appl. Phys. 58 (3), 1210–1222.
Grady, D.E., Kipp, M.E., 1987. The growth of unstable thermoplastic shear with application to steady-wave shock compression in solids. J. Mech. Phys. Solids 35 (1), 95–119.
Grady, D.E., Kipp, M.E., 1995. Experimental measurement of dynamic failure and fragmentation properties of metals. Int. J. Solids Struct. 32 (17–18), 2779–2781, 2783–2791.
Grady, D.E., Kipp, M.E., 1997. Fragmentation properties of metals. Int. J. Impact Eng. 20 (1–5), 293–308.
Grady, D.E., Olsen, M.L., 2003. A statistics and energy based theory of dynamic fragmentation. Int. J. Impact Eng. 29 (1–10), 293–306.
Grady, D.E., Kipp, M.E., Benson, D.A., 1984. Energy and Statistical Effects in the Dynamic Fragmentation of Metal Rings. Sandia National Labs., Albuquerque, NM (USA) (No. SAND-83-0859C; CONF-840411-7).
Grady, D.E., 1973. Experimental analysis of spherical wave propagation. J. Geophys. Res. 78, 1299–1307.
Grady, D.E., 1981. Application of survival statistics to the Impulsive fragmentation of ductile rings. In: Meyers, M.A., Murr, L.E. (Eds.), Shock Waves and High-Strain-Rate Phenomena in Metals. Plenum, New York, pp. 181–192.
Grady, D.E., 1982. Local inertial effects in dynamic fragmentation. J. Appl. Phys. 53, 322–325.
Grady, D.E., 2006. Fragmentation of Rings and Shells, the Legacy of N F Mott. Springer.
Gray, G.T., Blumenthal, W.R., 2000. Split-Hopkinson pressure bar testing of soft materials. In: Kuhn, H., Medlin, D. (Eds.), ASM Handbook: Mechanical Testing and Evaluation, Materials, vol. 8. ASM International, Park, OH, pp. 488–496.
Green, A.E., Rivlin, R.S., 1957. The mechanics of nonlinear materials with memory, Part 1. Arch. Ration. Mech. Anal. 1, 1–21.
Greenman, W.F., Vreeland Jr., T., Wood, D.S., 1967. Dislocation mobility in copper. J. Appl. Phys. 38 (9), 3595–3603.
Griffith, A.A., 1920. The phenomena of rupture and flow in solids. Phil. Trans. Roy. Soc. Lond. A221 (582–583), 163–198.
Griffith, A.A., 1924. The theory of rupture. In: Proc First Inter Congr Appl Mech, Delft, pp. 55–63.
Grüneisen, E., 1912. Theorie des festen Zustandes einatomiger Elemente. Ann. Phys. 344 (12), 257–306.
Gutmanas, E.Y., Nadgornyi, E.M., Stepanov, A.V., 1963a. Dislocation movement in sodium chloride crystals. Sov. Phys. Solid State 3, 743–747.
Gutmanas, E.U., Nadgomi, E.M., Stepanov, A.V., 1963b. Dislocation motion in sodium chloride crystals. Sov. Phys. Solid State 5 (4), 743–747.
Harding, J., Wood, E.O., Campbell, J.D., 1960. Tensile testing of materials at impact rates of strain. J. Mech. Eng. Sci. 2 (2), 88–96.
Hawke, R.S., Brooks, A.C., Mitchell, A.C., Fowler, C.M., Peterson, D.R., Shaner, J.W., 1981. Railguns for EOS Research. UCRL-85298.

Hertzberg, R.W., 1976. Deformation and Fracture Mechanics of Engineering Materials. John Wiley & Sons, New York.
Hild, F., Denoual, C., Forquin, P., et al., 2003. On the probabilistic-deterministic transition involved in a fragmentation process of brittle materials. Comput. Struct. 81 (12), 1241–1253.
Hiroe, T., Fujiwara, K., Hata, H., et al., 2008. Deformation and fragmentation behaviour of exploded metal cylinders and the effects of wall materials, configuration, explosive energy and initiated locations. Int. J. Impact Eng. 35 (12), 1578–1586.
Hiroe, T., Fujiwara, K., Hata, H., et al., 2010. Explosively driven expansion and fragmentation behavior for cylinders, spheres and rings of 304 stainless steel. Mater. Sci. Forum 638, 1035–1040.
Hoge, K.G., Mukherjee, A.K., 1977. The temperature and strain rate dependence of the flow stress of tantalum. J. Mater. Sci. 12 (8), 1666–1672.
Hoggatt, C.R., Recht, R.F., 1968. Fracture behavior of tubular bombs. J. Appl. Phys. 39 (3), 1856–1862.
Hoggatt, C.R., Recht, R.F., 1969. Stress-strain data obtained at high rates using an expanding ring. Exp. Mech. 9 (10), 441–448.
Holmquist, T.J., Johnson, G.R., Cook, W.H., 1993. A computational constitutive model for concrete subjected large strain, high strain rates and high pressure. In: Proc. 7th Int. Symp. Ballistics, Québec, Canada, pp. 591–600.
Holt, D.L., Babcock, S.G., Green, S.J., Maiden, C.J., 1967. The strain-rate dependence of the flow stress in some aluminum alloys. Trans. Am. Soc. Metals 60 (2), 152–159.
Hopkinson, J., 1872. On the rupture of iron wire by a blow. Proc. Manchester Literary Philosophical Society. 11, 40-45; Further experiments on the rupture of iron wire. Proc. Manch. Lit. Philos. Soc. 11, 119–121.
Hopkinson, B., 1905. The effect of momentary stresses in metals. Proc. Roy. Soc. A74, 498.
Hopkinson, B., 1910. Scientific Papers. Cambridge University Press, London.
Hopkinson, B., 1914. A method of measuring the pressure produced in the detonation of high explosives or by the impact of bullets. In: Containing Papers of a Mathematical or Physical Character. London: Philosophical Transactions of the Royal Society, A213, pp. 437–456.
Huang, D.-J., Shi, S.-Q., Wang, L.-L., 2001. Studies on rate-dependent evolution of damage and its effects on dynamic constitutive response by using a random fuse network model. In: Chiba, A., Tanimura, S., Hokamoto, K. (Eds.), Impact Engineering and Application. Elsevier Science Ltd., Tokyo, pp. 743–748.
Huang, K.-Z., Yu, S.-W., 1985. Elastic-Plastic Fracture Mechanics. Tsinghua University Press, Beijing (in Chinese).
Huang, Z.-P., Yang, L.-M., Pan, K.-L., 1993. Dynamic damage and failure of materials. Adv. Mech. 23 (4), 433–467.
Huang, K., 1966. Solid State Physics. People's Education Press (in Chinese).
Hummeltenberg, A., Curbach, M., 2012. Design and construction of a biaxial Split-Hopkinson-Bar. Beton- Stahlbetonbau 107 (6), 394–400.
Inaoka, H., Toyosawa, E., Takayasu, H., 1997. Aspect ratio dependence of impact fragmentation. Phys. Rev. Lett. 78 (18), 3455–3458.
Inglis, C.E., 1913. Stresses in a plate due to the presence of cracks and sharp corners. Trans. Inst. Nav. Archit. 55, 219–230.
Inoue, H., Harrigan, J.J., Reid, S.R., 2001. Review of inverse analysis of indirect measurement of impact force. Appl. Mech. Rev. 54 (6), 503–524.
Irwin, G.R., 1948. Fracture Dynamics, Fracturing of Metals. American Society of Metals, pp. 147–166.
Irwin, G.R., 1957. Analysis of stresses and strains near the end of a crack traversing a plate. J. Appl. Mech. 24 (3), 361–364.

Ishii, T., Matsushita, M., 1992. Fragmentation of long thin glass rods. J. Phys. Soc. Jpn. 61 (10), 3474–3477.

Ivanov, A.G., Novikov, S.A., 1961. Rarefaction shock waves in iron and steel. J. Exp. Theor. Phys. 40, 1880–1882.

Jiang, F., Vecchio, K.S., 2007. Experimental investigation of dynamic effects in a two-bar/three-point bend fracture test. Rev. Sci. Instrum. 78, 063903.

Jiang, F., Vecchio, K.S., 2009. Hopkinson bar loaded fracture experimental technique: a critical review of dynamic fracture toughness tests (Review). Appl. Mech. Rev. 62 (6), 1–39.

Jing, F.-Q., 1999. Introduction to Experimental Equation of State, second ed. Science Press, Beijing (in Chinese).

Johnson, G.R., Cook, W.H., 1983. A constitutive model and data for metals subjected to large strains, high strain rates and high temperature. In: Proc. 7thInt. Symp. Ballistics, Netherlands, pp. 541–547.

Johnson, P.C., Stein, B.A., Davis, R.S., 1963. Measurement of dynamic plastic flow properties under uniform stress. In: Symposium on the Dynamic Behavior of Materials. ASTM Special Publication.

Johnson, G.R., Hoegfeldt, J.M., Lindholm, U.S., Nagy, A., 1983. Response of various metals to large torsional strains over a large range of strain rates – Part 1: ductile metals. Trans. ASME J. Eng. Mater. Technol. 105, 42–47.

Johnston, W.G., Gilman, J.J., 1959. Dislocation velocities, dislocation densities, and plastic flow in lithium fluoride crystals. J. Appl. Phys. 30 (2), 129–144.

Jones, N., 2011. Structural Impact, 2nd Revised edition. Cambridge University Press, Cambridge.

Kachanov, L.M., 1958. Time of the rupture process under creep conditions. Izv AN SSSR Otd Tekhn Nauk 8, 26–31 (in Russia).

Kachanov, L.M., 1986. Introduction to Continuum Damage Mechanics. Matinus Nijhoff, Dordrecht.

Kadono, T., 1997. Fragment mass distribution of platelike objects. Phys. Rev. Lett. 78 (8), 1444–1447.

Kalthoff, J.F., 1986. Fracture behavior under high rates of loading. Eng. Fract. Mech. 23, 289–298.

Kalthoff, J.F., 1986. Fracture behavior under high rates of loading. Eng. Fract. Mech. 23.

Kalthoff, J.F., 2000. Modes of dynamic shear failure in solids. Int. J. Fract. 101, 1–31.

Kalthoff, J.F., Winkler, S., 1987. Failure mode transition at high rates of shear loading. In: Chen, C.Y., et al. (Eds.), Impact Loading and Dynamic Behavior of Materials. DGM, Bremen, pp. 185–195.

Kalthoff, J.F., Beinert, J., Winkler, S., 1977a. Measurement of dynamic stress intensity factors for fast running and arresting cracks in double cantilever beam specimens. In: Hahn, G.T., Kanninen, M.F. (Eds.), Fast Fracture and Crack Arrest, ASTM STP 627. ASTM, Philadelphia, pp. 161–176.

Kalthoff, J., Beinert, J., Winkler, S., Blauel, J., 1977b. On the determination of the crack arrest toughness. In: ICF4, Waterloo (Canada) 1977.

Kanel, G.I., Petrova, E.N., 1981. The strength of titanium VT6 at shock wave loading. In: Dremin, A.N., et al. (Eds.), Workshop on Detonation. Chernogolovka, Russia, pp. 136–142.

Ke, F.-J., Bai, Y.-L., Xia, M.-F., 1990. The evolution characteristics of the ideal micro crack system. Chin. Sci. 33, 1447–1459 (in Chinese).

Kinslow, R. (Ed.), 2012. High-Velocity Impact Phenomena. Elsevier.

Kipp, M.E., Grady, D.E., 1985. Dynamic fracture growth and interaction in one dimension. J. Mech. Phys. Solids 33 (4), 399–415.

Klepaczko, J., 1968. Strain rate history effects for polycrystalline aluminum and theory of intersections. J. Mech. Phys. Solids 16, 255–266.
Klepaczko, J., 1975. Thermally activated now and strain rate history effects for some polycrystalline f.c.c. metals. Mater. Sci. Eng. 18, 121–135.
Klepaczko, J.R., 1979. Application of the split hopkinson pressure bar to fracture dynamic. In: Harding, J. (Ed.), Mechanical Properties at High Rates of Strain. The Institute of Physics, Bristol, London, pp. 201–214.
Klepaczko, J.R., 1990. Dynamic crack initiation, some experimental methods and modeling. In: Klepaczko, J.R. (Ed.), Crack Dynamics in Metallic Materials. Springer, Vienna, pp. 255–454.
Knott, J.F., 1973. Fundamentals of Fracture Mechanics. Butterworth, London.
Kobayashi, A.S., Ramalu, M., 1985. A dynamic fracture analysis of crack curving and branching. J. Phys. 46. C5-197-206.
Kobayashi, A., Wang, L., 2001. Quest for Dynamic Deformation and Fracture of Visco-elastic Solids. Ryoin Publishers, Japan.
Kocks, U.F., Argon, A.S., Ashby, M.F., 1975. Thermodynamics and kinetics of slip. Prog. Mater. Sci. 19, 1–5.
Kolsky, H., 1949. An investigation of the mechanical properties of materials at very high strain rates of loading. Proc. Phys. Soc. B 62 (11), 676.
Kolsky, H., 1953. Stress Waves in Solids. Clarendous Press, Oxford.
Kou, S.-Q., 1979. Adiabatic shear – a material failure problem to be worth paying attention. Adv. Mech. 3, 52–59 (in Chinese).
Lai, H.W., Wang, L.L., 2011. Studies on dynamic behavior of Nylon through modified Lagrangian analysis based on particle velocity profiles measurements. J. Exp. Mech. 26 (2), 221–226 (in Chinese).
Lankford, J., Blanchard, C.R., 1991. Fragmentation of brittle materials at high rates of loading. J. Mater. Sci. 26 (11), 3067–3072.
Lawn, B.R., Wilshaw, T.R., 1975. Fracture of Brittle Solids. Cambridge University Press, Cambridge.
Lee, Y.-S., Yoon, Y.-K., Yoon, H.-S., 2002. Dynamic fracture toughness of Chevron-Notch ceramic specimen measured in split hopkinson pressure bar. Int. J. Korean Soc. Precis. Eng. 3, 69–75.
Lee, E.H., 1971. Plastic-Wave propagation analysis and elastic-plastic theory at finite deformation. In: Burke, J.J., Weiss, V. (Eds.), Shock Wave and the Mechanical Properties of Solids, vol. 3. Syracuse Univ. Press.
Lemaitre, J., 1992. A Course on Damage Mechanics. Spring-Verlag.
Lenski (Ленский), B.C., 1951. метод построения динамической зависимости между напряжениями и дефомациями по распределению остаточныхдеф омации. Becmник МГУ 5, 13 (in Russian).
Lewis, J.L., Goldsmith, W., 1973. A biaxial split Hopkinson bar for simultaneous torsion and compression. Rev. Sci. Instrum. 44 (7), 811–813.
Li, Y.-C., Tan, F.-L., Guo, Y., Yuan, F.-P., 2003a. Determination of the damage evolution equation and spallation criterion of metals. J. Ningbo Univ. (NSEE) 16 (4), 443–447 (in Chinese).
Li, J.R., Yu, J.L., Wei, Z.G., 2003b. Influence of specimen geometry on adiabatic shear instability of tungsten heavy alloys. Int. J. Impact Eng. 28 (3), 303–314.
Li, X.-Q., Ma, L., 1996. Neural Network, Engineering Application. Chongqing University Publishing House, Chongqing (in Chinese).
Lienau, C.C., 1936. Random fracture of a brittle solid. J. Frankl. Inst. 221, 485–494, 674–686, 769–787.
Lindholm, U.S., Yeakley, L.M., 1968. High strain-rate testing: tension and compression. Exp. Mech. 8 (1), 1–9.

Lindholm, U.S., 1968. Some experiments in dynamic plasticity under combined stress. In: Mechanical Behavior of Materials under Dynamic Loads. Springer Verlag, New York, pp. 77–95.

Lindholm, U.S., 1971. High Strain-Rate Tests. In: Techniques in Metals Research, vol. 5. Interscience, New York, p. 199.

Lindholm, U.S., 1974. Review of dynamic testing techniques and material behavior. In: Institute of Physics Conference Series, vol. 21, pp. 3–70.

Liu Shuyi, 1954. On the realistic stress space of solids. Acta Phys. Sin. 10 (1), 13–34 (in Chinese).

Liu, X.M., Hu, S.S., 2000. Two-dimensional numerical analysis for the dispersion of stress waves in large-diameter-SHPB. J. Exp. Mech. 15 (4), 371–376 (in Chinese).

Liu, X.M., Hu, S.S., Chen, Z., 2002. The wave propagation attenuation and dispersion in a viscoelastic Hopkinson pressure bar. Acta Mech. Solida Sin. 23 (1), 81–86 (in Chinese).

Liu, X.-H., Huang, X.-C., Chen, Y.-Z., Su, X.-Y., Zhu, J.-S., 2007. A review on constitutive models for plastic deformation of metal materials under dynamic loading. Adv. Mech. 37 (3), 361–374 (in Chinese).

Lockett, F.J., 1972. Nonlinear Viscoelastic Solids. Academic Press, London.

Lu, W.-X., Lu, Z.-Q., 1985. The microscopic analysis of adiabatic shear for a titanium alloy in the stress-wave-riveting. Explos. Shock Waves 5 (1), 67–72 (in Chinese).

Lu, W.-X., Wang, L.-L., Lu, Z.-Q., 1986. The adiabatic shear of a β-titanium alloy under high strain rates. Acta Metall. Sin. 22 (4), A317–A320 pA41-42 (in Chinese).

Lüdwik, P., 1909. Elemente der technologischen Mechanik. Springer-Verlag, Berlin.

Maiden, C.J., Green, S.J., 1966. Compressive strain-rate tests on six selected materials at strain rates from 10^{-3} to 10^{4} in./in./sec. J. Appl. Mech. 33, 496–504.

Makin, M.J., 1968. The obstacles responsible for the hardening of neutron irradiated copper crystals. Phil. Mag. 18 (156), 1245–1255.

Malvern, L.E., 1951. The propagation of longitudinal waves of plastic deformation in a bar of material exhibiting strain-rate effect. J. Appl. Mech. 18, 203–208.

Marsh, K.J., Campbell, J.D., 1963. The effect of strain rate on the post-yield flow of mild steel. J. Mech. Phys. Solids 11 (1), 49–63.

Marsili, M., Zhang, Y.C., 1996. Probabilistic fragmentation and effective power law. Phys. Rev. Lett. 77 (17), 3577–3580.

Matthewson, M.J., Field, J.E., 1980. An improved strength-measurement technique for brittle materials. J. Phys. 13, 355–*359*.

Maudlin, P.J., Davidson, R.F., Henninger, R.J., 1990. Implementation and Assessment of the Mechanical-Threshold-Stress Model Using the EPIC2 and PINON Computer Codes. Los Alamos National Lab., NM (USA). No. LA-11895-MS.

McGregor, C.W., Fisher, J.C., 1946. A velocity-modified temperature for the plastic flow of metals. J. Appl. Mech. 13, 11–16.

McQueen, R.C., Marsh, S.R., 1962. Ultimate yield strength of copper. J. Appl. Phys. 33, 654–665.

McNaughton, I.I., Chisman, S.W., Booker, J.D., 1970. Hail impact studies. In: Fyall, A.A., King, R.B. (Eds.), Proc 3rd Int Conf on Rain Erosion and Associated Phenomena. Royal Aircraft Establishment, Farnborough, UK.

McQueen, R.G., Fritz, J.N., Marsh, S.P., 1967. Hugoniot equation of state of twelve rocks. J. Geophys. Res. 72 (20), 4999–5036.

Meibom, A., Balslev, I., 1996. Composite power laws in shock fragmentation. Phys. Rev. Lett. 76 (14), 2492–2494.

Messerschmidt, U., 2010. Dislocation Dynamics during Plastic Deformation. Springer Science & Business Media.

Meyer, L.W., Manwaring, S., 1986. Critical adiabatic shear strength of low alloyed steel under compressive loading. In: Murr, L.E., Staudhammer, K.P., Meyers, M.A. (Eds.),

Metallurgical Applications of Shock Wave and High-Strain-Rate Phenomena. Marcel Dekker Inc., New York, pp. 657–674.
Meyer, L.W., 2004. Material behaviour at high strain rates. In: Proc. 1st Int. Conf. on High Speed Forming- CHSF, Dortmund, pp. 45–56.
Meyers, M.A., Aimone, C.T., 1983. Dynamic fracture (spalling) of metals. Prog. Mater. Sci. 28, 1.
Meyers, M.A., Chawla, K.K., 1984. Mechanical Metallurgy, Principles and Applications. Prentice-Hall, Englewood Cliffs, NJ.
Meyers, M.A., Chawla, K.K., 2009. Mechanical Behavior of Materials. Cambridge University Press, Cambridge.
Meyers, M.A., 1994. Dynamic Behavior of Materials. Wiley-Interscience.
Mie, G., 1903. Zur kinetischen Theorie der einatomigen Körper. Ann. Phys. 316 (8), 657–697.
Migault, A., 1971. J. Phys. 32, 437.
Migault, A., 1972. J. Phys. 33, 707.
Miklowitz, T., 1978. The Theory of Elastic Waves and Wave Guides. North-Holland Publ. Co, Amsterdam.
Mines, R.A.W., Ruiz, C., 1985. The dynamic behavior of the instrumented charpy-test. J. Phys. 46, 187–196.
Molinari, A., Clifton, R.J., 1983. Localization et la déformation viscoplastique en cisaillement simple: résultats exacts en théorie non linéaire. C. R. Acad. Sci. Paris 296 (II-I), 1–4.
Molinari, A., Clifton, R.J., 1988. Analytical characterization of shear localization in thermoviscoplastic solids. J. Appl. Mech. 54, 806–812.
Molinari, A., Mercie, S., Jacques, N., 2014. Dynamic failure of ductile materials, 23rd international congress of theoretical and applied mechanics. Procedia IUTAM 10, 201–220.
Mott, N.F., Linfoot, E.H., 1943. A Theory of Fragmentation. Extra-Mural Research No F72/80. Ministry of Supply. AC 3348.
Mott, N.F., 1943. A Theory of the Fragmentation of Shells and Bombs. Ministry of Supply. AC4035, May.
Mott, N.F., 1947. Fragmentation of shell cases. Proc. R. Soc. Lond. Ser. A 189 (1018), 300–308.
Mott, N.F., 1948. Brittle fracture in mild steel plates. Engineering 165, 16–18.
Mott, N.F., 1956. LVII Creep in metal crystals at very low temperatures. Phil. Mag. 1 (6), 568–572.
Nabarro, F.R.N., 1947. Dislocations in a simple cubic lattice. Proc. Phys. Soc. 59 (2), 256–272.
Nicholas, T., 1981. Tensile testing of materials at high rates of strain. Exp. Mech. 21 (5), 177–188.
Nilsson, F., 1984. Crack growth initiation and propagation under dynamic loading, mechanical properties at high rates of strain. Inst. Phys. Conf. Ser. 70, 185–204.
Niordson, F.I., 1965. A unit for testing materials at high strain rates. Exp. Mech. 5 (1), 29–32.
Novikov, S.A., Divnov, I.I., Ivanov, A.G., 1966. The study of fracture of steel, aluminum and copper under explosive loading. Phys. Met. Met. Sci. (USSR) 21 (4), 608–615.
Oddershede, L., Dimon, P., Bohr, J., 1993. Self-organized criticality in fragmenting. Phys. Rev. Lett. 71 (19), 3107–3110.
Ogawa, K., 1984. Impact-tension compression test by using a split-Hopkinson bar. Exp. Mech. 24 (2), 81–86.
Orowan, E., 1934. Plasticity of crystals. Z. Phys. 89 (9–10), 605–659.
Orowan, E., 1940. Problems of plastic gliding. Proc. Phys. Soc. 52 (1), 8–22.

Orowan, E., 1949. Fracture and strength of solids. Rep. Prog. Phys. 12 (1), 185–232.

Oscarson, J.H., Graff, K.F., 1968. Spall Fracture and Dynamic Response of Materials. Battelle Memorial Institute. BAT-197A-4-3(AD 669440).

Pastine, D.J., 1965. Formulation of the Gruneisen parameter on monoatomic cubic crystals. Phys. Rev. 138, A767.

Peierls, R.E., 1940. The size of a dislocation. Proc. Phys. Soc. 52 (1), 34–37.

Perzyna, P., 1966. Fundamental problems in viscoplasticity. Adv. Appl. Mech. 9, 935–950.

Piekutowski, A.J., 1995. Fragmentation of a sphere initiated by hypervelocity impact with a thin sheet. Int. J. Impact Eng. 17, 627–638.

Piekutowski, A.J., 1996. Formation and Description of Debris Cloud Produced by Hypervelocity Impact, pp. 98–99. NASA CR-4707, 1996.

Piekutowski, A.J., 2001. Debris clouds produced by the hypervelocity impact of nonspherical projectiles. Int. J. Impact Eng. 26 (1–10), 613–624.

Piekutowski, A.J., 2003. Fragmentation-Initiation threshold for spheres impacting at hypervelocity. Int. J. Impact Eng. 29, 563–574.

Polanyi, M., 1934. Lattice distortion which originates plastic flow. Z. Phys. 89 (9–10), 660–662.

Prieto, F.E., Renero, C., 1970. Equation for the shock adiabat. J. Appl. Phys. 41 (9), 3876–3883.

Qian, X.-S., 1962. Notes on Physical Mechanics. Science Press (in Chinese).

Ramalu, M., Kobayashi, A.S., 1983. Dynamic crack curving and crack branching. In: Material Behavior under High Stress and Ultrahigh Loading Rates, New York, p. 241.

Randall, M.R.D., Campbell, J.D., 1974. Effect of Strain Rate on Plastic Deformation of Materials under Combined Loading, 1972, University of Oxford Report No. 1047/22; Mechanical Properties of Materials at High Rates of Strain. Proc. Univ. of Oxford.

Ravi-Chandar, K., Knauss, W.G., 1984a. An experimental investigation into dynamic fracture-I. Crack initiation and crack arrest. Int. J. Fract. 25, 247–262.

Ravi-Chandar, K., Knauss, W.G., 1984b. An experimental investigation into dynamic fracture-II. Microstructural aspects. Int. J. Fract. 26, 65–80.

Ravi-Chandar, K., Knauss, W.G., 1984c. An experimental investigation into dynamic fracture-111. On steady state crack propagation and branching. Int. J. Fract. 26, 141–154.

Ravi-Chandar, K., 2004. Dynamic Fracture. Elsevier Ltd, Oxford.

Ravichandran, G., Clifton, R.J., 1989. Dynamic fracture under plane wave loading. Int. J. Fract. 40, 157–201.

Ravichandran, G., Subhash, G., 1994. Critical appraisal of limiting strain rates for compression testing of ceramics in a split Hopkinson pressure bar. J. Am. Ceram. Soc. 77 (1), 263–267.

Rayleigh, L., 1885. On waves propagated along the plan surface of an elastic solid. Proc. Lond. Math. Soc. 1 (1), 4–11.

Recht, R.F., 1964. Catastrophic thermoplastic shear. J. Appl. Mech. 31, 189–193.

Rice, J.R., 1968. Mathematical analysis in the mechanics of fracture, Chapter 3, fracture: an advanced treatise. In: Liebowitz, H. (Ed.), Mathematical Fundamentals, vol. 2. Academic Press, New York, pp. 191–311.

Rice, J.R., 1970. On the structure of stress-strain relations for time-dependent plastic deformation in metals. J. Appl. Mech. 37 (3), 728–737.

Rickerby, D.G., 1977. High Velocity Liquid Impact and Fracture Phenomena. University of Cambridge. Ph D thesis.

Rinehart, J.S., 1951. Some quantitative data bearing on the scabbing of metals under explosive attack. J. Appl. Phys. 22 (5), 555–560.

Rinehart, J.S., 1952. Scabbing of metals under explosive attack: multiple scabbing. J. Appl. Phys. 23 (11), 1229–1233.

Rinehart, J.S., 1975. Stress Transients in Solids. Hyper Dynamics, New Mexico.
Rittel, D., Maigre, H., Bui, H.D., 1992. A new method for dynamic fracture toughness testing. Scripta Metall. Mater. 26, 1593–1598.
Rogers, H.C., 1979. Adiabatic plastic deformation. Annu. Rev. Mater. Sci. 9, 283–311.
Rome, J., Isaacs, J., Nemat-Nasser, S., 2004. Hopkinson Techniques for Dynamic Triaxial Compression Tests. Recent Advances in Experimental Mechanics. Springer, Netherlands, pp. 3–12.
Rosakis, A.J., Ravichandran, G., 2000. Dynamic failure mechanics. Int. J. Solids Struct. 37, 331–348.
Rosakis, A.J., Owen, D.M., Zaharieva, R., Haberman, K., Vardoulakis, I., Exadaktylos, G., 1999. Explosion at the Parthenon: Can We Pick Up the Pieces? California Institute of Technology, Pasadena. Report No. CalCIT SM report, 99-3.
Royce, E.B., 1971. High pressure eq. of state from shock wave data. In: Course XWII, Physics of High Energy Density, School of Physics, Enrico Fermi Varenna, Italy, July 14–26, 1969. Academic Press, pp. 80–95.
Royce, E.B., 1974. Lawrence Livermore Laboratory. UCRL-51121.
Ruiz, C., Mines, R.A.W., 1985. The Hopkinson pressure bar: an alternative to the instrumented pendulum for Charpy tests. Int. J. Fract. 29, 101–109.
Schardin, H., 1941. Jahrbuch Der Deutsche Akademie Der Luftfahrtforschung, p. 314.
Schoeck, G., 1962. Correlation between dislocation length and density. J. Appl. Phys. 33 (5), 1745–1747.
Seaman, L., Barbee, T.W., Crewdson, R.C., February 1972. Dynamic Fracture Criteria of Homogeneous Materials. AFWL-TR-71-156. Air Force Weapons Laboratory, Kirtland AFB, Albuquerque, New Mexico.
Seaman, L., Curran, D.R., Shockey, D.A., 1976. Computational models for ductile and brittle fracture. J. Appl. Phys. 47 (11), 4814–4826.
Seaman, L., Curran, D.R., Crewdson, R.C., 1978. Transformation of observed crack traces on a section to true crack density for fracture calculations. J. Appl. Phys. 49 (10), 5221–5229.
Seaman, L., 1974. Lagrange analysis for multiple stress or velocity gages in attenuating waves. J. Appl. Phys. 45 (10), 4303–4314.
Seeger, A., 1955. The generation of lattice defects by moving dislocations, and its application to the temperature dependence of the flow stress of face-centered cubic crystals. Phil. Mag. 46, 1194–1217.
Semiatin, S.L., Lahoti, G.D., Oh, S.I., 1983. The occurrence of shear bands in metalworking. In: Mescall, J., Weiss, V. (Eds.), Material Behavior under High Stress and Ultra High Loading Rates. Plenum Press, New York, pp. 119–160.
Semiatin, S.L., Staker, M.R., Jonas, J.J., 1984. Plastic instability and flow localization in shear at high rates of deformation. Acta Met. 32 (9), 1347–1354.
Shi, S.Q., Wang, L.L., 2000. Passive confined pressure SHPB test method for materials under quasi-one dimensional strain state. J. Exp. Mech. 15 (4), 377–384 (in Chinese).
Shi, S.-Q., Gan, S., Wang, L.-L., 1990. The thermo-viscoelastic mechanical behaviour of an aeronautical PMMA under impact loading. J. Ningbo Univ. (Nat. Sci. Eng. Ed.) 3 (2), 66–75 (in Chinese).
Shi, S.-Q., Yu, B., Wang, L.-L., 2007a. The thermo-viscoelastic constitutive equation of PP/PA blends and its rate temperature equivalency at high strain rates. Explos. Shock Waves 27 (3), 210–216 (in Chinese).
Shi, S.-Q., Yu, B., Wang, L.-L., 2007b. The thermo-viscoelastic constitutive equation of PP/PA blends and its rate temperature equivalency at high strain rates. Macromol. Symp. 247, 28–34.

Shi, S.Q., Yu, B., Wang, L.L., 2009. The dynamic impact experiments under active confining pressure and the constitutive equation of PP/PA blends at multi-axial compressive stress state. Macromol. Symp. 286, 53–59.

Shockey, D.A., Curran, D.R., Seaman, L., 1985. Development of Improved Dynamic Failure Models. Sri International Menlo Park, CA.

Sih, G.C., Embley, G.T., Ravera, R.S., 1972. Impact response of a finite crack in plane extension. Int. J. Solids Struct. 8 (7), 977–993.

Sih, G.C. (Ed.), 1977. Mechanics of Fracture 4, Elastodynamic Crack Problem. Noordhoff International Publishing, Leyden.

Skidmore, I.C., 1965. An introduction to shock waves in solids. Appl. Mater. Res. 4 (3), 131.

Slater, J.C., 1939. Introduction to Chemical Physics. McGraw-Hill Book Co., New York.

Slepyan, L.I., 1981. Crack Mechanics. Sudostroenie, Leningrad first ed., 1981; second ed., 1990.

Song, L., Hu, S.S., 2006. A new technique for testing soft material using a modified Hopkinson pressure bar. Eng. Mech. 23 (5), 24–28 (in Chinese).

Stein, D.F., Low Jr., J.R., 1960. Mobility of edge dislocations in silicon-iron crystals. J. Appl. Phys. 31 (2), 362.

Steinberg, D., Chau, H., Dittbenner, G., Weingart, R., 1978. Electric Gun: A New Method for Generating Shock Pressures in Excess of 1 TPa, No. UCID-17943. California Univ.

Steinberg, D.J., Cochran, S.G., Guinan, M.W., 1980. A constitutive model for metals applicable at high strain rate. J. Appl. Phys. 51 (3), 1498.

Stelly, M., Legrand, J., Dormeval, R., 1981. Some metallurgical aspects of the dynamic expansion of shells. In: Murr, L.E., Meyers, M.A. (Eds.), Shock Waves and High-Strain-Rate Phenomena in Metals, pp. 113–125.

Stroppe, H., Clos, R., Schreppel, U., 1992. Determination of the dynamic fracture toughness using a new stress pulse loading method. Nucl. Eng. Des. 137, 315–321.

Sun, Z.-J., Wang, L.-L., 2005. Dynamic damage evolution described by modules change for PP/PA polymer blends with two different ompatibilizersm. J. Ningbo Univ. (Nat. Sci. Eng. Ed.) 18 (2), 159–162 (in Chinese).

Sun, Z.J., Wang, L.L., 2006. Studies on impact constitutive behavior and damage evolution for PP/PA polymer blends at large deformation. J. Phys. IV 134, 117–124.

Sun, Z.J., Wang, L.L., 2008. Studies on dynamic damage evolution for PP/PA polymer blends under high strain rates. Int. J. Mod. Phys. B 22 (9–11), 1409–1416.

Sweeney, M.A., Perry, F.C., Asay, J.R., 1981. Shock effects in particle-beam fusion targets. Bull. APS 26 (4), 650.

Symonds, P.S., Chon, C.T., 1974. Approximation Techniques for Impulsive Loading of Structures of Time-dependent Plastic Behaviour with Finite-Deflections. In: Mechanical Properties of Materials at High Strain Rates, pp. 299–316. Institute of Physics Conference Series No.21.

Symonds, P.S., 1965. Viscoplastic behavior in response of structures to dynamic loading. In: Huffington, N.J. (Ed.), Behavior of Materials under Dynamic Loading. ASME, pp. 106–124.

Symonds, P.S., 1967. Survey of Methods of Analysis for Plastic Deformation of Structures under Dynamic Loading. Brown University. Division of Engineering Report BU/NSRDC/1–67.

Tang, Z.-P., Tian, L.-Q., Chu, C.-H., Wang, L.-L., 1981. Mechanical behavior of epoxy resin under high strain rates. In: Second National Conference of Explosive Mechanics, Chinese Society of Theoretical and Applied Mechanics, Yangzhou (in Chinese).

Taylor, G.I., Quinney, H., 1934. The latent energy remaining in a metal after cold working. Proc. Roy. Soc. 143, 307–326.

Taylor, G.I., 1934. The mechanism of plastic deformation of crystals, part I theoretical. Proc. Roy. Soc. A145, 362–387.

Taylor, G.I., 1946. The testing of materials at high rates of loading. J. ICE 26 (8), 486–519.

Taylor, G.I., 1948. The use of flat ended projectiles for determining dynamic yield stress. I: theoretical considerations. Proc. Roy. Soc. Lond. A 194 (1038), 289–299.

Tegart, W.J.M., 1966. Elements of Mechanical Metallurgy. Macmillan, New York.

Thurstan, R.S., Mudd, W.L., 1968. Spallation Criterion for Numerical Computational Data. LA-4013, TID-4500.

Tillotson, J.H., 1962. Metallic Equations of State for Hypervelocity Impact. General Atomic, San Diego, CA. General Atomic Report GA-3216.

Tong, W., Clifton, R.J., Huang, S., 1992. Pressure-shear impact investigation of strain rate history effects in oxygen – free high-conductivity copper. J. Mech. Phys. Solids 40, 1251–1294.

Trainor, R.J., Graboske, H.C., Long, K.S., Shaner, J.W., 1978. Application of High-Power Lasers to Equation-of-State Research at Ultrahigh Pressures. California Univ. No. UCRL-52562.

Tsien, H.-S., 1962. Notes on Physical Mechanics. Science press, Beijing (in Chinese).

Tuler, F.R., Butcher, B.M., 1968. A criterion for the time dependence of dynamic fracture. Int. J. Fract. Mech. 4 (4), 431–437.

Tyler, W.W., Dash, W.C., 1957. Dislocation arrays in germanium. J. Appl. Phys. 28 (11), 1221–1224.

Urtiew, P.A., Grover, R.J., 1977. Appl. Phys. 48, 1122.

Vohringer, O., 1989. Deformation behavior of metallic materials. In: Chiem, C.Y. (Ed.), Int. Summer School on Dynamic Behaviour of Materials for Europ. Engineers and Scientists, EMSN, Nantes, September 11–15, 1989, pp. 7–50.

Volterra, E., 1948. Alcuni risultati di prove dinamiche sui materiali (Some results on the dynamic testing of materials). Riv. Nuovo Cimento 4, 1–28.

Walley, S.M., Field, J.E., 2005. The contribution of the Cavendish Laboratory to the understanding of solid particle erosion mechanisms. Wear 258, 552–566.

Wang, L.L., 1984. A thermo-viscoplastic constitutive equation based on hyperbolic shape thermo-activated barriers. Trans. ASME J. Eng. Mat. Technol. 106, 331–336.

Wang, L.-L., 1986. A criterion of thermo-viscoplastic instability for adiabatic shearing. In: Zheng, Z.-M., et al. (Eds.), Proc. Int. Symp. on Intense Dynamic Loading and its Effects. Science Press, Beijing, pp. 787–792.

Wang, L., 1992. Adiabatic shearing—the constitutive instability of materials under impact loading. In: Wang, L., Yu, T., Li, Y. (Eds.), Progress in Impact Dynamics. the Press of University of Science and Technology of China, Hefei, pp. 3–33 (1992).

Wang, L.-L., 2003. On constitutive modeling in numerical simulation of explosion mechanics. Explos. Shock Waves 23 (2), 97–104 (in Chinese).

Wang, L., 2005. Foundations of Stress Waves. National Defense Industry Press, Beijing first ed., 1985, second ed., English ed., Elsevier, Amsterdam, 2007.

Wang, L., 2007a. Foundations of Stress Waves. National Defense Industry Press, Beijing first ed., 1985, second ed., 2005 (in Chinese), English Edition, Elsevier, Amsterdam.

Wang, L.L., 2007b. Foundations of Stress Waves. Elsevier Science Ltd, Amsterdam.

Wang, L., Bao, H., 1991. A strain-localization analysis for adiabatic shear band at different environmental temperatures (Invited Lecture). In: Joho, M., Inoue, T. (Eds.), Mechanical Behavior of Materials-VI, Proceedings of the Sixth International Conference on Mechanical Behavior of Materials, Kyoto, Japan, 29 July–2 August, vol. 1. Pergamon Press, pp. 479–486.

Wang, L., Hu, S., 1985. The dynamic stress-strain relation of Ti-alloy TB-2 under high strain rates. Explos. Shock Waves 5 (1), 9–15 (in Chinese).

Wang, L.-L., Hu, S.-S., 1986. Dynamic Stress-Strain Relation of Al-Alloy LF6R and Al L4R under High Strain Rates. Acta Mechnica Solida Sinica No. 2, pp. 163–166 (in Chinese).

Wang, Y.G., Wang, L.L., 2004. Stress wave dispersion in large diameter SHPB and its manifold manifestations. J. Beijing Inst. Technol. (Soc. Sci. Ed.) 13 (3), 247–253.

Wang, C.Y., Xia, Y.M., 1998. Application of Fourier analysis of the dispersion effect to split Hopkinson pressure testing system and bar-bar tensile impact testing system. Explos. Shock Waves 18 (3), 213–219 (in Chinese).

Wang, L.-L., Yang, L.-M., 1992. Non-linear viscoelastic constitutive relation of solid polymeric materials. In: Wang, L., Yu, T., Li, Y. (Eds.), Progress in Impact Dynamics. the Press of University of Science and Technology of China, Hefei, pp. 88–116 (in Chinese).

Wang, Y.-G., Zhou, F.-H., 2008. Experimental study on the dynamic tensile fragmentations of Al_2O_3 rings under radial expansion. Acta Mech. Solida Sin. 29 (3), 245–249 (in Chinese).

Wang, L., Hu, S.-S., Wang, X.-J., 1983. The attenuation of plane shock waves propagation in an elasto-plastic medium. J. China Univ. Sci. Technol. 13 (1), 90–100 (in Chinese).

Wang, L. L., Lu, W. X., Hu, S.S., Tang, Z.P., 1985. Study on the initiation and development of adiabatic shear bands for a titanium alloy under high strain rates. In: IUTAM Symposium on MMMHVDF, Aug. 12–15, 1985, Tokyo, Japan; in Kawata, K. and Shioiri, J. (Eds.) Macro- and Micro-mechanics of High-Velocity Deformation and Fracture. Springer-Verlag, Berlin, 395-406.

Wang, L.-L., Bao, H.-S., Lu, W.-X., 1988. The dependence of adiabatic shear banding on strain-rate, strain and temperature. J. Phys. 49 (9), 207–214. C3.

Wang, L.L., Labibes, K., Azari, Z., Pluvinage, G., 1992a. On the use of a viscoelastic bar in the split Hopkinson bar technique. In: Maekawa, I. (Ed.), Proceedings of the International Symposium on Impact Engineering. ISIE, Sendai, Japan, pp. 532–537.

Wang, L., Yu, T., Li, Y., 1992b. Progress in Impact Dynamics. the Press of University of Science and Technology of China, Hefei (in Chinese).

Wang, L.-L., Zhu, X.-X., Shi, S.-Q., Gan, S., Bao, H.-S., 1992c. An impact dynamics investigation on some problems in bird strike on windshields of high speed aircrafts. Chin. J. Aeronaut. 5 (3), 205–213.

Wang, R., Huang, W.-B., Huang, Z.-P., 1992d. Introduction to Plastic Mechanics, Revised Edition. Peking University Press, Beijing (in Chinese).

Wang, L., Dong, X., Hu, S., Yu, J., 1994a. A macro-and microscopic study of adiabatic shearing extension of mode-II crack at dynamic loading. J. Phys. IV 4 (C8), C8-465–C8-470.

Wang, L.-L., Bao, H.-S., Lu, W.-X., 1993. The inverse strain-rate effect induced by damage and its influence on constitutive relation and thermo-viscoplastic instability. Explos. Shock Waves 13 (1), 1–8 (in Chinese).

Wang, L., Field, J.E., Sun, Q., Liu, J., 1994b. Surface damage of PMMA plates by ice and nylon ball impacts. In: Zhemin, Z., Qingming, T. (Eds.), IUTAM Symposium on Impact Dynamics. Peking University Press, Beijing, China, pp. 417–433.

Wang, L.L., Labibes, K., Azari, Z., Pluvinage, G., 1994c. Generalization of split Hopkinson bar technique to use viscoelastic bars. Int. J. Impact Eng. 15, 669–686.

Wang, L.-L., Field, J.E., Sun, Q., Liu, J., 1995a. Surface damage of polymethylmethacrylate plates by ice and nylon ball impacts. J. Appl. Phys. 78 (3), 1643–1649.

Wang, L.L., Pluvinage, G., Labibes, K., 1995b. The influence of dynamic constitutive relations of polymers at impact loading on the viscoelastic wave propagation character. J. Ningbo Univ. (Nat. Sci. Eng. Ed.) 8 (3), 30–57 (in Chinese).

Wang, L.-L., Jiang, Z.-B., Chen, J.-Y., 1996. Micro-damage evolution in high velocity deformation and its influence on rate-dependent constitutive relation of materials. J. Ningbo Univ. (Nat. Sci. Eng. Ed.) 9 (3), 47–55 (in Chinese).
Wang, L.L., Jiang, Z.B., Chen, J.Y., 1999. Studies on rheological relation of materials by taking account of rate-dependent evolution of internal defects at high strain rates. In: Wang, R. (Ed.), Rheology of Bodies with Defects. Kluwer Academic Publishers, Dordrecht, pp. 167–178.
Wang, L.-L., Shi, S.-Q., Chen, J.-Y., Huang, D.-J., 2000a. Studies on ZWT non-linear thermo-viscoelastic constitutive relation and its application. J. Ningbo Univ. (Nat. Sci. Eng. Ed.) 13 (2000), 141–149 (in Chinese).
Wang, L.-L., Lai, H.-W., Sun, Z.-J., Xu, M.-Q., Huang, D.-J., Shi, S.-Q., 2003a. Dynamic deformation and fracture of polymers taking account of damage evolution (invited paper). In: Sino-France Workshop on Dynamic Fracture, Nov. 18-20, Ningbo, China; Jour, vol. 16. Ningbo Univ. (Natural Sci. & Engr. Edition), pp. 372–380 (6).
Wang, L.-L., Xu, M.-Q., Shi, S.-Q., 2003b. Application of BP neural network to SHPB technique for the investigation of impact response of polymers. In: Proc. 2003 SEM Annual Conference on Experimental and Applied Mechanics, June 2–4, 2003, Charlotte, North Carolina, USA, 167/1-8.
Wang, L.-L., Dong, X.-L., 2004. Influences of rate-dependent damage evolution upon constitutive response and fracture of materials at high strain rates. In: Proceedings of International Conference on Fracture and Damage of Advanced Material (FDAM 2004), Aug. 10–14, 2004, Hangzhou, China, pp. 69–78.
Wang, L., Xu, M., Zhu, J., Hu, S., 2006a. A method of combined SHPB technique and by neural network to study impact response of materials. Strain 42 (3), 149–158.
Wang, L.-L., Dong, X.-L., Sun, Z.-J., 2006b. Dynamic constitutive behavior of materials at high strain rate taking account of damage evolution. Explos. Shock Waves 26 (3), 193–198 (in Chinese).
Wang, L., Yang, L., Huang, D., Zhang, Z., Chen, G., 2008. An impact dynamics analysis on a new crashworthy device against ship-bridge collision. Int. J. Impact Eng. 35 (8), 895–904.
Wang, L.L., Zhou, F.H., Sun, Z.J., Wang, Y.Z., Shi, S.Q., 2010. Studies on rate-dependent macro-damage evolution of materials at high strain rates. Int. J. Damage Mech. 19, 805–820.
Wang, L.L., Zhu, J., Lai, H.W., 2011. A new method combining Lagrangian analysis with HPB technique. Strain 47, 173–182.
Wang, L.L., Ding, Y.Y., Yang, L.M., 2013a. Experimental investigation on dynamic constitutive behavior of aluminum foams by new inverse methods from wave propagation measurements. Int. J. Impact Eng. 62, 48–59.
Wang, L.-L., Ren, H.-Q., Yu, J.-L., Zhou, F.-H., Wu, X.-Y., Tang, Z.-P., Hu, S.-S., Yang, L.-M., Dong, X.-L., 2013b. Development and application of the theory of nonlinear stress wave propagation. Chin. J. Solid Mech. 34 (3), 217–240 (in Chinese).
Wang, L.-L., 1982a. Tensile stress regions caused by reverse plastic deformation in one dimensional strain elastic-plastic compressive wave propagation. Explos. Shock Waves 2, 39–44 (in Chinese).
Wang, L.-L., 1982b. Tensile stress zones caused by reverse plastic deformation in the propagation of one-dimensional strain elasto-plastic compression waves. Explos. Shock Waves 2 (2), 39–44 (in Chinese).
Wang, L.-L., 2004. Unloading waves and unloading failures in structures under impact loading. Int. J. Impact Eng. 30 (8/9), 889–900.
Wang, L.L., Dong, X.L., 2001. Positive and negative strain-rate effect for materials with damage and/or phase transformation at high strain rates. In: Staudhammer, K.P.,

Murr, L.E., Meyers, M.A. (Eds.), Fundamental Issues and Applications of Shock-Wave and High-Strain-Rate Phenomena. Elsevier Science Ltd., Amsterdam, pp. 193–200.

Wang, L.L., Hu, S.S., Yang, L.M., Sun, Z.J., Zhu, J., Lai, H.W., Ding, Y.Y., 2014. Development of experimental methods for impact testing by combining hopkinson pressure bar with other techniques. Acta Mech. Solida Sin. 27 (4), 331–344.

Wang, L.-L., Shi, S.-Q., Chen, J.-Y., Huang, D.-J., Li, D.-H., 2000b. Rate-Dependent constitutive relation for cement mortar at high strain-rates by taking account of damage evolution. In: Xu, B.-Y., Huang, Z.-P. (Eds.), Advances in Plasticity and Geodynamics (80th Anniversary Volume in Honor of Academician Ren Wang). International Academic Publishers, Beijing, pp. 257–262 (in Chinese).

Ward, I.M., 1983. Mechanical Properties of Solid Polymers, second ed. Wiley-Interscience.

Warnes, R.H., Duffey, T.A., Karpp, R.R., et al., 1981. An improved technique for determining dynamic material properties using the expanding ring. In: Shock Waves and High–Strain-Rate Phenomena in Metals. Springer US, pp. 23–36.

Wei, Z.-G., Yu, J.-L., Hu, S.-S., Li, Y.-C., 2000. Influence of microstructure on adiabatic shear localization of pre-twisted tungsten heavy alloys. Int. J. Impact Eng. 24 (6–7), 747–758.

Wei, Z.-G., Yu, J.-L., Li, J.-R., Li, Y.-C., Hu, S.-S., 2001. Influence of stress condition on adiabatic shear localization of tungsten heavy alloys. Int. J. Impact Eng. 26 (1–10), 843–852.

Weisbrod, G., Rittel, D., 2000. A method for dynamic fracture toughness determination using short beams. Int. J. Fract. 104, 89–103.

Whiffin, A.C., 1948. The use of flat – ended projectiles for determining dynamic yield stress, II, tests on various metallic materials. Proc. Roy. Soc. A194, 300.

Whiteman, P., 1962. Preliminary Report on the Effect of Stress Rate on the Dynamic Fracture of Steel, Brass and Aluminum. Atomic Weapons Research Establishment, Great Britain. Report No. UNDEX 445.

Wilson, M.L., Hawley, R.H., Duffy, J., 1980. The effect of loading rate and temperature on fracture initiation in 1020 hot-rolled steel. Eng. Fract. Mech. 13 (2), 371–385.

Wingrove, A.L., 1973. The influence of projectile geometry on adiabatic shear and target failure. Metall. Trans. 4 (8), 1829–1833.

Wright, T.W., Batra, R.C., 1985. The initiation and growth of adiabatic shear bands. Int. J. Plast. 1, 202–212.

Wright, T.W., Walter, J.W., 1987. On stress collapse in adiabatic shear bands. J. Mech. Phys. Solids 35, 701–720.

Xia, M.-F., Han, W.-S., Ke, F.-J., Bai, Y.-L., 1995a. Statistical mesoscopic damage mechanics and damage evolution induced mutation (I). Adv. Mech. 25 (1), 1–23 (in Chinese).

Xia, M.-F., Han, W.-S., Ke, F.-J., Bai, Y.-L., 1995b. Statistical mesoscopic damage mechanics and damage evolution induced mutation (II). Adv. Mech. 25 (2), 145–159 (in Chinese).

Xie, X., 1961. Fang Junxin, Solid Physics, vol. I. Shanghai Science and Technology Press (in Chinese).

Xing, W.-X., Xie, J.-X., 1999. Modern Optimization Calculation Method. Tsinghua University Press, Beijing (in Chinese).

Xu, M.-Q., Wang, L.-L., 2006. A new method for studying the dynamic response and damage evolution of polymers at high strain rates. Mech. Mater. 38 (1–2), 68–75.

Xu, T.-P., Wang, L.-L., Lu, W.-X., 1987. The thermo-viscoplasticity and adiabatic shear deformation for titanium alloy ti-6al-4v under high strain rates. Explos. Shock Waves 7, 1–8 (in Chinese).

Xu, J., Leonhardt, T., Farrell, J., Effgen, M., Zhai, T., 2008. Anomalous strain-rate effect on plasticity of a Mo-Re alloy at room temperature. Mater. Sci. Eng. A479, 76–82.

Yang, L.-M., Wang, L.-L., 1991. An investigation on the linear viscoelastic constitutive relation for composites on the basis of Eshelby's Theory. Explos. Shock Waves 11 (3), 244–251 (in Chinese).

Yang, L., Wang, L.-L., Zhu, Z.-X., 1993. A micromechanical study on the nonlinear viscoelastic constitutive relation of high rigidity particulate-filled polymers. Acta Mech. Sin. 25 (5), 606–614 (in Chinese).

Yang, L., Wang, L., Zhu, Z., 1994. A micromechanical analysis of the nonlinear elastic and viscoelastic constitutive relation of a polymer filled with rigid particles. Acta Mech. Sin. 10 (2), 176–185.

Yang, L.M., Shim, V.P.W., 2005a. Characterizing viscoelastic behavior of materials – applications of ZWT model. In: Bai, Y.-L., Huang, Z.-P., Yu, J.-L., Chen, D.-N. (Eds.), Dynamic Response of Structures and Materials. the Press of University of Science and Technology of China, Hefei, pp. 206–228 (in Chinese).

Yang, L.M., Shim, V.P.W., 2005b. An analysis of stress uniformity in split Hopkinson bar test specimens. Int. J. Impact Eng. 31 (2), 129–150.

Yang, L.-M., Chu, C.-H., Wang, L.-L., 1986. Effects of short-glass-fiber reinforcement on nonlinear viscoelastic behavior of polycarbonate. Explos. Shock Waves 6 (1), 1–9 (in Chinese).

Yang, T.-Q., Luo, W.-B., Xu, P., Wei, Y.-T., Gang, Q.-G., 2004. Visco-elasticity Theory and Application. Science Press, Beijing (in Chinese).

Yoffe, E., 1951. The moving Griffith crack. Philos. Mag. A 42, 739–750.

Yokoyama, T., Kishida, K., 1989. A novel impact three-point bend test method for determining dynamic fracture–initiation toughness. Exp. Mech. 29, 188–194.

Yokoyama, T., 1993. Determination of dynamic fracture-initiation toughness using a novel impact bend test procedure. J. Press. Vessel Technol. 115 (l), 389–397.

Zehnder, A.T., Rosakis, A.J., 1990. Dynamic fracture initiation and propagation in 4340 steel under impact loading. Int. J. Fract. 43, 271–285.

Zener, C., Hollomon, J.H., 1944. Effect of strain rate upon the plastic flow of steel. J. Appl. Phys. 15 (1), 22–32.

Zerilli, F.J., Armstrong, R.W., 1987. Dislocation-mechanics-based constitutive relations for material dynamics calculations. J. Appl. Phys. 61 (5), 1816–1825.

Zerilli, F.J., Armstrong, R.W., 1990. Description of tantalum deformation behavior by dislocation mechanics based constitutive relations. J. Appl. Phys. 68 (4), 1580–1591.

Zhang, Q., Huang, F., 2000. Introduction of Hypervelocity Impact Dynamics. Science Press, Beijing (in Chinese).

Zhang, H., Ravi-Chandar, K., 2006. On the dynamics of necking and fragmentation—I. Real-time and post-mortem observations in Al 6061-O. Int. J. Fract. 142 (3–4), 183–217.

Zhang, Q.B., Zhao, J., 2014. A review of dynamic experimental techniques and mechanical behaviour of rock materials. Rock Mech. Rock Eng. 47, 1411–1478.

Zhao, H., Gary, G., 1995. A three dimensional analytical solution of the longitudinal wave propagation in an infinite linear viscoelastic cylindrical bar: application to experimental techniques. J. Mech. Phys. Solids 43, 1335–1348.

Zhao, H., Gary, G., Klepaczko, J.R., 1997. On the use of a viscoelastic split Hopkinson pressure bar. Int. J. Impact Eng. 19, 319–330.

Zheng, Z.-M., Xie, B.-M., 2004. A proposed solution on the calculation model of underground explosion, Report on the research of science and technology of the Institute of Mechanics of the Chinese Academy of Sciences, No.00756, May, 1965. In: Zhe-min, Zheng (Ed.), Anthology. Science Press, Beijing, pp. 166–190 (in Chinese).

Zheng, Y.-X., Zhou, F.-H., Hu, S.-S., 2014. A SHPB based experimental technique for dynamic fragmentations of expanding rings. Explos. Shock Waves 34 (4), 483–488 (in Chinese).

Zhou, G.-Q., Liu, X.-M., 1996. Theory of Viscoelasticity. University of Science & Technology China press, Hefei (in Chinese).
Zhou, F.-H., Wang, Y.-G., 2008. Factors controlling sizes of brittle fragments due to impact loading. Explos. Shock Waves 28 (4), 298–303 (in Chinese).
Zhou, F., Molinari, J.-F., 2004. Stochastic fracture of ceramics under dynamic tensile loading. Int. J. Solids Struct. 41 (22–23), 6573–6596.
Zhou, F., Molinari, J.F., Ramesh, K.T., 2005. A cohesive-model based fragmentation analysis: effects of strain rate and initial defects distribution. Int. J. Solids Struct. 42, 5181–5207.
Zhou, F.-H., Wang, L.-L., 2010. Growth of an array of cohesive crack points in brittle solids and the influence of crack spacing. Chin. J. Theor. Appl. Mech. 42 (4), 691–701 (in Chinese).
Zhou, F.-H., Wang, L.-L., Hu, S.-S., 1992a. A damage-modified nonlinear viscoelastic constitutive relation and failure criterion of PMMA at high stain rates. Explos. Shock Waves 12 (4), 333–342 (in Chinese).
Zhou, F.H., Wang, L.L., Hu, S.S., 1992b. On the effect of stress nonuniformness in polymer specimen of SHPB tests. J. Exp. Mech. 7 (1), 23–29 (in Chinese).
Zhou, F., Molinari, J.F., Ramesh, K.T., 2006a. Effects of material properties on the fragmentation of brittle materials. Int. J. Fract. 139, 169–196.
Zhou, F., Molinari, J.-F., Ramesh, K.T., 2006b. Characteristic fragment size distributions in dynamic fragmentation. Appl. Phys. Lett. 88 (26), 261918.
Zhou, F.-H., Guo, L.-N., Wang, L.-L., 2010. The rapidest unloading characteristics in the fragmentation process of brittle solids. Chin. J. Solid Mech. 31 (3), 286–295 (in Chinese).
Zhu, Z.-X., Li, Y.-C., Wang, X.-J., 1981. Numerical analysis of the spallation of steel target under explosive loading. Appl. Math. Mech. 2 (4), 383–402.
Zhu, Z.-X., Xu, D.-B., Wang, L.-L., 1988. Thermo-Viscoelastic constitutive equation and time-temperature equivalence of epoxy resin at high strain rates. J. Ningbo Univ. (Nat. Sci. Eng. Ed.) 1 (1), 58–68 (in Chinese).
Zhu Jue, On SHPB Technique and Lagrangian Analysis Used for Studying the Impact Response of Concrete-Like Materials, Ph. D Dissertation (Supervisor: Hu S. and Wang L.), Department of Mechanics and Mechanical Engineering, University of Science and Technology of China, May 2006.
Zhu, J., Hu, S.S., Wang, L.L., 2006. An analysis of stress uniformity for viscoelastic materials during SHPB tests. Lat. Am. J. Solid. Struct. 3 (2), 125–148.
Zhu, J., Hu, S.S., Wang, L.L., 2009. An analysis of stress uniformity for concrete-like specimens during SHPB Tests. Int. J. Impact Eng. 36 (1), 61–72.
Zhu, Z.-X., 2015. Lectures on Material Constitutive Relation Theory. Science Press, Beijing (in Chinese).
Zhurkov, S.N., 1965. Kinetic concept of strength of solids. Int. J. Fract. Mech. 1 (4), 311–323. Int. J. Fract, 1984, 26 (4), 295–307.
Zhurkov, S.N., Sanfirova, T.P., 1955. The temperature and time dependence of the strength of pure metals. Dokl. Akad. Nauk SSSR 101, 237.
Zukas, J.A., Nicholas, T., Swift, H.F., Greszczuk, L.B., Curran, D.R., 1982. Impact Dynamics. J. Wiley & Sons, New York.
Zurek, A.K., Frantz, C.E., Gray, G.T., 1992. In: Meyers, M.A., Murr, L.E., Staudhammer, K.P. (Eds.), Shock-Wave and High-Strain-Rate Phenomena in Materials. Marcel Dekker, New York, p. 759.
Zwaag, S Van Der, Field, J.E., 1983. Rain erosion damage in brittle materials. Eng. Fract. Mech. 17 (4), 367–379.

Further reading

Aberson, J.A., Anderson, J.M., King, W.W., 1977. Dynamic analysis of cracked structures using singularity finite elements. In: Sih, G.C. (Ed.), Mechanics of Fracture 4, Elastodynamic Crack Problem. Noordhoff International Publishing, Leyden, pp. 249–294.

Albertini, C., Cadoni, E., Labibes, K., 1997. Impact fracture process and mechanical properties of plain concrete by means of an Hopkinson bar bundle. J. Phys. IV Colloq. 07 (C3), C3-915–C3-920. https://doi.org/10.1051/jp4:19973154 jpa-00255443.

Barbee, T.W., Seaman, L., Crewdson, R., Curran, D., 1972. Dynamic fracture criteria for ductile and brittle metals. J. Mater. 7 (3), 393–401.

Cadoni, E., Albertini, C., 2011. Modified Hopkinson bar technologies applied to the high strain rate rock tests. In: Zhou, Y.X., Zhao, J. (Eds.), Advances in Rock Dynamics and Applications. CRC Press, USA, pp. 79–104.

Chandar, K.R., Knauss, W.G., 1982. Dynamic crack-tip stresses under stress wave loading—a comparison of theory and experiment. Int. J. Fract. 20 (3), 209–222.

Chen, Y.M., Wilkins, 1977. Numerical analysis of dynamic crack problems. In: Sih, G.C. (Ed.), Mechanics of Fracture 4, Elastodynamic Crack Problem. Noordhoff International Publishing, Leyden, pp. 295–345.

Chen, G.-Y., Wang, L.-L., Yang, L.-M., Chen, M.-D., 2013. Theory and Design of Crashworthy Device against Ship-Bridge Collision. People's Communications Press, Beijing (in Chinese).

Coleman, B.D., Noll, W., 1957. Foundation of linear viscoelasticity. Rev. Mod. Phys. 33, 239, 1961.

Curran, D.R., Seaman, L., Shockey, D.A., 1977. Dynamic failure of solids. Phys. Today 46–55.

Field, J.E., Lesser, M.B., Dear, J.P., 1985. Studies of two-dimensional liquid wedge impact and their relevance to liquid-drop impact problems. Proc. Roy. Soc. Lond. A401, 225–249.

Jones, O.E., 1972. Metal Response under Explosive Loading. Sandia National Lab (SNL-NM), Albuquerque, NM (United States) (No. SC-DC-721202; CONF-720303-2).

Orowan, E., 1950. Fundamentals of brittle behaviour in metals. In: Proceedings of the Symposium on Fatigue and Fracture of Metals. J. Wiley and Sons, New York, pp. 139–167.

Index

Note: 'Page numbers followed by "f" indicate figures and "t" indicate tables'.

E

F

G

H

I

T

W

Y

Z

CPI Antony Rowe
Eastbourne, UK
September 25, 2019